Selected Works in Probability and Statistics

For other titles published in this series, go to
www.springer.com/series/8556

Ross Maller · Ishwar Basawa
Peter Hall · Eugene Seneta
Editors

Selected Works of C.C. Heyde

Springer

Editors
Ross Maller
Centre for Mathematics and its Applications,
and School of Finance, Actuarial Studies
and Applied Statistics
Australian National University
ACT, Australia
ross.maller@anu.edu.au

Peter Hall
Department of Mathematics and Statistics
University of Melbourne
Parkville Campus
Melbourne Victoria
Australia
halpstat@ms.unimelb.edu.au

Ishwar Basawa
Department of Statistics
University of Georgia
Athens Georgia
USA
ishwar@stat.uga.edu

Eugene Seneta
School of Mathematics and Statistics
University of Sydney
Sydney New South Wales
Australia
eseneta@maths.usyd.edu.au

ISBN 978-1-4419-5822-8 e-ISBN 978-1-4419-5823-5
DOI 10.1007/978-1-4419-5823-5
Springer New York Dordrecht Heidelberg London

Library of Congress Control Number: 2010932621

Printed on acid-free paper

Springer is part of Springer Science+Business Media (www.springer.com)

Preface to the Series

Springer's Selected Works in Probability and Statistics series offers scientists and scholars the opportunity of assembling and commenting upon major classical works in statistics, and honors the work of distinguished scholars in probability and statistics. Each volume contains the original papers, original commentary by experts on the subject's papers, and relevant biographies and bibliographies.

Springer is committed to maintaining the volumes in the series with free access on SpringerLink, as well as to the distribution of print volumes. The full text of the volumes is available on SpringerLink with the exception of a small number of articles for which links to their original publisher is included instead. These publishers have graciously agreed to make the articles freely available on their websites. The goal is maximum dissemination of this material.

The subjects of the volumes have been selected by an editorial board consisting of Anirban DasGupta, Peter Hall, Jim Pitman, Michael Sörensen, and Jon Wellner.

Christopher Charles Heyde

Preface

In 1945, very early in the history of the development of a rigorous analytical theory of probability, Feller (1945) wrote a paper called "The fundamental limit theorems in probability" in which he set out what he considered to be "the two most important limit theorems in the modern theory of probability: the central limit theorem and the recently discovered ... 'Kolmogoroff's celebrated law of the iterated logarithm' ". A little later in the article he added to these, via a charming description, the "little brother (of the central limit theorem), the weak law of large numbers", and also the strong law of large numbers, which he considers as a close relative of the law of the iterated logarithm. Feller might well have added to these also the beautiful and highly applicable results of renewal theory, which at the time he himself together with eminent colleagues were vigorously producing. Feller's introductory remarks include the visionary: "The history of probability shows that our problems must be treated in their greatest generality: only in this way can we hope to discover the most natural tools and to open channels for new progress. This remark leads naturally to that characteristic of our theory which makes it attractive beyond its importance for various applications: a combination of an amazing generality with algebraic precision."

No better description of, or prescription for, Chris Heyde's work in each of these fundamental areas, as well as in the other areas to which he made significant and original contributions, could be found.

The selection of papers in the volume to follow amply displays Chris's commitment to these principles, as it traces the growth of his interest in the classical theory of probability and stochastic processes from very early days to a full flowering decades later, as well as the broadening of his interests and expertise from its early origins in probability theory.

Chris realised early on that the theory of probability was inspired and invigorated, and should always continue to be refreshed, by the real-world practical problems from which it stems. While never diminishing or neglecting the role that theoretical investigations play, his research quickly moved on from them to encompass a remarkably wide repertoire; to mention just some of it, his work ranged from applications in applied probability, especially to branching processes and their genetic implications, to statistics, in areas ranging from times series analysis and estimating equations theory through to very applied subjects—and on to his later interests in and contributions to financial modeling and analysis.

In addition to these, Chris nurtured and maintained a deep interest in the history of probability and statistics. He returned, and contributed substantially to this subject, throughout his entire career. And above all, convinced of its utmost importance, Chris lobbied and promoted his subject, probability and statistics, as hard as he could to those who had influence, throughout his life; thus, we even see papers addressing the role of statistics, and statistical science, in his publications list.

In all that he attempted, Chris's research was up with or ahead of the times in which he worked, and at all times his work was characterised by its elegance and a striving for completeness, as well as by a willingness to attack the most difficult and fundamental areas.

The volume is arranged as follows. First, is given an "author's pick" section, comprised of comments by Chris himself on a selection of papers he considered interesting and important in the development of his, and the profession's, probabilistic and statistical thinking. Then follow four articles by colleagues and friends of Chris: Ishwar Basawa, Peter Hall, Eugene Seneta, and myself. These articles concentrate on four quite different strands of Chris's thinking, giving in all a wide ranging review of Chris's interests over a broad area of probability and statistics.

Sadly, Chris died while this volume was in preparation. His mind remained clear and focussed till the end, and he wrote a draft of his "author's pick" section in February 2008. It was typed and forwarded to me by his wife Beth just a few days before his death, and I needed only to edit it lightly into its present form. Knowing that time was short, the other contributors, Ishwar, Peter, Eugene and myself, encouraged by our series editor Anirban DasGupta, worked hard to get our pieces together in time, hopefully, to present the finished volume to Chris. Unfortunately that was not to be.

An appreciation of Chris's life and works, and the honours accorded to him, written by his long-time friends and colleagues Joe Gani and Eugene Seneta, appeared as an introduction to Gani and Seneta (2004). This was a special issue of the Journal of Applied Probability, which contained a collection of articles by his colleagues and friends. It includes a list of his publications up till 2004; that list is included, and updated, herein. Beyond these, there are papers yet to appear, as coauthors finalise them for publication. No doubt a complete list will be made available at some stage.

A published interview of Chris appeared in Glasserman and Kou (2006). An obituary by Gani and Seneta (2008) extends the introduction mentioned above. A brief biographical sketch by Joe Gani of Chris's life is included on page xxi of this volume.

In closing I would like to add my personal appreciation for Chris's support and guidance over many years, following his role as the initial supervisor of my PhD in the 1970s. He made a deep impression on me in ways I did not realise for many decades. His contribution was immense, as is the gap he leaves.

Canberra
February 2010

Ross Maller

Preface Bibliography

— W. Feller. The fundamental limit theorems in probability. *Bull. Amer. Math. Soc.*, 1945.

I2 Christopher Charles Heyde, AM, DSc, FAA, FASSA. In J. Gani and E. Seneta, editors, *Stochastic Methods and Their Applications: Papers in Honour of Chris Heyde*, J. Appl. Probab. Special Volume 41A, pages vii–x. Applied Probability Trust, Sheffield, 2004.

I3 P. Glasserman and S. Kou. A conversation with Chris Heyde. *Statist. Sci.*, 21(2):286–298, 2006.

I5 J. Gani and E. Seneta. Obituary: Christopher Charles Heyde AM, DSc, FAA, FASSA. *J. Appl. Probab.*, 45(3):587–592, 2008.

Acknowledgements

This series of selected works is possible only because of the efforts and cooperation of many people, societies, and publishers. The series editors originated the series and directed its development. The volume editors spent a great deal of time organizing the volumes and compiling the previously published material. The contributors provided comments on the significance of the papers. The societies and publishers who own the copyright to the original material made the volumes possible and affordable by their generous cooperation:

American Institute of Physics
American Mathematical Society
American Statistical Association
Applied Probability Trust
Australian Mathematical Society
Bernoulli Society
Cambridge University Press
Canadian Mathematical Society
Danish Society for Theoretical Statistics
Elsevier
Finnish Statistical Society
Indian Statistical Institute
Institute of Mathematical Statistics
International Chinese Statistical Association
International Statistical Institute
John Wiley and Sons
New Zealand Statistical Association
Norwegian Statistical Society
Oxford University Press
Princeton University and the Institute for Advanced Studies
Royal Statistical Society
Statistical Society of Australia
Swedish Statistical Society
University of California Press
University of Illinois, Department of Mathematics
University of North Carolina Press

Contents

Contributors

Ishwar Basawa Department of Statistics, University of Georgia, Athens, Georgia 30602, USA, Ishwar@stat.uga.edu

Joe Gani Mathematical Sciences Institute, Australian National University, Canberra ACT 0200 AUSTRALIA, joe.gani@anu.edu.au

Peter Hall Department of Mathematics and Statistics, University of Melbourne, Melbourne, VIC 3010, AUSTRALIA halpstat@ms.unimelb.edu.au

Ross Maller Centre for Mathematics and its Applications, and School of Finance, Actuarial Studies and Applied Statistics, Australian National University, ACT 0200 AUSTRALIA, Ross.Maller@anu.edu.au

Emeritus Professor Eugene Seneta School of Mathematics and Statistics, FO7, University of Sydney, NSW 2006, AUSTRALIA, eseneta@maths.usyd.edu.au

Introduction

A Biographical Sketch of C. C. Heyde

Christopher Charles Heyde (Chris) was born in Sydney, Australia on 29 April 1939, and went to school at Barker College, Hornsby where he excelled in sports. In his last year at school, he became interested in mathematics, and after completing his secondary studies, he enrolled at the University of Sydney in 1957 to study mathematical statistics. He graduated in 1961 with a first class Honours degree, and was awarded the University Medal.

In 1962, Chris won a Commonwealth Postgraduate Research Scholarship in Pat Moran's Department of Statistics at the Australian National University (ANU). He was awarded his PhD in 1965 for a thesis on "Results related to first passage time problems and some of their applications". In 1963, his paper "On a property of the lognormal distribution" appeared in J. Roy. Statist. Soc. B 25, 392–393; in this, he showed that the lognormal distribution was not determined by its moments, a result which was to make him famous.

Chris was married to Elizabeth (Beth) James in 1965, and they had two sons, Neil born in 1967 and Eric born in 1969. Both are now married, each with two children who have been the great delight of their grandparents.

Chris had a very international career: his first job was at Michigan State University, East Lansing in 1964–1965, followed by the University of Sheffield in 1965–1967, the University of Manchester in 1967–1968, the Department of Statistics (headed by E. J. Hannan) at the ANU in 1968–1974. and the CSIRO Division of Mathematics and Statistics in 1974-1983. He was appointed Professor and Chairman of the Department of Statistics at the University of Melbourne in 1983, but returned to the ANU as Professor of Statistics in 1986. When the ANU School of Mathematical Sciences (now the Mathematical Sciences Institute) was established, he became its Foundation Dean. From 1993 until his death in 2008, he divided his time between the ANU and Columbia University, New York, where he was a Professor in the Department of Statistics. He founded and acted as Director of its Center in Applied Probability. He served at Columbia between September and December each Fall semester, and was at the ANU for the rest of the year.

Chris was the author of over 200 papers and wrote or edited a dozen books, all of which are recorded in this volume. Among his many contributions, Chris considered the following to be his most significant: (a) the result that the lognormal distribution was not determined by its moments (J. Roy. Statist. Soc. B 25, 392–393); (b) the observation that the best linear predictor in a discrete time series is the best predictor if the innovations are martingale differences (Ann. Math. Statist. 43 (1972) 2058–2066, jointly with E. J. Hannan); (c) the clarification of the concept of long-range dependence (J. Appl. Prob. 34 (1997) 939-944, jointly with Y. Yang); and (d) the formulation of the fractal activity time geometric Brownian motion model for a risky asset (J. Appl. Prob. 36 (1999) 1234–1239).

Chris acted as the Editor or Associate Editor of several journals. For 18 years between 1990 and 2007, he was Editor-in-Chief of both the Journal of and Advances in Applied Probability and initiated the section on "Stochastic Geometry" in the latter. His judgment was always sound, and he was greatly valued as one of the Trustees of the Applied Probability Trust.

Chris received well-deserved recognition for his professional contributions. He was elected a Fellow of the Institute of Mathematical Statistics in 1972, the Australian Academy of Science in 1977, and the Academy of Social Sciences in Australia in 2003. He was awarded the Academy of Science's Hannan Medal in 1994 and its Lyle Medal in 1995. He served as a Council Member of the Australian Mathematical Society in 1980–1983 and was its Vice-President in 1981. He was Vice-President of the International Statistical Institute in 1985–1987 and again in 1993–1995, as well as President of its Bernoulli Society in 1985–1987. He was Federal President of the Statistical Society of Australia in 1985–1986, and was awarded its Pitman Medal in 1988. In recognition of his services to mathematics, he was awarded Membership of the Order of Australia by the Australian Government in 2003.

Chris was an eminent scientist with wide statistical interests; he was also a kind and thoughtful human being, with a strong sense of justice. He died in Canberra on 6 March 2008 from the effects of metastatic melanoma. He is sorely missed by his family, and his many friends and colleagues. His life and work has enriched us all.

Canberra
29 January 2010

Joe Gani

Complete Author Bibliography

Books Authored

B1 C. C. Heyde and E. Seneta. *I. J. Bienaymé. Statistical theory anticipated*, volume 3 of *Studies in the History of Mathematics and Physical Sciences*. Springer-Verlag, New York, 1977.

B2 P. Hall and C. C. Heyde. *Martingale limit theory and its application.* Probability and Mathematical Statistics. Academic Press Inc. [Harcourt Brace Jovanovich Publishers], New York, 1980. Probability and Mathematical Statistics.

B3 C. C. Heyde. *Quasi-likelihood and its application: A general approach to optimal parameter estimation.* Springer Series in Statistics. Springer-Verlag, New York, 1997.

Books Edited

E1 C. C. Heyde, editor. *Studies in Modelling and Statistical Science. Papers in honour of J. Gani.* Austral. J. Statist. Special Volume 30A. Australian Statistical Pub. Association, 1988.

E2 C. C. Heyde and E. Seneta, editors. *Bicentennial history issue.* Austral. J. Statist. Special Vol. 30B. Australian Statistical Publishing Association, Canberra, 1988.

E3 W. Dunsmuir, C. C. Heyde, and I. McRae, editors. *Youth Employment and Unemployment.* Austral. J. Statist. Special Volume 31B. Australian Statistical Publishing Association, 1989.

E4 C. C. Heyde, Yu. V. Prohorov, R. Pyke, and S. Rachev, editors. *Athens Conference on Applied Probability and Time Series Analysis. Vol. I*, volume 114 of *Springer Lecture Notes in Statistics*. Springer-Verlag, New York, 1996. Applied probability, In honor of J. M. Gani, Papers from the conference held in Athens, March 22–26, 1995.

E5 L. Accardi and C. C. Heyde, editors. *Probability towards 2000*, volume 128 of *Lecture Notes in Statistics*. Springer-Verlag, New York, 1998. Papers from the symposium held at Columbia University, New York, October 2–6, 1995.

E6 V. V. Anh and C. C. Heyde, editors. *Special Issue on Long-range Dependence*, volume 80. J. Statist. Plann. Inference, 1999.

E7 I. V. Basawa, C. C. Heyde, and R. L. Taylor, editors. *Selected Proceedings of the Symposium on Inference for Stochastic Processes*, volume 37 of *Institute of Mathematical Statistics Lecture Notes—Monograph Series*. Inst. Math. Statist., Beachwood, OH, 2001. Held at the University of Georgia, Athens, GA, May 10–12, 2000.

E8 C. C. Heyde and E. Seneta, editors. *Statisticians of the centuries*. Springer-Verlag, New York, 2001.

Mathematical Articles

M1 C. C. Heyde. On a property of the lognormal distribution. *J. R. Stat. Soc. Ser. B Stat. Methodol.*, 25:392–393, 1963.

M2 C. C. Heyde. Some remarks on the moment problem. I. *Quart. J. Math. Oxford. 2nd Series*, 14:91–96, 1963.

M3 C. C. Heyde. Some remarks on the moment problem. II. *Quart. J. Math. Oxford. 2nd Seri es*, 14:97–105, 1963. (2nd series) Oxford.

M4 C. C. Heyde. On the stationary waiting time distribution in the queue *GI/G/*1. *J. Appl. Probab.*, 1:173–176, 1964.

M5 C. C. Heyde. Two probability theorems and their application to some first passage problems. *J. Aust. Math. Soc.*, 4:214–222, 1964.

M6 C. C. Heyde. Some renewal theorems with application to a first passage problem. *Ann. Math. Statist.*, 37:699–710, 1966.

M7 C. C. Heyde. Some results on small-deviation probability convergence rates for sums of independent random variables. *Canad. J. Math.*, 18:656–665, 1966.

M8 C. C. Heyde. Asymptotic renewal results for a natural generalization of classical renewal theory. *J. R. Stat. Soc. Ser. B Stat. Methodol.*, 29:141–150, 1967.

M9 C. C. Heyde. A contribution to the theory of large deviations for sums of independent random variables. *Z. Wahrsch. Verw. Gebiete.*, 7:303–308, 1967.

M10 C. C. Heyde. A limit theorem for random walks with drift. *J. Appl. Probab.*, 4:144–150, 1967.

M11 C. C. Heyde. On large deviation problems for sums of random variables which are not attracted to the normal law. *Ann. Math. Statist.*, 38:1575–1578, 1967.

M12 C. C. Heyde. On the influence of moments on the rate of convergence to the normal distribution. *Z. Wahrsch. Verw. Gebiete.*, 8:12–18, 1967.

Indicates article found in this volume

M13 C. C. Heyde. Some local limit results in fluctuation theory. *J. Aust. Math. Soc.*, 7:455–464, 1967.

M14 C. C. Heyde and V. K. Rohatgi. A pair of complementary theorems on convergence rates in the law of large numbers. *Math. Proc. Cambridge Philos. Soc.*, 63:73–82, 1967.

M15 C. C. Heyde. An extension of the Hájek-Rényi inequality for the case without moment conditions. *J. Appl. Probab.*, 5:481–483, 1968.

M16 C. C. Heyde. A further generalization of the arc-sine law. *J. Aust. Math. Soc.*, 8:369–372, 1968.

M17 C. C. Heyde. On almost sure convergence for sums of independent random variables. *Sankhyā Ser. A*, 30:353–358, 1968.

🕮 M18 C. C. Heyde. On large deviation probabilities in the case of attraction to a non-normal stable law. *Sankhyā Ser. A*, 30:253–258, 1968.

🕮 M19 C. C. Heyde. On the converse to the iterated logarithm law. *J. Appl. Probab.*, 5:210–215, 1968.

M20 C. C. Heyde. On the growth of a random walk. *Ann. Inst. Statist. Math.*, 20:315–321, 1968.

M21 C. C. Heyde. Variations on a renewal theorem of Smith. *Ann. Math. Statist.*, 39:155–158, 1968.

M22 C. C. Heyde. A derivation of the ballot theorem from the Spitzer-Pollaczek identity. *Math. Proc. Cambridge Philos. Soc.*, 65:755–757, 1969.

🕮 M23 C. C. Heyde. A note concerning behaviour of iterated logarithm type. *Proc. Amer. Math. Soc.*, 23:85–90, 1969.

M24 C. C. Heyde. On a fluctuation theorem for processes with independent increments. II. *Ann. Math. Statist.*, 40:688–691, 1969.

🕮 M25 C. C. Heyde. On extended rate of convergence results for the invariance principle. *Ann. Math. Statist.*, 40:2178–2179, 1969.

M26 C. C. Heyde. On extremal factorization and recurrent events. *J. R. Stat. Soc. Ser. B Stat. Methodol.*, 31:72–79, 1969.

🕮 M27 C. C. Heyde. On the maximum of sums of random variables and the supremum functional for stable processes. *J. Appl. Probab.*, 6:419–429, 1969.

M28 C. C. Heyde. Some properties of metrics in a study on convergence to normality. *Z. Wahrsch. Verw. Gebiete.*, 11:181–192, 1969.

M29 C. C. Heyde and E. Heyde. A stochastic approach to a one substrate one product enzyme reaction in the initial velocity phase. *J. Theoret. Biol.*, 25:159–172, 1969.

M30 C. C. Heyde. Characterization of the normal law by the symmetry of a certain conditional distribution. *Sankhyā Ser. A*, 32:115–118, 1970.

🕮 M31 C. C. Heyde. Extension of a result of Seneta for the super-critical Galton-Watson process. *Ann. Math. Statist.*, 41:739–742, 1970.

M32 C. C. Heyde. On some mixing sequences in queuing theory. *Operations Res.*, 18:312–315, 1970.

🕮 M33 C. C. Heyde. On the implication of a certain rate of convergence to normality. *Z. Wahrsch. Verw. Gebiete.*, 16:151–156, 1970.

🕮 M34 C. C. Heyde. A rate of convergence result for the super-critical Galton-Watson process. *J. Appl. Probab.*, 7:451–454, 1970.

🕮 M35 C. C. Heyde and B. M. Brown. On the departure from normality of a certain class of martingales. *Ann. Math. Statist.*, 41:2161–2165, 1970.

🕮 Indicates article found in this volume

M36 C. C. Heyde. On the growth of the maximum queue length in a stable queue. *Operations Res.*, 19:447–452, 1971.

M37 C. C. Heyde. Some almost sure convergence theorems for branching processes. *Z. Wahrsch. Verw. Gebiete.*, 20:189–192, 1971.

M38 C. C. Heyde. Some central limit analogues for supercritical Galton-Watson processes. *J. Appl. Probab.*, 8:52–59, 1971.

M39 C. C. Heyde and B. M. Brown. An invariance principle and some convergence rate results for branching processes. *Z. Wahrsch. Verw. Gebiete.*, 20:271–278, 1971.

M40 C. C. Heyde and E. Heyde. Stochastic fluctuations in a one substrate one product enzyme system: are they ever relevant? *J. Theoret. Biol.*, 30(2):395–404, 1971.

M41 C. C. Heyde and J. R. Leslie. Improved classical limit analogues for Galton-Watson processes with or without immigration. *Bull. Austral. Math. Soc.*, 5:145–155, 1971.

M42 C. C. Heyde and E. Seneta. Analogues of classical limit theorems for the supercritical Galton-Watson process with immigration. *Math. Biosci.*, 11:249–259, 1971.

M43 E. J. Hannan and C. C. Heyde. On limit theorems for quadratic functions of discrete time series. *Ann. Math. Statist.*, 43:2058–2066, 1972.

M44 C. C. Heyde. Martingales: a case for a place in the statistician's repertoire. *Austral. J. Statist.*, 14:1–9, 1972. invited paper.

M45 C. C. Heyde and J. R. Leslie. On the influence of moments on approximations by portion of a Chebyshev series in central limit convergence. *Z. Wahrsch. Verw. Gebiete.*, 21:225–268, 1972.

M46 C. C. Heyde and E. Seneta. Estimation theory for growth and immigration rates in a multiplicative process. *J. Appl. Probab.*, 9:235–256, 1972.

M47 C. C. Heyde and E. Seneta. Studies in the history of probability and statistics. XXXI. The simple branching process, a turning point test and a fundamental inequality: a historical note on I. J. Bienaymé. *Biometrika*, 59:680–683, 1972.

M48 C. C. Heyde. An iterated logarithm result for martingales and its application in estimation theory for autoregressive processes. *J. Appl. Probab.*, 10:146–157, 1973.

M49 C. C. Heyde. On the uniform metric in the context of convergence to normality. *Z. Wahrsch. Verw. Gebiete.*, 25:83–95, 1973.

M50 C. C. Heyde. Revisits for transient random walk. *Stochastic Processes Appl.*, 1:33–51, 1973.

M51 C. C. Heyde and D. J. Scott. Invariance principles for the law of the iterated logarithm for martingales and processes with stationary increments. *Ann. Probab.*, 1:428–436, 1973.

M52 C. C. Heyde. An iterated logarithm result for autocorrelations of a stationary linear process. *Ann. Probab.*, 2:328–332, 1974.

M53 C. C. Heyde. Limit theory for stationary processes via approximating martingales. *Adv. in Appl. Probab.*, 6:196–197, 1974. Abstract of invited paper given to the 3rd Conference on Stochastic Processes and their Applications, Sheffield, August 1973.

M54 C. C. Heyde. On estimating the variance of the offspring distribution in a simple branching process. *Adv. in Appl. Probab.*, 6:421–433, 1974.

Indicates article found in this volume

M55 C. C. Heyde. On martingale limit theory and strong convergence results for stochastic approximation procedures. *Stochastic Processes Appl.*, 2(4):359–370, 1974.

M56 C. C. Heyde. On the central limit theorem for stationary processes. *Z. Wahrsch. Verw. Gebiete.*, 30:315–320, 1974.

M57 C. C. Heyde and E. Seneta. Notes on "Estimation theory for growth and immigration rates in a multiplicative process" (J. Appl. Probability 9 (1972), 235–256). *J. Appl. Probab.*, 11:572–577, 1974.

M58 C. C. Heyde. Kurtosis and departure from normality. In S. K. G. P. Patil and J. K. Ord, editors, *Statistical Distributions in Scientific Work*, volume 1 (Models and Structures), pages 193–202. D. Reidel Publ. Co., Dordrecht, 1975.

M59 C. C. Heyde. Martingale methods in estimation theory. *Adv. in Appl. Probab.*, 7:235–237, 1975. Abstract of invited paper given to the 4th Conference on Stochastic Processes and their Applications, Toronto, August 1974.

M60 C. C. Heyde. A nonuniform bound on convergence to normality. *Ann. Probab.*, 3(5):903–907, 1975.

M61 C. C. Heyde. On the central limit theorem and iterated logarithm law for stationary processes. *Bull. Austral. Math. Soc.*, 12:1–8, 1975.

M62 C. C. Heyde. Remarks on efficiency in estimation for branching processes. *Biometrika*, 62:49–55, 1975.

M63 C. C. Heyde. A supplement to the strong law of large numbers. *J. Appl. Probab.*, 12:173–175, 1975.

M64 C. C. Heyde and P. D. Feigin. On efficiency and exponential families in stochastic process estimation. In S. K. G.P. Patil and J. Ord, editors, *Statistical Distributions in Scientific Work*, volume 1 (Models and Structures), pages 227–240. D. Reidel Publ. Co., Dordrecht, 1975.

M65 C. C. Heyde and E. Seneta. Bienaymé. In *Proceedings of the 40th Session of the International Statistical Institute (Warsaw, 1975), Book 2. Invited papers*, Bull. Inst. Internat. Statist. volume 46, pages 318–331. International Statistical Institute, 1975. With discussion.

M66 C. C. Heyde and E. Seneta. The genetic balance between random sampling and random population size. *J. Math. Biol.*, 1(4):317–320, 1975.

M67 I. V. Basawa, P. D. Feigin, and C. C. Heyde. Asymptotic properties of maximum likelihood estimators for stochastic processes. *Sankhy Ser. A*, 38(3):259–270, 1976.

M68 P. G. Hall and C. C. Heyde. On a unified approach to the law of the iterated logarithm for martingales. *Bull. Austral. Math. Soc.*, 14(3):435–447, 1976.

M69 C. C. Heyde. On asymptotic behavior for the Hawkins random sieve. *Proc. Amer. Math. Soc.*, 56:277–280, 1976.

M70 C. C. Heyde and J. R. Leslie. On moment measures of departure from the normal and exponential laws. *Stochastic Processes Appl.*, 4(3):317–328, 1976.

M71 C. C. Heyde. The effect of selection on genetic balance when the population size is varying. *Theoret. Population Biol.*, 11:249–251, 1977.

M72 C. C. Heyde. On central limit and iterated logarithm supplements to the martingale convergence theorem. *J. Appl. Probab.*, 14(4):758–775, 1977.

Indicates article found in this volume

M73 C. C. Heyde. An optimal property of maximum likelihood with application to branching process estimation. In *Proceedings of the 41st Session of the International Statistical Institute (New Delhi, 1977)*, volume 47 book 2, pages 407–417, 446–448. International Statistical Institute, 1977.

M74 C. C. Heyde. A log log improvement to the Riemann hypothesis for the Hawkins random sieve. *Ann. Probab.*, 6(5):870–875, 1978.

M75 C. C. Heyde. On an explanation for the characteristic clutch size of some bird species. *Adv. in Appl. Probab.*, 10:723–725, 1978.

M76 C. C. Heyde. On an optimal asymptotic property of the maximum likelihood estimator of a parameter from a stochastic process. *Stochastic Process. Appl.*, 8(1):1–9, 1978.

M77 C. C. Heyde and H. J. Schuh. Uniform bounding of probability generating functions and the evolution of reproduction rates in birds. *J. Appl. Probab.*, 15(2):243–250, 1978.

M78 C. C. Heyde and E. Seneta. Bienaymé, Irénée Jules. In C. C. Gillispie, editor, *Dictionary of Scientific Biography*, volume 15, pages 30–33. Scribner, 1978.

M79 C. C. Heyde. Applications of stochastic processes: some general principles and their illustration. *Math. Sci.*, 4:1–8, 1979.

M80 C. C. Heyde. On assessing the potential severity of an outbreak of a rare infectious disease: a Bayesian approach. *Austral. J. Statist.*, 21(3):282–292, 1979.

M81 C. C. Heyde and I. M. Johnstone. On asymptotic posterior normality for stochastic processes. *J. R. Stat. Soc. Ser. B Stat. Methodol.*, 41(2):184–189, 1979.

M82 C. C. Heyde. On a probabilistic analogue of the Fibonacci sequence. *J. Appl. Probab.*, 17(4):1079–1082, 1980.

M83 P. Hall and C. C. Heyde. Rates of convergence in the martingale central limit theorem. *Ann. Probab.*, 9(3):395–404, 1981.

M84 C. C. Heyde. Invariance principles in statistics. *Internat. Statist. Rev.*, 49(2):143–152, 1981.

M85 C. C. Heyde. Looking forward into the 1980s: a personal view of the problems and prospects for the statistical profession. *Austral. J. Statist.*, 23:1–14, 1981. Presidential Address to the Statistical Society of Australia, August 1980.

M86 C. C. Heyde. On Fibonacci (or lagged Bienaymé-Galton-Watson) branching processes. *J. Appl. Probab.*, 18(3):583–591, 1981.

M87 C. C. Heyde. On the survival of a gene represented in a founder population. *J. Math. Biol.*, 12(1):91–99, 1981.

M88 C. C. Heyde. Trends in the statistical sciences. *Austral. J. Statist.*, 23:273–286, 1981. The Belz Lecture for 1980.

M89 D. J. Daley, P. Hall, and C. C. Heyde. Further results on the survival of a gene represented in a founder population. *J. Math. Biol.*, 14(3):355–363, 1982.

M90 C. C. Heyde. The Australian Journal of Statistics. In N. L. Johnson and S. Kotz, editors, *Encyclopedia of Statistical Sciences*, volume 1, pages 147–148. Wiley, 1982.

M91 C. C. Heyde. The effect of differential reproductive rates on the survival of a gene represented in a founder population. In J. Gani and E. J. Han-

Indicates article found in this volume

nan, editors, *Essays in statistical science*, J. Appl. Probab. Special Vol. 19A, pages 19–25. Applied Probability Trust, 1982.

M92 C. C. Heyde. Estimation in the presence of a threshold theorem: principles and their illustration for the traffic intensity. In P. K. G. Kallianpur and J. Ghosh, editors, *Statistics and probability: essays in honor of C. R. Rao*, pages 317–323. North-Holland, Amsterdam, 1982.

M93 C. C. Heyde. On the asymptotic behavior of random walks on an anisotropic lattice. *J. Statist. Phys.*, 27(4):721–730, 1982.

M94 C. C. Heyde and J. Gani. Statistics. In *Mathematical Sciences in Australia 1981*, pages 60–70. Australian Academy of Science, Canberra, 1982.

M95 C. C. Heyde, M. Westcott, and E. R. Williams. The asymptotic behavior of a random walk on a dual-medium lattice. *J. Statist. Phys.*, 28(2):375–380, 1982.

M96 D. Najock and C. C. Heyde. On the number of terminal vertices in certain random trees with an application to stemma construction in philology. *J. Appl. Probab.*, 19(3):675–680, 1982.

M97 A. G. Pakes and C. C. Heyde. Optimal estimation of the criticality parameter of a supercritical branching process having random environments. *J. Appl. Probab.*, 19(2):415–420, 1982.

M98 C. C. Heyde. An alternative approach to asymptotic results on genetic composition when the population size is varying. *J. Math. Biol.*, 18:163–168, 1983.

M99 C. C. Heyde. The effect on humans of exposure to herbicides: a contentious contemporary problem in statistical inference. *Math. Sci.*, 8:63–73, 1983.

M100 C. C. Heyde. Invariance principles and functional limit theorems. In N. L. Johnson and S. Kotz, editors, *Encyclopedia of Statistical Sciences*, volume 4, pages 225–228. Wiley, New York, 1983.

M101 C. C. Heyde. Laws of large numbers. In N. L. Johnson and S. Kotz, editors, *Encyclopedia of Statistical Sciences*, volume 4, pages 566–568. Wiley, New York, 1983.

M102 C. C. Heyde. Laws of the iterated logarithm. In N. L. Johnson and S. Kotz, editors, *Encyclopedia of Statistical Sciences*, volume 4, pages 528–530. Wiley, New York, 1983.

M103 C. C. Heyde. Limit theorems, central. In N. L. Johnson and S. Kotz, editors, *Encyclopedia of Statistical Sciences*, volume 4, pages 651–655. Wiley, New York, 1983.

M104 C. C. Heyde. The sociology of discovery in pre-20th century probability and statistics. I. Period pre-1830. *Math. Sci.*, 8:1–10, 1983.

M105 C. C. Heyde. On limit theorems for gene survival. In P. Revesz, editor, *Limit theorems in probability and statistics, Vol. II (Veszprém, 1982)*, volume 36 of *Colloq. Math. Soc. János Bolyai*, pages 573–586. North-Holland, Amsterdam, 1984.

M106 C. C. Heyde and T. Nakata. On the asymptotic equivalence of *Lp* metrics for convergence to normality. *Z. Wahrsch. Verw. Gebiete.*, 68(1):97–106, 1984.

M107 C. C. Heyde. An asymptotic representation for products of random matrices. *Stochastic Process. Appl.*, 20(2):307–314, 1985.

Indicates article found in this volume

M108 C. C. Heyde. The multidimensional central limit theorem. In N. L. Johnson and S. Kotz, editors, *Encyclopedia of Statistical Sciences*, volume 5, pages 643–646. Wiley, New York, 1985.

M109 C. C. Heyde. On inference for demographic projection of small populations. In *Proceedings of the Berkeley conference in honor of Jerzy Neyman and Jack Kiefer, Vol. I (Berkeley, Calif., 1983)*, Wadsworth Statist./Probab. Ser., pages 215–223, Belmont, CA, 1985. Wadsworth.

M110 C. C. Heyde. On macroscopic stochastic modelling of systems subject to criticality. *Math. Sci.*, 10(1):3–8, 1985.

M111 C. C. Heyde. On some new probabilistic developments of significance to statistics: martingales, long range dependence, fractals, and random fields. In A. Atkinson and S. Fienberg, editors, *A Celebration of Statistics. The ISI Centenary Volume*, pages 355–368. Springer, New York, 1985.

M112 C. C. Heyde and J. E. Cohen. Confidence intervals for demographic projections based on products of random matrices. *Theoret. Population Biol.*, 27(2):120–153, 1985.

M113 C. C. Heyde. On the use of time series representations of population models. In J. M. Gani and M. B. Priestley, editors, *Essays in time series and allied processes*, J Applied Prob. Special Vol. 23A, pages 345–353. Applied Probability Trust, 1986.

M114 C. C. Heyde. Probability theory (outline). In N. L. Johnson and S. Kotz, editors, *Encyclopedia of Statistical Sciences*, volume 7, pages 249–252. Wiley, New York, 1986.

M115 C. C. Heyde. Quantile transformation methods. In N. L. Johnson and S. Kotz, editors, *Encyclopedia of Statistical Sciences*, volume 7, pages 432–433. Wiley, New York, 1986.

M116 C. C. Heyde. Random matrices. In N. L. Johnson and S. Kotz, editors, *Encyclopedia of Statistical Sciences*, volume 7, pages 549–551. Wiley, New York, 1986.

M117 C. C. Heyde. Random sum distributions. In N. L. Johnson and S. Kotz, editors, *Encyclopedia of Statistical Sciences*, volume 7, pages 565–567. Wiley, New York, 1986.

📖 M118 V. P. Godambe and C. C. Heyde. Quasi-likelihood and optimal estimation. *Internat. Statist. Rev.*, 55(3):231–244, 1987.

M119 C. C. Heyde. On combining quasilikelihood estimating functions. *Stochastic Process. Appl.*, 25(2):281–287, 1987.

M120 C. C. Heyde. Optimality in estimation for stochastic processes under both fixed and large sample conditions. In Yu. V. Prohorov, V. A. S. cius, V. V. Sazonov, and B. Griegelionis, editors, *Probability theory and mathematical statistics, Vol. I (Vilnius, 1985)*, pages 535–541. VNU Sci. Press, Utrecht, 1987.

M121 P. M. Kulkarni and C. C. Heyde. Optimal robust estimation for discrete time stochastic processes. *Stochastic Process. Appl.*, 26(2):267–276, 1987.

M122 C. C. Heyde. Asymptotic efficiency results for the method of moments with application to estimation for queueing processes. In *Queueing theory and its applications*, volume 7 of *CWI Monogr.*, pages 405–412. North-Holland, Amsterdam, 1988. Liber Amicorum for J.W. Cohen.

📖 Indicates article found in this volume

M123 C. C. Heyde. Fixed sample and asymptotic optimality for classes of estimating functions. In *Statistical inference from stochastic processes (Ithaca, NY, 1987)*, volume 80 of *Contemp. Math.*, pages 241–247. Amer. Math. Soc., Providence, RI, 1988.

M124 C. C. Heyde. Official statistics in the late colonial period leading on to the work of the first Commonwealth Statistician, G. H. Knibbs. In C. C. Heyde and E. Seneta, editors, *Bicentennial history issue*, Austral. J. Statist. Special Vol. 30B, pages 23–43. Australian Statistical Publishing Association, 1988.

M125 C. C. Heyde. Some thoughts on stationary processes and linear time series analysis. In J. M. Gani, editor, *A Celebration of Applied Probability*, J. Appl. Probab. Special Vol. 25A, pages 309–318. 1988.

M126 C. C. Heyde. On efficiency for quasi-likelihood and composite quasi-likelihood methods. In Y. Dodge, editor, *Statistical data analysis and inference (Neuchâtel, 1989)*, pages 209–213. North-Holland, Amsterdam, 1989.

M127 C. C. Heyde. Quasi-likelihood and optimality for estimating functions: some current unifying themes. In *Proceedings of the 47th Session of the International Statistical Institute, Book 1 (Paris, 1989)*, volume 53, pages 19–29, 1989. The Fisher Lecture.

M128 C. C. Heyde and R. Gay. On asymptotic quasi-likelihood estimation. *Stochastic Process. Appl.*, 31(2):223–236, 1989.

M129 N. G. Becker and C. C. Heyde. Estimating population size from multiple recapture experiments. *Stochastic Process. Appl.*, 36(1):77–83, 1990.

M130 R. Gay and C. C. Heyde. On a class of random field models which allows long range dependence. *Biometrika*, 77(2):401–403, 1990.

M131 C. C. Heyde. A view from the mathematical sciences. In *Proceedings of the Forum on the Profile of Australian Science, ASTEC, Canberra*, pages 41–44, 1990.

M132 C. C. Heyde. Patrick Alfred Pierce Moran 1917–1988. *Biographical Memoirs of Fellows of the Royal Society*, 37:366–379, 1991. Also in *Historical Records of Australian Science,* 9:17–30, 1992.

M133 C. C. Heyde. The promotion and development of Applied Probability: a note on the contributions of J. Gani. In I. V. Basawa and R. L. Taylor, editors, *Selected Proceedings of the Symposium on Applied Probability*, volume 18 of *IMS Monograph Series*, pages 9–11. Inst. Math. Statist., Hayward, Calif., 1991.

M134 C. C. Heyde and Y.-X. Lin. Approximate confidence zones in an estimating function context. In V. P. Godambe, editor, *Estimating Functions*, volume 7 of *Oxford Statist. Sci. Ser.*, pages 161–168. Oxford Univ. Press, New York, 1991.

M135 J. Gani and C. C. Heyde. On the retirement of Mavis Hitchcock. *J. Appl. Probab.*, 29:484–486, 1992.

M136 C. C. Heyde. New developments in inference for temporal stochastic processes. *J. Statist. Plann. Inference*, 33:121–129, 1992.

M137 C. C. Heyde. On best asymptotic confidence intervals for parameters of stochastic processes. *Ann. Statist.*, 20(1):603–607, 1992.

M138 C. C. Heyde. Some results on inference for stationary processes and queueing systems. In U. N. Bhat and I. V. Basawa, editors, *Queue-*

Indicates article found in this volume

ing and related models, volume 9 of *Oxford Statist. Sci. Ser.*, pages 337–345. Oxford Univ. Press, New York, 1992.

M139 C. C. Heyde and R. Gay. Thoughts on the modelling and identification of random processes and fields subject to possible long-range dependence. In L. H. Y. Chen, K. P. Choi, K. Hu, and J.-H. Lou, editors, *Probability theory (Singapore, 1989)*, pages 75–81. de Gruyter, Berlin, 1992.

M140 C. C. Heyde and Y.-X. Lin. On quasi-likelihood methods and estimation for branching processes and heteroscedastic regression models. *Austral. J. Statist.*, 34(2):199–206, 1992.

M141 C. C. Heyde. Allan, Frances Elizabeth (1905–1952). In *Australian Dictionary of Biography*, volume 13, pages 26–27. Melbourne Univ. Press, 1993.

M142 C. C. Heyde. Asymptotics for two-dimensional anisotropic random walks. In R. L. K. S. Cambanis, J. K. Ghosh and P. K. Sen, editors, *Stochastic Processes: A Festschrift in Honour of Gopinath Kallianpur*, pages 125–130. Springer, New York, 1993.

M143 C. C. Heyde. Quasi-likelihood and general theory of inference for stochastic processes. In A. Obretenov and V. T. Stefanov, editors, *7th International Summer School on Probability Theory and Mathematical Statistics, Lecture Notes*, pages 122–152. Science Culture Technology Publishing, Singapore, 1993.

M144 C. C. Heyde. Societies, membership and the future of ISI. *Bull. Int. Statist. Inst.*, 49, book 1:79–84, 1993.

M145 C. C. Heyde. Weak convergence of probability measures. In *Soviet Encyclopedia of Mathematics*, volume 9, pages 448–449. Kluwer, Dordrecht, 1993.

M146 C. C. Heyde and R. Gay. Smoothed periodogram asymptotics and estimation for processes and fields with possible long-range dependence. *Stochastic Process. Appl.*, 45(1):169–182, 1993.

M147 C. C. Heyde and R. Morton. On constrained quasi-likelihood estimation. *Biometrika*, 80(4):755–761, 1993.

M148 Y.-X. Lin and C. C. Heyde. Optimal estimating functions and Wedderburn's quasi-likelihood. *Comm. Statist. Theory Methods*, 22(8):2341–2350, 1993.

M149 C. C. Heyde. Edward James Hannan (29 January 1921–7 January 1994). *Austral. Math. Soc. Gaz.*, 21(4):121–122, 1994.

M150 C. C. Heyde. On the effects of noise in systems involving products of random matrices. In M. L. Puri and J. P. Vilaplana, editors, *Recent Advances in Statistics and Probability*, pages 277–284. VSP International Science Publishers, Utrecht, 1994.

🕮 M151 C. C. Heyde. A quasi-likelihood approach to estimating parameters in diffusion-type processes. In J. Galambos and J. Gani, editors, *Studies in applied probability*, J. Appl. Probab. Special Volume 31A, pages 283–290. Applied Probability Trust, 1994.

M152 C. C. Heyde. A quasi-likelihood approach to the REML estimating equations. *Statist. Probab. Lett.*, 21(5):381–384, 1994.

M153 K. Chen and C. C. Heyde. On asymptotic optimality of estimating functions. *J. Statist. Plann. Inference*, 48(1):101–112, 1995.

🕮 Indicates article found in this volume

M154 W. Dai and C. C. Heyde. On the robustness of limit theorems. In *Bull. Int. Statist. Inst. 50th session*, volume 56, book 2, pages 549–555, 1995.

M155 C. Heyde. A conversation with Joe Gani. *Statist. Sci.*, 10(2):214–230, 1995.

M156 W. Dai and C. C. Heyde. Itô's formula with respect to fractional Brownian motion and its application. *J. Appl. Math. Stoch. Anal.*, 9(4):439–448, 1996.

M157 C. C. Heyde. On the use of quasi-likelihood for estimation in hidden-Markov random fields. *J. Statist. Plann. Inference*, 50(3):373–378, 1996.

M158 C. C. Heyde and W. Dai. On the robustness to small trends of estimation based on the smoothed periodogram. *J. Time Ser. Anal.*, 17(2):141–150, 1996.

M159 C. C. Heyde and R. Morton. Quasi-likelihood and generalizing the EM algorithm. *J. R. Stat. Soc. Ser. B Stat. Methodol.*, 58(2):317–327, 1996.

M160 C. C. Heyde. Asymptotic optimality. In M. Hazewinkel, editor, *Encyclopedia of Mathematics, Suppl. Vol. I*, pages 69–70. Kluwer, Dordrecht, 1997.

M161 C. C. Heyde. Avoiding the likelihood. In I. V. Basawa, V. P. Godambe, and R. L. Taylor, editors, *Selected Proceedings of the Symposium on Estimating Functions (Athens, GA, 1996)*, volume 32 of *IMS Lecture Notes Monogr. Ser.*, pages 35–42. Inst. Math. Statist., 1997.

M162 C. C. Heyde and Y. Yang. On defining long-range dependence. *J. Appl. Probab.*, 34(4):939–944, 1997.

M163 Y.-X. Lin and C. C. Heyde. On spaces of estimating functions. *J. Statist. Plann. Inference*, 63(2):255–264, 1997.

M164 C. C. Heyde. The chasm between the scientist and the media. In A. M. Herzberg and I. Krupka, editors, *Statistics, Science and Public Policy. Proceedings of a Conference held at Herstmonceaux Castle, Hailsham, UK, April 10-13, 1996*, chapter 5, pages 31–34. Queen's University, Kingston, Ontario, Canada, 1998.

M165 C. C. Heyde. Risk assessment for catastrophic events. In *Statistics, Science and Public Policy II, Hazards and Risks. Proceedings of a Conference held at Queen's University, Kingston, Ontario, Canada, April 23–25, 1997*, chapter 13, pages 85–87. Queens's University, 1998.

M166 C. C. Heyde. Role of national academies. In A. M. Herzberg and I. Krupka, editors, *Statistics, Science and Public Policy. Proceedings of a Conference held at Herstmonceaux Castle, Hailsham, UK, April 10-13, 1996*, chapter 11, pages 65–68. Queen's University, Kingston, Ontario, Canada, 1998.

M167 C. C. Heyde and R. Morton. Multiple roots in general estimating equations. *Biometrika*, 85(4):954–959, Dec. 1998.

M168 V. V. Anh, C. C. Heyde, and Q. Tieng. Stochastic models for fractal processes. *J. Statist. Plann. Inference*, 80(1-2):123–135, 1999.

M169 C. C. Heyde. A risky asset model with strong dependence through fractal activity time. *J. Appl. Probab.*, 36(4):1234–1239, 1999.

M170 A. Thavaneswaran and C. C. Heyde. Prediction via estimating functions. *J. Statist. Plann. Inference*, 77(1):89–101, 1999.

Indicates article found in this volume

M171 J. Gao, V. Anh, C. Heyde, and Q. Tieng. Parameter estimation of stochastic processes with long-range dependence and intermittency. *J. Time Ser. Anal.*, 22(5):517–535, 2001.

M172 C. C. Heyde. Agner Krarup Erlang. In C. C. Heyde and E. Seneta, editors, *Statisticians of the centuries*, pages 328–330. Springer-Verlag, New York, 2001.

M173 C. C. Heyde. George Handley Knibbs. In C. C. Heyde and E. Seneta, editors, *Statisticians of the centuries*, pages 257–260. Springer-Verlag, New York, 2001.

M174 C. C. Heyde. John Graunt. In C. C. Heyde and E. Seneta, editors, *Statisticians of the centuries*, pages 14–16. Springer-Verlag, New York, 2001.

M175 C. C. Heyde. The changing scene for higher education in science: a view from the statistical profession. In *Statistics, Science and Public Policy V, Society, Science and Education. Proceedings of a Conference held at Herstmonceaux Castle, Hailsham, UK, April, 26–29, 2000*, chapter 2, pages 13–23. Queens's University, Kingston, Ontario, Canada, 2001.

M176 C. C. Heyde. On option pricing with the FATGBM risky asset model. In *Bull. Int. Statist. Inst. 53rd session*, volume 59, book 3, pages 313–314, 2001.

M177 C. C. Heyde. An overview of the Symposium on Inference for Stochastic Processes. In I. V. Basawaa, C. C. Heyde, and R. Taylor, editors, *Selected Proceedings of the Symposium on Inference for Stochastic Processes: Held at the University of Georgia, Athens, GA, May 10–12, 2000*, volume 37 of *IMS Lecture Notes-Monograph Series*, pages 3–7. Institute of Mathematical Statistics, 2001.

M178 C. C. Heyde. Probabilistic models. In A. H. El-Shaarawi and W. W. Piegorsch, editors, *Encyclopedia of Environmetrics*, volume 3, pages 1637–1638. Wiley, Chichester, 2001.

M179 C. C. Heyde. Shifting paradigms in inference. In I. V. Basawaa, C. C. Heyde, and R. Taylor, editors, *Selected Proceedings of the Symposium on Inference for Stochastic Processes (Athens, GA, 2000)*, volume 37 of *IMS Lecture Notes Monogr. Ser.*, pages 9–21. Inst. Math. Statist., Beachwood, OH, 2001. Held at the University of Georgia, Athens, GA, May 10–12, 2000.

M180 C. C. Heyde, S. Liu, and R. Gay. Fractal scaling and Black-Scholes: the full story. A new view of long-range dependence in stock prices. *Journal of the Australian Society of Security Analysts*, Issue 1:29–32, 2001.

M181 C. C. Heyde and S. Liu. Empirical realities for a minimal description risky asset model. The need for fractal features. *J. Korean Math. Soc.*, 38(5):1047–1059, 2001. Principal Invited Paper, Mathematics in the New Millenium Conference, Yonsei University, Seoul, Korea, October 2000.

M182 A. Thavaneswaran and C. C. Heyde. A note on filtering for long memory processes. In S. Mittnik, editor, *Stable non-Gaussian models in finance and econometrics*, volume 34 of *Math. Comput. Modelling*, pages 1139–1144. Pergamon, 2001.

M183 V. V. Anh, C. C. Heyde, and N. N. Leonenko. Dynamic models of long-memory processes driven by Lévy noise. *J. Appl. Probab.*, 39(4):730–747, 2002.

📖 M184 J. Gao, V. Anh, and C. Heyde. Statistical estimation of nonstationary Gaussian processes with long-range dependence and intermittency. *Stochastic Process. Appl.*, 99(2):295–321, 2002.

M185 C. C. Heyde. Challenges facing statistics for the 21st century. In S. M. Chandra Gulati, Yan-Xia Lin and J. Rayner, editors, *Advances in statistics, combinatorics and related areas*, pages 238–244. World Scientific, River Edge, NJ, 2002. Selected papers from the International Conference on Statistics, Combinatorics and Related Areas and the 8th International Conference of the Forum for the Interdisciplinary Mathematics (SCRA2001-FIMVIII) held at the University of Wollongong, Wollongong, December 19–21, 2001.

M186 C. C. Heyde. On modes of long-range dependence. *J. Appl. Probab.*, 39(4):882–888, 2002.

M187 S. Liu and C. C. Heyde. Some efficiency comparisons for estimators from quasi-likelihood and generalized estimating equations. In M. Moore, S. Froda, and C. Léger, editors, *Mathematical statistics and applications: Festschrift for Constance van Eeden*, volume 42 of *IMS Lecture Notes Monogr. Ser.*, pages 357–371. Inst. Math. Statist., Beachwood, OH, 2003.

M188 C. C. Heyde. Asymptotics and criticality for a correlated Bernoulli process. *Aust. N. Z. J. Stat.*, 46(1):53–57, 2004. Festschrift in honour of Daryl Daley.

M189 C. C. Heyde and A. Irle. On subordinated market models. In B. M. de Silva and N. Mukhopadhyay, editors, *Proceedings of the International Sri Lankan Statistical Conference: Visions of Futuristic Methodologies, December 2004*, pages 1–15. Postgraduate Inst. Science, Univ. Peradenya, Sri Lanka, 2004.

M190 C. C. Heyde and S. G. Kou. On the controversy over tailweight of distributions. *Oper. Res. Lett.*, 32(5):399–408, 2004.

M191 B. Wong and C. C. Heyde. On the martingale property of stochastic exponentials. *J. Appl. Probab.*, 41(3):654–664, 2004.

M192 C. C. Heyde and N. N. Leonenko. Student processes. *Adv. in Appl. Probab.*, 37(2):342–365, 2005.

M193 K. Au and C. C. Heyde. On the problem of discriminating between the tails of distributions. In S. L. P. Brown and D. Sharma, editors, *Contributions to Probability and Statistics: Applications and Challenges, Proceedings of the University of Canberra International Statistical Workshop, April 2005*, pages 246–258. World Scientific, Singapore, 2006.

M194 B. Wong and C. C. Heyde. On changes of measure in stochastic volatility models. *J. Appl. Math. Stoch. Anal.*, 2006. Art. ID 18130.

M195 C. C. Heyde and K. Au. A cautionary note on model choice and the Kullback-Leibler information. *J. Stat. Theory Pract.*, 2:221–232, 2008.

M196 C. C. Heyde and A. Sly. A cautionary note on modeling with fractional Lévy flights. *Phys. A*, 387(21):5024–5032, Sep. 1 2008.

M197 A. Sly and C. Heyde. Nonstandard limit theorem for infinite variance functionals. *Ann. Probab.*, 36(2):796–805, 2008.

📖 Indicates article found in this volume

M198 C. C . Heyde. Scaling issues for risky asset modelling. *Mathematical Methods of Operations Research*, 69:593–603, 2009.

M199 C. C. Heyde and D. Wang. Finite-time ruin probability with an exponential Lévy process investment return and heavy-tailed claims. *Adv. in Appl. Probab.*, 41:206–224, 2009.

M200 C. C. Heyde. The central limit theorem. In J. Teugels and B. Sundt, editors, *Encyclopedia of Actuarial Science*. Wiley, Chichester, to appear.

Abstracts

A1 C. C. Heyde. Estimation of parameters for stochastic processes. *Math. Sci.*, 1:3–5, 1976. Abstract of invited paper given to the First Conference of the CSIRO Division of Mathematics and Statistics, Sydney, February 1975.

A2 C. C. Heyde. On the rate of convergence in the martingale convergence theorem. *Adv. in Appl. Probab.*, 9:196, 1977. Abstract of invited paper given to the 6th Conference on Stochastic Processes and their Applications, Tel Aviv, June 1976.

A3 C. C. Heyde. Some rate of convergence results for the martingale convergence theorem. *Math. Sci.*, 2:141–143, 1977. Abstract of invited paper given to the 3rd Conference of the Statistical Society of Australia, Melbourne, August 1976.

A4 C. C. Heyde. On central limit and iterated logarithm results for subadditive processes. *Adv. in Appl. Probab.*, 11:283–284, 1979. Abstract of paper given to the 8th Conference on Stochastic Processes and their Applications, Canberra, July 1978.

Book Reviews Authored

R1 C. C. Heyde. Review of *Bounds on the rate of convergence of moments in the central limit theorem*, by P. Hall. *Mathematical Reviews*, 84a, 1984. MR672300 (84a #60035).

R2 C. C. Heyde. Review of *The central limit theorem around 1935* by L. Le Cam. *Mathematical Reviews*, 87h, 1987. MR833276 (87h #60054).

R3 L. Le Cam. Review of *La statistique mathématique depuis 1950* by L. Le Cam. *Mathematical Reviews*, 2002a, 2002. MR1796857 (2002a #62001).

Discussion Contributions

D1 C. C. Heyde. Discussion of *Some statistical applications of Poisson's work*, by I. J. Good. *Statist. Sci.*, 1:176–177, 1986.

D2 C. C. Heyde. Comments on the discussion of the paper *An extension of quasi-likelihood estimation*, by V. P. Godambe and M. E. Thompson. *J. Statist. Plann. Inference*, 22(2):160, 1989.

D3 C. C. Heyde. Discussion of *Eliminating multiple root problems in estimation*, by C. G. Small, J. Wang and Z. Yang. *Statist. Sci.*, 15(4):334–335, 2000.

D4 C. C. Heyde. Discussion of *Evidence functions and the optimality of the law of likelihood*, by S. Lele. In *The Nature of Scientific Evidence: Empirical, Statistical and Philosophical Considerations*, pages 203–205. Univ. Chicago Press, Chicago, IL, 2004.

Other Contributions

C1 C. C. Heyde. *Results Related to First Passage Time Problems and Some of Their Applications.* Ph.d. dissertation, Australian National University, 1964.

C2 C. C. Heyde. A limit theorem for random walks with drift. Technical report, Aarhus Universitet Matematisk Institut, 1966. (A seven page article with the same title, published in J. App. Probab., also appears in this bibliography.).

C3 W. J. Ewens and C. C. Heyde. Obituary: Motoo Kimura. *J. Appl. Probab.*, 32(4):1142–1144, 1995.

C4 C. C. Heyde, editor. *Branching processes*, volume 99 of *Lecture Notes in Statistics*, New York, 1995. Springer-Verlag.

C5 C. C. Heyde. Fate more than roll of the dice. *The Canberra Times*, page 14, Aug. 6 1998.

C6 C. C. Heyde. Obituary: Richard L. Tweedie. Born 22 August 1947–died 7 June 2001. *Appl. Stoch. Models Bus. Ind.*, 18(1):1–2, 2002.

C7 C. C. Heyde. Method and system for modeling financial markets and assets using fractal activity time. US patent 6643631, Nov. 4 2003.

C8 S. Liu and C. C. Heyde. On estimation in conditional heteroskedastic time series models under non-normal distributions, 2008.

C9 C. C. Heyde, S. G. Kou, and X. H. Peng. What is a good external risk measure: Bridging the grips between robustness, subadditivity and insurance risk measures. Submitted.

Interviews and Biographies

I1 Pitman Medal awarded to C. C. Heyde. *Austral. J. Statist.*, 32(3):259–260, 1990.

I2 Christopher Charles Heyde, AM, DSc, FAA, FASSA. In J. Gani and E. Seneta, editors, *Stochastic Methods and Their Applications: Papers in Honour of Chris Heyde*, J. Appl. Probab. Special Volume 41A, pages vii–x. Applied Probability Trust, Sheffield, 2004.

I3 P. Glasserman and S. Kou. A conversation with Chris Heyde. *Statist. Sci.*, 21(2):286–298, 2006.

I4 J. Gani. Life and work of Christopher Charles Heyde: In memoriam. *J. Stat. Theory Pract.*, 2:327–329, 2008.

I5 J. Gani and E. Seneta. Obituary: Christopher Charles Heyde AM, DSc, FAA, FASSA. *J. Appl. Probab.*, 45(3):587–592, 2008.

I6 L. Nash. Chris Heyde: an appreciation. *J. Appl. Probab.*, 45(3):593–594, 2008.

Miscellaneous

Z1 Publications of Christopher Charles Heyde. In J. Gani and E. Seneta, editors, *Stochastic Methods and Their Applications: Papers in Honour of Chris Heyde*, J. Appl. Probab. Special Volume 41A, pages xi–xviii. Applied Probability Trust, Sheffield, 2004.

Z2 J. Gani and E. Seneta, editors. *Stochastic Methods and Their Applications: Papers in Honour of Chris Heyde*. J. Appl. Probab. Special Volume 41A. Applied Probability Trust, Sheffield, 2004.

Author's Pick

C.C. Heyde[1]

Paper [M1] shows by example that the lognormal distribution is not determined by its moments. It is referenced in Feller's second volume (Feller 1971) and has become a widely cited result.

Paper [M43] with Ted Hannan has the key time series result that the best linear predictor is the best predictor if the innovations are martingale differences. This has major implications for the linear/nonlinear debate, and in this respect has received much attention, especially from econometricians. This was one of the very early papers which made it clear that martingales would play an important role in statistics.

Paper [M44] illustrates a realization of the emerging role which martingales were to play in probability and statistics. The derivative of the log likelihood is a martingale. A general Central Limit Theorem for martingales emerged. This research was influenced by the first edition of Billingsley's then recent book (Billingsley (1999)).

Paper [M162] with Y. Yang was a serious attempt to bridge the gap between long range dependence as a property of processes and the inflexible and restricted formal definitions. "Long range dependence" is really a misnomer. The term persistence is better at catching the idea of large values following large values and small values following small values much more than under independence. Correlation plays no role in the concept, nor is there need for stationarity or finite variances, so mathematical definitions involving these restrictions are unhelpful. In this paper a self-normalised limit criterion is included to characterize the concept. None of the correlation related restrictions apply, but if they do happen to hold, the definitions of this paper are almost equivalent to the usual forms. The framework of the paper easily allows for treatment of the persistence of self-similar infinite variance processes such as fractional Lévy noises, which it had hitherto not been possible to integrate with that of fractional Brownian motion.

Paper [M169] introduces the fractal activity time geometric Brownian motion (FATGBM) risky asset model (also see [M180], [M181], [M189]). This is a minimal description model aimed at capturing the essentially empirical properties which the Geometric Brownian Motion (GBM) lacks. These are leptokurtic returns distributions (higher peaks and fatter tails than normal), persistent strong dependence of absolute values of functions of the returns and evident stochastic volatility effects. The efficient market hypothesis applies to the FATGBM model, the log returns being martingale differences. Empirical scales of the FATGBvM model strongly suggest fractal scaling of the activity time and empirical volatility, and provide considerable support for the need for power tailed return distributions rather than the somewhat more popular (tractable) exponentially tailed ones.

References

[1] Billingsley, P. (1999) *Convergence of Probability Measures*, 2nd ed., Wiley, New York
[2] Feller, W. (1971) *An Introduction to Probability Theory and its Applications*, Vol. II, 3rd ed., Wiley, New York

[1]Centre for Mathematical Analysis, Australian National University, Canberra, ACT

R. Maller et al. (eds.), *Selected Works of C.C. Heyde*, Selected Works in Probability and Statistics,
DOI 10.1007/978-1-4419-5823-5_1, © Springer Science+Business Media, LLC 2010

Chris Heyde's Contribution to Inference in Stochastic Processes

Ishwar Basawa[1]

This overview of Heyde's research on inference in stochastic processes is based on a subjective sampling of his extensive publications in this area.

The question of asymptotic efficiency of the maximum likelihood (ML) estimator was first addressed by R.A. Fisher (1925). Fisher's criterion of efficiency was further explored by C.R. Rao (1962). The Fisher-Rao approach seeks to establish the asymptotic optimality of the maximum likelihood estimator in the class of consistent and asymptotically normal estimators by showing that the ML estimator has the minimum asymptotic variance, under some broad regularity conditions. LeCam (1960) provided a general framework for asymptotics via the local asymptotic normal (LAN) family which formalizes the theoretical basis needed for the Fisher-Rao approach. Even though most of the early work on asymptotic inference was developed for independent and identically distributed (i.i.d.) observations, the theory has since been extended to stochastic processes. The extension to dependent observations mainly involves replacing the classical law of large numbers and central limit theorem for the likelihood score function by their analogues for martingales (Hall and Heyde (1980)).

In [M62], Heyde's early work on asymptotic efficiency in estimation for branching processes led to some unexpected foundational questions about the classical Fisher-Rao-LeCam approach. Suppose $L_n(\theta)$ is the likelihood function based on n dependent observations $(X_1, X_2, \ldots, X_n)$, and

$$\xi_n(\theta) = -\frac{d^2 \ln L_n(\theta)}{d\theta\, d\theta^T}.$$

The Fisher information matrix is denoted by $I_n(\theta) = E(\xi_n(\theta))$. For a large class of processes (including the classical i.i.d. model), typically one can verify that $I_n^{-1}(\theta)\xi_n(\theta)$ converges in probability to a non-random matrix (usually, the identity matrix). When this happens, the classical Fisher-Rao-LeCam asymptotics usually can be extended to dependent observations. However, for branching processes (and many other stochastic processes) the normalized information $I_n^{-1}(\theta)\xi_n(\theta)$ has a non-degenerate limit. For such processes, the Fisher-Rao-LeCam asymptotic optimality criterion no longer applies. In particular, the ML estimator is not asymptotically normal, and the LAN property fails. In [M62], Heyde provided a remarkably simple modification of Rao's (1962) asymptotic efficiency criterion by replacing $I_n(\theta)$ in Rao's criterion by $\xi_n(\theta)$. In [M62], an estimator T_n is defined to be asymptotically efficient if it satisfies:

$$\text{(H)} \qquad \xi_n^{\frac{1}{2}}(\theta)\left[(T_n - \theta) - \sigma_T(\theta)\xi_n^{-1}(\theta)\frac{d \ln L_n(\theta)}{d\theta}\right] \xrightarrow{P} 0,$$

[1]Department of Statistics, University of Georgia, Athens, Georgia 30602, USA

R. Maller et al. (eds.), *Selected Works of C.C. Heyde*, Selected Works in Probability and Statistics,
DOI 10.1007/978-1-4419-5823-5_2, © Springer Science+Business Media, LLC 2010

for some positive number σ_T. [M76] shows that the ML estimator $\hat{\theta}_{ML}$ satisfies (H) with σ_T equal to 1, and it maximizes the limiting probability of concentration

$$\text{(W)} \qquad \lim_{n\to\infty} P[I_n^{\frac{1}{2}}(\theta)\,|\,T_n - \theta\,| \le a]$$

for all $a>0$ (see also, Wolfowitz (1965)). Note that the criterion (W) does not require the asymptotic normality of T_n, nor does it need the LAN framework. Moreover, in [M137], Heyde demonstrates the advantages of using the random norm $\xi_n(\theta)$ in constructing asymptotically optimal confidence intervals.

Heyde and Johnstone in [M81] established the asymptotic normality of the posterior density for stochastic processes. Interestingly, this result is applicable for branching processes, thus avoiding the problem of a threshold parameter (the offspring mean of a branching process) drastically affecting the asymptotic inference. Bayesian asymptotic inference can then be based on the asymptotic posterior normality.

In the context of stochastic processes, it is often the case that complete specification of the likelihood function may not be available. It is then not possible to use the ML estimation or Bayesian approach. [B3] gives a comprehensive review of the theory of quasi-likelihood estimation. Let $\mu_t(\theta) = E(X_t \mid \mathcal{F}_{t-1})$ and $V_t(\theta) = \mathrm{Var}(X_t \mid \mathcal{F}_{t-1})$ denote the conditional mean and variance respectively. A typical quasi-likelihood (QL) estimating equation is defined by

$$\sum_{t=1}^{n} V_t^{-1}(\theta)\left(\frac{d\mu_t(\theta)}{d(\theta)}\right)(X_t - \mu_t(\theta)) = 0.$$

This method of estimation requires specification only of the conditional mean and variance. In [M118] Godambe and Heyde, and in [M127], Heyde, discussed both finite sample and asymptotic optimality properties of the QL estimating functions. See also [M151] for quasi-likelihood estimation for diffusion processes.

Stochastic processes with long-range dependence are becoming increasingly important tools for modeling financial, meteorological and diverse other time series data. Heyde's joint work with Anh, Gao, et al. in [M171] and [M184], provides an excellent example of path-breaking research in this important area.

In [M171] estimation for a zero-mean stationary Gaussian process $\{Y(t)\}$ with spectral density

$$\phi(\omega,\theta) = \eta\,|\,\omega\,|^{-2\beta}\,(1+\omega^2)^{-\alpha},\ \omega \in (-\infty,\infty),$$

where $\theta = (\alpha,\beta,\eta) \in [\frac{1}{2},\infty)\times(0,\frac{1}{2})\times(0,\infty)$ is a vector of unknown parameters, is discussed. This process has long range dependence (LRD) and is known as the fractional Riesz-Bessel motion (fRBm). A continuous version of the Gauss-Whittle contrast function is defined as

$$L_N(\theta) = \frac{1}{4\pi}\int_{-\infty}^{\infty} \{\log\phi(\omega,\theta) + I_N(\omega)\phi^{-1}(\omega,\theta)\}(1+\omega^2)^{-1}d\omega$$

where

$$I_N(\omega) = \frac{1}{2\pi N}\left|\int_0^N e^{-i\omega t}Y(t)dt\right|^2$$

is the periodogram of $\phi(\omega,\theta)$, N being the upper bound of the interval $[0,N]$ on which $Y(t)$ is observed. The minimum contrast estimator of θ is defined by

$$\hat{\theta}_N = \arg\min_\theta L_N(\theta).$$

Consistency and asymptotic normality of the estimator $\hat{\theta}_N$ are established in [M171].

[M184] considers the nonstationary Gaussian process $\{Y(t)\}$ with spectral density given by

$$\phi^*(\omega,\theta) = \phi(\omega,\theta)\,\pi(\omega),\ \omega \in (-\infty,\infty),$$

where $\pi(\omega)$ is a positive continuous function satisfying a regularity condition, and $\theta \in [0,\infty) \times \left(\frac{1}{2}, \frac{3}{2}\right) \times (0,\infty)$ is the unknown parameter. Under regularity conditions, the consistency and asymptotic normality of the minimum contrast estimator $\hat{\theta}_N$ of θ are established.

The two papers [M171] and [M184] solve the problem of parameter estimation for both stationary and nonstationary Gaussian processes with long range dependence. The method of estimation used in both the papers is based on a natural Gauss-Whittle contrast function. These models are illustrated with an application to real data sets on time series of maximum daily wind speed [M171], and data on air pollution [M184].

References

[1] Fisher, R.A. (1925). Theory of statistical estimation. *Proc. Cambridge Phil. Soc.*, 11, 700–725.

[2] Hall, P.G. and Heyde, C.C. (1980). *Martingale Limit Theory and Its Applications*, Academic Press, NY.

[3] LeCam, L. (1960). Locally asymptotically normal families of distributions. *Univ. Calif. Pub. Prob. Statist.*, 3, 27–98.

[4] Rao, C.R. (1962). Efficient estimates and optimum inference procedures in large samples. *J. Roy. Statist. Soc. B*, 24, 46–72.

[5] Wolfowitz, J. (1965). Asymptotic efficiency of the maximum likelihood estimator. *Theor. Prob. Appl.*, 10, 247–260.

Chris Heyde's Work on Rates of Convergence in the Central Limit Theorem

Peter Hall[1]

The best-known result on rates of convergence in the central limit theorem is undoubtedly that of A.C. Berry and C.-G. Esseen, which describes the rate in the case of finite third moments. In particular, if $X, X_1, X_2, \ldots$ are independent and identically distributed random variables for which

$$E|X|^3 < \infty, \tag{1}$$

if $E(X)=0$ and $E(X^2)=1$, and if we define

$$S_n = \sum_{i=1}^{n} X_i,$$

then

$$\sup_{-\infty<x<\infty} | P(S_n \le n^{1/2} x) - \Phi(x) | \le A\, E|X|^3\, n^{-1/2}, \tag{2}$$

where Φ is the standard normal distribution function and A denotes an absolute constant.

The bound at (2) is the best possible, in the sense that for many distributions the convergence rate is exactly that given by (2). For example, Esseen also showed that if the distribution of X is non-lattice, and if (1) holds, then

$$P(S_n \le n^{1/2} x) = \Phi(x) + n^{-1/2} \frac{1}{6} E(X^3)\,(1-x^2)\,\phi(x) + o(n^{-1/2}), \tag{3}$$

uniformly in x as $n \to \infty$, where $\phi = \Phi'$ is the standard normal density function. However, the simplicity of (2) and (3) tends to obscure some of the more delicate, and therefore more interesting, properties of rates of convergence. Chris Heyde played a central role in revealing and exploring those important subtleties.

To access these issues we should mention first that the Berry-Esseen bound, at (2), is readily generalised to the case of moments lower than the third. In particular, if we replace condition (1) by the less stringent assumption that

$$E|X|^{2+\delta} < \infty, \tag{4}$$

where $0 < \delta \le 1$, then (2) admits the following more general form:

$$\sup_{-\infty<x<\infty} | P(S_n \le n^{1/2} x) - \Phi(x) | \le A_\delta\, E|X|^{2+\delta}\, n^{-\delta/2}, \tag{5}$$

---

[1]Department of Mathematics and Statistics, The University of Melbourne, Melbourne, VIC 3130, Australia
e-mail: p.hall@ms.unimelb.edu.au

R. Maller et al. (eds.), *Selected Works of C.C. Heyde*, Selected Works in Probability and Statistics,
DOI 10.1007/978-1-4419-5823-5_3, © Springer Science+Business Media, LLC 2010

where the constant A_δ depends only on δ, not on the distribution of X or on n. We know from expansions such as (3) that we cannot expect (5) to hold for $\delta>1$, and of course (5) is degenerate when $\delta=0$. However, when $0<\delta<1$ result (5) is rather conservative. In fact, in this case it can be shown that if (4) holds then

$$\sup_{-\infty<x<\infty} |P(S_n \le n^{1/2}x) - \Phi(x)| = o(n^{-\delta/2}). \tag{6}$$

From some viewpoints it is perhaps disappointing that the inequality (5) is so conservative for $\delta<1$. Moment conditions such as (4) are very important to statistics, and the architecture of probability theory would have been simpler had there been an elementary, concise interpretation of (4) in terms of convergence rates. One of the attractions of the Berry-Esseen theorem is that it gives (in many cases) an exact convergence rate under an elementary moment assumption. However, as Einstein once remarked, "If you are out to describe the truth, leave elegance to the tailor." The conservatism evidenced by (6) begs the question of what the truth might be on this occasion; what rates of convergence are characterised by the moment assumption (4)?

Heyde gave a beautiful answer to this question, so exquisite and memorable that it will endure while ever moment conditions are discussed in connection with the central limit theorem. He showed in [M12] that if $0<\delta<1$ then (4) holds if and only if

$$\sum_{n=1}^{\infty} n^{(\delta/2)-1} \sup_{-\infty<x<\infty} |P(S_n \le n^{1/2}x) - \Phi(x)| < \infty. \tag{7}$$

Results of [M7] are related. In [M45], Heyde and Leslie derived analogous properties connecting finiteness of moments to approximations by Edgeworth series.

Condition (4) is rather uninformative if $\delta=0$. In that instance it tells us that X has finite variance, and so the constants $n^{1/2}$ give an appropriate standardisation in the central limit theorem, but the amount of additional information it provides seems marginal, at least on the surface. Nevertheless, (7) is meaningful when $\delta=0$, and we are bound to ask what moment conditions are equivalent to that property, i.e. to:

$$\sum_{n=1}^{\infty} n^{-1} \sup_{-\infty<x<\infty} |P(S_n \le n^{1/2}x) - \Phi(x)| < \infty. \tag{8}$$

In [M12] Heyde solved this problem, too, showing that (8) holds if and only if

$$E\{X^2 \log(1+|X|)\} < \infty. \tag{9}$$

In [M60] Chris extended these results to rates of convergence in L_p metrics, showing, among other results, that if $0<\delta<1$ then (4) is equivalent to

$$\sum_{n=1}^{\infty} n^{(\delta/2)-1} \|F_n - \Phi\|_p < \infty \tag{10}$$

for $1\le p\le\infty$, where $F_n(x)=P(S_n\le n^{1/2}x)$ and, with $\Delta=F_n-\Phi$,

$$\|\Delta\|_p = \begin{cases} \left\{\int_{-\infty<x<\infty} |\Delta(x)|^p\, dx\right\}^{1/p} & \text{if } 1\le p<\infty, \\ \sup_{-\infty<x<\infty} |\Delta(x)| & \text{if } p=\infty. \end{cases}$$

Moreover, [M60] shows that if $\delta=0$ then (9) and (10) are equivalent. Heyde and Nakata in [M106] obtained related results addressing the effects of choice of p on convergence rates.

Theory for convergence rates in the central limit theorem is exceptionally delicate, and so it is unlikely that results such as these, striking though they may be, probe the very depths of the convergence-rate problem. Indeed, one is left wondering just what kinds of convergence rates are achievable under the condition of finite variance alone. In this circumstance, faster rates are sometimes possible if the sum S_n is centred not at its expected value but at a truncated moment, or is rescaled using a truncated measure of variance. Examples of such location and scale constants include

$$\mu_n = nE\{XI(|X| \le n^{1/2})\},\ \sigma_n^2 = nE\{x^2 I(|X| \le n^{1/2})\}. \tag{11}$$

In [M33] and [M49], Heyde proved that if centring and scaling constants are chosen to give the fastest possible convergence rate then the series at (8) converges under solely the assumption of finite variance. Indeed, if $X_1, X_2, \ldots$ come from a distribution in the domain of partial attraction of the normal distribution, then

$$E(X^2) < \infty \tag{12}$$

if and only if

$$\sum_{n=1}^{\infty} n^{-1} \inf_{c>0,-\infty<d<\infty} \sup_{-\infty<x<\infty} |P(S_n \le cx+d) - \Phi(x)| < \infty. \tag{13}$$

This equivalence continues to hold if we fix $d=0$ in (13), and seek only the infimum over $c>0$. Moreover, the series in (13) converges if we take either $(c,d)=(\sigma_n,\mu_n)$ or $(c,d)=(\sigma_n,0)$, for each n, where μ_n and σ_n are given by (11). A further investigation of convergence rates shows that choosing $(c,d)=(\mu_n,\sigma_n)$ can give improved convergence rates, relative to $(c,d)=(0,\sigma_n)$, although the improvements are not expressible in terms of convergence of the series at (13).

The equivalence of (12) and (13) was proved independently by Egorov (1973), although under the slightly stronger condition that the distribution of X is in the domain of attraction, not the domain of partial attraction, of the normal law. In [M28] Chris had previously derived a similar result for rates of convergence in the L_1 metric. Egorov (1973) extended this to L_p metrics for $1 \le p \le 2$, and Rozovskiĭ (1978) treated the case $p \ge 1$. Ibragimov (1966), Osipov and Petrov (1967) and many others, working from the 1960s to the 1980s, developed a variety of related results.

This multiplicity of authors indicates the considerable interest attached to this line of research at the time. Heyde was competing with some of the strongest probabilists anywhere, and obtained definitive results. His work was fundamental to our understanding of the connections between moment conditions and convergence rates in the standard central limit theorem, in cases where that theorem is only barely valid; that is, where little more than finite variance is assumed.

References

[1] Egorov, V.A. (1973). The rate of convergence to the normal law that is equivalent to the existence of the second moment. (In Russian.) *Teor. Verojatnost. i Primenen.* **18**, 180–185.

[2] Ibragimov, I.A. (1966). On the accuracy of approximation by the normal distribution of distribution functions of sums of independent random variables. (In Russian.) *Teor. Verojatnost. i Primenen* **11**, 632–655.

[3] Osipov, L.V. and Petrov, V.V. (1967). On the estimation of the remainder term in the central limit theorem. (In Russian.) *Teor. Verojatnost. i Primenen* **12**, 322–329.

[4] Rozovskiĭ, L.V. (1978). The accuracy of an estimate of the remainder term in the central limit theorem. (In Russian.) *Teor. Veroyatnost. i Primenen.* **23**, 744–761.

Chris Heyde's Work in Probability Theory, with an Emphasis on the LIL

Ross Maller*,[1]

In one of his first few papers, [M5], Chris gives necessary and sufficient conditions for the convergence of the series

$$\sum_{n=1}^{\infty} n^k P(S_n \le x), \quad -\infty < x < \infty, \tag{1}$$

for integers $k=0,1,2,\ldots$, and for a similar series with n^k replaced by e^{rn}, where $r>0$. Here $S_n = X_1 + X_2 + \cdots + X_n$ is a random walk comprised of independent and identically distributed summands $(X_i)_{i=1,2,\ldots,n}$. Series of the kind in (1) are fundamental objects in renewal theory. Under the assumptions $E|X|<\infty$ and $EX>0$ (X is any random variable having the same distribution as the X_i), Chris shows that the series in (1) converges (for the designated value of k) if and only if $E|X^-|^{k+2}$ is finite, and the series with n^k replaced by e^{rn} converges for some $r>0$ if and only if X^- has an analytic characteristic function, or, equivalently, if $Ee^{-\theta X}$ is finite for all θ in an interval $[0,K]$, for some $K>0$. (X^- is defined as $\min(0,X)$.) Thus, the series in (1) converges if and only if the negative tail of the distribution of the X_i is not too heavy. Two aspects are worth noting; first, the X_i are allowed to take either sign (early work in renewal theory restricted S_n to be a "renewal process", i.e., comprised of non-negative summands); and, secondly, Chris obtains complete characterisations of the convergence of the series, subject only to the X_i having a finite and positive expectation. Only much later was this latter restriction relaxed, in connection with (1) (in Kesten and Maller (1996)).

In a follow-up paper to [M5], Chris in [M6] extends the study to look at the *magnitudes* (rather than just the convergence or not) of the series in (1) (and with n^k replaced by e^{rn}, $r>0$), considered as functions of x. Actually, slightly more general series of the form $\sum_n a_n P(S_n \le x)$, where $a_n>0$ is a deterministic sequence satisfying a certain regularity condition (related to regular variation), are studied. If k is the "index" of the sequence (the least real k such that $a_n = O(n^k)$ as $n\to\infty$) then, when it converges, the asymptotic behaviour of $\sum_n a_n P(S_n \le x)$, as $x\to\infty$, can be specified in terms of the asymptotic behaviour of a_n. Another theorem shows that the same result holds with S_n replaced by the *maximal* sum, $\max_{1\le j\le n} S_j$. The latter result is closely connected with the large sample behaviour (as $x\to\infty$) of moments of the first passage time, $M(x) := \inf\{n\ge 1 : S_n > x\}$, $x>0$; again, an object of primary interest in renewal theory and its applications.

[M5] and [M6] deal with the convergence of series of probabilities. A very natural next step, given such a convergence, is to apply the Borel-Cantelli lemma and make some kind of statement about the a.s. behaviour of the random walk (where "a.s." stands for "almost surely"). [M6] contains some information on the a.s. convergence of $\max_{1\le j\le n} S_j$, analogous to the Marcinkiewicz-Zygmund strong law for S_n, but Chris soon turned his attention to the more delicate issue of the law of the iterated logarithm

*This research was partially supported by ARC grant DP0664603.

[1]Centre for Mathematical Analysis, and School of Finance and Applied Statistics, Australian National University, Canberra, ACT, e-mail: Ross.Maller@anu.edu.au

R. Maller et al. (eds.), *Selected Works of C.C. Heyde*, Selected Works in Probability and Statistics,
DOI 10.1007/978-1-4419-5823-5_4, © Springer Science+Business Media, LLC 2010

(LIL). Hartman and Wintner (1941) had given this in its ultimate form: provided the increments of the random walk have finite variance σ^2 and expectation 0, we have

$$-1 = \liminf_{n\to\infty} \frac{S_n}{\sigma\sqrt{2n\log\log n}} < \limsup_{n\to\infty} \frac{S_n}{\sigma\sqrt{2n\log\log n}} = +1, \text{ a.s.} \tag{2}$$

As part of his deep investigations into functional versions of the LIL (of which, more below), Strassen (1966) proved a converse: (2) can hold only if S_n has finite variance σ^2 and expectation 0. Strassen's proof relied on a representation of Skorokhod, whereby S_n has the same distribution as a Brownian motion evaluated at a certain sequence of random times. Chris, in [M19], gives a completely different proof of the converse result by modifying a clever and highly technical method developed just previously by Stone (1966), to investigate recurrence properties of $S_n/\sqrt{n}$.

This is Chris's first foray into LIL territory. Prior to this, however, in [M18], Chris proves an interesting result which will have some significant repercussions later. He is aware of course, that central limit theorems with remainder (covered in Peter Hall's article in the present volume), can prove useful in establishing LIL and related results. In [M9] and [M27], Chris has already made contributions to various aspects of the convergence theory of random walks to of stable laws, that is, to situations where the normed and centered sum converges in distribution, but *not* to the normal distribution. [M18] proves the interesting result, that, for any such walks, the "large deviation" probability, $P(|S_n|>x_n)$, where x_n is a deterministic sequence such that $S_n/x_n \xrightarrow{P} 0$, must satisfy

$$0 < \liminf_{n\to\infty} \frac{P(|S_n|>x_n)}{nP(|X|>x_n)} \le \limsup_{n\to\infty} \frac{P(|S_n|>x_n)}{nP(|X|>x_n)} < \infty \tag{3}$$

(and the same with $nP(|X|>x_n)$ replaced by $P(\max_{1\le k\le n}|X_k|>x_n)$). More precisely, in [M18] Chris proves a much more general result than this: (3) holds provided S_n is *not* in the *domain of partial attraction of the normal law*; that is, provided no subsequence of (S_n), after norming and centering, converges in distribution to the normal distribution.

This thread is taken up again in [M23], but now applied directly to "generalised iterated logarithm results". This terminology is used to describe a result of the form

$$0 < \limsup_{n\to\infty} \frac{|S_n|}{b_n} < \infty, \text{ a.s.}, \tag{4}$$

where now the $\sqrt{n\log\log n}$ norming used in (2) is replaced by a more general deterministic sequence, $b_n>0$. Chris shows that if $\limsup_{n\to\infty}|S_n|/b_n<\infty$ a.s. for such a b_n, and S_n is *not* in the domain of partial attraction of the normal law, then it must be the case that $\lim_{n\to\infty}(S_n-med(S_n))/b_n=0$ a.s., where $med(S_n)$ is a median of S_n. Thus, if for example the X_i are symmetrically distributed around zero, a generalised LIL as in (4) can hold only if S_n *is* in the domain of partial attraction of the normal law. This result was published soon after a closely related one of Rogozin (1968); Chris was at the cutting edge of a quite "hot" area of research into understanding the law of the iterated logarithm, and its generalisations.

The results just cited might seem obscure or even cryptic, since the class of random walks belonging to the domain of partial attraction of the normal is extremely narrowly defined; but not so–they provide the key for Kesten (1972) to prove the most general possible form of the LIL: (4) holds, after allowing certain kinds of centering, if and only if S_n is in the domain of partial attraction of the normal law.

Papers on the LIL continue to appear even in the 2000's, showing that the subject retains its appeal and vigour. In Kesten and Maller (2004), Harry Kesten and I reconsidered, and extended somewhat, Chris's 1968 and 1969 papers – a long lead time from when I first read them.

After [M6] and [M23], as the next logical extension of his research, Chris turns his attention to functional limit laws and invariance principles. The note in [M25] gives a rate of convergence result for functionals of the interpolated or "broken line" process $\xi_n(t)$ defined by: $\xi_n(t) = S_k / (\sigma\sqrt{n})$ if $t = k/n$; and

$$\xi_n(t) = \xi_n(k/n) + n\left(\xi_n((k+1)/n) - \xi_n(k/n)\right)(t - k/n), \tag{5}$$

if $k/n \le t \le (k+1)/n$, $1 \le k \le n$, $n = 1, 2, \ldots$. Here the X_i have mean 0 and finite variance σ^2, and convergence is to the corresponding functional of Brownian motion.

The paper [M25] deals with a *weak* invariance principle. Chris is also concerned with *strong* invariance principles. Rather than work these out just for random walks, however, his research takes a notable new direction, broadening its range of applicability tremendously, by considering limit laws for *martingales.* In collaboration with long-time colleagues Peter Hall and Bruce Brown, and (at the time) PhD student David Scott, Chris publishes a series of papers, and a book, that do much to demonstrate and publicise the possible applications of martingales.

The first step, as for random walks, is to look at rates of convergence to normality. Heyde and Brown in [M35] do this in a very general way. The normalisation of the martingale is by its (deterministic) standard deviation. A uniform bound for the difference between the cumulative distribution function (cdf) of the normalised martingale, and the standard normal cdf, is given in terms of the conditional variances of the increments and their moments of order great than 2. The method of proof is via the Skorokhod representation, as in Strassen (1967).

Next, we have an iterated logarithm result in [M48]. The main theorem in this paper is a direct analogue of the classical LIL in (2), but where the summands are now required only to form a stationary martingale difference sequence, satisfying a kind of strong law condition on their conditional variances. This provides a generalisation of a slightly earlier result of Stout (1970), who required that the summands be ergodic, and used a random normalisation of the martingale in terms of the conditional variances of the increments. Its usefulness is demonstrated in [M48] by an application, giving a LIL for autoregressive processes.

Heyde and Scott in [M51] carry on this line of research by giving first a strong invariance principle for martingales, then generalising it to a wide class of processes with stationary increments. The resulting theorems are versions of the Strassen (1964) functional law for the interpolated process, whereby its almost sure cluster set is shown to be precisely the class of absolutely continuous functions $\xi(t)$ on $[0,1]$, with $\xi(0) = 0$ and $\int_0^1 (\mathrm{d}\xi(t)/\mathrm{d}t)^2 \mathrm{d}t \le 1$. The martingale difference sequence in [50] is not in general required to be stationary, so Lindberg-type conditions are imposed. The norming for the martingale is of the form $\sqrt{2s_n \log\log s_n}$, where s_n^2 is the variance of the martingale. Thus, we have a *constant norming* LIL and invariance principle. By contrast, Strassen's (1964) functional law uses a *random norming*. Chris in [M52] applies his results to get a theorem for stationary linear processes, in a somewhat similar way as he did in [M48], for autoregressive processes.

The threads are drawn together in Hall and Heyde [M68], where the previous constant and random norming versions of the LIL and functional law are replaced by a more general random norming version. The Strassen law is shown to hold when the norming is of the form $W_n\sqrt{2\log\log W_n}$, where W_n^2 is asymptotic to a conditional variance of the martingale increments, truncated in a certain (in general, random) way. The resulting theorem is, as the title of the paper suggests, a unification of previous results which can be obtained from it as special cases.

[M72] gives a further application of LIL invariance principles, this time using a clever trick of Barbour (1974) to transfer both weak and strong laws for martingales to convergent tail series of martingale difference sequences. An application to the Pólya urn scheme is given.

Finally in this section, we have [M74], again demonstrating Chris's mastery of a wide range of problem areas, by applying his martingale LIL to obtain an a.s. upper bound for the Hawkins (1958) random sieve. This is a probabilistic device which "produces a random sequence with asymptotic properties similar in many ways to the primes". This neat paper turns us full circle to Feller (1945), who drew attention in the introduction to his paper to an intimate connection between number-theory and the law of the iterated logarithm, in respect to "a problem in Diophantine approximations treated by Hardy and Littlewood (1914)". The particular stream of Chris's research outlined in this section has followed classical lines, but in thoughtful, innovative and useful ways.

References

[1] Barbour, A.D. (1974) Tail sums of convergent series of independent random variables. Proc. Camb. Phil. Soc. 75, 361–364.

[2] Feller, W. (1945) The fundamental limit theorems in probability. Bull. Amer. Math, Soc., 800–832.

[3] Hardy, G.H. and Littlewood, J.E. (1914) Some problems of Diophantine approximation, Acta Math. 37, 155–239.

[4] Hartman, P. and Wintner, A. (1941) On the law of the iterated logarithm, Amer. J. Math., 63, 169–176.

[5] Hawkins, D. (1958) The random sieve. Math. Mag. 31, 1–3.

[6] Kesten, Harry (1972) Sums of independent random variables—without moment conditions. Ann. Math. Statist., 43, 701–732.

[7] Kesten, Harry, and R. A. Maller (1996) Two renewal theorems for general random walks tending to infinity. Prob. Theor. Rel. Fields 106, (1996), 1–38.

[8] Harry Kesten and R. A. Maller (2004) The effect of trimming on the law of the iterated logarithm. In: *Stochastic Methods and Their Applications: Papers in Honour of Chris Heyde*, Eds J. Gani and E. Seneta, J. Appl. Prob. Special Vol. 41A, (2004), 253–272.

[9] Rogozin, B.A. (1968) On the existence of exact upper sequences. Theor. Prob. Appl., 13, 667–672.

[10] Stout, W.F (1970) The Hartman-Wintner law of the iterated logarithm for martingales. Ann. Math. Statist., 41, 2158–2160.

[11] Strassen, V. (1964) An invariance principle for the law of the iterated logarithm. Z. Wahrscheinlichkeitstheorie und Verw. Gebiete 3, 211–226.

[12] Strassen, V. (1967) Almost sure behaviour of sums of independent random variables and martingales. Proc. V Berk. Symp. Math. Statist. Prob., 2, 315–343. Univ. of Calif. Press.

Chris Heyde on Branching Processes and Population Genetics

Eugene Seneta[1]

When Chris Heyde returned to Australia in September 1968 to take up a position as Reader in Ted Hannan's teaching Department of Statistics in the-then School of General Studies at the Australian National University, there was still a strong trend of research into population genetics emanating from Pat Moran's purely research department in the Institute of Advanced Studies, a direction synthesized in Moran (1962). There was also strong activity in both departments in the theory of branching processes. With its common focus with population genetics on reproduction and extinction, it was the primary interest at the time of the author of this commentary. A strong impetus to branching process theory had occurred with the appearance of the book of Harris (1963).

The dominant theme of Chris's work prior to his arrival had been refinement of classical limit theory for sums of independent random variables. A principal theme of his work since 1970 has been the theory and use of martingale methods in situations of statistical dependence. This seems to have its origin in his paper [M31], in the context of a simple (Bienaymé-Galton-Watson) branching process $\{Z_n\}, n \geq 0; Z_0 = 1$. Suppose the non-degenerate offspring distribution has probability generating function

$$F(s) = \sum_{j=0}^{\infty} s^j P(Z_1 = j),\ s \in [0,1].$$

When $1 < m = E(Z_1) < \infty$ the process is called supercritical, and in this case the probability of extinction, q, is the unique real number in $[0,1)$ satisfying $F(q) = q$. The random sequence $\{Z_n/m^n\}$ is clearly a non-negative martingale, which in the supercritical case under the-then standard condition $E(Z_1^2) < \infty$ was known to converge almost surely (a.s) to a random variable W for which $P(W=0) = q$ and which has a continuous density on the set of positive real numbers. After a number of refinements of this convergence result, it was shown in Seneta (1968) that in the non-degenerate supercitical case without further conditions, the choice $c_n = 1/h_n(s_0)$, where s_0 is any fixed number in $(0, -\log q)$, and $h_n(s_0)$ is the n-th functional iterate of $h(s)$, which is the inverse function of $k(s) = -\log F(e^{-s})$, results in convergence in distribution of $\{Z_n/c_n\}$ to a proper non-degenerate random variable, W. While it was noticed that $\{W_n = h_n(s_0)Z_n\}$ was a submartingale, a.s. convergence could not be deduced from the usual theorem since $E(W_n)$ is bounded if and only if $E(Z_1 \log Z_1) < \infty$. Chris's insight was that $\{\exp(-h_n(s_0)Z_n)\}$ is a martingale, with (obviously) bounded mean, thus giving the sharpening to a.s. convergence. Reciprocally, this strengthening immediately led to an analogous result for the supercritical branching process with immigration $\{X_n\}$ (Seneta, 1970), which then led to collaboration with Chris in [M42] on analogues of the Central Limit Theorem and the Law of the Iterated Logarithm as refinements of the results

$$X_n / m^n \to V, \text{ a.s.},$$

[1] School of Mathematics and Statistics, FO7, University of Sydney, N.S.W. 2006, Australia
e-mail: eseneta@maths.usyd.edu.au

R. Maller et al. (eds.), *Selected Works of C.C. Heyde*, Selected Works in Probability and Statistics,
DOI 10.1007/978-1-4419-5823-5_5,

and

$$X_{n+r} / X_n \to m^r, \text{a.s.}$$

The note [M31] stimulated Chris to develop theory for the supercritical process analogous to the classical rate of convergence theorems for the difference $\{Z_n/m^n - W\}$ under the classical condition $E(Z_1^2) < \infty$. The central limit theorem analogue developed in [M34] and [M38] is beautifully attained with random norming and the necessary conditioning:

$$\lim_{n\to\infty} Pr\{(m^2 - m)^{1/2}\sigma^{-1}Z_n^{-1/2}(m^nW - Z_n) \le x \mid Z_n > 0\} = \Phi(x), \qquad (1)$$

where $\sigma^2 = \mathrm{Var}Z_1$. This result is followed in [M37] by a corresponding iterated logarithm analogue, using an intermediate result from [M39], under the condition $E(Z_1^3) < \infty$, later relaxed in [M41] to $E(Z_1^2) < \infty$. These papers display a finely-honed appreciation of, and ability to adapt in a non-trivial way, and to a quite different context, tools of the classical limit theory.

The note [M31] had several other important aspects. The chief of these was the reinvigoration of the theory of statistical estimation of m on the basis of a single finite observed realization, including a focus on estimation theory for a stochastic process in which extinction may occur. In [M62] Chris returns to the investigation of m under the second moment constraint, and under the simplifying assumption that $P(Z_1=0)=0$ so that $P(W>0)=1$. The starting point is a comparison of the maximum-likelihood estimator $\widehat{m}_n = (Y_n - 1)Y_{n-1}^{-1}$, where $Y_n = 1 + Z_1 + \cdots + Z_n$, and the simple ratio estimator $\widetilde{m}_n = Z_n Z_{n-1}^{-1}$. There are different non-random normings of each of $\widehat{m}_n - m$, $\widetilde{m}_n - m$, which result in a limit law of form $W^{-1/2}Z$, where $Z \sim \mathcal{N}(0,1)$, on the basis of which he concludes that asymptotic relative efficiency of $\widetilde{m}_n$ relative to $\widehat{m}_n$ is $1 - m^{-1}$. Thus the limit law is normal variance-mixing. Chris then shows that under a power-series condition, $\widehat{m}_n$ is asymptotically efficient (according to an extension of Rao's concept of efficiency). Here a central limit theorem for martingales is a dominant tool, arising from the fact that $\{Z_n/m^n\}$ is a martingale.

Chris's inclination towards a framework of time-series type results in which his martingale work could be applied was hugely influential, and the extensive use of martingale central limit theory as in [M54] and [M62] for inferential purposes, was at least partly stimulated by appearance of a paper of (Bruce) Brown (1971). An early manifestation in [M46] stems from noticing that $\{Y_n/m^n\}$ is a martingale, where $Y_n = X_n - \mu$, and $\{X_n\}$ is a subcritical $(0<m<1)$ and hence (under moment conditions) an asymptotically stationary branching process with immigration. Thus $Y_{n+1} = mY_n + \delta_n$, $n = 0, 1, 2, \ldots$, where $E\delta_n = 0$, and δ_n is uncorrelated with Y_n, in analogy to the first-order autoregressive process. There is certainly a line-of-descent on estimation theory for m in branching processes with immigration, which is traced in Section 1 of Qi (2007). A related line of descent, which also continues, can be traced from Ku and Seneta (1998).

Chris's interest in population genetics models was reinvigorated by a note of the author (Seneta, 1974) which extended Malécot's Markov chain formulation $\{X_n\}$ of the Fisher-Wright diploid model for fluctuation of gene frequency, say of allele a out of the two possible alleles A, a, at a single diploid locus, to randomly fluctuating population size $\{N_n\}$, so that the proportion of A alleles in the population in generation n is given by $Y_n = X_n/(2N_n)$. The extension assumes the bivariate process $\{X_n, N_n\}$ is bivariate Markov, but retains the binomial sampling structure in the manner that

$$X_n \mid \{N_{n+1}, N_n, X_n\} \sim \mathcal{B}(2N_{n+1}, Y_n).$$

Since $\{Y_n\}$ is an (obviously bounded) martingale, by the martingale convergence theorem $Y_n \to Y$ a.s., with $0 \le Y \le 1$. The distribution of Y, unlike the fixed population-size model, is not necessarily concentrated at the two points 0, 1. A handle on its actual concentration is provided by the fact that $h_n = E(Y_n(1 - Y_n)) \to E(Y(1-Y))$, equals h, say. For a deterministically varying population size, with $N_n > 1$, $n \ge 1$, we have $h > 0$ if and only if $\Sigma_{i=1}^{\infty} 1/N_i < \infty$. This result emanates from the fact that:

$$h_{n+1} = (1 - (2N_{n+1})^{-1})h_n. \qquad (2)$$

The themes of almost sure convergence of relative frequency $\{Y_n\}$, and the concentration of probability of Y via the study of the evolution of $\{h_n\}$ by relations such as (2), using martingale methods, are the dominant themes of Chris's subsequent work. Thus, in [M66] almost immediately, two theorems, focussed on when $h=0$ or not, are based on investigation of the zero-mean martingale with bounded increments $\{U_n, \mathcal{F}_n, n \geq 1\}$, where $\mathcal{F}_n$ is the σ-field generated by the random vectors $\{(X_0,N_0),(X_1,N_1),\ldots,(X_n,N_n)\}$, and

$$U_n = \Sigma_{k=1}^{n}[Y_k(1-Y_k) - E(Y_k(1-Y_k) \mid \mathcal{F}_{k-1})], n \geq 1. \tag{3}$$

The device underlying (3) for constructing a martingale is used repeatedly in later papers.

In [M71] the model is extended to permit selection via a parameter $\sigma \geq 0$, with $\sigma = 0$ corresponding to the selective neutrality of [M66]. In this extension $\{Y_n, \mathcal{F}_n\}$ is a submartingale in general, so the martingale convergence theorem can still be used; and there is the pleasing conclusion that $\sigma > 0$ implies that $Y(1-Y)=0$, a.s., just as in the case of fixed population size.

In [M87], Chris addresses such issues by generalizing Moran's overlapping generation model, which is also a Markov chain. Here the transition time-points are times of death of single haploid individuals, of type a or A. The evolution of population structure is again studied through the movement of the (haploid) proportion $Z_n = X_n/N_n$, $n \geq 0$, which is again a bounded martingale, so $Z_n \to Z$, $0 \leq Z \leq 1$. The quantity $Z_n(1-Z_n)$ is now randomly normed to be a non-negative martingale relative to the σ-field $\mathcal{F}_n$, and martingale theory provides complete resolution of when $h = E(Z(1-Z)) > 0$, inasmuch as a necessary and sufficient condition is that $1 < m < \infty$, where m is the mean of the offspring distribution. This insight, progressive assimilation and evolution of earlier techniques to resolve a question completely, is characteristic of Chris's work.

A concluding paper [M98] to the sequence [M66], [M71], [M87], [M91], [M89], [M105], reviews difficulties in studying effects of varying population size focussed on *relative* frequencies such as $\{Z_n\}$. To remedy these, Chris characteristically uses a perceived unifying feature based on the growth of *actual* frequencies $\{X_n\}$:

$$E(X_{n+1} \mid \mathcal{F}_n) = X_n(1+f_n), \text{ a.s.}, \tag{4}$$

where f_n is $\mathcal{F}_n$ -measurable, and $1+f_n \geq 0$, a.s. There is, for the first time, no attention paid to the limiting behavior of $E(Z_n(1-Z_n))$ as a means of information on the limiting distribution of Z.

Curiously, in a slightly earlier paper in the same year and in the same journal, Buckley and Seneta (1983) generalized the selectively neutral context of Seneta (1974) to, in essence, conditional branching processes, with the focus then being on the behavior of the (diploid) sequence $\{Y_n(1-Y_n)\}$ in terms of the conditional expectation:

$$E(Y_{n+1}(1-Y_{n+1}) \mid \mathcal{F}_n) = (1-W_{n+1})Y_n(1-Y_n), \tag{5}$$

where

$$W_{n+1} = (V_n/(2N_n - 1))(N_n / N_{n+1})^2,$$

and V_n is a conditional variance of an offspring distribution. This somewhat resembles both (2) and (4). Once the expression (5) is established, the theorems giving sufficient conditions for each of $E(Y(1-Y))=0$ or $\neq 0$ follow precisely as in [M66]. For example, the former holds if $\Sigma_{n=0}^{\infty} W_{n+1} = \infty$, a.s., a condition later mentioned by Donnelly (1986) as equivalent to his own necessary and sufficient condition for fixation. The very recent paper of Prince and Weber (2007) achieves an ultimate tidying up of such issues. This separate line of descent from [M66] further testifies to the enduring vitality of Chris's ideas.

References

[1] Brown, B.M. (1971) Martingale central limit theorems. *Ann. Math. Statist.*, **42**,59–66.

[2] Buckley, M.J. and Seneta, E. (1983) The genetic balance between varying population size and selective neutrality. *J. Mathematical Biology*, **17**, 217–222.

[3 Donnelly, P. (1986) A genealogical approach to variable-population-size models in population genetics. *J. Appl. Prob.,* **23**, 283–296.

[4] Harris, T.E. (1963) *The Theory of Branching Processes.* Springer, Berlin.

[5] Ku, S. and Seneta, E. (1998) Practical estimation from the sum of AR(1) processes. *Commun. Statist. – Simula.,* **27**, 981–998.

[6] Moran, P.A.P. (1962) *The Statistical Processes of Evolutionary Theory.* Clarendon Press, Oxford.

[7] Prince, T. and Weber, N. (2007) Fixation in conditional branching process models in population genetics. *J. Appl. Prob.,* **44**, 1103–1110.

[8] Qi, Y. (2007) On asymptotic normality of sequential estimators for branching processes with immigration. *Journal of Statistical Planning and Inference,* **137**, 2892–2902.

[9] Seneta, E. (1968) On recent theorems concerning the supercritical Galton-Watson process.*Ann. Math. Statist.,* **39**, 2098–2102.

[10] Seneta, E. (1970) On the supercritical Galton-Watson process with immigration. *Math. Biosciences,* **7**, 9–14.

[11] Seneta, E. (1974) A note on the balance between random sampling and population size. (On the 30th anniversary of G. Malécot's paper.) *Genetics,* **77**, 607–610.

On a Property of the Lognormal Distribution

BY

C. C. HEYDE

Reprinted from

THE JOURNAL OF THE ROYAL STATISTICAL SOCIETY

SERIES B (METHODOLOGICAL)

Volume 25, No. 2, 1963

pp. 392 – 393

PRINTED FOR PRIVATE CIRCULATION

1963

On a Property of the Lognormal Distribution

By C. C. Heyde

The Australian National University

[Received May 1963]

Summary

In this paper it is established that the lognormal distribution is not determined by its moments. Some brief comments are made on the set of distributions having the same moments as a lognormal distribution.

1. Introduction

It seems to be generally accepted that all the commonly used distributions in statistics, which possess moments of all orders, are determined by their moments. I will show, however, that this is not the case for the lognormal distribution.

Let the random variable X have the general lognormal distribution. We take $\log(X-a)$ as normally distributed with mean m and variance σ^2. Then X has frequency function

$$f(x) = \begin{cases} \{\sigma(x-a)(2\pi)^{\frac{1}{2}}\}^{-1}\exp[-\tfrac{1}{2}\sigma^{-2}\{\log(x-a)-m\}^2] & (a<x<\infty), \\ 0 \quad (x \leqq a). \end{cases}$$

It can readily be shown that this distribution possesses the same moments, namely

$$\alpha_n = \sum_{r=0}^{n} \binom{n}{r} a^{n-r}\exp(rm+\tfrac{1}{2}r^2\sigma^2) \quad (n=0,1,2,\ldots), \tag{1}$$

as the distribution with frequency function

$$g(x) = \begin{cases} \{\sigma(x-a)(2\pi)^{\frac{1}{2}}\}^{-1}\exp[-\tfrac{1}{2}\sigma^{-2}\{\log(x-a)-m\}^2] \\ \qquad \times(1+\epsilon\sin[2\pi k\sigma^{-2}\{\log(x-a)-m\}]) & (a<x<\infty), \\ 0 \quad (x \leqq a), \end{cases}$$

where $0<\epsilon<1$ and k is a positive integer. Thus the lognormal distribution is not determined by its moments. It is clear that distributions can be chosen possessing the same moments as a lognormal distribution and such that they are indistinguishable from the lognormal for all practical purposes.

2. Remarks on the Moment Problem

The theory of the indeterminate moment problem enables us to make some rather interesting remarks.

Applying Theorem 2.13 of Shohat and Tamarkin (1943) to the case of the moment sequence (1), we see that for any real x there exists a purely discrete distribution having x as a point of discontinuity and moment sequence (1).

We now define P as the set of all points of increase of a distribution function F. Applying Theorem 10.42 of Stone (1932) to the case of the moment sequence (1), we see that there exists a constant $K>a$ such that there are cardinal c distributions, possessing the moment sequence (1) and for which $\inf P$ assumes an arbitrary value μ on the range $-\infty \leqq \mu < K$. There is, however, just one distribution with the moment sequence (1) and for which $\inf P = K$.

3. A Special Case

It is worth remarking that in the case $a = 0$, $m = 0$, the various determinants which arise out of the theory of the indeterminate moment problem may be expressed in terms of alternants (see for example Muir, 1960, Chapter XI).

Using the notation of Shohat and Tamarkin (1943, p. 6) we have the determinants

$$\Delta_n = \left|\exp\{\tfrac{1}{2}\sigma^2(i+j)^2\}\right|_{i,j=0}^{n} \quad (n \geqslant 1)$$
$$= \exp\{\tfrac{1}{2}n^2\sigma^2(n+1)\} \prod_{r=1}^{n} (e^{r\sigma^2}-1)^{n-r+1},$$

and

$$\Delta_n^{(1)} = \left|\exp\{\tfrac{1}{2}\sigma^2(i+j+1)^2\}\right|_{i,j=0}^{n} \quad (n \geqslant 1)$$
$$= \exp\{\tfrac{1}{2}\sigma^2(n+1)(2n+1)\}\,\Delta_n.$$

The corresponding continued fraction (Shohat and Tamarkin, 1943, p. 73) has coefficients l_i $(i = 1, 2, 3, \ldots)$ given by

$$l_1 = 1,$$
$$l_{2n+1} = \{\Delta_{n-1}^{(1)}\}^2/\{\Delta_{n-1}\Delta_n\} \quad (n = 1, 2, 3, \ldots)$$
$$= \exp\{\tfrac{1}{2}n(n-1)\sigma^2\} \Big/ \prod_{r=1}^{n} (e^{r\sigma^2}-1).$$
$$l_2 = -\exp(\tfrac{1}{2}\sigma^2),$$
$$l_{2n} = -\Delta_{n-1}^2/\{\Delta_{n-1}^{(1)}\Delta_{n-2}^{(1)}\} \quad (n = 2, 3, 4, \ldots)$$
$$= \exp\{\tfrac{1}{2}(-n^2-n+1)\sigma^2\} \prod_{r=1}^{n} (e^{r\sigma^2}-1).$$

Cases where these may be computed explicitly from the moments are particularly rare.

It is possible, in this case, to obtain the Stieltjes transforms of all the distributions possessing the same moments as the lognormal. I do not include either the calculation or the result here as the calculation is routine and tedious, a particular case of analysis is given in Shohat and Tamarkin (1943, Chapter II), and the result is not in a compact or useful form.

References

MUIR, T. (1960), *A Treatise on the Theory of Determinants* (revised and enlarged by W. H. Metzler). New York: Dover.

SHOHAT, J. A. and TAMARKIN, J. D. (1943), *The Problem of Moments*. New York: American Mathematical Society.

STONE, M. H. (1932), *Linear Transformations in Hilbert Space and their Applications to Analysis*. New York: American Mathematical Society.

HEYDE C. C.
1964
THE JOURNAL OF THE AUSTRALIAN
MATHEMATICAL SOCIETY
Vol. IV, Part. 2, p.p. 214-222

TWO PROBABILITY THEOREMS AND THEIR APPLICATION TO SOME FIRST PASSAGE PROBLEMS

BY

C. C. HEYDE

Reprinted from
THE JOURNAL OF THE AUSTRALIAN
MATHEMATICAL SOCIETY
Volume IV, Part 2, p.p. 214-222
1964

TWO PROBABILITY THEOREMS AND THEIR APPLICATION TO SOME FIRST PASSAGE PROBLEMS

C. C. HEYDE

(received 23 July 1963)

1. Introduction

Let X_i, $i = 1, 2, 3, \cdots$ be a sequence of independent and identically distributed random variables and write $S_n = X_1 + X_2 + \cdots + X_n$. If the mean of X_i is finite and positive, we have $Pr(S_n \leqq x) \to 0$ as $n \to \infty$ for all x, $-\infty < x < \infty$ using the weak law of large numbers. It is our purpose in this paper to study the rate of convergence of $Pr(S_n \leqq x)$ to zero. Necessary and sufficient conditions are established for the convergence of the two series

$$\sum_{n=1}^{\infty} n^k \, Pr(S_n \leqq x), \qquad -\infty < x < \infty$$

where k is a non-negative integer, and

$$\sum_{n=1}^{\infty} e^{rn} \, Pr(S_n \leqq x), \qquad -\infty < x < \infty$$

where $r > 0$. These conditions are applied to some first passage problems for sums of random variables. The former is also used in correcting a queueing Theorem of Finch [4].

2. Two Probability Theorems

Let X_i, $i = 1, 2, 3, \cdots$ be independent and identically distributed random variables. We write $S_n = X_1 + X_2 + \cdots + X_n$, $X_i^- = \min(0, X_i)$ and $X_i^+ = X_i + X_i^-$.

We shall establish the following two Theorems:

THEOREM A[1]. *Suppose* $E|X| < \infty$, $EX > 0$ *and let* k *be a non-negative*

[1] My attention has been drawn to a statement of Theorem A without proof in Smith, W. L. "On the elementary renewal theorem for non-identically distributed random variables" Univ. North Carolina Mimeographed Notes No. 352 (Feb. 1963). Professor Smith states that a proof of this result will appear in a paper entitled "On functions of characteristic functions and their applications to some renewal-theoretic random walk problems".

integer. A necessary and sufficient condition for the convergence of the series

$$\sum_{n=1}^{\infty} n^k \, Pr(S_n \leqq x), \qquad -\infty < x < \infty,$$

is that $E|X^-|^{k+2} < \infty$.

THEOREM B. *Suppose* $E|X| < \infty$ *and* $EX > 0$. *A necessary and sufficient condition for the convergence of the series*

$$\sum_{n=1}^{\infty} e^{rn} \, Pr(S_n \leqq x), \qquad -\infty < x < \infty,$$

for some $r > 0$ *is that* X^- *has an analytic characteristic function* [2].

(It is clear that analogous Theorems will hold in the case $EX < 0$). We defer the proofs of Theorems A and B until some Lemmas have been established.

LEMMA 1. *If* $E|X|^r < \infty$ *for some integer* $r \geqq 1$ *and* $EX > 0$, *then*

$$\sum n^{r-2} \, Pr(S_n \leqq x) < \infty, \qquad -\infty < x < \infty.$$

PROOF: Write $EX = \mu$. Using Katz [5], Theorem 1, we have

$$\sum n^{r-2} \, Pr\{|S_n - n\mu| \geqq n\varepsilon\} < \infty, \text{ every } \varepsilon > 0$$

from which we obtain, in particular

$$\sum n^{r-2} \, Pr(S_n \leqq (\mu - \varepsilon)n) < \infty, \text{ every } \varepsilon > 0. \tag{1}$$

Now we choose ε so small that $\varepsilon < \mu$. We then have, for n sufficiently large,

$$Pr\,(S_n \leqq x) \leqq Pr(S_n \leqq (\mu - \varepsilon)n)$$

and the result follows immediately from (1).

LEMMA 2. *Let* $E|X| < \infty$ *and* $EX > 0$ *or else* $E|X| = \infty$ *and, in either case,* $E|X^-|^r < \infty$ *for some integer* $r \geqq 1$. *Then*

$$\sum n^{r-2} \, Pr(S_n \leqq x) < \infty, \qquad -\infty < x < \infty.$$

PROOF. We define a new random variable Y as follows

$$\begin{aligned} Y &= X \text{ if } X < K \\ &= 0 \text{ otherwise,} \end{aligned}$$

where the constant $K (> 0)$ is chosen so that $EY > 0$. Then, $Y \leqq X$ and $E|Y|^r < \infty$. It follows from Lemma 1 that

[2] The term "analytic characteristic function" is used for a characteristic function which is analytic in a strip containing the origin as an interior point.

$$\sum n^{r-2} Pr(Y_1+Y_2+\cdots+Y_n \leqq x) < \infty, \qquad -\infty < x < \infty.$$

Also, if $X_1+X_2+\cdots+X_n \leqq x$, then $Y_1+Y_2+\cdots+Y_n \leqq x$ so that

$$Pr(Y_1+Y_2+\cdots+Y_n \leqq x) \geqq Pr(X_1+X_2+\cdots+X_n \leqq x) = Pr(S_n \leqq x)$$

and hence

$$\sum n^{r-2} Pr(S_n \leqq x) < \infty, \qquad -\infty < x < \infty.$$

This completes the proof.

LEMMA 3. *Let* $E|X| < \infty$, $EX = \mu > 0$ *and*

$$\sum n^k Pr(S_n \leqq x) < \infty, \qquad -\infty < x < \infty$$

for some non-negative integer k. *Then* $E|X^-|^{k+2} < \infty$.

Our proof relies heavily on techniques used by Erdös [3].

PROOF. Write $X_i^* = X_i - \mu$ and $Z_n = \sum_{i=1}^n X_i^*$. We then have

$$Pr(S_n \leqq x) = Pr(Z_n \leqq x - n\mu) \geqq Pr(Z_n \leqq -nc)$$

for $c > \mu$ and n sufficiently large.

Now from the fact that $E|X| < \infty$, it follows by a simple rearrangement that

$$\sum_{n=1}^{\infty} Pr(X_i^* < -(c+\varepsilon)n) < \infty,$$

for arbitrary $\varepsilon > 0$. Also, since the terms in this series are non-increasing, we have

(2) $$nPr(X_i^* < -(c+\varepsilon)n) \to 0 \text{ as } n \to \infty.$$

Write $\{E\}$ for the event E and $\{\bar{E}\}$ for the complement of $\{E\}$. We have

$$\{Z_n \leqq -nc\} \supseteqq \bigcup_{i=1}^{n} [\{X_i^* < -n(c+\varepsilon)\} \cap \{|X_1^*+\cdots+X_{i-1}^*+X_{i+1}^*+\cdots +X_n^*| < (n-1)\varepsilon\}],$$

and for the sake of brevity, we put

$$A_i = \{X_i^* < -n(c+\varepsilon)\},$$
$$B_i = \{|X_1^*+\cdots+X_{i-1}^*+X_{i+1}^*+\cdots+X_n^*| < (n-1)\varepsilon\},$$

$i = 1, 2, 3, \cdots, n$. Thus,

$$Pr(Z_n \leqq -nc) \geqq Pr[\bigcup_{i=1}^{n} (A_i \cap B_i)]$$
$$= Pr[\bigcup_{i=1}^{n} \{(\overline{A_1 \cap B_1}) \cap (\overline{A_2 \cap B_2}) \cap \cdots \cap (\overline{A_{i-1} \cap B_{i-1}}) \cap (A_i \cap B_i)\}]$$
$$= \sum_{i=1}^{n} Pr[(\overline{A_1 \cap B_1}) \cap \cdots \cap (\overline{A_{i-1} \cap B_{i-1}}) \cap (A_i \cap B_i)]$$
$$\geqq \sum_{i=1}^{n} Pr(\bar{A}_1 \cap \bar{A}_2 \cap \cdots \cap \bar{A}_{i-1} \cap A_i \cap B_i) \tag{3}$$
$$\geqq \sum_{i=1}^{n} [Pr(A_i \cap B_i) - Pr\{(A_1 \cup A_2 \cup \cdots \cup A_{i-1}) \cap A_i\}]$$
$$\geqq \sum_{i=1}^{n} [Pr(B_i) - (i-1) Pr(A_1)] Pr(A_i)$$
$$\geqq \sum_{i=1}^{n} [Pr(B_i) - n\, Pr(A_1)] Pr(A_i).$$

Now take arbitrary ρ, $0 < \rho < 1$ and $\delta > 0$ such that $1 - 2\delta \geqq \rho$. It follows from the weak law of large numbers that we can find an integer N_1 such that

$$Pr(B_i) > 1 - \delta \text{ for } n \geqq N_1.$$

Also, from (2), we can find an integer N_2 such that

$$n\, Pr(A_1) < \delta \text{ for } n \geqq N_2.$$

Thus, for $n \geqq \max(N_1, N_2)$, we have from (3)

$$Pr(Z_n \leqq -nc) \geqq n\rho\, Pr(X_i^* < -(c+\varepsilon)n), \tag{4}$$

and hence

$$\sum n^{k+1} Pr(X_i^* < -(c+\varepsilon)n) < \infty.$$

We now introduce the random variable Y defined by

$$Y = X^* \text{ if } X^* < 0$$
$$= 0 \text{ otherwise,}$$

and obtain

$$\sum n^{k+1} Pr(Y < -(c+\varepsilon)n) < \infty.$$

It follows from this, by a simple rearrangement, that $E|Y|^{k+2} < \infty$ and hence that $E|X^-|^{k+2} < \infty$. This completes the proof.

Proof of Theorem A. Theorem A follows immediately from Lemmas 2 and 3.

We now go on to give two Lemmas leading up to a proof of Theorem B.

The development of the proof is similar to that used above in the proof of Theorem A.

LEMMA 4. *Suppose X^- has an analytic characteristic function and either $E|X| < \infty$ and $EX > 0$ or $E|X| = \infty$. There exists a constant $R > 0$ such that*

$$\sum e^{rn} Pr(S_n \leqq x) < \infty$$

for every x, $-\infty < x < \infty$ and every r, $0 < r < R$.

PROOF. Since X^- has an analytic characteristic function, there exists a constant $K > 0$ such that

$$\Phi(\theta) = E(e^{-\theta X}) < \infty$$

for all θ in $0 \leqq \theta < K$. Now for such a θ, a well-known Chebyshev type inequality gives

$$Pr(S_n \leqq x) \leqq e^{\theta x} E(e^{-\theta S_n}) = e^{\theta x}\{\Phi(\theta)\}^n.$$

Also, in view of our assumption that $E|X^-| < \infty$ and either $E|X| < \infty$ and $EX > 0$ or $E|X| = \infty$, we must have $\Phi(\theta) < 1$ for sufficiently small positive θ. We then choose R so small that $e^R\Phi(\theta) < 1$ and for all r, $0 < r < R$,

$$\sum e^{rn} Pr(S_n \leqq x) < \infty.$$

This completes the proof of the Lemma. I am indebted to the referee for this direct proof. My original proof was based on Baum, Katz and Read [1], Theorem 2, 190.

Lemma 4 is a generalization of the well-known result of Stein [9] which deals with the case $X^- = 0$. It should be noted that although Stein's result is correct, his proof is invalidated by a misinterpretation of the Markov chain property of the sequence $\{S_n\}$ of sums.

LEMMA 5. *Let $E|X| < \infty$. Suppose $EX = \mu > 0$ and*

$$\sum e^{rn} Pr(S_n \leqq x) < \infty$$

for all r, $0 < r < R$ and all x, $-\infty < x < \infty$. Then X^- has an analytic characteristic function.

PROOF. We retain the notation of Lemma 3. Proceeding precisely as in Lemma 3, we obtain (4)

$$Pr(Z_n \leqq -nc) \geqq n\rho\, Pr(X_i^* < -(c+\varepsilon)n),$$

so that certainly

$$\sum e^{rn} Pr(X_i^* < -(c+\varepsilon)n) < \infty.$$

We now introduce the random variable Y defined by

$$Y = X^* \text{ if } X^* < 0$$
$$= 0 \text{ otherwise,}$$

and obtain

$$\sum e^{rn}\, Pr(Y < -(c+\varepsilon)n) < \infty.$$

It follows immediately, using Lukacs [7], Theorem 7.2.1, 137, that Y and hence X^- has an analytic characteristic function. This completes the proof of thc Lcmma.

PROOF OF THEOREM B: Theorem B follows immediately from Lemmas 4 and 5.

It is worth remarking that it is quite likely that in Theorems A and B the condition $E|X| < \infty$, $EX > 0$ can be replaced by the condition $E|X| < \infty$, $EX > 0$ or $E|X^-| < \infty$, $E|X| = \infty$.

3. Application to some first passage problems

Let X_i, $i = 1, 2, 3, \cdots$ be independent and identically distributed random variables and write $S_n = X_1+X_2+ \cdots +X_n$. Consider a single boundary at $A (\geqq 0)$ so that if

$$F_0(x) = \begin{cases} 1 & x \geqq 0 \\ 0 & x < 0, \end{cases}$$

$$F_1(x) = Pr(S_1 \leqq x),$$

$$F_n(x) = Pr(S_n \leqq x;\ \max_{1 \leqq k \leqq n-1} S_k \leqq A),\ n > 1,$$

the probability p_n that the first passage time out of the interval $(-\infty, A]$ for the process S_n is n is given by

$$p_n = F_{n-1}(A) - F_n(A), \qquad n \geqq 1.$$

This passage problem in the case $A = 0$ arises, for example, in the busy period distribution of the queue $GI/G/1$ which has been considered by various authors such as Finch [4].

We introduce the probability generating function $P(\lambda) = \sum_{r=1}^{\infty} \lambda^r p_r$ for the first passage time distribution (henceforth abbreviated F.P.T.D.) $Pr(N = n) = p_n$. We have formally

$$P'(1) = E(N) = 1 + \sum_{r=1}^{\infty} F_r(A)$$

$$P''(1) = \sum_{r=2}^{\infty} r(r-1)p_r = E(N^2) - E(N) = 2 \sum_{r=1}^{\infty} rF_r(A),$$

and in general for $k > 1$,

$$P^{(k)}(1) = (\alpha)_k \text{ (the } k\text{-th factorial moment of } N)$$
$$= k \sum_{r=k-1}^{\infty} (r)_k F_r(A) = \sum_{r=0}^{k} s(k, r) E(N^r)$$

where $(r)_k = r(r-1)(r-2) \cdots (r-k+1)$ and $s(k, r)$ are the Stirling numbers of the first kind. It is thus clear that $E(N^r) < \infty$ for some positive integer r if and only if $\sum n^{r-1} F_n(A) < \infty$. Also, the random variable N has an analytic characteristic function if and only if the radius of convergence of $P(\lambda)$ is greater than unity or equivalently if $\sum e^{rn} F_n(A) < \infty$ for some $r > 0$.

Now we write

$$q_n = Pr\,(\max_{1 \leqq k \leqq n} S_k \leqq 0), \qquad n \geqq 1; \qquad q_0 = 1.$$

Spitzer [8], 332, shows that

$$\sum_{n=0}^{\infty} q_n t^n = \exp\left\{\sum_{n=1}^{\infty} \frac{t^n}{n} Pr(S_n \leqq 0)\right\}, \tag{6}$$

a result originally due, in a slightly different form, to E. Sparre Andersen. From this we obtain

$$q_n \geqq \frac{1}{n} Pr(S_n \leqq 0).$$

Thus,

$$Pr(S_n \leqq A) \geqq Pr\,(\max_{1 \leqq k \leqq n} S_k \leqq A) = F_n(A) \geqq q_n \geqq \frac{1}{n} Pr(S_n \leqq 0),$$

and we readily obtain from Theorem B:

Theorem 1. *The F.P.T.D. generated by the random variable* X *with* $E|X| < \infty$ *and* $EX > 0$ *has an analytic characteristic function if and only if* X^- *has an analytic characteristic function.*

Further, we obtain immediately from Theorem A:

Theorem 2. *Let* $r > 1$ *be a positive integer. Consider the F.P.T.D. generated by the random variable* X *with* $E|X| < \infty$ *and* $EX > 0$. *If the F.P.T.D. has a finite r-th moment, then* $E|X^-|^r < \infty$. *If, on the other hand,* $E|X^-|^r < \infty$, *then the F.P.T.D. has finite moments at least up to the* $(r-1)$*th.*

In the particular case where $A = 0$, we can improve this Theorem by virtue of the relation (6). In fact, formally differentiating (6) $(r-1)$ times, we see that

$$\sum n^{r-2} Pr(S_n \leqq 0) < \infty \text{ if and only if } \sum n^{r-1} q_n < \infty.$$

Thus, in view of our comments above, the r-th moment of the F.P.T.D. exists if and only if

$$\sum n^{r-2} Pr(S_n \leqq 0) < \infty.$$

We therefore obtain immediately from Theorem A:

THEOREM 3. *Let $r > 1$ be a positive integer. The zero-barrier F.P.T.D. generated by the random variable X with $E|X| < \infty$ and $EX > 0$ has a finite r-th moment if and only if X^- has a finite r-th moment.*

Before ending this section, it is worth remarking that Derman and Robbins [2] show that it is possible to have $E|X^+| = \infty$, $E|X^-| = \infty$ and $Pr(S_n > 0 \text{ i.o.}) = 1$, $Pr(S_n \leqq 0 \text{ i.o.}) = 0$ and hence, following Kemperman [6], Theorem 15.2, 81, $\sum 1/n\, Pr(S_n > 0) = \infty$, $\sum 1/n\, Pr(S_n \leqq 0) < \infty$. This provides us with a limitation on eventual improvements of the Theorems given above.

4. Correction to a theorem of Finch [4]

Let η be the difference between the inter-arrival and service time in a $GI/G/1$ queue. We refrain from stating the usual queueing assumptions for the sake of brevity. Let Π_n be the probability that n customers are served in a busy period. Then, as is well known,

$$\Pi_1 = Pr(\eta_1 > 0)$$
$$\Pi_n = Pr\,(\max_{1 \leqq k \leqq n-1} \eta_1 + \eta_2 + \cdots + \eta_k \leqq 0,\ \eta_n > 0), \qquad n > 1,$$

so that $Pr(T = n) = \Pi_n$ is a zero-barrier F.P.T.D.

Finch [4] gives the following Theorem (his Theorem 2, 223).

THEOREM *Suppose that $E|\eta| < \infty$. Write $\Pi = \sum_{j=1}^{\infty} \Pi_j$, $N = \sum_{j=1}^{\infty} j\Pi_j$, and $a_n = Pr(\eta_1 + \eta_2 + \cdots + \eta_n > 0)$. Then,*

$$\Pi = \begin{cases} 1 & \text{if } E\eta \geqq 0 \\ 1 - \exp\{-\sum_{n=1}^{\infty} n^{-1} a_n\} & \text{if } E\eta < 0 \end{cases}$$

$$N = \begin{cases} \exp\{\sum_{n=1}^{\infty} n^{-1}(1-a_n)\} & \text{if } E\eta > 0 \\ \infty & \text{if } E\eta = 0 \\ \sum_{n=1}^{\infty} a_n \exp\{-\sum_{n=1}^{\infty} n^{-1} a_n\} & \text{if } E\eta < 0. \end{cases}$$

It is the final part of the statement of this Theorem that is incorrect, namely that

$$N = \sum_{n=1}^{\infty} a_n \exp\{-\sum_{n=1}^{\infty} n^{-1} a_n\} < \infty \text{ if } E\eta < 0.$$

In fact, under the condition $E\eta < 0$, we see from a negative mean analogue of Theorem A that $\sum_{n=1}^{\infty} a_n < \infty$ if and only if $E|\eta^+|^2 < \infty$. Thus, $N = \infty$ if $E\eta < 0$ and $E|\eta^+|^2 < \infty$. Finch's error arises from an invalid application of the Borel zero-one criterion which yields $Pr(S_n > 0 \text{ i.o.}) = 0$ or 1 according as $\sum a_n < \infty$ or $= \infty$. Actually, using Kemperman [6], Theorem 15.2, 81, $Pr(S_n > 0 \text{ i.o.}) = 0$ or 1 according as $\sum n^{-1} a_n < \infty$ or $= \infty$. Finch's Theorem and his proof of it can easily be repaired in terms of these comments. A correct statement of the Theorem is as follows:

THEOREM. *Suppose that* $E|\eta| < \infty$. *Write* $\Pi = \sum_{j=1}^{\infty} \Pi_j$, $N = \sum_{j=1}^{\infty} j\Pi_j$, *and* $a_n = Pr(\eta_1+\eta_2+\cdots+\eta_n) > 0$. *Then*

$$\Pi = \begin{cases} 1 & \text{if } E\eta \geqq 0 \\ 1-\exp\{-\sum_{n=1}^{\infty} n^{-1} a_n\} & \text{if } E\eta < 0 \end{cases}$$

$$N = \begin{cases} \exp\{\sum_{n=1}^{\infty} n^{-1}(1-a_n)\} & \text{if } E\eta > 0 \\ \infty & \text{if } E\eta = 0 \text{ or } E\eta < 0 \text{ and } E|\eta^+|^2 = \infty \\ \sum_{n=1}^{\infty} a_n \exp\{-\sum_{n=1}^{\infty} n^{-1} a_n\} & \text{if } E\eta < 0 \text{ and } E|\eta^+|^2 < \infty. \end{cases}$$

References

[1] Baum, L. E., Katz, M. L., and Read, R. R., Exponential convergence rates for the law of large numbers, *Trans. Amer. Math. Soc.*, **102** (1962), 187–199.

[2] Derman, C., and Robbins, H., The strong law of large numbers when the first moment does not exist, *Proc. Nat. Acad. Sci. U.S.A.*, **41** (1955), 586–587.

[3] Erdös, P., On a theorem of Hsu and Robbins, *Ann. Math. Statist.*, **20** (1949), 286–291.

[4] Finch, P. D., On the busy period in the queueing system *GI/G/*1, *J. Austral. Math. Soc.*, **2** (1961), 217–228.

[5] Katz, M. L., The probability in the tail of a distribution, *Ann. Math. Statist.*, **34** (1963), 312–318.

[6] Kemperman, J. H. B., *The passage problem for a stationary Markov chain*, Univ. of Chicago Press (1961).

[7] Lukacs, E., *Characteristic functions*, Griffin, London (1960).

[8] Spitzer, F., A combinatorial lemma and its applications to probability theory, *Trans. Amer. Math. Soc.*, **82** (1956), 323–339.

[9] Stein, C., A note on cumulative sums, *Ann. Math. Statist.*, **17** (1946), 498–499.

The Australian National University,
Canberra

SOME RENEWAL THEOREMS WITH APPLICATION TO A FIRST PASSAGE PROBLEM

BY C. C. HEYDE
Michigan State University

Reprinted from THE ANNALS OF MATHEMATICAL STATISTICS
Vol. 37, No. 3, June, 1966
Printed in U.S.A.

Reprinted from The Annals of Mathematical Statistics
Vol. 37, No. 3, June, 1966
Printed in U.S.A.

SOME RENEWAL THEOREMS WITH APPLICATION TO A FIRST PASSAGE PROBLEM[1]

By C. C. Heyde

Michigan State University

1. Introduction. Let X_i, $i=1, 2, 3, \cdots$ be a sequence of independent and identically distributed random variables with $E|X_i| < \infty$, $EX_i = \mu > 0$. Write $X_i^- = -\min(0, X_i)$, $S_n = \sum_{i=1}^{n} X_i$ and $M_n = \max_{1 \leqq k \leqq n} S_k$. In this paper we shall discuss the asymptotic behaviour as $x \to \infty$ of the sums $\sum_{n=1}^{\infty} a_n \Pr(S_n \leqq x)$ and $\sum_{n=1}^{\infty} a_n \Pr(M_n \leqq x)$ for certain classes of positive coefficient sequences $\{a_n\}$ and use the results on the latter sums to investigate the behaviour of the first passage time out of the interval $(-\infty, x]$ for the process S_n as $x \to \infty$.

The analysis that we shall use in obtaining the theorems on asymptotic behaviour follows closely on that of Smith [6] who discussed sums $\sum_{n=1}^{\infty} a_n \Pr(S_n \leqq x)$ for a class of coefficient sequences that we shall also discuss and for non-identically distributed random variables. In fact, our Theorem 1 follows directly from a specialization of the analysis of Smith. One of the particularly interesting characteristics of this technique is that it enables us to study the asymptotic behaviour of the sums $\sum_{n=1}^{\infty} a_n \Pr(S_n \leqq x)$ and $\sum_{n=1}^{\infty} a_n \Pr(M_n \leqq x)$ in the one operation in spite of essential differences in their behaviour.

2. Renewal theorems. For the first set of positive term coefficient sequences $\{a_n\}$ that we consider we shall suppose (as in [6]) that there exist real numbers $\alpha > 0$, $\gamma \geqq 0$ and some non-negative function of slow growth $L(x)$ such that

$$(1) \qquad \sum_{n=1}^{\infty} a_n x^n \sim [\alpha/(1-x)^\gamma] L(1-x)^{-1}, \qquad \text{as} \qquad x \to 1^-.$$

This is satisfied, for example, if

$$a_n \sim [\alpha/\Gamma(\gamma)] n^{\gamma-1} L(n) \qquad \text{as} \qquad n \to \infty$$

using an Abelian theorem of Doetsch [3], 460.

In the subsequent work we shall need the following definition:

Definition. The index k of the sequence $\{a_n\}$ is the least real k such that $a_n = O(n^k)$.

Consideration will be restricted to cases where $\sum a_n$ diverges.

Theorem 1. *Suppose* $E|X| < \infty$, $EX = \mu > 0$. *Let* k *be the index of the sequence* $\{a_n\}$ *and* m *be non-negative. In order that*

$$\sum_{n=1}^{\infty} a_n \Pr(S_n \leqq x) \sim [\alpha L(x)/\Gamma(1+\gamma)](x/\mu)^\gamma \qquad \textit{as} \qquad x \to \infty$$

for each sequence $\{a_n\}$ *such that* $k \leqq m$ *it is necessary and sufficient that* $E|X^-|^{m+2} < \infty$.

Received 2 June 1965; revised 6 January 1966.

[1] Research supported in part by the National Institutes of Health.

THEOREM 2. *Suppose* $E|X| < \infty$, $EX = \mu > 0$. *Let* k *be the index of the sequence* $\{a_n\}$ *and* m *be non-negative. In order that*

$$\textstyle\sum_{n=1}^{\infty} a_n \Pr(M_n \leqq x) \sim [\alpha L(x)/\Gamma(1+\gamma)](x/\mu)^{\gamma} \qquad as \qquad x \to \infty$$

for each sequence $\{a_n\}$ *such that* $k \leqq m$ *it is necessary that* $E|X^-|^{m+1} < \infty$ *and sufficient that* $E|X^-|^{m+2} < \infty$.

I conjecture that the condition $E|X^-|^{m+1} < \infty$ is both necessary and sufficient in Theorem 2. It is certainly known that in the particular case where $a_n = 1$ for all n so that $k = 0$ we only need $E|X^-| < \infty$ (see for example Chow and Robbins [2]).

The proof of the two theorems shall be deferred until we have given four lemmas. The first of these is given in a form which is more general than we shall need subsequently as it has some independent interest.

LEMMA 1. *Suppose the random variables* X_i, $i = 1, 2, 3, \cdots$ *are independent and identically distributed.* Write $S_n = \sum_{i=1}^{n} X_i$, $M_n = \max_{1 \leqq k \leqq n} S_k$, *and, in the case where* $E|X_i| < \infty$, $EX_i = \mu$. *If* $E|X_i|^r < \infty$ *with* $1 \leqq r < 2$ *and* $\mu \geqq 0$, *then*

$$n^{-1/r}(M_n - n\mu) \to_{\text{a.s.}} 0.$$

If $E|X_i|^r < \infty$ *with* $0 < r < 1$ *or* $E|X_i|^r < \infty$ *with* $1 \leqq r < 2$ *and* $\mu < 0$, *then*

$$n^{-1/r} M_n \to_{\text{a.s.}} 0.$$

("a.s." *denotes almost sure convergence*).

The corresponding almost sure convergence versions for the sums S_n have been given by Kolmogorov ($r = 1$) and Marcinkiewicz ($r \neq 1$) (see for example Loève [5] 242, 243).

PROOF. Suppose $E|X_i|^r < \infty$. Let $c_r = \mu$ if $1 \leqq r < 2$; $c_r = 0$ if $0 < r < 1$. If $c_r \geqq 0$, we have

$$\begin{aligned} M_n - nc_r &= \max_{1 \leqq j \leqq n} S_j - nc_r \\ &\leqq \max_{1 \leqq j \leqq n} (S_j - jc_r) \\ &\leqq \max_{1 \leqq j \leqq n} j^{1/r} a_j, \end{aligned}$$

where $a_j = j^{-1/r}|S_j - jc_r|$. Therefore, if $M_n \geqq nc_r$, $0 \leqq M_n - nc_r \leqq \max_{1 \leqq j \leqq n} j^{1/r} a_j$, and if $M_n < nc_r$, $0 > M_n - nc_r \geqq S_n - nc_r = -n^{1/r} a_n \geqq -\max_{1 \leqq j \leqq n} a_j$, so that $0 \leqq n^{-1/r}|M_n - nc_r| \leqq \max_{1 \leqq j \leqq n} n^{-1/r} j^{1/r} a_j$. In order to obtain the desired result in this case it suffices to show that for arbitrary $\epsilon > 0$,

$$\Pr\left(\textstyle\bigcup_{k \geqq n} \{\max_{1 \leqq j \leqq k} j^{1/r} k^{-1/r} a_j \geqq \epsilon\}\right) \to 0 \quad \text{as} \quad n \to \infty.$$

Now the strong laws of Kolmogorov and Marcinkiewicz ([5], 242, 243) imply that for arbitrary $\epsilon > 0$,

$$\Pr(\max_{j \geqq n} a_j \geqq \epsilon) \to 0 \qquad \text{as} \qquad n \to \infty.$$

Therefore, given $\eta > 0$ arbitrarily small we can choose an N so large that

$$\Pr(\max_{j > N} a_j \geqq \tfrac{1}{2}\epsilon) \leqq \tfrac{1}{2}\eta \tag{2}$$

and then select an $n_0(>N)$ so large that for $n \geqq n_0$,

(3) $$\Pr(N^{1/r}n^{-1/r} \max_{1 \leqq j \leqq N} a_j \geqq \tfrac{1}{2}\epsilon) \leqq \tfrac{1}{2}\eta.$$

Then, for $n \geqq n_0$,

$$\begin{aligned}
\Pr(\textstyle\bigcup_{k \geqq n} \{\max_{1 \leqq j \leqq k} j^{1/r}k^{-1/r}a_j \geqq \epsilon\})
&= \Pr(\textstyle\bigcup_{k \geqq n} \{\max_{1 \leqq j \leqq N} j^{1/r}k^{-1/r}a_j + \max_{N<j \leqq k} j^{1/r}k^{-1/r}a_j \geqq \epsilon\}) \\
&\leqq \Pr(\textstyle\bigcup_{k \geqq n} \{\max_{1 \leqq j \leqq N} j^{1/r}k^{-1/r}a_j + \max_{N<j \leqq k} a_j \geqq \epsilon\}) \\
&\leqq \Pr(N^{1/r}n^{-1/r} \max_{1 \leqq j \leqq N} a_j + \max_{j>N} a_j \geqq \epsilon) \\
&\leqq \Pr(N^{1/r}n^{-1/r} \max_{1 \leqq j \leqq N} a_j \geqq \tfrac{1}{2}\epsilon) + \Pr(\max_{j>N} a_j \geqq \tfrac{1}{2}\epsilon) \\
&\leqq \eta,
\end{aligned}$$

using (2) and (3). This completes the proof of this part of the lemma. It remains to consider the case $c_r < 0$.

In the case $c_r < 0$, $M_n^+ = \max(0, M_n)$ actually has a proper limiting distribution (finite with probability one), $M^+ = \lim_{n\to\infty} M_n^+$.

We have

$$\begin{aligned}
\Pr(\textstyle\bigcup_{k \geqq n} \{k^{-1/r}|M_k| \geqq \epsilon\})
&= \Pr(\textstyle\bigcup_{k \geqq n} \{k^{-1/r}M_k \geqq \epsilon\} \cup \bigcup_{k \geqq n} \{k^{-1/r}M_k \leqq -\epsilon\}) \\
&\leqq \Pr(\textstyle\bigcup_{k \geqq n} \{k^{-1/r}M_k \geqq \epsilon\}) + \Pr(\bigcup_{k \geqq n} \{k^{-1/r}M_k \leqq -\epsilon\}) \\
&\leqq \Pr(\textstyle\bigcup_{k \geqq n} \{M^+ \geqq k^{1/r}\epsilon\}) + \Pr(\bigcup_{k \geqq n} \{X_1 \leqq -k^{1/r}\epsilon\}) \\
&= \Pr(M^+ \geqq n^{1/r}\epsilon) + \Pr(X_1 \leqq -n^{1/r}\epsilon)
\end{aligned}$$

and both these terms approach zero as $n \to \infty$. Therefore, $n^{-1/r}M_n \to_{\text{a.s.}} 0$ as required. This completes the proof of the lemma.

In the subsequent work we shall write $F_n(x) = \Pr(S_n \leqq x)$, $G_n(x) = \Pr(M_n \leqq x)$ and $H_n(x)$ to mean either $F_n(x)$ or $G_n(x)$ (so that if a property holds for both $F_n(x)$ and $G_n(x)$ it holds for $H_n(x)$ and conversely).

Lemma 2. $\int_\mu^\infty \{1 - H_n(nx)\}\, dx \to 0$ *as* $n \to \infty$.

Proof. Introduce the new random variables $Y_i = \max(0, X_i)$, $i = 1, 2, 3, \cdots$. Suppose $EY_i = \alpha$. If $M_n > x$ then $\sum_{i=1}^n Y_i > x$ so that $\Pr(\sum_{i=1}^n Y_i > x) \geqq \Pr(M_n > x) \geqq \Pr(S_n > x)$ and hence

(4) $$1 - F_n(x) \leqq 1 - G_n(x) \leqq 1 - K_n(x),$$

where we have written $K_n(x) = \Pr(\sum_{i=1}^n Y_i \leqq x)$.

We show firstly that $\int_\alpha^\infty \{1 - K_n(nx)\}\, dx \to 0$ as $n \to \infty$. We have, by a simple integration by parts,

$$\begin{aligned}
\alpha &= \textstyle\int_0^\infty \{1 - K_n(nx)\}\, dx \\
&= \textstyle\int_0^\alpha \{1 - K_n(nx)\}\, dx + \int_\alpha^\infty \{1 - K_n(nx)\}\, dx.
\end{aligned}$$

Further, by the law of large numbers, $K_n(nx) \to 0$ as $n \to \infty$ for $x < \alpha$. It follows from the mean value theorem that $\int_0^\alpha \{1 - K_n(nx)\}\, dx \to \alpha$ as $n \to \infty$ and hence that $\int_\alpha^\infty \{1 - K_n(nx)\}\, dx \to 0$ as $n \to \infty$.

Then, making use of (4),

$$0 \leqq \int_\mu^\infty \{1 - F_n(nx)\}\, dx \leqq \int_\mu^\infty \{1 - G_n(nx)\}\, dx$$
$$\leqq \int_\mu^\alpha \{1 - G_n(nx)\}\, dx + \int_\alpha^\infty \{1 - K_n(nx)\}\, dx.$$

Now, in view of Lemma 1 (case $r = 1, \mu > 0$), $G_n(nx) \to 1$ as $n \to \infty$ for $x > \mu$ so that by the mean value theorem $\int_\mu^\alpha \{1 - G_n(nx)\}\, dx \to 0$ as $n \to \infty$. We have shown that $\int_\alpha^\infty \{1 - K_n(nx)\}\, dx \to 0$ as $n \to \infty$ so the proof is complete.

LEMMA 3. (Smith [6]). *If the non-negative constants* $\{a_n\}$ *satisfy* (1) *then as* $s \to 0+$,

$$\sum_{n=1}^\infty a_n e^{-\mu s n} \sim \alpha(\mu s)^{-\gamma} L(s^{-1}).$$

LEMMA 4. *If the non-negative constants* $\{a_n\}$ *satisfy* (1) *then as* $s \to 0+$,

$$\sum_{n=1}^\infty n a_n e^{-\mu s n} \sim \alpha(\mu s)^{-\gamma-1} L(s^{-1}) \quad \text{if} \quad \gamma > 0$$
$$\limsup_{s\to 0+} s[L(s^{-1})]^{-1} \sum_{n=1}^\infty n a_n e^{-\mu s n} < \alpha \mu^{-1} \quad \text{if} \quad \gamma = 0.$$

Smith [6] has established the former result, ignoring the possibility of the case $\gamma = 0$. The result given above for $\gamma = 0$ can be readily extracted from Smith's proof and is adequate in the present context.

PROOF OF THEOREMS 1 AND 2. We follow the methods of [6] but work the proof in terms of $H_n(x)$.

Suppose firstly that $E|X^-|^{m+2} < \infty$. Take β arbitrary with $0 < \beta < \mu$. Consider

$$0 \leqq K_n = \int_{n\beta}^{n\mu} e^{-sx} H_n(x)\, dx \qquad (s \geqq 0)$$
$$= n \int_\beta^\mu e^{-nsx} H_n(nx)\, dx$$
$$\leqq n e^{-n\beta s} \int_\beta^\mu H_n(nx)\, dx.$$

Now using the law of large numbers and the inequality $H_n(y) \leqq F_n(y)$ (or alternatively referring to Lemma 1 as well), we see that $H_n(nx) \to 0$ as $n \to \infty$ for all $x < \mu$. Hence, using the mean value theorem, we may write

$$K_n = n e^{-n\beta s} \delta_n', \tag{5}$$

where $\delta_n' \to 0$ as $n \to \infty$, uniformly in $s \geqq 0$.

Now consider

$$0 \leqq L_n = \int_{n\mu}^\infty e^{-sx} \{1 - H_n(x)\}\, dx \qquad (s \geqq 0)$$
$$= n \int_\mu^\infty e^{-nsx} \{1 - H_n(nx)\}\, dx$$
$$\leqq n e^{-ns\beta} \int_\mu^\infty \{1 - H_n(nx)\}\, dx.$$

In view of Lemma 2, we may write

$$L_n = n e^{-n\beta s} \delta_n'' \tag{6}$$

where $\delta_n'' \to 0$ as $n \to \infty$, uniformly in $s \geqq 0$.

Combining (5) and (6) and putting $\delta_n = \delta_n' - \delta_n''$, we obtain

(7) $$\sum_{n=1}^{\infty} a_n(L_n - K_n) = -\sum_{n=1}^{\infty} na_n\delta_n e^{-n\beta s}.$$

Now given arbitrary $\epsilon > 0$ we can choose an integer $n_0(\epsilon)$ so large that $|\delta_n| < \epsilon$ for all $n > n_0$. Then,

$$\left|\sum_{n=1}^{\infty} na_n\delta_n e^{-n\beta s}\right| < \sum_{n=1}^{n_0} na_n|\delta_n|e^{-n\beta s} + \epsilon\sum_{n=1}^{\infty} na_n e^{-n\beta s},$$

so that by Lemma 4,

$$\limsup_{s\to 0+} [(\mu s)^{\gamma+1}/L(s^{-1})]\left|\sum_{n=1}^{\infty} na_n\delta_n e^{-n\beta s}\right| \leqq \epsilon\gamma\alpha \text{ if } \gamma > 0$$
$$\leqq \epsilon\alpha \text{ if } \gamma = 0$$

and hence

(8) $$[(\mu s)^{\gamma+1}/L(s^{-1})]\sum_{n=1}^{\infty} a_n(L_n - K_n) \to 0 \quad \text{as} \quad s \to 0+.$$

Next write

(9) $$\Phi_\beta(x) = \sum_{n=1}^{\infty} a_n H_n(x) U(x - n\beta)$$
$$= \sum_{n=1}^{\infty} a_n U(x - n\mu) - \sum_{n=1}^{\infty} a_n\{U(x - n\mu) - H_n(x)\}U(x - n\beta)$$

where

$$U(x) = 1, \qquad x \geqq 0,$$
$$= 0, \qquad x < 0.$$

If we denote the Laplace transform of a function $A(x)$ by $A^0(s) = \int_0^\infty e^{-sx}A(x)\,dx$ we have

(10) $$\Phi_\beta^0(s) = s^{-1}\sum_{n=1}^{\infty} a_n e^{-n\mu s} - \sum_{n=1}^{\infty} a_n(L_n - K_n);$$

the term by term integration being justified by monotone convergence.

From (8), (10) and Lemma 3 it follows that

(11) $$[(\mu s)^{\gamma+1}/L(s^{-1})]\Phi_\beta^0(s) \to \alpha\mu \quad \text{as} \quad s \to 0+.$$

Appealing to a Tauberian theorem of Doetsch [3], 511, we then obtain

(12) $$[1/t^{\gamma+1}L(t)]\int_0^t \Phi_\beta(x)\,dx \to \alpha/\mu^\gamma\Gamma(\gamma + 2).$$

Now for $t > 0$ and $0 < \theta < 1$,

$$\Phi_\beta(\theta t)(t - \theta t) \leqq \int_{\theta t}^t \Phi_\beta(x)\,dx \leqq \Phi_\beta(t)(t - \theta t)$$

so that

$$[1/t^\gamma L(t)]\Phi_\beta(\theta t) \leqq [1/(1 - \theta)]\{[1/t^{\gamma+1}L(t)]\int_0^t \Phi_\beta(x)\,dx$$
$$- [1/t^{\gamma+1}L(t)]\int_0^{\theta t}\Phi_\beta(x)\,dx\} \leqq [1/t^\gamma L(t)]\Phi_\beta(t).$$

Then, for fixed θ and $t \to \infty$, $[1/t^{\gamma+1}L(t)]\int_0^t \Phi_\beta(x)\,dx \to \alpha/\mu^\gamma\Gamma(\gamma + 2)$, $[1/(\theta t)^{\gamma+1}L(t)]\int_0^{\theta t}\Phi_\beta(x)\,dx \to [1/(\theta t)^{\gamma+1}L(\theta t)]\int_0^{\theta t}\Phi_\beta(x)\,dx \to \alpha/\mu^\gamma\Gamma(\gamma + 2)$ by

(12) so that

$$(13)\quad \limsup_{t\to\infty} [1/t^\gamma L(t)]\Phi_\beta(\theta t) \leqq [\alpha/\mu^\gamma\Gamma(\gamma+2)](1-\theta^{\gamma+1})/(1-\theta) \leqq \liminf_{t\to\infty} [1/t^\gamma L(t)]\Phi_\beta(t).$$

Taking $\theta \to 1$ in the right hand part of inequality (13) gives

$$(14)\quad \liminf_{t\to\infty} [1/t^\gamma L(t)]\Phi_\beta(t) \geqq \alpha/\mu^\gamma\Gamma(\gamma+1).$$

Further, the left hand part of the inequality (13) can be written

$$\limsup_{t\to\infty} [1/(\theta t)^\gamma L(\theta t)]\Phi_\beta(\theta t) \leqq [-\theta\alpha/\mu^\gamma\Gamma(\gamma+2)](1-\theta^{-(\gamma+1)})/(1-\theta)$$

and the left hand side of this is equal to $\limsup_{t\to\infty} [1/t^\gamma L(t)]\Phi_\beta(t)$ for θ fixed. Then, taking $\theta \to 1$ in the right hand side, we obtain

$$(15)\quad \limsup_{t\to\infty} [1/t^\gamma L(t)]\Phi_\beta(t) \leqq \alpha/\mu^\gamma\Gamma(\gamma+1),$$

so that combining (14) and (15),

$$(16)\quad \lim_{t\to\infty} [1/t^\gamma L(t)]\Phi_\beta(t) = \alpha/\mu^\gamma\Gamma(\gamma+1).$$

Now under the condition $E|X^-|^{m+2} < \infty$ we certainly have for all x, $-\infty < x < \infty$, $\sum n^m H_n(x) < \infty$ in view of Theorem A of Heyde [4] and the inequality $H_n(x) \leqq F_n(x)$. Hence, $\sum a_n H_n(x) < \infty$, since the sequence $\{a_n\}$ has index $k \leqq m$. Write

$$(17)\quad \sum_{n=1}^\infty a_n H_n(x) = \Phi_\beta(x) + \Psi_\beta(x),$$

where

$$(18)\quad \Psi_\beta(x) = \sum_{n=1}^\infty a_n H_n(x)\{1 - U(x - n\beta)\}.$$

We shall go on to show that $\Psi_\beta(x)/x^\gamma L(x) \to 0$ as $x \to \infty$.

Define a new sequence of random variables Y_i, $i = 1, 2, 3, \cdots$ by

$$Y_i = X_i - \beta.$$

Then, $EY > 0$ and $E|Y^-|^{m+2} < \infty$ since $E|X^-|^{m+2} < \infty$. It follows from Theorem A of [4] applied to the Y's that for $k \leqq m$, $\sum n^k F_n(n\beta) < \infty$, and since $H_n(n\beta) \leqq F_n(n\beta)$,

$$(19)\quad \sum n^k H_n(n\beta) < \infty.$$

Also, it is clear from (18) that $\Psi_\beta(x) \leqq \sum_{n=1}^\infty a_n H_n(n\beta)$ and (19) ensures that this upper bound is finite since k is the index of the sequence $\{a_n\}$. We therefore have $\Psi_\beta(x)/x^\gamma L(x) \to 0$ as $x \to \infty$ and hence, using (16) and (17),

$$\sum_{n=1}^\infty a_n H_n(x) \sim [\alpha L(x)/\Gamma(1+\gamma)](x/\mu)^\gamma \quad \text{as} \quad x \to \infty.$$

This result is true for all sequences $\{a_n\}$ with index $k \leqq m$ and hence establishes the sufficiency parts of both Theorems 1 and 2.

The necessity parts of both the theorems are easy to establish. It suffices to note that in particular $\sum n^m H_n(x) < \infty$ for all x, $-\infty < x < \infty$, and hence by Theorem A of [4] in the case $H_n(x) = F_n(x)$ we obtain $E|X^-|^{m+2} < \infty$ and by Theorem A of [4] together with the well known inequality $G_n(0) = \Pr(M_n \leqq 0) \geqq n^{-1} \Pr(S_n \leqq 0)$ in the case $H_n(x) = G_n(x)$ we obtain $E|X^-|^{m+1} < \infty$. This completes the proof of both Theorems 1 and 2.

Some remarks on the possibility that the condition $E|X^-|^{m+1} < \infty$ might be both necessary and sufficient in Theorem 2 would seem in order. To establish that this is the case, it would be adequate to show that if $EX = \mu > 0$ and $E|X^-|^{k+1} < \infty$, then for $0 < \beta < \mu$,

$$\sum n^k G_n(n\beta) = \sum n^k \Pr(M_n \leqq n\beta) < \infty.$$

The nearest approach to this that I have obtained is summarized in the following theorem:

THEOREM 3. *Suppose* $E|X| < \infty$, $EX > 0$, $\Pr(X < 0) > 0$, *and let* k *be a non-negative integer. A necessary and sufficient condition for the convergence of the series*

$$\sum_{n=1}^{\infty} n^k \Pr(M_n \leqq x), \qquad -\infty < x < \infty,$$

is that $E|X^-|^{k+1} < \infty$.

PROOF. It follows from the work of Heyde [4] that a necessary and sufficient condition for the convergence of the series $\sum n^k \Pr(M_n \leqq 0)$ is that $E|X^-|^{k+1} < \infty$. Therefore, in order to complete the proof it is only necessary to show that the convergence of $\sum n^k \Pr(M_n \leqq 0)$ implies that of $\sum n^k \Pr(M_n \leqq x)$, $0 < x < \infty$.

To accomplish this we define a new sequence of random variables:

$$N_0 = S_0 = 0,$$

$$N_1 = \max(S_0, S_1) = X_1^+,$$

$$N_2 = \max(S_0, S_1, S_2) = (X_1 + X_2^+)^+,$$

$$\cdots$$

$$N_n = \max(S_0, S_1, S_2, \cdots, S_n) = (X_1 + (X_2 + \cdots + (X_{n-1} + X_n^+)^+ \cdots)^+$$

$$\cdots$$

Since the X_i are independent and identically distributed we may write $N_0 = 0$, $N_{n+1} \sim (X + N_n)^+$, $n \geqq 1$, where for two random variables X and Y, we write $X \sim Y$ if they have the same distribution. Clearly, $N_n = M_n^+ = \max(0, M_n)$, so that

$$\Pr(N_n \leqq x) = \Pr(M_n \leqq x), \qquad x > 0,$$

$$\Pr(N_n = 0) = \Pr(M_n \leqq 0).$$

Now for $A > 0$, $n \geqq 1$, we have

$$\Pr(M_{n+1} \leqq A) = \int_{-\infty}^{A} \Pr(M_n \leqq A - y)\, d\Pr(X \leqq y)$$

so that for arbitrary $B > 0$,

$$\begin{aligned}\sum_{n=1}^{N} n^{\alpha} \Pr(M_{n+1} \leqq A) &= \int_{-\infty}^{A} \{\sum_{n=1}^{N} n^{\alpha} \Pr(M_n \leqq A - y)\}\, d\Pr(X \leqq y) \\ &\geqq \int_{-\infty}^{-B} \{\sum_{n=1}^{N} n^{\alpha} \Pr(M_n \leqq A - y)\}\, d\Pr(X \leqq y) \\ &\geqq \{\sum_{n=1}^{N} n^{\alpha} \Pr(M_n \leqq A + B)\} \Pr(X \leqq -B).\end{aligned}$$

Then, choosing B so that $\Pr(X \leqq -B) > 0$, we see that if $\sum n^{\alpha} \Pr(M_n \leqq A)$ converges then $\sum n^{\alpha} \Pr(M_n \leqq A + B)$ converges. Let $A \to 0+$; we see that the convergence of $\sum n^{\alpha} \Pr(M_n \leqq 0)$ implies that of $\sum n^{\alpha} \Pr(M_n \leqq rB)$ for all positive integral r and hence that of $\sum n^{\alpha} \Pr(M_n \leqq x)$ for all x, $0 < x < \infty$. This completes the proof of the theorem.

The restriction to non-negative integral k in Theorem 3 is unfortunate. This comes about from the use of the derivatives of the generating function

$$\sum_{n=0}^{\infty} \Pr(M_n \leqq 0)t^n = \exp\{\sum_{n=1}^{\infty} n^{-1} t^n \Pr(S_n \leqq 0)\}$$

in [4].

We now go on to consider coefficient sequences of the form $a_n = e^{rn}$, $r > 0$, and shall establish the following two theorems:

THEOREM 4. *Suppose* $E|X| < \infty$, $EX = \mu > 0$. *In order that*

$$\sum_{n=1}^{\infty} e^{rn} \Pr(S_n \leqq x) \sim r^{-1} e^{xr\mu^{-1}} \quad \text{as} \quad x \to \infty$$

for any r *in some interval* $0 < r < R$, *it is necessary and sufficient that* X^- *should have an analytic characteristic function.*

(The term "analytic characteristic function" is used for a characteristic function which is analytic in a strip containing the origin as an interior point.)

THEOREM 5. *Suppose* $E|X| < \infty$, $EX = \mu > 0$. *In order that*

$$\sum_{n=1}^{\infty} e^{rn} \Pr(M_n \leqq x) \sim r^{-1} e^{xr\mu^{-1}} \quad \text{as} \quad x \to \infty$$

for any r *in some interval* $0 < r < R$, *it is necessary and sufficient that* X^- *should have an analytic characteristic function.*

The proofs of these theorems, of course, follow markedly similar lines to the proofs of Theorems 1 and 2. A generalized coefficient sequence form along the lines of Theorems 1 and 2 is not, however, convenient in this case.

PROOF OF THEOREMS 4 AND 5. We construct the proofs in parallel fashion as we did with Theorems 1 and 2.

Suppose X^- has an analytic characteristic function. Take β arbitrary with $0 < \beta < \mu$. We consider to begin with

$$\begin{aligned}K_n &= \int_{n\beta}^{n\mu} e^{-sx} H_n(x)\, dx \qquad (s > 0) \\ &\leqq n e^{-n\beta s} \int_{\beta}^{\mu} H_n(nx)\, dx \\ &= n e^{-n\beta s}(\mu - \beta) H_n(n\xi)\end{aligned}$$

for some $\xi = \xi(n)$ in $\beta < \xi < \mu$. $(X - \xi)^-$ has an analytic characteristic function and $E(X - \xi) > 0$. It follows from Theorem B of [4] and the inequality

$H_m \leqq F_m$ that for any r in some interval $0 < r < R$,

$$\sum e^{rm} H_m(m\xi) < \infty, \tag{20}$$

and hence, considering the totality of series (20) as n varies,

$$H_n(n\xi) = o(e^{-rn}) \quad \text{as} \quad n \to \infty. \tag{21}$$

Therefore,

$$\sum_{n=1}^{\infty} e^{rn} K_n \leqq (\mu - \beta) \sum_{n=1}^{\infty} n e^{-n\beta s} H_n(n\xi) e^{rn}$$
$$< \infty \quad \text{as} \quad s \to \mu^{-1} r+$$

in view of (21) and so

$$\lim_{s \to \mu^{-1} r+} (s - \mu^{-1} r) \sum_{n=1}^{\infty} e^{rn} K_n = 0. \tag{22}$$

Next consider

$$L_n = \int_{n\mu}^{\infty} e^{-sx} \{1 - H_n(x)\}\, dx \qquad (s > 0).$$

Using the mean value theorem and the fact that $H_n(nx) \to 1$ as $n \to \infty$ for $x > \mu$, we may write

$$L_n = \delta_n \int_{n\mu}^{\infty} e^{-sx}\, dx = \delta_n s^{-1} e^{-n\mu s},$$

where $\delta_n > 0$ and $\delta_n \to 0$ as $n \to \infty$ uniformly in $s > 0$. Now given arbitrary $\epsilon > 0$ we can choose an integer $n_0(\epsilon)$ so large that $\delta_n < \epsilon$ for all $n > n_0$. Then, for $s > \mu^{-1} r$,

$$\sum_{n=1}^{\infty} e^{rn} L_n = s^{-1} \sum_{n=1}^{\infty} \delta_n e^{-n\mu(s - \mu^{-1} r)}$$
$$< s^{-1} \sum_{n=1}^{n_0} \delta_n e^{-n\mu(s - \mu^{-1} r)} + \epsilon s^{-1} \sum_{n=1}^{\infty} e^{-n\mu(s - \mu^{-1} r)},$$

and since for $s > \mu^{-1} r$,

$$\sum_{n=1}^{\infty} e^{-n\mu(s - \mu^{-1} r)} = e^{-\mu(s - \mu^{-1} r)} / (1 - e^{-\mu(s - \mu^{-1} r)}) \sim 1/\mu(s - \mu^{-1} r) \quad \text{as} \quad s \to \mu^{-1} r+,$$

it follows that

$$\lim_{s \to \mu^{-1} r+} (s - \mu^{-1} r) \sum_{n=1}^{\infty} e^{rn} L_n = 0, \tag{23}$$

so that combining (22) and (23),

$$\lim_{s \to \mu^{-1} r+} (s - \mu^{-1} r) \sum_{n=1}^{\infty} e^{rn} (L_n - K_n) = 0. \tag{24}$$

Now consider the function

$$\Phi_\beta(x) = \sum_{n=1}^{\infty} e^{rn} H_n(x) U(x - n\beta)$$
$$= \sum_{n=1}^{\infty} e^{rn} U(x - n\mu) - \sum_{n=1}^{\infty} e^{rn} \{U(x - n\mu) - H_n(x)\} U(x - n\beta).$$

Taking Laplace transforms, we have

$$\Phi_\beta^0(s) = s^{-1} \sum_{n=1}^{\infty} e^{-n\mu(s - \mu^{-1} r)} - \sum_{n=1}^{\infty} e^{rn} (L_n - K_n),$$

from which it follows, using (24), that $(s - \mu^{-1}r)\Phi_\beta{}^0(s) \to r^{-1}$ as $s \to \mu^{-1}r+$. Now let

$$\Theta_\beta{}^0(s - \mu^{-1}r) = \Phi_\beta{}^0(s); \tag{25}$$

we have $w\Theta_\beta{}^0(w) \to r^{-1}$ as $w \to 0+$, and appealing to a Tauberian theorem of Widder [7], Theorem 4.3, 192, we obtain

$$t^{-1} \int_0^t \Theta_\beta(x)\,dx \to r^{-1} \quad \text{as} \quad t \to \infty. \tag{26}$$

Then, following through exactly the analysis that we used in the case of Theorems 1 and 2 (we are here dealing with the case $\gamma = 0$, $L(t) = 1$, $\alpha\mu^{-\gamma}[\Gamma(\gamma + 2)]^{-1} = r^{-1}$) we find from (26) that

$$\lim_{t\to\infty} \Theta_\beta(t) = r^{-1}. \tag{27}$$

However, using (25) and the uniqueness theorem for Laplace transformations, $\Theta_\beta(t) = e^{-r\mu^{-1}t}\Phi_\beta(t)$, so that (27) gives

$$\lim_{t\to\infty} e^{-r\mu^{-1}t}\Phi_\beta(t) = r^{-1}. \tag{28}$$

Now write

$$\sum_{n=1}^{\infty} e^{rn} H_n(x) = \Phi_\beta(x) + \Psi_\beta(x), \tag{29}$$

where

$$\Psi_\beta(x) = \sum_{n=1}^{\infty} e^{rn} H_n(x)\{1 - U(x - n\beta)\}, \tag{30}$$

$\sum e^{rn} H_n(x)$ being finite for r in $0 < r < R$ by (20). Also it is clear from (30) and (20) that for all x, $\Psi_\beta(x) \leqq \sum_{n=1}^{\infty} e^{rn} H_n(n\beta) < \infty$. We therefore have $e^{-r\mu^{-1}x}\Psi_\beta(x) \to 0$ as $x \to \infty$, and hence by (28), $\sum_{n=1}^{\infty} e^{rn} H_n(x) \sim r^{-1} e^{r\mu^{-1}x}$ as $x \to \infty$. This establishes the sufficiency parts of Theorems 4 and 5.

The necessity parts of the theorems follow readily from Theorem B of [4]. We have $\sum e^{rn} H_n(x) < \infty$ for all x, $-\infty < x < \infty$ and Theorem B gives the result directly in the case $H_n(x) = F_n(x)$. In the case $H_n(x) = G_n(x)$, we take $x > 0$ and make use of the well known relations $G_n(x) = \Pr(M_n \leqq x) \geqq \Pr(M_n \leqq 0) \geqq n^{-1} \Pr(S_n \leqq 0)$. It is clear that the convergence of $\sum e^{rn} \Pr(M_n \leqq x)$ implies the convergence of $\sum n^{-1} e^{rn} \Pr(S_n \leqq 0)$ and hence that of $\sum e^{sn} \Pr(S_n \leqq 0)$ for $s < r$ and making use of Theorem B again we see that X^- must have an analytic characteristic function. This completes the proofs of both Theorems 4 and 5.

3. Application to a first passage problem. Let X_i, $i = 1, 2, 3, \cdots$ be independent and identically distributed random variables with $E|X| < \infty$, $EX = \mu > 0$. Write $S_n = \sum_{i=1}^{n} X_i$ and $M_n = \max_{1\leqq k\leqq n} S_k$. Consider a single boundary at x ($\geqq 0$) so that if

$$G_0(x) = 1,$$

$$G_n(x) = \Pr(M_n \leqq x), \qquad n \geqq 1,$$

the probability p_n that the first passage time, $M(x)$, out of the interval $(-\infty, x]$ for the process S_n is n is given by $p_n = G_{n-1}(x) - G_n(x)$, $n \geqq 1$.

We introduce the probability generating function $P(\lambda) = \sum_{r=1}^{\infty} \lambda^r p_r$ for the first passage time distribution $\Pr(M(x) = n) = p_n$.

Formally differentiating,

$$P^{(1)}(1) = E[M(x)] = 1 + \sum_{r=1}^{\infty} G_r(x),$$

and for $k > 1$,

$$P^{(k)}(1) = (\alpha)_k \quad \text{(the } k\text{th factorial moment of } M(x))$$
$$= k \sum_{r=k-1}^{\infty} (r)_k G_r(x) = \sum_{r=0}^{k} s(k, r) E\{[M(x)]^r\},$$

where $(r)_k = r(r-1)(r-2) \cdots (r-k+1)$ and $s(k, r)$ are the Stirling numbers of the first kind. It is thus clear that $E\{[M(x)]^r\} < \infty$ for some positive integer r if and only if $\sum n^{r-1} G_r(x) < \infty$. Also, the random variable $M(x)$ has an analytic characteristic function if and only if the radius of convergence of $P(\lambda)$ is greater than unity or equivalently if $\sum e^{rn} G_n(x) < \infty$ for some $r > 0$.

As a particular case of Theorem 2 we obtain for integral $k \geqq 1$,

$$kx^{-k} \sum_{n=1}^{\infty} n^{k-1} G_n(x) \to \mu^{-k} \quad \text{as} \quad x \to \infty,$$

so long as $E|X^-|^{k+1} < \infty$. Therefore, in view of the above comments, we see that as $x \to \infty$, $E\{[x^{-1}M(x)]^r\} \to \mu^{-r}$ for integral $r \geqq 1$ so long as $E|X^-|^{r+1} < \infty$. (The $r = 1$ and $r = 2$ cases are included in the results of [2] and [1] respectively). If $E|X^-|^r = \infty$ then Theorem 3 shows us that $\sum n^{r-1} G_n(x)$ diverges and hence that $E\{[x^{-1}M(x)]^r\} = \infty$. If we have the condition that X^- possesses an analytic characteristic function, then it follows from Theorem B of [4] and the inequality, $\Pr(S_n \leqq x) \geqq \Pr(M_n \leqq x) = G_n(x)$, that $M(x)$ possesses an analytic characteristic function. We have, in fact, for $r > 0$ sufficiently small,

$$E[e^{rM(x)}] = \sum_{n=1}^{\infty} e^{rn} P_n$$
$$= e^r + (e^r - 1) \sum_{n=1}^{\infty} e^{rn} G_n(x),$$

so that by Theorem 5,

$$E[e^{rM(x)}] \sim r^{-1}(e^r - 1) e^{r\mu^{-1}x} \quad \text{as} \quad x \to \infty.$$

The function $r^{-1}(e^r - 1)e^{r\mu^{-1}x}$ is the Laplace-Stieltjes transform of the convolution of a rectangular distribution on the interval $(0, 1)$ and a degenerate distribution at $x\mu^{-1}$. We have therefore obtained the following theorem:

THEOREM 6. *Suppose* $EX = \mu > 0$. *If for some integral* $r \geqq 1$, $E|X^-|^{r+1} < \infty$, *then*

$$E\{[x^{-1}M(x)]^r\} \to \mu^{-r} \quad as \quad x \to \infty.$$

If X^- *possesses an analytic characteristic function then for* $r > 0$ *sufficiently small*

$$E[e^{rM(x)}] \sim r^{-1}(e^r - 1)e^{r\mu^{-1}x} \quad as \quad x \to \infty.$$

One can further use a method due to Doob based on the strong law of large numbers to obtain the following result:

THEOREM 7. *Suppose* $EX = \mu > 0$. *Then,*

$$x^{-1}M(x) \rightarrow_{\text{a.s.}} \mu^{-1} \quad as \quad x \rightarrow \infty.$$

PROOF. According to the strong law of large numbers, we have for $0 < \epsilon < \mu$ and sufficiently large n, $(\mu - \epsilon)n \leqq S_n \leqq (\mu + \epsilon)n$ with probability one. In particular, if $n = M(x)$ the left hand side implies $M(x)/x \leqq 1/(\mu - \epsilon)$ and if $n = M(x) + 1$ the right side implies $M(x) + 1 \geqq x/(\mu + \epsilon)$. Thus, for large x, $1/(\mu + \epsilon) - 1/x \leqq M(x)/x \leqq 1/(\mu - \epsilon)$ with probability one. The result follows.

Acknowledgments. I am indebted to J. F. Hannan for a helpful discussion on the subject of Lemma 1 and to Y. S. Chow for the observation which led to Theorem 7.

REFERENCES

[1] CHOW, Y. S. (1966). On the moments of some one-sided stopping rules. *Ann. Math. Statist.* **37** 382–387.

[2] CHOW, Y. S. and ROBBINS, H. (1963). A renewal theorem for random variables which are dependent or non-identically distributed. *Ann. Math. Statist.* **34** 390–395.

[3] DOETSCH, G. (1950). *Handbuch der Laplace-Transformation* **1**. Verlag-Birkhauser, Basel.

[4] HEYDE, C. C. (1964). Two probability theorems and their application to some first passage problems. *J. Austral. Math. Soc.* **4** 214–222.

[5] LOÈVE, M. (1960). *Probability Theory* (2nd edition). Van Nostrand, New York.

[6] SMITH, W. L. (1964). On the elementary renewal theorem for nonidentically distributed random variables. *Pacific J. Math.* **14** 673–699.

[7] WIDDER, D. V. (1946). *The Laplace Transform.* Princeton Univ. Press.

HEYDE, C. C.
1966
CAN. J. MATH.
Vol. 18, pp. 656–665

SOME RESULTS ON SMALL-DEVIATION PROBABILITY CONVERGENCE RATES FOR SUMS OF INDEPENDENT RANDOM VARIABLES

C. C. HEYDE

Reprinted from

The Canadian Journal of Mathematics

R. Maller et al. (eds.), *Selected Works of C.C. Heyde*, Selected Works in Probability and Statistics, DOI 10.1007/978-1-4419-5823-5_9, © Springer Science+Business Media, LLC 2010

SOME RESULTS ON SMALL-DEVIATION PROBABILITY CONVERGENCE RATES FOR SUMS OF INDEPENDENT RANDOM VARIABLES

C. C. HEYDE

1. Introduction. Let $\{X_j, j = 1, 2, 3, \ldots\}$ be a sequence of independent, non-degenerate random variables and write

$$S_n = \sum_{j=1}^{n} X_j.$$

Under quite a diverse variety of conditions we may obtain

$$\Pr(S_n < n^p x) \to \tfrac{1}{2}$$

as $n \to \infty$ for all x, $-\infty < x < \infty$, and some real $p \geqslant 0$. For example, suppose the $\{X_j\}$ happen to be distributed identically and belong to the domain of normal attraction of a symmetric stable law with characteristic exponent α, $0 < \alpha \leqslant 2$, $\alpha \neq 1$. If $EX_j = 0$ whenever $E|X_j| < \infty$, then

$$\Pr(S_n < n^p x) \to \tfrac{1}{2}$$

as $n \to \infty$ for all x, $-\infty < x < \infty$, as long as $0 \leqslant p < 1/\alpha$, in view of the central limit theory. It is the purpose of this paper to establish some results on the rate of convergence of $\Pr(S_n < n^p x)$ to $\frac{1}{2}$ by investigating convergence properties of the series

$$\sum_{n=1}^{\infty} n^{-\beta} |\Pr(S_n < n^p x) - \tfrac{1}{2}|, \qquad -\infty < x < \infty, \beta \leqslant 1.$$

Problems of this type could usefully be called small-deviation problems (following terminology that appears to have been introduced by Borovkov **2**).

This work is another link in the chain of development of a problem that arose in connection with the study of certain random walks. These previous investigations were restricted to the case of identically distributed random variables. In the random walk context, Spitzer **(8)** showed that if $EX^2 < \infty$ and $EX = 0$, then

$$\sum_{n=1}^{\infty} n^{-1} (\Pr(S_n < 0) - \tfrac{1}{2}) < \infty.$$

Rosén **(7)** showed that this series is absolutely convergent, his proof being based essentially on his rediscovery of Gil-Pelaez' **(3)** inversion formula. Baum

Received May 12, 1965.

and Katz (**1**) sharpened the methods of (**7**) and showed that if $E|X|^{2+\beta} < \infty$ for some $0 \leqslant \beta < 1$ and $EX = 0$, then

$$\sum n^{-1+\frac{1}{2}\beta}|\Pr(S_n < 0) - \tfrac{1}{2}| < \infty.$$

We go on to put all this theory in a broader context from two points of view. First, we generalize to certain broad classes of non-identically distributed random variables and secondly we include the case $\Pr(S_n < n^p x)$ for $p \geqslant 0$, $-\infty < x < \infty$.

2. Preliminaries. Suppose $\{X_j, j = 1, 2, 3, \ldots\}$ is a sequence of independent random variables with corresponding characteristic functions

$$\{\phi_j(t), j = 1, 2, 3, \ldots\}.$$

If there exists a non-degenerate random variable with characteristic function $\phi(t)$ and a constant $\delta > 0$ such that

$$\max_j|\phi_j(t)| \leqslant |\phi(t)| \qquad \text{for } |t| \leqslant \delta,$$

we shall say that the sequence $\{X_j, j = 1, 2, 3, \ldots\}$ satisfies the condition L. If, in addition, there exists a characteristic function with imaginary part $I(t)$ such that if $I_j(t)$ is the imaginary part of $\phi_j(t)$, $j = 1, 2, 3, \ldots$, there is a constant $\eta > 0$ such that

$$\max_j|I_j(t)| \leqslant |I(t)| \qquad \text{for } |t| \leqslant \eta,$$

then we shall say that the sequence $\{X_j, j = 1, 2, 3, \ldots\}$ satisfies the condition L'.

In the case where the X_j are identically distributed, the conditions L and L' are automatically satisfied. A simple non-identically distributed case in which they are satisfied is that in which some of the $\phi_j(t)$ are of the form $f(t)\phi(t)$ where $f(t)$ and $\phi(t)$ are characteristic functions, $f(t)$ being real, while others are of the form $\phi(t)$. We then have

$$|\phi_j(t)| = |f(t)|\,|\phi(t)| \leqslant |\phi(t)|$$

or

$$|\phi_j(t)| = |\phi(t)|$$

and, similarly,

$$|I_j(t)| = |f(t)|\,|I(t)| \leqslant |I(t)|$$

or

$$|I_j(t)| = |I(t)|.$$

LEMMA 1 (**5**, 61; **7**, Lemma 1). *Let X be a non-degenerate random variable with characteristic function $\phi(t)$. There exist constants $\delta > 0$ and $c > 0$ such that*

$$|\phi(t)| \leqslant 1 - ct^2 \qquad \textit{for } |t| \leqslant \delta.$$

LEMMA 2. *Let* $\{X_j, j = 1, 2, 3, \ldots\}$ *be a sequence of independent, non-degenerate random variables satisfying the condition L. Write*

$$S_n = \sum_{j=1}^{n} X_j.$$

I_n *is an interval on the x-axis and* $l(I_n)$ *is its length. c is a constant that is independent of n and* I_n.

(a) *If* $l(I_n) \leqslant n^p$, $0 < p < \frac{1}{2}$, *then* $\Pr(S_n \in I_n) \leqslant c/n^{\frac{1}{2}-p}$

(b) *If* $l(I_n) \leqslant \epsilon\sqrt{n}$, $\epsilon > 0$, *then* $\Pr(S_n \in I_n) \leqslant \epsilon(c + \xi(\epsilon, n))$ *where* $\xi(\epsilon, n) \to 0$ *for every fixed* $\epsilon > 0$.

(c) *If* $l(I_n) \leqslant M$ *(constant), then* $\Pr(S_n \in I_n) \leqslant c/\sqrt{n}$.

(d) $\max_a \Pr(S_n = a) \leqslant c/\sqrt{n}$.

These results cannot, in general, be improved.

This lemma is an extension of **(7**, Theorem 1**)** from the case of identically distributed random variables to the case of random variables satisfying the condition L. Only minor modifications are necessary to the proof given in **(7)** to make it applicable in this case. It is necessary to replace $\phi^n(t)$ by

$$\prod_{j=1}^{n} \phi_j(t)$$

($\phi_j(t)$ being the characteristic function of X_j). Then, in view of the condition L, we have the relation

$$\prod_{j=1}^{n} |\phi_j(t)| \leqslant |\phi(t)|^n$$

in $|t| \leqslant \delta_1$ (say) for some characteristic function $\phi(t)$. Thus, by virtue of Lemma 1,

$$\begin{aligned}\int_{|t|\leqslant\delta} \prod_{j=1}^{n} |\phi_j(t)|\, dt &\leqslant \int_{|t|\leqslant\delta} |\phi(t)|^n\, dt \\ &\leqslant \int_{|t|\leqslant\delta} (1 - ct^2)^n\, dt \\ &\leqslant \int_{|t|\leqslant\delta} e^{-ct^2n}\, dt \leqslant c_1/\sqrt{n};\end{aligned}$$

here $\delta = \min(\delta_1, \delta_2)$ where δ_2 is the δ of Lemma 1. The equation (2.2) of **(7)** then follows and the remainder of the proof is identical.

LEMMA 3 (Gil-Pelaez' inversion formula **(3**; see also **7**, Theorem 2 and **10))**. *Suppose the random variable X has distribution function* $F(x)$ *and characteristic function* $\phi(t)$. *Then*

$$\begin{aligned}\frac{1}{2}[F(x-0) + F(x+0)] &= \frac{1}{2} + \lim_{A\to 0}\lim_{B\to\infty} \frac{1}{2\pi i}\int_A^B t^{-1}\{e^{ixt}\phi(-t) - e^{-ixt}\phi(t)\}\, dt \\ &= \frac{1}{2} + \lim_{A\to 0}\frac{1}{2\pi i}\int_A^{\delta} t^{-1}\{e^{ixt}\phi(-t) - e^{-ixt}\phi(t)\}\, dt + R(1, x, \delta)\end{aligned}$$

where $\delta > A > 0$ *and*

$$R(1, x, \delta) = \frac{1}{\pi} \int_{-\infty}^{\infty} dF(y) \int_{\delta}^{\infty} \frac{\sin (x - y)t}{t} dt.$$

The use of the improper lower limit in the range of integration enables us to remove the restriction

$$\int_{-\infty}^{\infty} \log(1 + |x|) \, dF(x) < \infty$$

which was imposed in **(7)**.

Note. The corresponding remainder term for the random variable S_n (an n-element convolution) will be denoted $R(n, x, \delta)$.

LEMMA 4. *Suppose the sequence* $\{X_j, j = 1, 2, 3, \ldots\}$ *of random variables satisfies the condition L. For any* ϵ, $0 < \epsilon < \frac{1}{2}$, *there exists a constant c, independent of n and x, such that for all* $\delta < \delta_0$,

$$|R(n, x, \delta)| \leqslant cn^{\epsilon - \frac{1}{2}}.$$

This lemma is an extension of the lemma of **(1)** from the case of identically distributed random variables to the case of random variables satisfying the condition L. δ_0 is the δ_0 of Lemma 2. The proof in **(1)** remains formally unchanged in view of the results of Lemma 2.

3. Results.

THEOREM 1. *Suppose* $\{X_j, j = 1, 2, 3, \ldots\}$ *is a sequence of independent non-degenerate random variables satisfying the condition* L'. *Then for* $-\infty < x < \infty$, $1 \geqslant \beta > \frac{1}{2} + p$, *and* $0 \leqslant p < \frac{1}{2}$,

$$\sum_{n=1}^{\infty} n^{-\beta} |\Pr(S_n < n^p x) - \tfrac{1}{2}| < \infty$$

as long as

$$\lim_{A \to 0} \int_A^{\delta} t^{2\beta - 5} |I(t)| \, dt < \infty$$

for some $\delta > 0$. ($I(t)$ *is the uniform bound from the definition of* L'.)

Proof. We shall establish first that when the sequence $\{X_j, j = 1, 2, 3, \ldots\}$ satisfies the condition L', $\beta > \frac{1}{2}$, and

$$\lim_{A \to 0} \int_A^{\delta} t^{2\beta - 5} |I(t)| \, dt < \infty,$$

then

$$\sum_{n=1}^{\infty} n^{-\beta} |\Pr(S_n < 0) - \tfrac{1}{2}| < \infty.$$

The proof of this result follows along lines similar to the corresponding work in (**1** and **7**).

Making use of Lemma 3, we have

$$\begin{aligned}\Pr(S_n < 0) - \tfrac{1}{2} &= \tfrac{1}{2}\{F_n(0+) - F_n(0-)\} - \tfrac{1}{2}\Pr(S_n = 0) - \tfrac{1}{2}\\ &= \frac{1}{2\pi i}\lim_{A\to 0}\int_A^{\delta} t^{-1}\left\{\prod_{j=1}^{n}\phi_j(-t) - \prod_{j=1}^{n}\phi_j(t)\right\}dt\\ &\qquad + R(n, 0, \delta) - \tfrac{1}{2}\Pr(S_n = 0)\\ &= \frac{1}{2\pi i}\lim_{A\to 0}\int_A^{\delta} t^{-1}\prod_{j=1}^{n}|\phi_j(t)|\sin\left(\sum_{j=1}^{n}\arg\phi_j(t)\right)dt\\ &\qquad + R(n, 0, \delta) - \tfrac{1}{2}\Pr(S_n = 0),\end{aligned}$$

where $\delta > 0$, which is yet to be completely specified, is chosen so small that the inequalities of conditions L, L', and Lemma 1 are all satisfied. Therefore, using the condition L,

$$\begin{aligned}(1)\quad &\sum_{n=1}^{\infty} n^{-\beta}|\Pr(S_n < 0) - \tfrac{1}{2}|\\ &\leqslant \frac{1}{\pi}\lim_{A\to 0}\int_A^{\delta} t^{-1}\sum_{n=1}^{\infty} n^{-\beta}\prod_{j=1}^{n}|\phi_j(t)|\sum_{j=1}^{n}|\arg\phi_j(t)|\,dt\\ &\qquad + \sum_{n=1}^{\infty} n^{-\beta}|R(n, 0, \delta)| + \frac{1}{2}\sum_{n=1}^{\infty} n^{-\beta}\Pr(S_n = 0)\\ &\leqslant \frac{1}{\pi}\lim_{A\to 0}\int_A^{\delta} t^{-1}\sum_{n=1}^{\infty} n^{-\beta}|\phi(t)|^n\sum_{j=1}^{n}|\arg\phi_j(t)|\,dt\\ &\qquad + \sum_{n=1}^{\infty} n^{-\beta}|R(n, 0, \delta)| + \frac{1}{2}\sum_{n=1}^{\infty} n^{-\beta}\Pr(S_n = 0).\end{aligned}$$

We consider separately the three parts on the right-hand side of (1).

In view of our choice of δ, it follows from Lemma 4 with ϵ chosen to be less than $\beta - \frac{1}{2}$ that

$$\sum_{n=1}^{\infty} n^{-\beta}|R(n, 0, \delta)| < \infty.$$

Also, it follows from Lemma 2, part (d) that

$$\sum_{n=1}^{\infty} n^{-\beta}\Pr(S_n = 0) < \infty.$$

Now we can write

$$\phi_j(t) = R_j(t) + iI_j(t)$$

where $R_j(t)$ and $I_j(t)$ are real functions, bounded on any finite interval. Then

$$\arg\phi_j(t) = \arctan\{I_j(t)[R_j(t)]^{-1}\}.$$

$R_j(t)$ $(=\frac{1}{2}(\phi_j(t) + \phi_j(-t))$ is itself a characteristic function and hence is continuous about $R_j(0) = 1$ in a neighbourhood of the origin. Therefore, given suitably small $\epsilon > 0$ we can find $\delta_j > 0$ so small that $|R_j(t) - 1| < \epsilon$ in $|t| \leqslant \delta_j$. Choose $\delta_0 = \min_j \delta_j$ (clearly $\delta_0 > 0$). Then, in view of the relation $|R_j(t) - 1| < \epsilon$ uniformly in j for $|t| < \delta_0$, we must have

$$|\arg \phi_j(t)| \leqslant c|I_j(t)| \leqslant c|I(t)| \tag{2}$$

for $0 \leqslant t \leqslant \delta_0$ and some constant $c > 0$, $I(t)$ being the I of condition L'. At this stage we further restrict the δ of the theorem to make it less than the above δ_0 as well. Then, using (2), we have

$$\begin{aligned} \lim_{A\to 0} \int_A^\delta t^{-1} \sum_{n=1}^\infty n^{-\beta} |\phi(t)|^n \sum_{j=1}^n |\arg \phi_j(t)|\, dt \\ \leqslant \lim_{A\to 0} c \int_A^\delta t^{-1} \sum_{n=1}^\infty n^{1-\beta} |\phi(t)|^n |I(t)|\, dt. \end{aligned} \tag{3}$$

By an Abelian theorem of Widder (**11**, p. 182, Corollary 1a), we see that

$$\lim_{u\to 1-} (1-u)^{2-\beta} \sum_{n=1}^\infty n^{1-\beta} u^n = \text{const.}$$

and thus, for $0 < u < 1$, there is a positive constant c such that

$$\sum_{n=1}^\infty n^{1-\beta} u^n < c(1-u)^{\beta-2}.$$

Putting $u = |\phi(t)|$ in this inequality, we obtain, for $t \neq 0$,

$$\sum_{n=1}^\infty n^{1-\beta} |\phi(t)|^n < c(1 - |\phi(t)|)^{\beta-2}. \tag{4}$$

Then, by (4) and Lemma 1, there is a positive constant k such that

$$\lim_{A\to 0} \int_A^\delta t^{-1} \sum_{n=1}^\infty n^{1-\beta} |\phi(t)|^n\, |I(t)|\, dt < k \lim_{A\to 0} \int_A^\delta t^{2\beta-5} |I(t)|\, dt$$

and the right-hand side of this inequality is finite by hypothesis. This establishes the first part of the theorem.

The remainder of the proof is now easy to complete. We have, for $x \geqslant 0$,

$$\begin{aligned} \sum_{n=1}^\infty n^{-\beta} |\Pr(S_n < n^p x) - \tfrac{1}{2}| \leqslant \sum_{n=1}^\infty n^{-\beta} |\Pr(S_n < 0) - \tfrac{1}{2}| \\ + \sum_{n=1}^\infty n^{-\beta} \Pr(|S_n| < n^p x). \end{aligned}$$

The first series on the right-hand side converges as we have just shown and the second series converges by Lemma 2(a) as long as $\beta > \frac{1}{2} + p$. This establishes the result in the case $x \geqslant 0$, and the case $x < 0$ follows similarly.

In general Theorem 1 cannot be extended to the case $\beta = \frac{1}{2} + p$, $0 \leqslant p < \frac{1}{2}$. Take, for example, each of the X_j as normally distributed with zero mean and unit variance. We have, for $x > 0$,

$$n^{-\beta}|\Pr(S_n < n^p x) - \tfrac{1}{2}| = n^{-\beta} \Pr(0 < S_n < n^p x) = \frac{1}{\sqrt{(2\pi)}n^\beta} \int_0^{n^{p-\frac{1}{2}}x} e^{-\frac{1}{2}u^2}\, du$$

$$\sim \frac{x}{\sqrt{(2\pi)}n^{\beta-p+\frac{1}{2}}} \sim \frac{x}{\sqrt{(2\pi)}n} \qquad \text{when } \beta = \tfrac{1}{2} + p$$

so that the series

$$\sum_{n=1}^{\infty} n^{-\beta}|\Pr(S_n < n^p x) - \tfrac{1}{2}|$$

diverges when $\beta = \frac{1}{2} + p$.

As an immediate corollary of Theorem 1 we obtain:

COROLLARY 1. *Suppose* $\{X_j, j = 1, 2, 3, \ldots\}$ *is a sequence of independent, non-degenerate, symmetric random variables satisfying the condition L. Then for* $-\infty < x < \infty$, $1 \geqslant \beta > \frac{1}{2} + p$, *and* $0 \leqslant p < \frac{1}{2}$,

$$\sum_{n=1}^{\infty} n^{-\beta}|\Pr(S_n < n^p x) - \tfrac{1}{2}| < \infty.$$

Even when $x = 0$, the result of Corollary 1 need not extend to the case $\beta = \frac{1}{2}$. We have

$$|\Pr(S_n < 0) - \tfrac{1}{2}| = \tfrac{1}{2}\Pr(S_n = 0)$$

and it is easy to find examples where $\Pr(S_n = 0) \sim cn^{-\frac{1}{2}}$ for some positive constant c as $n \to \infty$. This is the case, for instance, when the X_j are identically distributed with a finite variance, zero mean, and characteristic function $\phi(t) = E(e^{itX})$ such that $|\phi(t)| = 1$ when t is any multiple of 2π; cf. (**9**, p. 75).

The following example shows that the condition L restriction in Theorem 1 (and Corollary 1) cannot be relaxed in general. Take X_j as normally distributed with zero mean and variance $j^{-(1-2p)} \log j$, $j = 1, 2, 3, \ldots$. Then

$$\phi_j(t) = \exp(-\tfrac{1}{2}j^{-(1-2p)} \log j t^2) = |\phi_j(t)|$$

and $|\phi_j(t)| \to 1$ as $j \to \infty$ for all t so the condition L is not satisfied. S_n is normally distributed with mean zero and variance

$$\sum_{k=1}^{n} k^{-(1-2p)} \log k = a_n^2 \sim n^{2p} \log n \qquad \text{as } n \to \infty.$$

We have, taking $x > 0$ for convenience,

$$\Pr(S_n < n^p x) - \tfrac{1}{2} = \frac{1}{\sqrt{(2\pi)}a_n} \int_0^{n^p x} \exp(-\tfrac{1}{2}u^2 a_n^{-2})\, du$$

$$= (2\pi)^{-\frac{1}{2}} a_n^{-1} n^p x (\exp -\tfrac{1}{2}x^2 a_n^{-2} n^{2p} \theta_n^2), \quad 0 < \theta_n < 1$$

using the mean value theorem. It follows that

$$|\Pr(S_n < n^p x) - \tfrac{1}{2}| = O(\log n)^{-\frac{1}{2}} \qquad \text{as } n \to \infty,$$

so that the results of neither Theorem 1 nor Corollary 1 can hold. The breakdown in this case comes in Lemma 2(a).

LEMMA 5. *Suppose the random variable X has distribution function $F(x)$ and $I(t)$ is the imaginary part of its characteristic function. If, whenever*

$$\int_{-\infty}^{\infty} |x|\, dF(x) < \infty$$

we have

$$\int_{-\infty}^{\infty} x\, dF(x) = 0,$$

then for $\gamma > 0$,

$$\int_0^1 |I(t)| t^{-(1+\gamma)}\, dt < \infty$$

as long as

$$\int_{-\infty}^{\infty} |x|^{\gamma}\, dF(x) < \infty.$$

This type of result seems to have been obtained first by Hsu **(4)**. Rosén **(7)** treats the case $\gamma = 2$ while Baum and Katz **(1)** treat the case $2 \leqslant \gamma < 3$. The proof is quite simple and will be omitted; whenever

$$\int_{-\infty}^{\infty} |x|\, dF(x) < \infty$$

and hence

$$\int_{-\infty}^{\infty} x\, dF(x) = 0$$

we use

$$I(t) = \int_{-\infty}^{\infty} (\sin xt - xt)\, dF(x),$$

and otherwise we use

$$I(t) = \int_{-\infty}^{\infty} \sin xt\, dF(x).$$

Lemma 5 is directly applicable in Theorem 1 only in the particular case considered by Baum and Katz **(1)**. For in order that it should be applicable, we must have $\gamma = 4 - 2\beta$ and $1 \geqslant \beta > \frac{1}{2}$ so that $3 > \gamma \geqslant 2$. It is not at all surprising that this is the best we can do in general. As pointed out previously, we cannot in general obtain convergence up to $\beta = \frac{1}{2}$ no matter how many moments we assume to be finite. On the other hand, if the X_j do not possess a finite variance, it is possible to obtain all sorts of pathological behaviour for the probabilities $\Pr(S_n < 0)$ despite the restriction $EX_j = 0$ when $E|X_j| < \infty$; cf. Spitzer **(9**, pp. 228–230**)**. He treats the case of identically distributed summands possessing zero mean but not a finite variance and shows that it is possible to find cases where the Césaro limit

$$\lim_{n\to\infty} n^{-1}\sum_{k=1}^{n} \Pr(S_k < 0)$$

takes on any value a, $0 \leqslant a \leqslant 1$, and also cases where the Césaro limit fails to exist altogether.

In the particular case where Lemma 5 is applicable in Theorem 1, we obtain:

THEOREM 2. *Suppose* $\{X_j, j = 1, 2, 3, \ldots\}$ *is a sequence of independent, non-degenerate random variables satisfying the condition* L'. *If there exists a uniform bound* $I(t)$ *(for the definition of* L'*) belonging to the characteristic function of a random variable* Y *for which* $E|Y|^{2+\alpha} < \infty$, $0 \leqslant \alpha < 1$, *and* $EY = 0$, *then for* $0 \leqslant p < \frac{1}{2}(1 - \alpha)$,

$$\sum_{n=1}^{\infty} n^{-(1-\frac{1}{2}\alpha)} |\Pr(S_n < n^p x) - \tfrac{1}{2}| < \infty .$$

Other, albeit rather less general, cases where Theorem 1 may be used are not difficult to find. In order to illuminate the situation slightly, let us mention another expression for $I(t)$ (the imaginary part of a characteristic function $\phi(t)$ which corresponds to a distribution function $F(x)$). This is the form

$$I(t) = t\int_{0+}^{\infty} [1 - F(x) - F(-x + 0)] \cos tx\, dx, \tag{5}$$

which is obtained by integrating $\phi(t)$ by parts. It is clear from (5) that whether an integral of the form

$$\lim_{A\to 0} \int_{A}^{\delta} t^{2\beta-5} |I(t)|\, dt$$

is finite or not depends in a critical way upon $K(x) = 1 - F(x) - F(-x + 0)$, which is in a sense a function measuring departure from symmetry. By considering certain classes of functions $K(x)$ it is quite easy to obtain applications of Theorem 1. For example, we could follow Pitman **(6)** who has studied cases where $K(x)$ has the property that for $\lambda > 0$,

$$\frac{K(\lambda x)}{K(x)} \to \frac{1}{\lambda^m} \qquad \text{as } x \to 0$$

for some fixed $m \geqslant 0$ and has obtained asymptotic estimates of the behaviour of $I(t)$ as $t \to 0+$. This is, of course, just what we need in order to be able to apply Theorem 1. It does not seem appropriate, however, to pursue this rather specialized matter further in the present context.

References

1. L. E. Baum and M. L. Katz, *On the influence of moments on the asymptotic distribution of sums of random variables*, Ann. Math. Statist., *34* (1963), 1042–1044.
2. A. A. Borovkov, *Limit theorems on the distribution of maxima of sums of bounded lattice random variables*. I, Theor. Probability Appl., *5* (1960), 125–155.
3. J. Gil-Pelaez, *Note on the inversion theorem*, Biometrika, *38* (1951), 481–482.
4. P. L. Hsu, *Absolute moments and characteristic functions*, J. Chinese Math. Soc., *1* (1951), 259–280.
5. E. Lukacz, *Characteristic functions* (London, 1960).
6. E. J. G. Pitman, *Some theorems on characteristic functions of probability distributions*, Proc. 4th Berkeley Symp. on Math. Statist. and Prob., *2* (1961), 393–402.
7. B. Rosén, *On the asymptotic distribution of sums of independent identically distributed random variables*, Ark. Mat., *4* (1962), 323–332.
8. F. Spitzer, *A Tauberian theorem and its probability interpretation*, Trans. Amer. Math. Soc., *94* (1960), 150–169.
9. ——— *Principles of random walk* (New York, 1964).
10. J. G. Wendel, *The non-absolute convergence of Gil-Pelaez' inversion integral*, Ann. Math. Statist., *32* (1961), 338–339.
11. D. V. Widder, *The Laplace transformation* (Princeton, 1964).

Michigan State University

Z. Wahrscheinlichkeitstheorie verw. Geb. 7, 303—308 (1967)

A Contribution to the Theory of Large Deviations for Sums of Independent Random Variables

C. C. Heyde

Received March 22, 1966

1. Introduction

Let the random variables X_i, $i = 1, 2, 3, \ldots$ be independent and identically distributed. Write $S_n = \sum_{i=1}^{n} X_i$ and $Z_n = S_n B_n^{-1} - A_n$ for normed and centered sums. The classical central limit formulation is for $\Pr(Z_n \leqq x_n)$ where $x_n = O(1)$ as $n \to \infty$ and thus gives only trivial information in the case where $x_n \to \infty$ as $n \to \infty$. However, we often require information on $\Pr(Z_n > x_n)$ under these circumstances. This type of problem is called a problem on the probability of large deviations and little in the way of comprehensive general results on such problems has so far been obtained. For references to the known results and also to various fields in which large deviation problems arise see, for example, Linnik [3] and Sethuraman [7].

Up till the work of Linnik [3], all the general large deviation results obtained had made use of the condition $E\{\exp(a|X_i|)\} < \infty$ for some $a > 0$ (that is, the X_i possess an analytic characteristic function) or something even stronger. The analytic characteristic function assumption was quite basic in much of the work as the results were obtained, in effect, from a use of the saddlepoint method in the theory of functions of a complex variable. Linnik broke away from the saddlepoint technique and the associated assumption of an analytic characteristic function. He has, however, only studied the problem of convergence to the normal distribution and, with the exception of one theorem, only cases where $x_n = o(\sqrt{n})$ as $n \to \infty$. In this paper we study, for the first time, large deviation problems associated with convergence to non-normal stable laws. A general expression is found for the order of magnitude of the large deviation probability in such cases.

2. Large Deviations in the Case of Attraction to a Non-normal Stable Law

Let $\{X_i, i = 1, 2, 3, \ldots\}$ be a sequence of independent and identically distributed random variables with law $\mathscr{L}(X)$ which belong, with normalizing constants B_n, to the domain of attraction of the stable law with characteristic function

$$\exp\left\{-a|t|^{\alpha}\left(1 + i\frac{t}{|t|}\frac{c_1 - c_2}{c_1 + c_2}\tan\frac{\pi\alpha}{2}\right)\right\},$$

$0 < \alpha < 2$, $\alpha \neq 1$, $a > 0$, $c_1 > 0$, $c_2 > 0$, and distribution function F. c_1, c_2 and a

R. Maller et al. (eds.), *Selected Works of C.C. Heyde*, Selected Works in Probability and Statistics,
DOI 10.1007/978-1-4419-5823-5_10, © Springer Science+Business Media, LLC 2010

are related according to

$$a = (c_1 + c_2)\,\alpha \int_0^\infty \frac{1 - \cos x}{x^{1+\alpha}}\,dx$$

(GNEDENKO and KOLMOGOROV [1], 169—171. We note in passing that an α is missing from the second and third terms on the right hand side of equation 12, 168 of [1]. This omission is perpetuated in the sequel). As a matter of convenience we shall exclude the case where $c_1(c_2)$ is zero although the problem is readily amenable to the same sort of treatment. Under these circumstances the stable law would be bounded on the left (right) (LUKACS [5], 106). If $\alpha > 1$ we suppose the X_i to be relocated so that $EX_i = 0$. Write $Z_n = B_n^{-1} \sum_{i=1}^{n} X_i = B_n^{-1} S_n$ and let $F_n(u) = \Pr(Z_n \leqq u)$. If $\alpha = 1$, the probability of interest cannot generally be put into the form $\Pr(Z_n > x_n)$ (GNEDENKO and KOLMOGOROV [1], 175). The problem in this case can be treated but it needs some special discussion and will be omitted here.

Under the above conditions we may write for $x > 0$, using GNEDENKO and KOLMOGOROV [1], Theorem 2, 175,

$$\Pr(X \leqq -x) = \frac{L_1(x)}{x^\alpha},$$

$$\Pr(X > x) = \frac{L(x)}{x^\alpha},$$

where $L_1(x)$ and $L(x)$ are slowly varying functions with

$$\frac{L_1(x)}{L(x)} \to \frac{c_1}{c_2} \quad \text{as} \quad x \to \infty .$$

$L(x)$ and the sequence B_n are related in such a way that

$$\frac{n L(B_n)}{B_n^\alpha} \to c_2 \quad \text{as} \quad n \to \infty .$$

We shall write $L_1(x) + L(x) = M(x)$ so that $M(x)$ is a slowly varying function, $M(x) \sim \left(1 + \frac{c_1}{c_2}\right) L(x)$ as $x \to \infty$ and

$$q(x) = \Pr(X \leqq -x) + \Pr(X > x) = \frac{M(x)}{x^\alpha}.$$

In the following discussion we shall restrict consideration to large deviations in the right hand tail. It is clear that corresponding results will hold for the left hand tail. We shall obtain the following theorem.

Theorem. *Let $\{x_n\}$ be a sequence with $x_n \to \infty$ as $n \to \infty$. Then,*

$$0 < \varliminf_{n\to\infty} \frac{x_n^\alpha L(B_n)}{L(x_n B_n)} \Pr(S_n > x_n B_n) \leqq \varlimsup_{n\to\infty} \frac{x_n^\alpha L(B_n)}{L(x_n B_n)} \Pr(S_n > x_n B_n) < \infty .$$

Proof. Denote by A_i and B_i the events $\{X_i > x_n B_n\}$ and $\left\{\sum_{\substack{j=1\\ j\neq i}}^{n} X_j > 0\right\}$

respectively, $i = 1, 2, \ldots, n$. If $\bar{E}$ is the complement of E, we then have

$$
\begin{aligned}
\Pr(S_n > x_n B_n) &\geqq \Pr\left[\bigcup_{i=1}^{n}(A_i \cap B_i)\right] \\
&= \Pr\left[\bigcup_{i=1}^{n}\left\{\bigcap_{j=1}^{i-1}(\overline{A_j \cap B_j}) \cap (A_i \cap B_i)\right\}\right] \\
&= \sum_{i=1}^{n}\Pr\left[\bigcap_{j=1}^{i-1}(\overline{A_j \cap B_j}) \cap (A_i \cap B_i)\right] \qquad (1)\\
&\geqq \sum_{i=1}^{n}\Pr\left[\bigcap_{j=1}^{i-1}\bar{A}_j \cap (A_i \cap B_i)\right] \geqq \\
&\geqq \sum_{i=1}^{n}\left\{\Pr(A_i \cap B_i) - \Pr\left(\bigcup_{j=1}^{i-1}A_j \cap A_i\right)\right\} \geqq \\
&\geqq \sum_{i=1}^{n}\Pr(A_i)\,[\Pr(B_i) - (i-1)\Pr(A_i)] \geqq \\
&\geqq n\Pr(A_i)\,[\Pr(B_i) - n\Pr(A_i)]\,.
\end{aligned}
$$

Now $\Pr(B_i) \to A > 0$ as $n \to \infty$ where A is the probability that the corresponding stable law takes a positive value. Thus, given $\delta > 0$ with $A - 2\delta > 0$ we can choose N_1 so large that $\Pr(B_i) > A - \delta$ for $n \geqq N_1$. Also

$$n\Pr(A_i) = n\Pr(X > x_n B_n) = \frac{n L(x_n B_n)}{x_n^{\alpha} B_n^{\alpha}} = \frac{n L(B_n)}{B_n^{\alpha}}\,\frac{L(x_n B_n)}{x_n^{\alpha} L(B_n)}$$

and since $L(x)$ is a non-negative function of slow variation, it possesses a representation of the form

$$L(x) = \frac{a(x)}{x}\exp\left\{\int_1^x \frac{a(u)}{u}\,du\right\},$$

where $a(x) \to 1$ as $x \to \infty$ (KARAMATA [2]). Therefore,

$$\frac{L(x_n B_n)}{L(B_n)} = \frac{a(x_n B_n)}{a(B_n)}\,\frac{1}{x_n}\exp\left\{\int_{B_n}^{x_n B_n} \frac{a(u)}{u}\,du\right\},$$

and given arbitrarily small $\eta > 0$ we can choose N so large that for $n > N$,

$$\frac{L(x_n B_n)}{L(B_n)} < (1+\eta)\,\frac{1}{x_n}\exp\left\{(1+\eta)\int_{B_n}^{x_n B_n}\frac{du}{u}\right\} = (1+\eta)\,x_n^{\eta}\,.$$

Take $\eta < \alpha$. Then, for $n > N$,

$$n\Pr(A_i) < \frac{n L(B_n)}{B_n^{\alpha}}(1+\eta)\,x_n^{\eta-\alpha} \to 0 \quad \text{as} \quad n \to \infty\,.$$

Thus, given $\delta > 0$ we can choose N_2 so large that $n\Pr(A_i) < \delta$ for $n \geqq N_2$ and hence, for $n \geqq \max(N_1, N_2)$, we have from (1)

$$
\begin{aligned}
\Pr(S_n > x_n B_n) &\geqq n(A - 2\delta)\Pr(X > x_n B_n) = n(A - 2\delta)\,\frac{L(x_n B_n)}{x_n^{\alpha} B_n^{\alpha}} \\
&= (A - 2\delta)\,\frac{n L(B_n)}{B_n^{\alpha}}\,\frac{L(x_n B_n)}{x_n^{\alpha} L(B_n)}
\end{aligned}
$$

22*

so that

$$\frac{x_n^\alpha L(B_n)}{L(x_n B_n)} \Pr(S_n > x_n B_n) > (A - 2\delta) \frac{n L(B_n)}{B_n^\alpha}$$

and

$$\underline{\lim}_{n\to\infty} \frac{x_n^\alpha L(B_n)}{L(x_n B_n)} \Pr(S_n > x_n B_n) > 0 .$$

In order to complete the proof we shall work in terms of the symmetrized random variables X_i^s, $i = 1, 2, 3, \ldots$, making continued use of the weak symmetrization inequalities (LOÈVE [4], 245) to transfer the results.

Define

$$X_{kn}^s = \begin{cases} X_k^s & \text{if } \quad |X_k^s| \leqq x_n B_n \\ 0 & \text{otherwise,} \end{cases}$$

and write $S_n^s = \sum_{k=1}^n X_k^s$, $S_{nn}^s = \sum_{k=1}^n X_{kn}^s$. We have

$$\Pr(|S_n^s| > x_n B_n) \leqq n \Pr(|X^s| > x_n B_n) + \Pr(|S_{nn}^s| > x_n B_n) \tag{2}$$

and, using the weak symmetrization inequalities,

$$\begin{aligned} \frac{n x_n^\alpha L(B_n)}{L(x_n B_n)} \Pr(|X^s| > x_n B_n) &\leqq 2 \frac{n x_n^\alpha L(B_n)}{L(x_n B_n)} \Pr\left(|X| > \frac{x_n B_n}{2}\right) \leqq \\ &\leqq 2^{1+\alpha} \frac{n L(B_n)}{L(x_n B_n)} \frac{M(\frac{1}{2} x_n B_n)}{B_n^\alpha} \\ &= 2^{1+\alpha} \frac{n L(B_n)}{B_n^\alpha} \frac{M(\frac{1}{2} x_n B_n)}{M(x_n B_n)} \frac{M(x_n B_n)}{L(x_n B_n)}, \end{aligned}$$

so that

$$\overline{\lim}_{n\to\infty} \frac{n x_n^\alpha L(B_n)}{L(x_n B_n)} \Pr(|X^s| > x_n B_n) < \infty . \tag{3}$$

We now turn our attention to $\Pr(|S_{nn}^s| > x_n B_n)$. From Markov's inequality we obtain

$$\begin{aligned} \Pr(|S_{nn}^s| > x_n B_n) &\leqq x_n^{-2} B_n^{-2} E(S_{nn}^s)^2 = n x_n^{-2} B_n^{-2} E(X_{kn}^s)^2 \\ &= n x_n^{-2} B_n^{-2} \int_{|x| \leqq x_n B_n} x^2 \, d \Pr(X^s \leqq x) \\ &= - n x_n^{-2} B_n^{-2} \int_0^{x_n B_n} x^2 \, d \Pr(|X^s| > x) \\ &\leqq 2 n x_n^{-2} B_n^{-2} \int_0^{x_n B_n} x \Pr(|X^s| > x) \, dx \leqq 4 n x_n^{-2} B_n^{-2} \int_0^{x_n B_n} x \, q(\tfrac{1}{2} x) \, dx \leqq \\ &\leqq 16 n x_n^{-2} B_n^{-2} \int_0^{\frac{1}{2} x_n B_n} M(x) \, x^{1-\alpha} \, dx \end{aligned}$$

using integration by parts to obtain the third last inequality and the weak symmetrization inequalities to obtain the second last. Thus,

$$\frac{x_n^\alpha L(B_n)}{L(x_n B_n)} \Pr(|S_{nn}^s| > x_n B_n) \leqq 16 \frac{n L(B_n)}{B_n^\alpha} \frac{1}{L(x_n B_n)(x_n B_n)^{2-\alpha}} \int_0^{\frac{1}{2} x_n B_n} M(x) \, x^{1-\alpha} \, dx$$

$$(4) \qquad = 16 \frac{nL(B_n)}{B_n^\alpha} \frac{1}{L(x_n B_n)(x_n B_n)^{2-\alpha}} \int_0^A M(x)\, x^{1-\alpha}\, dx +$$

$$+ 16 \frac{nL(B_n)}{B_n^\alpha} \frac{M(\frac{1}{2} x_n B_n)}{M(x_n B_n)} \frac{M(x_n B_n)}{L(x_n B_n)} \frac{1}{M(\frac{1}{2} x_n B_n)(x_n B_n)^{2-\alpha}} \int_A^{\frac{1}{2} x_n B_n} M(x)\, x^{1-\alpha}\, dx$$

for fixed A. The first term on the right hand side becomes arbitrarily small for sufficiently large n since L is a slowly varying function and $\alpha < 2$. Further, making use of the representation result for non-negative slowly varying functions, we have for $A \leq x \leq \frac{1}{2} x_n B_n$,

$$\frac{M(x)}{M(\frac{1}{2} x_n B_n)} = \frac{b(x)}{b(\frac{1}{2} x_n B_n)} \frac{x_n B_n}{2x} \exp\left\{- \int_x^{\frac{1}{2} x_n B_n} \frac{b(u)}{u}\, du\right\}$$

where $b(u) \to 1$ as $u \to \infty$, and given $0 < \delta < 2 - \alpha$ we can choose A so large that

$$\frac{M(x)}{M(\frac{1}{2} x_n B_n)} < (1+\delta) \frac{x_n B_n}{2x} \exp\left\{-(1-\delta) \int_x^{\frac{1}{2} x_n B_n} \frac{du}{u}\right\} = (1+\delta) \left(\frac{x_n B_n}{2x}\right)^\delta.$$

Therefore, for suitably large A,

$$\frac{1}{M(\frac{1}{2} x_n B_n)(x_n B_n)^{2-\alpha}} \int_A^{\frac{1}{2} x_n B_n} M(x)\, x^{1-\alpha}\, dx < \frac{1+\delta}{2^\delta (x_n B_n)^{2-\alpha-\delta}} \int_A^{\frac{1}{2} x_n B_n} x^{1-\alpha-\delta}\, dx <$$

$$< \frac{1}{2^{2-\alpha}} \frac{1+\delta}{2-\alpha-\delta} < \infty,$$

and it follows from (4) that

$$(5) \qquad \overline{\lim_{n \to \infty}} \frac{x_n^\alpha L(B_n)}{L(x_n B_n)} \Pr(|S_{nn}^s| > x_n B_n) < \infty.$$

Using (3) and (5) in (2) we obtain

$$\overline{\lim_{n \to \infty}} \frac{x_n^\alpha L(B_n)}{L(x_n B_n)} \Pr(|S_n^s| > x_n B_n) < \infty$$

which, using the weak symmetrization inequalities, implies in particular that

$$(6) \qquad \overline{\lim_{n \to \infty}} \frac{x_n^\alpha L(B_n)}{L(x_n B_n)} \Pr(S_n - \operatorname{med} S_n > x_n B_n) < \infty.$$

Now $B_n^{-1} S_n$ converges in distribution so $x_n^{-1} B_n^{-1} S_n$ converges in probability to zero. Hence, $x_n^{-1} B_n^{-1} \operatorname{med} S_n \to 0$ as $n \to \infty$. It follows then from (6) by a simple transformation that

$$(7) \qquad \overline{\lim_{n \to \infty}} \frac{x_n^\alpha L(B_n)}{L(x_n B_n)} \Pr(S_n > x_n B_n) < \infty$$

and the proof of the theorem is complete.

In large deviation problems it is usual to compare the large deviation behaviour of the normed sums with that of the corresponding stable law. In this context, that amounts to looking at

$$\frac{1 - F_n(x_n)}{1 - F(x_n)} = \frac{x_n^\alpha}{c_2 + \alpha_2(x_n)} \Pr(S_n > x_n B_n)$$

where $\alpha_2(x) \to 0$ as $x \to \infty$ ([1], Theorem 5, 181—182), and we shall have

$$0 < \underline{\lim}_{n\to\infty} \frac{1 - F_n(x_n)}{1 - F(x_n)} \leqq \overline{\lim}_{n\to\infty} \frac{1 - F_n(x_n)}{1 - F(x_n)} < \infty$$

if and only if

$$0 < \underline{\lim}_{n\to\infty} \frac{L(B_n)}{L(x_n B_n)} \leqq \overline{\lim}_{n\to\infty} \frac{L(B_n)}{L(x_n B_n)} < \infty .$$

This is the case, for example, when the random variables $\{X_i,\ i = 1, 2, 3, \ldots\}$ actually belong to the domain of normal attraction of the particular stable law. In that case,

$$L(x) = c_2 + \beta_2(x)$$

where $\beta_2(x) \to 0$ as $x \to \infty$. In general, however, different distributions belonging to the same domain of attraction can possess quite different asymptotic behaviour.

It seems plausible that in the theorem the result

$$\underline{\lim}_{n\to\infty} \frac{x_n^\alpha L(B_n)}{L(x_n B_n)} \Pr(S_n > x_n B_n) = \overline{\lim}_{n\to\infty} \frac{x_n^\alpha L(B_n)}{L(x_n B_n)} \Pr(S_n > x_n B_n)$$

should hold. In this connection it should be mentioned that McLAREN [6] has shown, using combinatorial arguments rather like those of LINNIK [3], 302, that in the case of normal attraction,

$$\lim_{n\to\infty} \frac{1 - F_n(x_n)}{1 - F(x_n)} = 1$$

provided $\frac{x_n}{n^{1+\delta}} \to \infty$ as $n \to \infty$ for some $\delta > 0$.

Acknowledgement. I am indebted to A. D. McLAREN for some stimulating correspondence along the lines mentioned above.

References

1. GNEDENKO, B. V., and A. N. KOLMOGOROV: Limit distributions for sums of independent random variables. Translated and annotated by K. L. CHUNG. Cambridge (Mass.): Addison Wesley 1954.
2. KARAMATA, J.: Sur un mode de croissance régulière des fonctions. Mathematica (Cluj) **4**, 38—53 (1930).
3. LINNIK, YU. V.: On the probability of large deviations for sums of independent random variables. Proc. 4th Berkeley Sympos. math. Statist. Probability, **2**, 289—306 (1961).
4. LOÈVE, M.: Probability theory. New York: Van Nostrand 1960.
5. LUKACS, E.: Characteristic functions. London: Griffin 1960.
6. McLAREN, A. D.: Unpublished private communication (1963).
7. SETHURAMAN, J.: On the probability of large deviations of families of sample means. Ann. math. Statistics **35**, 1304—1316 (1964).

Department of Probability and Statistics,
University of Sheffield,
Sheffield 10, Great Britain

Reprinted from THE ANNALS OF MATHEMATICAL STATISTICS
Vol. 38, No. 5, October, 1967
Printed in U.S.A.

ON LARGE DEVIATION PROBLEMS FOR SUMS OF RANDOM VARIABLES WHICH ARE NOT ATTRACTED TO THE NORMAL LAW

BY C. C. HEYDE

University of Sheffield

1. Introduction. Let X_i, $i = 1, 2, 3, \cdots$, be a sequence of independent and identically distributed random variables with law $\mathcal{L}(X)$, and write $S_n = \sum_{i=1}^{n} X_i$. Let x_n, $n = 1, 2, 3, \cdots$, be a monotone sequence of positive numbers with $x_n \to \infty$ as $n \to \infty$ such that $x_n^{-1} S_n \to_P 0$. ("P" stands for convergence in probability). We shall call the probability $\Pr(|S_n| > x_n)$, or either of its component tail probabilities, a large deviation probability. This constitutes an extension of the ordinary concept of large deviation probabilities where the X_i belong to some domain of attraction, $Z_n = B_n^{-1}S_n - A_n$ are the normed and centered sums, and the probabilities $\Pr(|Z_n| > x_n)$ are called large deviation ones for $x_n \to \infty$ as $n \to \infty$.

In this paper, we shall consider random variables which do not belong to the domain of partial attraction of the normal distribution. That is, those for which (Lévy [4], 113)

$$\liminf_{u\to\infty} [u^2 \Pr(|X| > u)/\textstyle\int_{|x| \leqq u} x^2\, dF(x)] > 0, \tag{1}$$

where $F(x) = \Pr(X \leqq x)$. We shall, in addition, make the restriction that for $\alpha > 1$ and $\{x_n\}$ such that $x_n^{-1}S_n \to_P 0$ as $n \to \infty$,

$$0 < \liminf_{n\to\infty} [\Pr(|X| > \alpha x_n)/\Pr(|X| > x_n)]. \tag{2}$$

The condition (2) is imposed to remove from consideration some laws whose distribution functions have widely spaced points of increase and for which the large deviation behaviour would be somewhat pathological.

Under conditions (1) and (2), we shall obtain a general expression for the order of magnitude of the large deviation probability in the form of (3) and (4) below. This provides a generalization of the work of Heyde [3] where the corresponding result (one sided version) was obtained for the case where the X_i belong to the domain of attraction of a non-normal stable law. Apart from [3], there appears to be no other work directly on large deviation problems outside the domain of attraction of the normal distribution.

It is worth noting that the condition (2) goes some way towards implying condition (1). In this connection, we mention the following result which is a minor modification of a lemma of Doeblin [1]: If $\alpha > 1$ and for $u > u_0$,

$$\Pr(|X| > \alpha u)/\Pr(|X| > u) \geqq c_\alpha > 0,$$

where $\alpha^2 c_\alpha > 1$, then

$$\liminf_{u\to\infty} [u^2 \Pr(|X| > u)/\textstyle\int_{|x| \leqq u} x^2\, dF(x) \geqq c_\alpha(1 - 1/c_\alpha \alpha^2) > 0.$$

Received 13 February 1967.

R. Maller et al. (eds.), *Selected Works of C.C. Heyde*, Selected Works in Probability and Statistics, DOI 10.1007/978-1-4419-5823-5_11,

2. Results.

LEMMA. *If* $\{x_n\}$ *is a monotone sequence of positive numbess with* $x_n \to \infty$ *as* $n \to \infty$, *then* $x_n^{-1}S_n \to_P 0$ *if and only if for every* $\epsilon > 0$,

(i) $n \Pr(|X| > \epsilon x_n) \to 0$ *as* $n \to \infty$,

(ii) $n x_n^{-2} \int_{|x| \leqq x_n} x^2 \, dF(x) \to 0$ *as* $n \to \infty$,

(iii) $n x_n^{-1} \int_{|x| \leqq x_n} x \, dF(x) \to 0$ *as* $n \to \infty$.

This result is stated essentially for the sake of completeness. It follows immediately from the degenerate convergence criterion (Loève [5], 317). We note that when the X_i do not belong to the domain of partial attraction of the normal distribution, condition (ii) is satisfied as a consequence of conditions (1) and (i).

THEOREM. *Suppose that* X_i, $i = 1, 2, 3, \cdots$, *is a sequence of independent and identically distributed random variables satisfying the conditions* (1) *and* (2). *Further,* $\{x_n\}$ *is a monotone sequence of positive numbers such that* $x_n \to \infty$ *as* $n \to \infty$ *and* $x_n^{-1}S_n \to_P 0$. *Then,*

$$(3) \quad 0 < \liminf_{n\to\infty} [\Pr(|S_n| > x_n)/n \Pr(|X| > x_n)] \leqq \limsup_{n\to\infty} [\Pr(|S_n| > x_n)/n \Pr(|X| > x_n)] < \infty,$$

or equivalently,

$$(4) \quad 0 < \liminf_{n\to\infty} [\Pr(|S_n| > x_n)/\Pr(\max_{k \leqq n} |X_k| > x_n)] \leqq \limsup_{n\to\infty} [\Pr(|S_n| > x_n)/\Pr(\max_{k \leqq n} |X_k| > x_n)] < \infty.$$

PROOF. Take $\epsilon > 0$ and denote by A_i and B_i the events $\{|X_i| > (1 + \epsilon)x_n\}$ and $\{|\sum_{j=1, j\neq i}^{n} X_j| < \epsilon x_n\}$ respectively, $i = 1, 2, \cdots, n$. If $\bar{E}$ is the complement of E, we then have

$$(5) \quad \begin{aligned} \Pr(|S_n| > x_n) &\geqq \Pr[\textstyle\bigcup_{i=1}^{n} (A_i \cap B_i)] \\ &= \textstyle\sum_{i=1}^{n} \Pr[\bigcap_{j=1}^{i-1} \overline{(A_j \cap B_j)} \cap (A_i \cap B_i)] \\ &\geqq \textstyle\sum_{i=1}^{n} \Pr[\bigcap_{j=1}^{i-1} \bar{A}_j \cap (A_i \cap B_i)] \\ &\geqq \textstyle\sum_{i=1}^{n} \{\Pr(A_i \cap B_i) - \Pr(\bigcup_{j=1}^{i-1} A_j \cap A_i)\} \\ &\geqq \textstyle\sum_{i=1}^{n} \Pr(A_i)[\Pr(B_i) - (i-1)\Pr(A_i)] \\ &\geqq n \Pr(A_1)[\Pr(B_1) - n \Pr(A_1)]. \end{aligned}$$

Now $x_n^{-1}S_n \to_P 0$ so $\Pr(B_i) \to 1$ as $n \to \infty$ and, given $\delta > 0$ with $1 - 2\delta > 0$, we can choose N_1 so large that $\Pr(B_i) > 1 - \delta$ for $n \geqq N_1$. Further, from the lemma we see that $n \Pr(A_i) \to 0$ as $n \to \infty$, so that we can choose N_2 so large that $n \Pr(A_i) < \delta$ for $n \geqq N_2$. Thus, for $n \geqq N = \max(N_1, N_2)$, we obtain from (5),

$$(6) \quad \Pr(|S_n| > x_n) \geqq n(1 - 2\delta) \Pr(|X| > (1 + \epsilon)x_n),$$

so that

$$\Pr(|S_n| > x_n)/n \Pr(|X| > x_n) \geqq (1 - 2\delta) \Pr(|X| > (1 + \epsilon)x_n)/\Pr(|X| > x_n)$$

and

$$\liminf_{n\to\infty} [\Pr(|S_n| > x_n)/n \Pr(|X| > x_n)]$$
$$\geq (1 - 2\delta) \liminf_{n\to\infty} [\Pr(|X| > (1 + \epsilon)x_n)/\Pr(|X| > x_n)] > 0$$

in view of condition (2).

In order to complete the proof we shall work in terms of the symmetrized random variables X_i^s, $i = 1, 2, 3, \cdots$, making continued use of the weak symmetrization inequalities (Loève, [5], 245) to transfer the results.

Define

$$X_{kn}^s = X_k^s \quad \text{if} \quad |X_k^s| \leq 2x_n$$
$$= 0 \qquad \text{otherwise,}$$

and write $S_n^s = \sum_{k=1}^n X_k^s$, $S_{nn}^s = \sum_{k=1}^n X_{kn}^s$. We have

$$\Pr(|S_n^s| > 2x_n) \leq n \Pr(|X^s| > 2x_n) + \Pr(|S_{nn}^s| > 2x_n) \tag{7}$$

and, using the weak symmetrization inequalities,

$$\Pr(|X^s| > 2x_n)/\Pr(|X| > x_n) \leq 2 \Pr(|X| > x_n)/\Pr(|X| > x_n) = 2.$$

We now turn our attention to $\Pr(|S_{nn}^s| > 2x_n)$. From Chebyshev's inequality, we obtain

$$\Pr(|S_{nn}^s| > 2x_n) \leq (2x_n)^{-2} E(S_{nn}^s)^2$$
$$= n(2x_n)^{-2} \int_{|x| \leq 2x_n} x^2 \, d \Pr(X^s \leq x),$$

so that

$$[\Pr(|S_{nn}^s| > 2x_n)/n \Pr(|X| > x_n)]$$
$$= [\Pr(|S_{nn}^s| > 2x_n)/n \Pr(|X^s| > 2x_n)][\Pr(|X^s| > 2x_n)/\Pr(|X| > x_n)]$$
$$\leq 2(2x_n)^{-2} \int_{|x| \leq 2x_n} x^2 \, d \Pr(X^s \leq x)/\Pr(|X^s| > 2x_n),$$

and $\limsup_{n\to\infty} [\Pr(|S_{nn}^s| > 2x_n)/n \Pr(|X| > x_n)] < \infty$ since X^s cannot belong to the domain of partial attraction of the normal distribution. We then have, from (7), $\limsup_{n\to\infty} [\Pr(|S_n^s| > 2x_n)/n \Pr(|X| > x_n)] < \infty$ and hence, using the weak symmetrization inequalities, $\limsup_{n\to\infty} [\Pr(|S_n - \text{med}\, S_n| > 2x_n)/ n \Pr(|X| > x_n)] < \infty$. However, as $x_n^{-1} S_n \to_P 0$, $x_n^{-1} \text{med}\, S_n \to 0$ as $n \to \infty$ and so, for $\epsilon > 0$ and n sufficiently large,

$$\Pr(|S_n - \text{med}\, S_n| > 2x_n)/n \Pr(|X| > x_n)$$
$$\geq \Pr(|S_n| > (2 + \epsilon)x_n)/n \Pr(|X| > x_n) \tag{8}$$
$$= [\Pr(|S_n| > (2 + \epsilon)x_n)/n \Pr(|X| > (2 + \epsilon)x_n)$$
$$\cdot [\Pr(|X| > (2 + \epsilon)x_n)/\Pr(|X| > x_n)].$$

Using condition (2) and a simple transformation, it follows from (8) that

$\limsup_{n\to\infty} [\Pr(|S_n| > x_n)/n \Pr(|X| > x_n)] < \infty$. This completes the proof of (3) and it just remains to establish the equivalence of (3) and (4). This, however, is easily deduced as, from Bonferroni's inequalities (Feller [2], 100), we have

$$n \Pr(|X| > x_n)[1 - \tfrac{1}{2}(n-1) \Pr(|X| > x_n)] \leqq \Pr(\max_{k \leqq n} |X_k| > x_n) \leqq n \Pr(|X| > x_n)$$

and hence, in view of condition (i) of the lemma,

$$\Pr(\max_{k \leqq n} |X_k| > x_n) \sim n \Pr(|X| > x_n) \quad \text{as} \quad n \to \infty.$$

The required result follows and the proof of the theorem is complete.

REFERENCES

[1] DOEBLIN, W. (1940). Sur l'ensemble de puissances d'une loi de probabilité. *Studia Math.* **9** 71–96.

[2] FELLER, W. (1957). *An Introduction to Probability Theory and its Applications* **1** (2nd edition). Wiley, New York.

[3] HEYDE, C. C. (1967). A contribution to the theory of large deviations for sums of independent random variables. *Z. Wahrscheinlichkeitstheorie verw. Gebiet.* **7** 303–308.

[4] LÉVY, P. (1954). *Théorie de l'addition des Variables Aléatoires* (Deuxième édition). Gauthier-Villars, Paris.

[5] LOÈVE, M. (1963). *Probability Theory* (3rd edition). Van Nostrand, New York.

Z. Wahrscheinlichkeitstheorie verw. Geb. 8, 12—18 (1967)

On the Influence of Moments on the Rate of Convergence to the Normal Distribution

C. C. Heyde

Received July 26, 1966

R. Maller et al. (eds.), *Selected Works of C.C. Heyde*, Selected Works in Probability and Statistics,
DOI 10.1007/978-1-4419-5823-5_12, © Springer Science+Business Media, LLC 2010

1. Introduction

Let X_i, $i = 1, 2, 3, \ldots$ be a sequence of independent and identically distributed random variables with var $X_i = \sigma^2 < \infty$ and $EX_i = 0$. Write

$$S_n = \sum_{i=1}^{n} X_i, \quad n \geqq 1.$$

Then,

$$F_n(x) = \Pr(S_n \leqq \sigma \sqrt{n}\, x) \to \Phi(x) = \frac{1}{\sqrt{2\pi}} \int_{-\infty}^{x} e^{-\frac{1}{2}u^2} du$$

as $n \to \infty$, and the rate of this convergence is a matter of considerable importance. In this paper we shall establish the following theorem:

Theorem. *Let X_i, $i = 1, 2, 3, \ldots$ be a sequence of independent and identically distributed random variables with $EX_i^2 = \sigma^2 < \infty$, $EX_i = 0$. Then,*

$$\sum_{n=1}^{\infty} n^{-1+\delta/2} \sup_x |F_n(x) - \Phi(x)| < \infty, \quad 0 \leqq \delta < 1, \tag{1}$$

if and only if $E|X_i|^{2+\delta} < \infty$, $0 < \delta < 1$, $EX_i^2 \log(1 + |X_i|) < \infty$, $\delta = 0$.

With the exception of the boundary case $\delta = 0$, this result provides a global generalization of a theorem of BAUM and KATZ [1] in which they established the convergence of the series $\sum_{1}^{\infty} n^{-1+\delta/2} |\Pr(S_n \leqq 0) - \frac{1}{2}|$ under the condition

$$E|X_i|^{2+\delta} < \infty, \quad 0 \leqq \delta < 1.$$

The theorem of this paper is, in a sense, complementary to Theorem 3.4.1 of LINNIK and IBRAGIMOV [7] in which they show that

$$|F_n(x) - \Phi(x)| = O(n^{-\delta/2}), \quad 0 < \delta < 1,$$

if and only if

$$\int_{|x| > z} x^2 \, dF(x) = O(z^{-\delta}),$$

where $F(x) = \Pr(X \leqq x)$ (a sketch of the proof of this result has also appeared in IBRAGIMOV [5]). The large literature on the BERRY-ESSEEN estimate is also closely related. See for example KATZ [6] and BIKJALIS [2] for recent contributions in this field. A comparison of the forms of the various results would seem to indicate that moment conditions are equivalent to series conditions on the terms $\sup_x |F_n(x) - \Phi(x)|$ rather than conditions on the individual elements.

2. Details

In view of our assumption of a finite variance, we can express the characteristic function $f(t)$ of the X_i in the form

$$f(t) = \exp\left\{-\frac{\sigma^2 t^2}{2}(1+\gamma(t))\right\} \tag{2}$$

where $\lim_{t\to 0}\gamma(t) = 0$. Furthermore, in a neighbourhood of the origin, the equation

$$f(t) = \exp\left\{-\frac{\sigma^2 t^2}{2}\right\}$$

can be satisfied at not more than a finite number of points ([7], § 6, Chapter 1) and hence in some suitably small interval, say $0 < t < \varepsilon$, we have $\gamma(t) \neq 0$.

Firstly, we shall prove the theorem for symmetric random variables X_i. In this case $f(t)$ and $\gamma(t)$ are real valued and symmetric. The proof is accomplished with the aid of two lemmas.

Lemma 1. *For symmetric random variables the series* (1) *converges if and only if for* $A > 0$,

$$\int_0^A \frac{|\gamma(u)|}{u^{1+\delta}}\,du < \infty. \tag{3}$$

Proof. Firstly we prove necessity. Let $f_n(t)$ be the characteristic function corresponding to $F_n(x)$. Then, integrating by parts in the equation

$$f_n(t) - e^{-t^2/2} = \int_{-\infty}^{\infty} e^{itx}\,d[F_n(x) - \Phi(x)],$$

we obtain

$$-\frac{f_n(t) - e^{-t^2/2}}{it} = \int_{-\infty}^{\infty} e^{itx}[F_n(x) - \Phi(x)]\,dx.$$

Further, it is seen that

$$it\,e^{-t^2/2} = \int_{-\infty}^{\infty} e^{itx}\frac{1}{\sqrt{2\pi}}\,x\,e^{-x^2/2}\,dx.$$

We observe that the conditions for the applicability of the Parseval identity are satisfied and hence

$$\int_{-\infty}^{\infty}[f_n(t) - e^{-t^2/2}]\,e^{-t^2/2}\,dt = \frac{1}{\sqrt{2\pi}}\int_{-\infty}^{\infty}[F_n(x) - \Phi(x)]\,x\,e^{-x^2/2}\,dx.$$

Thus from (1)

$$\sum_1^\infty n^{-1+\delta/2}\left|\int_{-\infty}^{\infty}[f_n(t) - e^{-t^2/2}]\,e^{-t^2/2}\,dt\right| = \sum_1^\infty n^{-1+\delta/2}\left|\int_{-\infty}^{\infty}[F_n(x) - \Phi(x)]\,x\,e^{-x^2/2}\,dx\right|$$
$$\leq \sum_1^\infty n^{-1+\delta/2}\sup_x|F_n(x) - \Phi(x)| < \infty.$$

That is,

$$\sum_1^\infty n^{-1+\frac{1}{2}\delta}\left|\int_{-\infty}^{\infty} e^{-t^2}\left(1 - e^{-\frac{1}{2}t^2\gamma\left(\frac{t}{\sigma\sqrt{n}}\right)}\right)dt\right| < \infty,$$

and since $\gamma(u)$ does not change sign in some interval $[0, \varepsilon]$, we must have

$$\sum_1^\infty n^{-1+\frac{1}{2}\delta}\int_0^1 e^{-t^2}\left|1 - e^{-\frac{1}{2}t^2\gamma(t/(\sigma\sqrt{n}))}\right|dt < \infty.$$

However, as $n \to \infty$,

$$|1 - e^{-\frac{1}{2}t^2\gamma(t/(\sigma\sqrt{n}))}| = \frac{1}{2}t^2\left|\gamma\left(\frac{t}{\sigma\sqrt{n}}\right)\right|(1+o(1)),$$

so that

$$\sum_1^\infty n^{-1+\frac{1}{2}\delta}\int_0^1 t^2\left|\gamma\left(\frac{t}{\sigma\sqrt{n}}\right)\right| dt < \infty,$$

and, upon making the transformation $u = t/\sigma\sqrt{n}$, this yields

$$\sum_1^\infty n^{\frac{1}{2}(1+\delta)}\int_0^{1/(\sigma\sqrt{n})} u^2|\gamma(u)|\,du < \infty. \tag{4}$$

Now $\int_0^{1/(\sigma\sqrt{x})} u^2|\gamma(u)|\,du$ is montone decreasing as x increases so, for $X > 1$,

$$\int_1^X x^{\frac{1}{2}(1+\delta)}\left\{\int_0^{1/(\sigma\sqrt{x})} u^2|\gamma(u)|\,du\right\}dx \leqq \sum_{n=1}^{[X]}\int_n^{n+1} x^{\frac{1}{2}(1+\delta)}\left\{\int_0^{1/(\sigma\sqrt{x})} u^2|\gamma(u)|\,du\right\}dx$$
$$\leqq \sum_{n=1}^{[X]}(n+1)^{\frac{1}{2}(1+\delta)}\int_0^{1/(\sigma\sqrt{n})} u^2|\gamma(u)|\,du,$$

where $[X]$ is the largest integer less than or equal to X. Furthermore, it is possible to choose a constant $c > 0$ such that $(n+1)^{\frac{1}{2}(1+\delta)} < cn^{\frac{1}{2}(1+\delta)}$ for all positive integral n and hence, for all $X > 1$,

$$\int_1^X x^{\frac{1}{2}(1+\delta)}\left\{\int_0^{1/(\sigma\sqrt{x})} u^2|\gamma(u)|\,du\right\}dx < c\sum_{n=1}^{[X]} n^{\frac{1}{2}(1+\delta)}\int_0^{1/(\sigma\sqrt{n})} u^2|\gamma(u)|\,du.$$

Consequently, using (4) we see that

$$\int_1^\infty x^{\frac{1}{2}(1+\delta)}\left\{\int_0^{1/(\sigma\sqrt{x})} u^2|\gamma(u)|\,du\right\}dx < \infty. \tag{5}$$

Now, in view of (5) we must have

$$\int_w^{2w} x^{\frac{1}{2}(1+\delta)}\left\{\int_0^{1/(\sigma\sqrt{x})} u^2|\gamma(u)|\,du\right\}dx \to 0 \quad \text{as } w \to \infty,$$

and

$$\int_w^{2w} x^{\frac{1}{2}(1+\delta)}\left\{\int_0^{1/(\sigma\sqrt{x})} u^2|\gamma(u)|\,du\right\}dx \geqq w^{\frac{1}{2}(3+\delta)}\int_0^{1/(\sigma\sqrt{2w})} u^2|\gamma(u)|\,du \geqq 0,$$

so that, putting $v = 1/\sigma\sqrt{2w}$, we conclude that

$$\frac{1}{v^{3+\delta}}\int_0^v u^2|\gamma(u)|\,du \to 0 \quad \text{as } v \to 0. \tag{6}$$

Then, upon making the transformation $v = 1/\sigma\sqrt{x}$, (5) becomes

$$\int_0^{\sigma^{-1}}\left\{\int_0^v u^2|\gamma(u)|\,du\right\}\frac{dv}{v^{4+\delta}} < \infty,$$

and in view of (6), the fact that condition (3) is satisfied follows immediately from an integration by parts.

It remains to prove the sufficiency of the condition (3). Making use of Theorem 1.5.2 of [7] we have for $T > 0$,

$$|F_n(x) - \Phi(x)| \leqq \frac{1}{\pi}\int_{-T}^{T}\left|\frac{f_n(t) - e^{-t^2/2}}{t}\right| dt + \frac{24}{\pi\sqrt{2\pi}\,T}. \tag{7}$$

Take $T = B\sigma\sqrt{n}$ where $2\sigma^{-1} > B > 0$ is chosen so small that

$$\max_{0\leqq t\leqq B}|\gamma(t)| \leqq \tfrac{1}{2}. \tag{8}$$

Then, using the inequality

$$|e^x - 1| \leqq |x|\, e^{|x|},$$

we have

$$\begin{aligned}|f_n(t) - e^{-\frac{1}{2}t^2}| &= e^{-\frac{1}{2}t^2}\left|1 - e^{-\frac{1}{2}t^2\gamma(t/(\sigma\sqrt{n}))}\right| \\ &\leqq \frac{1}{2}t^2\left|\gamma\left(\frac{t}{\sigma\sqrt{n}}\right)\right|\exp\left\{-\frac{t^2}{2} + \frac{t^2}{2}\left|\gamma\left(\frac{t}{\sigma\sqrt{n}}\right)\right|\right\}.\end{aligned}$$

Hence for $|t| \leqq T$, using (8),

$$|f_n(t) - e^{-\frac{1}{2}t^2}| \leqq \frac{t^2}{2}\left|\gamma\left(\frac{t}{\sigma\sqrt{n}}\right)\right| e^{-\frac{1}{4}t^2},$$

so that we have

$$\begin{aligned}\sum_1^\infty n^{-1+\frac{1}{2}\delta}\int_{-T}^{T}\left|\frac{f_n(t) - e^{-\frac{1}{2}t^2}}{t}\right| dt &\leqq \sum_1^\infty n^{-1+\frac{1}{2}\delta}\int_0^{B\sigma\sqrt{n}} t\left|\gamma\left(\frac{t}{\sigma\sqrt{n}}\right)\right| e^{-\frac{1}{4}t^2}\,dt \\ = \sigma^2\sum_1^\infty n^{\frac{1}{2}\delta}\int_0^B u\,|\gamma(u)|\, e^{-\frac{1}{4}\sigma^2 nu^2}\,du &= \sigma^2\int_0^B u\,|\gamma(u)|\left\{\sum_1^\infty n^{\frac{1}{2}\delta}e^{-\frac{1}{4}\sigma^2 nu^2}\right\} du.\end{aligned} \tag{9}$$

Now, making use of a standard Abelian theorem (e.g. FELLER [3], 423) we see that

$$\lim_{S\uparrow 1}(1 - S)^{1+\frac{1}{2}\delta}\sum_{n=1}^\infty n^{\frac{1}{2}\delta}S^n = \Gamma(1 + \tfrac{1}{2}\delta),$$

and hence, for $u \neq 0$, it is possible to choose a constant $K > 0$ such that

$$\sum_{n=1}^\infty n^{\frac{1}{2}\delta}e^{-\frac{1}{4}\sigma^2 nu^2} \leqq K(1 - e^{-\frac{1}{4}\sigma^2 u^2})^{-(1+\frac{1}{2}\delta)}.$$

Furthermore, for $0 \leqq u \leqq B$ we have $\frac{1}{4}\sigma^2 u^2 < 1$, so that

$$1 - e^{-\frac{1}{4}\sigma^2 u^2} > \tfrac{1}{4}\sigma^2 u^2(1 - \tfrac{1}{8}\sigma^2 u^2),$$

and hence

$$\sum_{n=1}^\infty n^{\frac{1}{2}\delta}e^{-\frac{1}{4}\sigma^2 nu^2} < \frac{K\,2^{2+\delta}}{\sigma^{2+\delta}u^{2+\delta}(1 - \frac{1}{8}\sigma^2 u^2)^{1+\frac{1}{2}\delta}}. \tag{10}$$

Then, using (9) and (10) we see that

$$\sum_{n=1}^\infty n^{-1+\frac{1}{2}\delta}\int_{-T}^{T}\left|\frac{f_n(t) - e^{-\frac{1}{2}t^2}}{t}\right| dt \leqq \frac{K\,2^{2+\delta}}{\sigma^\delta}\int_0^B \frac{|\gamma(u)|}{u^{1+\delta}(1 - \frac{1}{8}\sigma^2 u^2)^{1+\frac{1}{2}\delta}}\,du$$

and this is finite in view of the condition (3). It follows from (7) that the condition (1) must be satisfied. This completes the proof of the lemma.

Lemma 2. *For symmetric random variables, the condition* (3) *is satisfied if and only if* $E|X_i|^{2+\delta} < \infty$, $0 < \delta < 1$, $EX_i^2 \log(1 + |X_i|) < \infty$, $\delta = 0$.

Proof. In a neighbourhood of the origin,

$$f(t) - 1 = -\frac{\sigma^2 t^2}{2}(1 + o(1)),$$

and then,

$$\log f(t) = \log\{1 + [f(t) - 1]\} = [f(t) - 1] + O(t^4). \tag{11}$$

Suppose, for the sake of definiteness, that in the interval $(0, C]$, $\gamma(t) < 0$; we obtain from (11),

$$\tfrac{1}{2}\sigma^2 \int_0^C \frac{|\gamma(t)|}{t^{1+\delta}}\, dt = \int_0^C \frac{\tfrac{1}{2}\sigma^2 t^2 + \log f(t)}{t^{3+\delta}}\, dt = \int_0^C \frac{\tfrac{1}{2}\sigma^2 t^2 + [f(t) - 1]}{t^{3+\delta}}\, dt + O(C^{2-\delta})$$

$$= \int_0^C \left\{\frac{\tfrac{1}{2}\sigma^2 t^2 + \int_{-\infty}^{\infty} (\cos tu - 1)\, dF(u)}{t^{3+\delta}}\right\} dt + O(C^{2-\delta})$$

$$= \int_{-\infty}^{\infty} \left\{\int_0^C \left(\frac{\cos tu - 1 + \tfrac{1}{2}t^2u^2}{t^{3+\delta}}\right) dt\right\} dF(u) + O(C^{2-\delta}).$$

Consequently, the condition (3) and the condition

$$\int_{-\infty}^{\infty} \left\{\int_0^C \left(\frac{\cos tu - 1 + \tfrac{1}{2}t^2u^2}{t^{3+\delta}}\right) dt\right\} dF(u) < \infty \tag{12}$$

are equivalent.

Now let us look at the integrand $\int_0^C \left(\frac{\cos tu - 1 + \frac{1}{2}t^2u^2}{t^{3+\delta}}\right) dt$. After two integrations by parts we obtain

$$\int_0^C \left(\frac{\cos tu - 1 + \tfrac{1}{2}t^2u^2}{t^{3+\delta}}\right) dt = -\frac{\cos Cu - 1 + \tfrac{1}{2}C^2u^2}{(2 + \delta)C^{2+\delta}} - \frac{u(Cu - \sin Cu)}{(1 + \delta)(2 + \delta)C^{1+\delta}} + \frac{u^2}{(2 + \delta)(1 + \delta)} \int_0^C \frac{1 - \cos tu}{t^{1+\delta}}\, dt,$$

so that the condition (12) is equivalent to the condition

$$\int_{-\infty}^{\infty} u^2 \left\{\int_0^C \frac{1 - \cos tu}{t^{1+\delta}}\, dt\right\} dF(u) < \infty,$$

which transforms to give

$$\int_{-\infty}^{\infty} |u|^{2+\delta} \left\{\int_0^{C|u|} \frac{1 - \cos v}{v^{1+\delta}}\, dv\right\} dF(u) < \infty. \tag{13}$$

(13) is obviously equivalent to the condition $E|X_i|^{2+\delta} < \infty$ when $0 < \delta < 1$.

On the other hand, when $\delta = 0$, we have for $|u| > 1$

$$\int_0^{C|u|} \frac{1-\cos v}{v}\,dv = \int_0^{C} \frac{1-\cos v}{v}\,dv + \int_C^{C|u|} \frac{dv}{v} - \int_C^{C|u|} \frac{\cos v}{v}\,dv$$
$$\sim \log|u| \quad \text{as} \quad |u| \to \infty,$$

so that (13) is equivalent to the condition $EX_i^2 \log(1 + |X_i|) < \infty$. This completes the proof of the lemma.

Lemmas 1 and 2 establish the theorem for the case of symmetric random variables. We go on to complete the proof of the theorem in the general case.

Firstly, we shall examine the necessity part. Consider the sequence Y_i, $i = 1, 2, 3, \ldots$ of independent symmetrized random variables; each Y_i having the distribution of the difference between two independent X_i's. Obviously, the characteristic function of the random variables Y_i is $|f(t)|^2$ and the distribution function of the sum

$$Z_n = \frac{Y_1 + \cdots + Y_n}{\sigma\sqrt{2n}}$$

is equal to $F_n(x\sqrt{2}) * (1 - F_n(-x\sqrt{2} - 0)) = G_n(x)$. Hence, if $F_n(x)$ satisfies the condition (1), then

$$\sum_1^\infty n^{-1+\frac{1}{2}\delta} \sup_x |G_n(x) - \Phi(x)|$$
$$= \sum_1^\infty n^{-1+\frac{1}{2}\delta} \sup_x |F_n(x\sqrt{2}) * (1 - F_n(-x\sqrt{2} - 0)) - \Phi(x\sqrt{2}) * (1 - \Phi(-x\sqrt{2}))|$$
$$\leqq \sum_1^\infty n^{-1+\frac{1}{2}\delta} \sup_x |F_n(x\sqrt{2}) * (1 - F_n(-x\sqrt{2} - 0)) - \Phi(x\sqrt{2}) * (1 - F_n(-x\sqrt{2} - 0))|$$
$$+ \sum_1^\infty n^{-1+\frac{1}{2}\delta} \sup_x |\Phi(x\sqrt{2}) * (1 - F_n(-x\sqrt{2} - 0)) - \Phi(x\sqrt{2}) * (1 - \Phi(-x\sqrt{2}))|$$
$$< \infty.$$

By Lemma 1 in such a case,

$$\int_0^A \frac{|\operatorname{Re}\gamma(t)|}{t^{1+\delta}}\,dt < \infty,$$

and proceeding exactly as with the proof of Lemma 2 we extract the information that $E|Y_i|^{2+\delta} < \infty$, $0 < \delta < 1$, $EY_i^2 \log(1 + |Y_i|) < \infty$, $\delta = 0$ and hence $E|X_i|^{2+\delta} < \infty$, $0 < \delta < 1$, $EX_i^2 \log(1 + |X_i|) < \infty$, $\delta = 0$.

In order to prove the sufficiency part, we shall make use of the result that if

$$\int_0^A \frac{|\gamma(t)|}{t^{1+\delta}}\,dt < \infty, \quad 0 \leqq \delta < 1, \tag{14}$$

then

$$\sum_1^\infty n^{-1+\frac{1}{2}\delta} \sup_x |F_n(x) - \Phi(x)| < \infty.$$

We note that $|\gamma(t)| = |\gamma(-t)|$. Then, the proof of this result is essentially the same as the proof of sufficiency in Lemma 1 and is hence omitted. We complete the proof by showing that the condition

$$E|X_i|^{2+\delta} < \infty\,, \quad 0 < \delta < 1\,, \quad E X_i^2 \log(1 + |X_i|) < \infty\,, \quad \delta = 0\,,$$

implies the condition (14).

Now exactly as with the proof of Lemma 2 we can show that $E|X_i|^{2+\delta} < \infty$, $0 < \delta < 1$, $E X_i^2 \log(1 + |X_i|) < \infty$, $\delta = 0$ implies

$$\int_0^A \frac{|\operatorname{Re}\gamma(t)|}{t^{1+\delta}}\, dt < \infty\,; \tag{15}$$

we shall show that the corresponding integral obtained by replacing $\operatorname{Re}\gamma(t)$ by $\operatorname{Im}\gamma(t)$ is also finite. Indeed, in a neighbourhood of the origin,

$$\begin{aligned} \tfrac{1}{2}\sigma^2 t^2 |\operatorname{Im}\gamma(t)| &= |\operatorname{Im}\log f(t)| \\ &= |\operatorname{Im} f(t)| + O(t^4)\,, \end{aligned}$$

so that

$$\tfrac{1}{2}\sigma^2 \int_0^A \frac{|\operatorname{Im}\gamma(t)|}{t^{1+\delta}}\, dt = \int_0^A \frac{\left|\int_{-\infty}^{\infty} (\sin tu - tu)\, dF(u)\right|}{t^{3+\delta}}\, dt + O(A^{2-\delta})$$

and the finiteness of the integral on the right hand side under the condition $E|X_i|^{2+\delta} < \infty$, $0 \leqq \delta < 1$, follows from Lemma 5 of HEYDE [4]. This result, together with (15), gives us (14) and thus completes the proof.

Addendum. Since the preparation of this paper there has appeared a related work "Convergence rates for the central limit theorem" by FRIEDMAN, KATZ and KOOPMANS, Proc. nat. Acad. Sci. USA **56**, 1062–1065 (1966). In this paper the convergence problem dealt with relates to a fixed x and not to the maximum over all x, $-\infty < x < \infty$, as in the present work. In so far as the results are connected, those of the present paper are superior. Different methods are employed in the two papers.

References

1. BAUM, L. E., and M. L. KATZ: On the influence of moments on the asymptotic distribution of sums of random variables. Ann. math. Statistics **34**, 1042–1044 (1963).
2. BIKJALIS, A.: An estimate for the remainder term in the central limit theorem (Russian, English summary). Litovsk. Mat. Sbornik **4**, 303–308 (1964).
3. FELLER, W.: An introduction to probability theory and its applications, Vol. II. New York: Wiley 1966.
4. HEYDE, C. C.: Some results on small deviation probability convergence rates for sums of independent random variables. Canadian J. Math. **18**, 656–665 (1966).
5. IBRAGIMOV, I. A.: The rate of convergence to the normal distribution. Soviet Math. Doklady **6**, 580–582 (1965).
6. KATZ, M. L.: Note on the Berry-Esseen theorem. Ann. math. Statistics **34**, 1107–1108 (1963).
7. LINNIK, YU. V., and I. A. IBRAGIMOV: Independent and stationary dependent random variables (Russian). Moscow 1965.

Department of Probability and Statistics
University of Sheffield, England,
and
Matematisk Institut,
Aarhus Universitet, Denmark

Reprinted from Sankhyā : The Indian Journal of Statistics, Series A, Vol. 30, Part 3, 1968

ON LARGE DEVIATION PROBABILITIES IN THE CASE OF ATTRACTION TO A NON-NORMAL STABLE LAW

By C. C. HEYDE
University of Sheffield

SUMMARY. In this paper, the precise asymptotic behaviour of the large deviation probability is found in the case where the random variables are attracted to a non-normal stable law. This extends previous work of the same author in which only the order of magnitude of the large deviation probability was found.

1. Introduction

Let X_i, $i = 1, 2, 3, \ldots$ be a sequence of independent and identically distributed random variables with law $\mathcal{L}(X)$ and write $S_n = \sum_{i=1}^{n} X_i$. Let y_n, $n = 1, 2, 3, \ldots$ be a monotone sequence of positive numbers with $y_n \to \infty$ as $n \to \infty$ such that $y_n^{-1} S_n \xrightarrow{P} 0$. ($P$ stands for convergence in probability). The probabilities $\Pr(|S_n| > y_n)$, or either of the one sided components, are called large deviation probabilities.

In the papers Heyde, (1967a and 1967b), the asymptotic behaviour of $\Pr(|S_n| > y_n)$ has been investigated for various types X which are not attracted to the normal law. The results contained therein, however, have a shortcoming in that only the order of magnitude of the larger deviation probability is found and not the precise behaviour. It is the object of this paper to remedy this deficiency in the case of the (1967a) paper where the X_i belong to the domain of attraction of a non-normal stable law. The context of the (1967b) paper, namely with the X_i not belonging to the domain of partial attraction of the normal distribution, is manifestly too general to allow for a corresponding complete answer.

2. Details

The formulation of the problem follows the same lines as that of Heyde (1967a) to which we refer for background details. Thus, the X_i belong, with normalizing constants B_n, to the domain of attraction of the stable law with characteristic function

$$\exp\left\{-a|t|^{\alpha}\left(1+i\frac{t}{|t|}\,\frac{c_1-c_2}{c_1+c_2}\tan\frac{\pi\alpha}{2}\right)\right\}, \qquad \ldots \quad (1)$$

$0 < \alpha < 2, \alpha \neq 1$, $a > 0$, $c_1 > 0, c_2 > 0$. Certain rather exceptional cases (c_1 or c_2 zero, $\alpha = 1$) are excluded as a matter of convenience. If $\alpha > 1$, we suppose that the X_i are located so that $EX_i = 0$. Consequently, we have $B_n^{-1}S_n$ converging in law to a stable distribution with characteristic function (1) and $x_n^{-1} B_n^{-1} S_n \xrightarrow{P} 0$ if x_n, $n = 1, 2, 3, \ldots$ is a monotone sequence of positive numbers with $x_n \to \infty$ as $n \to \infty$.

R. Maller et al. (eds.), *Selected Works of C.C. Heyde*, Selected Works in Probability and Statistics,
DOI 10.1007/978-1-4419-5823-5_13, © Springer Science+Business Media, LLC 2010

Under the above conditions we may write for $x > 0$, using Gnedenko and Kolmogorov (1954, Theorem 2, 175)

$$\Pr(|X| > x) = \frac{M(x)}{x^{\alpha}}, \qquad \dots \quad (2)$$

where $M(x)$ is a slowly varying function which satisfies the relation

$$\frac{nM(B_n)}{B_n^{\alpha}} \to (c_1+c_2) \quad \text{as } n \to \infty. \qquad \dots \quad (3)$$

We shall establish the following theorem concerning the two sided large deviation probability. Corresponding one sided versions can probably be obtained in much the same manner but there are some technical difficulties to overcome.

Theorem : *Let $\{x_n\}$ be a monotone sequence of positive numbers with $x_n \to \infty$ as $n \to \infty$. Then,*

$$\lim_{n \to \infty} \frac{Pr(|S_n| > x_n B_n)}{n\, Pr(|X| > x_n B_n)} = 1, \qquad \dots \quad (4)$$

or equivalently,

$$\lim_{n \to \infty} \frac{Pr(|S_n| > x_n B_n)}{Pr\left(\max\limits_{1 \leqslant k \leqslant n} |X_k| > x_n B_n\right)} = 1. \qquad \dots \quad (5)$$

This result extends that of Heyde (1967a) where it is shown, in essence, that

$$0 < \varliminf_{n \to \infty} \frac{\Pr(|S_n| > x_n B_n)}{n \Pr(|X| > x_n B_n)} \leqslant \varlimsup_{n \to \infty} \frac{\Pr(|S_n| > x_n B_n)}{n \Pr(|X| > x_n B_n)} < \infty.$$

Proof : Firstly we note that the analysis of the theorem of Heyde (1967b) can be used and we deduce immediately that

$$\varliminf_{n \to \infty} \frac{\Pr(|S_n| > x_n B_n)}{n \Pr(|X| > x_n B_n)} \geqslant 1.$$

It therefore just remains to show that

$$\varlimsup_{n \to \infty} \frac{\Pr(|S_n| > x_n B_n)}{n \Pr(|X| > x_n B_n)} \leqslant 1$$

and the result (4) will follow. The equivalence of the forms (4) and (5) is deduced precisely as in Heyde (1967b) and will not be repeated here.

For $1 \leqslant k \leqslant n$ and $n = 1, 2, 3, \dots$, define random variables

$$X_{kn} = \begin{cases} X_k & \text{if } |X_k| \leqslant z_n B_n \\ 0 & \text{if } |X_k| > z_n B_n, \end{cases}$$

where $z_n = x_n^{\gamma}$, $1 > \gamma > 1/2$, and write $S_{nn} = \sum_{k=1}^{n} X_{kn}$. Take $0 < \varepsilon < 1$ and define the events

$$E_n = \{|X_k| > (1-\varepsilon)x_n B_n \text{ for at least one } k \leqslant n\},$$

$$F_n = \{|X_k| > z_n B_n \qquad \text{for at least two } k\text{'s} \leqslant n\},$$

$$G_n = \{|S_{nn}| > \varepsilon x_n B_n\}.$$

Then, we have

$$\{|S_n| > x_n B_n\} \subset E_n U F_n U G_n,$$

so that

$$\begin{aligned}\Pr(|S_n| > x_n B_n) &\leqslant \Pr(E_n)+\Pr(F_n)+\Pr(G_n)\\ &\leqslant n \Pr(|X| > (1-\varepsilon)x_n B_n)+n^2[\Pr(|X| > z_n B_n)]^2\\ &\qquad +\Pr(|S_{nn}| > \varepsilon x_n B_n). \quad \ldots \quad (6)\end{aligned}$$

We shall deal separately with each of the three terms on the right hand side of (6).

From (2), we have

$$\frac{n\Pr(|X| > (1-\varepsilon)x_n B_n)}{n \Pr(|X| > x_n B_n)} = \frac{M[(1-\varepsilon)x_n B_n]}{M[x_n B_n]} \cdot \frac{x_n^{\alpha} B_n^{\alpha}}{x_n^{\alpha} B_n^{\alpha}(1-\varepsilon)^{\alpha}} \to \frac{1}{(1-\varepsilon)^{\alpha}} \text{ as } n \to \infty. \quad \ldots \quad (7)$$

Also,

$$\begin{aligned}\frac{n^2[\Pr(|X| > z_n B_n)]^2}{n \Pr(|X| > x_n B_n)} &= \frac{nM(B_n)}{B_n^{\alpha}} \cdot \frac{x_n^{\alpha}}{z_n^{2\alpha}} \frac{[M(z_n B_n)]^2}{M(B_n)M(x_n B_n)}\\ &\leqslant C \frac{x_n^{\alpha}}{z_n^{2\alpha}} \left(\frac{x_n}{z_n}\right)^{\eta_1} z_n^{\eta_2}\\ &= C\, x_n^{\alpha+\eta_1-\gamma(2\alpha+\eta_1-\eta_2)}\end{aligned}$$

for n sufficiently large using estimates obtained in Heyde (1967a), where C is a positive constant and η_1, η_2 are arbitrarily small and positive. Since $1 > \gamma > 1/2$, we can choose η_1 and η_2 such that

$$\alpha+\eta_1-\gamma(2\alpha+\eta_1-\eta_2) < 0,$$

and then

$$\frac{n[\Pr(|X| > z_n B_n)]^2}{\Pr(|X| > x_n B_n)} \to 0 \text{ as } n \to \infty. \quad \ldots \quad (8)$$

Finally, it remains to consider $\Pr(|S_{nn}| > \varepsilon x_n B_n)$. We have, using Chebyshev's inequality,

$$\begin{aligned}\Pr(|S_{nn}| > \varepsilon x_n B_n) &\leqslant \varepsilon^{-2} x_n^{-2} B_n^{-2} E(S_{nn}^2)\\ &= \varepsilon^{-2} x_n^{-2} B_n^{-2} [n\, EX_{kn}^2 + \tfrac{1}{2} n(n-1)(EX_{kn})^2]. \quad \ldots \quad (9)\end{aligned}$$

Then,

$$\frac{EX_{kn}^2}{x_n^2 B_n^2 \Pr(|X| > x_n B_n)} = \frac{\int_{|x| \leqslant z_n B_n} x^2 d \Pr(X \leqslant x)}{x_n^2 B_n^2 \Pr(|X| > x_n B_n)}$$

$$\leqslant \frac{2 \int_0^{z_n B_n} x \Pr(|X| > x) dx}{x_n^2 B_n^2 \Pr(|X| > x_n B_n)}$$

$$= \frac{2 \int_0^{z_n B_n} x \Pr(|X| > x) dx}{z_n^2 B_n^2 \Pr(|X| > z_n B_n)} \cdot \frac{z_n^2 \Pr(|X| > z_n B_n)}{x_n^2 \Pr(|X| > x_n B_n)}$$

where we have used integration by parts to obtain the second last line. Furthermore, using standard properties of slowly varying functions (Feller, 1966, 273), we have

$$\frac{2 \int_0^{z_n B_n} x \Pr(|X| > x) dx}{z_n^2 B_n^2 \Pr(|X| > z_n B_n)} \to \frac{2}{2-\alpha} \text{ as } n \to \infty,$$

while

$$\frac{z_n^2 \Pr(|X| > z_n B_n)}{x_n^2 \Pr(|X| > x_n B_n)} = \frac{z_n^{2-\alpha}}{x_n^{2-\alpha}} \frac{M(z_n B_n)}{M(x_n B_n)} \to 0 \text{ as } n \to \infty,$$

and therefore,

$$\frac{EX_{kn}^2}{x_n^2 B_n^2 \Pr(|X| > x_n B_n)} \to 0 \text{ as } n \to \infty. \qquad \ldots \quad (10)$$

Also, if $\alpha < 1$,

$$|EX_{kn}| \leqslant E|X_{kn}| = \int_{|x| \leqslant z_n B_n} |x| \, d \Pr(X \leqslant x)$$

$$\leqslant \int_0^{z_n B_n} \Pr(|X| > x) dx,$$

using integration by parts, so that

$$\frac{n(EX_{kn})^2}{x_n^2 B_n^2 \Pr(|X| > x_n B_n)} \leqslant \left[\frac{\int_0^{z_n B_n} \Pr(|X| > x) dx}{z_n B_n \Pr(|X| > z_n B_n)} \right]^2 \frac{n z_n^2 [\Pr(|X| > z_n B_n)]^2}{x_n^2 \Pr(|X| > x_n B_n)}.$$

In this case,

$$\frac{\int_0^{z_n B_n} \Pr(|X| > x) dx}{z_n B_n \Pr(|X| > z_n B_n)} \to \frac{1}{1-\alpha},$$

while

$$\frac{n z_n^2 [\Pr(|X| > z_n B_n)]^2}{x_n^2 \Pr(|X| > x_n B_n)} < \frac{n [\Pr(|X| > z_n B_n)]^2}{\Pr(|X| > x_n B_n)} \to 0 \text{ as } n \to \infty,$$

according to (8) above, so that for $\alpha < 1.$,

$$\frac{n(EX_{kn})^2}{x_n^2 B_n^2 \Pr(|X| > x_n B_n)} \to 0 \text{ as } n \to \infty. \qquad \dots \quad (11)$$

If, on the other hand, $1 < \alpha < 2$, we have since $EX = 0$,

$$\begin{aligned} |EX_{kn}| &= \left| \int_{|x| \leqslant z_n B_n} x d \Pr(X \leqslant x) \right| \\ &= \left| \int_{|x| > z_n B_n} x d \Pr(X \leqslant x) \right| \\ &\leqslant \int_{|x| > z_n B_n} |x| d \Pr(X \leqslant x) \\ &= - \int_{z_n B_n}^{\infty} x d \Pr(|X| > x) \\ &= \int_{z_n B_n}^{\infty} \Pr(|X| > x) dx + z_n B_n \Pr(|X| > z_n B_n), \end{aligned}$$

and

$$\frac{\int_{z_n B_n}^{\infty} \Pr(|X| > x) dx}{z_n B_n \Pr(|X| > z_n B_n)} \to \frac{1}{\alpha - 1},$$

so that

$$\frac{n(EX_{kn})^2}{x_n^2 B_n^2 \Pr(|X| > x_n B_n)} \leqslant \left[\frac{\int_{z_n B_n}^{\infty} \Pr(|X| > x) dx}{z_n B_n \Pr(|X| > z_n B_n)} + 1 \right]^2 \cdot \frac{n z_n^2 [\Pr(|X| > z_n B_n)]^2}{x_n^2 \Pr(|X| > x_n B_n)}$$

$$\to 0 \text{ as } n \to \infty, \qquad \dots \quad (12)$$

again using (8). Then, from (9), (10), (11), and (12), we see that

$$\frac{\Pr(|S_{nn}| > \varepsilon x_n B_n)}{n \Pr(|X| > x_n B_n)} \to 0 \text{ as } n \to \infty. \qquad \dots \quad (13)$$

Finally, using (7), (8) and (13) in (6), we see that

$$\overline{\lim_{n \to \infty}} \frac{\Pr(|S_n| > x_n B_n)}{n \Pr(|X| > x_n B_n)} \leqslant \frac{1}{(1-\varepsilon)^2}$$

and since ε, $0 < \varepsilon < 1$ is arbitrary,

$$\overline{\lim_{n \to \infty}} \frac{\Pr(|S_n| > x_n B_n)}{n \Pr(|X| > x_n B_n)} \leqslant 1.$$

This furnishes the proof of the theorem.

REFERENCES

FELLER, W. (1966) : *An Introduction to Probability Theory and its Applications*, Vol. II, Wiley, New York.

GNEDENKO, B. V. and KOLMOGOROV, A. N. (1954) : *Limit Distributions for Sums of Independent Random Variables*, Translated and annotated by K. L. Chung, Addison Wesley, Cambridge (Mass.).

HEYDE, C. C. (1967a) : A contribution to the theory of large deviations for sums of independent random variables. *Z. Wahrscheinlichkeitstheorie*, **7**, 303–308.

——— (1967b) : On large deviation problems for sums of random variables which are not attracted to the normal law. *Ann. Math. Stat.*, **38**, 1575–1578.

Paper received : June, 1967.

ON THE CONVERSE TO THE ITERATED LOGARITHM LAW

BY

C. C. HEYDE

Reprinted from the
JOURNAL OF APPLIED PROBABILITY
Vol. 5, No. 1, pp. 210–215
April 1968

R. Maller et al. (eds.), *Selected Works of C.C. Heyde*, Selected Works in Probability and Statistics,
DOI 10.1007/978-1-4419-5823-5_14, © Springer Science+Business Media, LLC 2010

J. Appl. Prob. **5**, 210–215 (1968)
Printed in Israel

ON THE CONVERSE TO THE ITERATED LOGARITHM LAW

C. C. HEYDE, *University of Sheffield*

1. Introduction

Let X_i, $i = 1,2,3,\cdots$ be a sequence of independent and identically distributed random variables with law $\mathscr{L}(X)$ and write $S_n = \sum_{i=1}^{n} X_i$. If $EX = 0$ and $EX^2 = \sigma^2 < \infty$, the law of the iterated logarithm (Hartman and Wintner [1]) tells us that

$$\Pr(\limsup_{n\to\infty} (2\sigma n \log\log n)^{-\frac{1}{2}} |S_n| = 1) = 1.$$

Furthermore, it has recently been shown by Strassen [4] that if

$$\Pr(\limsup_{n\to\infty} (n \log\log n)^{-\frac{1}{2}} |S_n| < \infty) = 1,$$

then $EX = 0$, $EX^2 < \infty$. The proof of Strassen, however, utilizes deep results of Skorokhod involving the use of an associated Wiener process and it is the object of the present paper to develop a direct approach to this result. As a by-product, it will be clear how the results of Stone [3] can be extended.

2. Results

Theorem. If $EX^2 = \infty$, then

$$\Pr(\limsup_{n\to\infty} (n \log\log n)^{-\frac{1}{2}} |S_n| = \infty) = 1.$$

Proof. Firstly suppose that $\sum \Pr[|X| > (n\log\log n)^{\frac{1}{2}}] = \infty$. Then, for $N > 0$,

$$\frac{|X_n|}{N[n\log\log n]^{\frac{1}{2}}} = \frac{|S_n - S_{n-1}|}{N[n\log\log n]^{\frac{1}{2}}} \leqq \frac{|S_n|}{N[n\log\log n]^{\frac{1}{2}}} + \frac{|S_{n-1}|}{N[n\log\log n]^{\frac{1}{2}}}$$

$$\leqq \frac{|S_n|}{N[n\log\log n]^{\frac{1}{2}}} + \frac{|S_{n-1}|}{N[(n-1)\log\log(n-1)]^{\frac{1}{2}}},$$

so that if $\Pr(|S_n| > N(n\log\log n)^{\frac{1}{2}} \text{ i.o.}) = 0$, then we must have

$$\Pr(|X_n| > 2N(n\log\log n)^{\frac{1}{2}} \text{i.o.}) = 0$$

Received 25 April 1967.

and hence, from the Borel-Cantelli Lemma (the X_i's being independent), $\sum \Pr[|X| > 2N(n\log\log n)^{\frac{1}{2}}] < \infty$. ("i.o." means infinitely often). This clearly contradicts the condition $\sum \Pr[|X| > (n\log\log n)^{\frac{1}{2}}] = \infty$ so, by appeal to the zero-one law, $\Pr(|S_n| > N(n\log\log n)^{\frac{1}{2}}\text{i.o.}) = 1$. Since N is arbitrarily large, the result of the theorem follows. Consequently, it remains to consider the case where $\sum \Pr[|X| > (n\log\log n)^{\frac{1}{2}}] < \infty$ but $\sum \Pr(|X| > n^{\frac{1}{2}}) = \infty$, this last condition being equivalent to $EX^2 = \infty$. We note, in particular, that $\sum\Pr(|X| > n^{1/(2-\delta)}) < \infty$; that is, $E|X|^{2-\delta} < \infty$, every $\delta > 0$. Furthermore, if $EX = \mu \neq 0$, the truth of the theorem is evident from the strong law of large numbers so we may subsequently restrict ourselves to the case $EX = 0$. We shall proceed by developing a modification of the argument of Stone [3].

Write F for the distribution function of X. Also, for a positive constant K, let H be the distribution function which results from truncating X at $-K$, $+K$ and then centering to make the mean zero. Then, there are positive constants a and b $(a + b = 1)$ such that $F = aG + bH$, G being a distribution function. Of course, $b = \Pr(|X| \leqq K)$. Let $Y_1, Y_2, \cdots$ be independent random variables with common distribution function G; $Z_1, Z_2, \cdots$ be independent random variables with common distribution function H and write $T_n = \sum_{i=1}^n Y_i$, $U_n = \sum_{i=1}^n Z_i$. Also, $\xi_1, \xi_2, \cdots$ are independent and identically distributed such that $\Pr(\xi_k = 1) = a$ and $\Pr(\xi_k = 0) = b$. The Y_k's, Z_k's and ξ_k's are taken as mutually independent. We write also, $j(n) = \sum_{i=1}^n \xi_i$, $k(n) = n - j(n)$ and $V_n = T_{j(n)} + U_{k(n)}$. Clearly V_n has the same probabilistic structure as S_n, so in order to obtain the desired result we have just to show that for any positive N,

$$\Pr[(n\log\log n)^{-\frac{1}{2}}|V_n| \geqq N \text{ i.o.}] = 1.$$

Now, since $EX = 0$ and $EZ_i = 0$, we have $EY_i = 0$ and consequently $\Pr(|T_n| \leqq \varepsilon(n\log\log n)^{\frac{1}{2}} \text{ i.o.}) = 1$ for any fixed $\varepsilon > 0$. Define events A_n, $n \geqq 3$, by

$$A_n = \{j(n) \leqq \tfrac{1}{2}(a+1)n;\ |T_{j(n)}| \leqq \varepsilon(n\log\log n)^{\frac{1}{2}}\}.$$

Then, $\Pr(A_n \text{ i.o.}) = 1$, for $n^{-1}j(n) \xrightarrow{\text{a.s.}} a$ as $n \to \infty$ and consequently,

$$\Pr(j(n) > \tfrac{1}{2}(a+1)n;\ |T_{j(n)}| \leqq \varepsilon(n\log\log n)^{\frac{1}{2}} \text{ i.o.}) = 0.$$

("a.s." denotes almost sure convergence). Furthermore,

$$\begin{aligned}\Pr[(n\log\log n)^{-\frac{1}{2}}|V_n| &\geqq N \text{ i.o.}]\\ &= \Pr(|T_{j(n)} + U_{k(n)}| \geqq N(n\log\log n)^{\frac{1}{2}} \text{ i.o.})\\ &= \Pr(A_n;\ |T_{j(n)} + U_{k(n)}| \geqq N(n\log\log n)^{\frac{1}{2}} \text{ i.o.})\\ &\geqq \Pr(A_n;\ |U_{k(n)}| \geqq (N+\varepsilon)(n\log\log n)^{\frac{1}{2}} \text{ i.o.}).\end{aligned}$$

Also, when A_n holds, $k(n) \geqq \frac{1}{2}nb$, and given a small positive η, we have $(1+\eta)\log\log k(n) > \log\log n$ for n sufficiently large. Thus,

$$\Pr(|V_n| \geqq N(n\log\log n)^{\frac{1}{2}} \text{ i.o.})$$
(1)
$$\geqq \Pr(A_n; |U_{k(n)}| \geqq (N+\varepsilon)[(1+\eta)2b^{-1}k(n)\log\log k(n)]^{\frac{1}{2}} \text{ i.o.}),$$

and the problem is to show that this right right hand probability is unity.

Let $\sigma^2 = \operatorname{var} Z_i = EZ_i^2$. Then, using the law of the iterated logarithm,

$$\Pr(|U_{k(n)}| \geqq (N+\varepsilon)[(1+\eta)2b^{-1}k(n)\log\log k(n)]^{\frac{1}{2}} \text{ i.o.}) = 1$$

provided $(N+\varepsilon)^2(1+\eta) < \sigma b$. In order to proceed further, we show that $(n\log\log n)^{-\frac{1}{2}}T_n \xrightarrow{P} 0$ as $n \to \infty$. ("P" denotes convergence in probability).

Firstly, we recall that $\sum \Pr[|X| > (n\log\log n)^{\frac{1}{2}}] < \infty$, and consequently $n\Pr[|X| > (n\log\log n)^{\frac{1}{2}}] \to 0$ as $n \to \infty$. Thus, using the degenerate convergence criterion (Loève [2], 317), we shall have $(n\log\log n)^{-\frac{1}{2}} S_n \xrightarrow{P} 0$ provided we can show that

(i) $$(\log\log n)^{-1}\int_{|x| \leqq (n\log\log n)^{\frac{1}{2}}} x^2\, dF(x) \to 0 \qquad \text{as } n \to \infty,$$

and

(ii) $$n^{\frac{1}{2}}(\log\log n)^{-\frac{1}{2}}\int_{|x| \leqq (n\log\log n)^{\frac{1}{2}}} x\, dF(x) \to 0 \qquad \text{as } n \to \infty.$$

For $n \geqq 3$ we have, using integration by parts,

$$\begin{aligned}
\int_{|x| \leqq (n\log\log n)^{\frac{1}{2}}} x^2 dF(x) &= -\int_0^{(n\log\log n)^{\frac{1}{2}}} x^2 d\Pr(|X| > x) \\
&\leqq 2\int_0^{(n\log\log n)^{\frac{1}{2}}} x\Pr(|X| > x)\,dx \\
&= 2\int_0^{(3\log\log 3)^{\frac{1}{2}}} x\Pr(|X| > x)\,dx \\
&\quad + 2\sum_{k=3}^{n-1}\int_{[k\log\log k]^{\frac{1}{2}}}^{[(k+1)\log\log(k+1)]^{\frac{1}{2}}} x\Pr(|X| > x)\,dx \\
&\leqq 2\int_0^{(3\log\log 3)^{\frac{1}{2}}} x\Pr(|X| > x)\,dx \\
&\quad + \sum_{k=3}^{n-1}\Pr[|X| > (k\log\log k)^{\frac{1}{2}}][(k+1)\log\log(k+1) - k\log\log k] \\
&\leqq 2\int_0^{(3\log\log 3)^{\frac{1}{2}}} x\Pr(|X| > x)dx + C\sum_{k=3}^{n-1}\log\log k\Pr[|X| > (k\log\log k)^{\frac{1}{2}}]
\end{aligned}$$

for some positive constant C. Consequently, since $\sum \Pr[|X| > (n\log\log n)^{\frac{1}{2}}) < \infty$, we have from the Kronecker Lemma (e.g. Loève [2], 238) that

$$(\log\log n)^{-1}\int_{|x|\leqq(n\log\log n)^{\frac{1}{2}}} x^2\,dF(x)\to 0 \quad \text{as} \quad n\to\infty,$$

and (i) is established.

In the case of (ii) we have, since $EX = 0$,

$$\left|\int_{|x|\leqq(n\log\log n)^{\frac{1}{2}}}\right| xdF(x) = \left|\int_{|x|>(n\log\log n)^{\frac{1}{2}}} xdF(x)\right|$$

$$\leqq \int_{|x|>(n\log\log n)^{\frac{1}{2}}} |x|\,dF(x)$$

$$= (n\log\log n)^{\frac{1}{2}}\Pr[|X|>(n\log\log n)^{\frac{1}{2}}] + \int_{(n\log\log n)^{\frac{1}{2}}}^{\infty} \Pr(|X|>x)dx$$

$$= (n\log\log n)^{\frac{1}{2}}\Pr[|X|>(n\log\log n)^{\frac{1}{2}}] + \sum_{k=n}^{\infty}\int_{[k\log\log k]^{\frac{1}{2}}}^{[(k+1)\log\log(k+1)]^{\frac{1}{2}}} \Pr(|X|>x)dx$$

$$\leqq (n\log\log n)^{\frac{1}{2}}\Pr[|X|>(n\log\log n)^{\frac{1}{2}}]$$

$$+\sum_{k=n}^{\infty}\Pr[|X|>(k\log\log k)^{\frac{1}{2}}]\{[(k+1)\log\log(k+1)]^{\frac{1}{2}}-[k\log\log k]^{\frac{1}{2}}\}$$

$$\leqq (n\log\log n)^{\frac{1}{2}}\Pr[|X|>(n\log\log n)^{\frac{1}{2}}]$$

$$+ C\sum_{k=n}^{\infty} k^{-\frac{1}{2}}(\log\log k)^{\frac{1}{2}}\Pr[|X|>(k\log\log k)^{\frac{1}{2}}]$$

$$\leqq (n\log\log n)^{\frac{1}{2}}\Pr[|X|>(n\log\log n)^{\frac{1}{2}}]$$

$$+ C'n^{-\frac{1}{2}}(\log\log n)^{\frac{1}{2}}\sum_{k=n}^{\infty}\Pr[|X|>(k\log\log k)^{\frac{1}{2}}],$$

C, C' being suitably chosen positive constants. Since $\sum\Pr[|X|>(n\log\log n)^{\frac{1}{2}}] < \infty$, it follows immediately that (ii) is satisfied. We therefore have $(n\log\log n)^{-\frac{1}{2}}S_n \xrightarrow{P} 0$ and consequently that $(n\log\log n)^{-\frac{1}{2}}T_n \xrightarrow{P} 0$.

Next we return to the main problem. Define events B_n by

$$B_n = \{|U_{k(n)}| \geqq (N+\varepsilon)[(1+\eta)2b^{-1}k(n)\log\log k(n)]^{\frac{1}{2}}\}.$$

Then from (1), our problem is to show that $\Pr(A_n\cap B_n \text{ i.o.}) = 1$, or equivalently, that $\lim_{n\to\infty}\Pr[\bigcup_{m=n}^{\infty}(A_m\cap B_m)] = 1$.

Let $\bar{A}$ denote the complement of A. We have

$$\Pr\left[\bigcup_{m=n}^{\infty}(A_m\cap B_m)\right] = \Pr\left[\bigcup_{m=n}^{\infty}\left\{\bigcap_{l=n}^{m-1}\overline{(A_l\cap B_l)}\cap(A_m\cap B_m)\right\}\right]$$

$$= \sum_{m=n}^{\infty}\Pr\left[\bigcap_{l=n}^{m-1}\overline{(A_l\cap B_l)}\cap(A_m\cap B_m)\right]$$

$$\geqq \sum_{m=n}^{\infty} \Pr\left[\bigcap_{l=n}^{m-1} \bar{B}_l \cap B_m \cap A_m\right]$$

$$= \sum_{m=n}^{\infty} \sum_{i=1}^{\frac{1}{2}(a+1)m} \Pr\left[\bigcap_{l=n}^{m-1} \bar{B}_l \cap B_m \cap A_m \cap \{j(m) = i\}\right]$$

$$\geqq \sum_{m=n}^{\infty} \sum_{i=\delta m}^{\frac{1}{2}(a+1)m} \Pr\left[\bigcap_{l=n}^{m-1} \bar{B}_l \cap B_m \cap \{j(m) = i\}\right] \Pr\left[A_m \cap \{j(m) = i\}\right],$$

for arbitrary δ, $0 < \delta < a$, the $\{Y_i\}$ and $\{Z_i\}$ being mutually independent. Now, since $(n \log\log n)^{-\frac{1}{2}} T_n \xrightarrow{P} 0$, $\Pr[|T_n| \leqq \varepsilon(n \log\log n)^{\frac{1}{2}}] \to 1$ as $n \to \infty$. Thus, we can find an integer I and a constant $K > 0$ such that

$$\Pr[A_m \cap \{j(m) = i\}] \geqq K \qquad \text{for all } m \geqq n \geqq I,\ \delta m \leqq i \leqq \tfrac{1}{2}(a+1)m.$$

Then, for $n \geqq I$,

$$\Pr\left[\bigcup_{m=n}^{\infty} (A_m \cap B_m)\right] \geqq K \sum_{m=n}^{\infty} \sum_{i=\delta m}^{\frac{1}{2}(a+1)m} \Pr\left[\bigcap_{l=n}^{m-1} \bar{B}_l \cap B_m \cap \{j(m) = i\}\right]$$

$$= K \sum_{m=n}^{\infty} \Pr\left[\bigcap_{l=n}^{m-1} \bar{B}_l \cap B_m \cap \{\delta m \leqq j(m) \leqq \tfrac{1}{2}(a+1)m\}\right] \tag{2}$$

$$\geqq K \sum_{m=n}^{\infty} \Pr\left[\bigcap_{l=n}^{m-1} \bar{B}_l \cap B_m\right] - K \sum_{m=n}^{\infty} \Pr\left[\overline{\{\delta m \leqq j(m) \leqq \tfrac{1}{2}(a+1)m\}}\right].$$

Now

$$\begin{aligned}
\Pr\left[\overline{\{\delta m \leqq j(m) \leqq \tfrac{1}{2}(a+1)m\}}\right] &= \Pr[j(m) < \delta m] + \Pr[j(m) > \tfrac{1}{2}(a+1)m] \\
&= \Pr[k(m) > (1-\delta)m] + \Pr[j(m) > \tfrac{1}{2}(a+1)m] \\
&\leqq e^{-(1-\delta)mt_1} E[e^{t_1 k(m)}] + e^{-\frac{1}{2}(a+1)mt_2} E[e^{t_2 j(m)}] \\
&= [e^{-t_1(1-\delta)}(ae^{t_1} + b)]^m + [e^{-\frac{1}{2}(a+1)t_2}(a - be^{t_2})]^m \\
&= [\xi(t_1)]^m + [\eta(t_2)]^m,
\end{aligned}$$

say, where t_1 and t_2 are positive and chosen so small that $\xi(t_1) < 1$, $\eta(t_2) < 1$. Then,

$$\sum_{m=n}^{\infty} \Pr\left[\overline{\{\delta m \leqq j(m) \leqq \tfrac{1}{2}(a+1)m\}}\right] \leqq \frac{[\xi(t_1)]^n}{1 - \xi(t_1)} + \frac{[\eta(t_2)]^m}{1 - \eta(t_2)}$$

$$\to 0 \text{ as } n \to \infty.$$

Consequently, from (2),

$$\lim_{n\to\infty} \Pr\left[\bigcup_{m=n}^{\infty} (A_m \cap B_m)\right] \geqq K \lim_{n\to\infty} \sum_{m=n}^{\infty} \Pr\left[\bigcup_{l=n}^{m-1} (\bar{B}_l \cap B_m)\right]$$

$$= K \lim_{n\to\infty} \Pr\left[\bigcup_{m=n}^{\infty} B_n\right] = K$$

for $(N+\varepsilon)^2(1+\eta) < \sigma b$ since $\Pr(B_n \text{ i.o.}) = 1$ under this condition. It then follows, from the zero-one law, that $\Pr(|S_n| > N(n \log\log n)^{\frac{1}{2}} \text{i.o.}) = 1$ provided that $(N+\varepsilon)^2(1+\eta) < \sigma b$. However, $EX^2 = \infty$, so σ can be made arbitrarily large by increasing K and, correspondingly, b tends towards unity. ε and η are arbitrarily small so in fact the relation $\Pr(|S_n| > N(n \log\log n)^{\frac{1}{2}} \text{i.o.}) = 1$ holds no matter how large be $N > 0$. This establishes the result of the theorem.

References

[1] Hartman, P. and Wintner, A. (1941) On the law of the iterated logarithm. *Amer. J. Math.* **63**, 169–176.

[2] Loève, M. (1963) *Probability Theory*. 3rd edition. Van Nostrand, New York.

[3] Stone, C. (1966) The growth of a recurrent random walk. *Ann. Math. Statist.* **37**, 1040–1041.

[4] Strassen, V. (1966) A converse to the law of the iterated logarithm. *Z. Wahrscheinlichkeitsth.* **4**, 265–268.

Reprinted from the
PROCEEDINGS OF THE AMERICAN MATHEMATICAL SOCIETY
Vol. 23, No. 1, October, 1969
pp. 85-90

A NOTE CONCERNING BEHAVIOUR OF ITERATED LOGARITHM TYPE

C. C. HEYDE

Let $\{X_i, i=1, 2, 3, \cdots\}$ be a sequence of independent and identically distributed random variables and write $S_n=\sum_{i=1}^n X_i$, $n\geqq 1$. It is well known that

$$\Pr(\limsup_{n\to\infty}(2n \log\log n)^{-1/2}S_n = 1) = 1$$

if and only if $EX_i=0$, $EX_i^2=1$ (Hartman and Wintner [6] obtained the sufficiency part and Strassen [9] the necessity) and the purpose of this note is to clarify the corresponding situation in the case where these moment conditions are violated. It turns out that the oscillation behaviour of normed sums rests essentially on whether or not the summands belong to the domain of partial attraction of the normal distribution.

Firstly, we remark that it has been shown by Freedman (see Strassen [9]) that there exist symmetric random variables X_i with $EX_i^2=\infty$ or even $E|X_i|=\infty$, for which it is possible to choose a monotone sequence $\{b_n, n=1, 2, 3, \cdots\}$ of positive constants with $b_n\to\infty$ as $n\to\infty$ such that

$$\Pr(\limsup_{n\to\infty} b_n^{-1}S_n = 1) = 1.$$

With this in mind, we go on to establish the following theorem.

THEOREM 1. *Suppose the X_i are such that there exists a monotone sequence of positive constants $\{b_n, n=1, 2, 3, \cdots\}$ with $b_n\to\infty$ as $n\to\infty$ for which*

$$\limsup_{n\to\infty} b_n^{-1}|S_n| < \infty \tag{1}$$

with positive probability. Then, at least one of the following two conditions is satisfied:

(i) $b_n^{-1}(S_n - \operatorname{med} S_n) \xrightarrow{\text{a.s.}} 0.$

(ii) *The X_i belong to the domain of partial attraction of the normal distribution. A necessary and sufficient condition for this is (Lévy* [7, p. 113])

Received by the editors November 1, 1968.

R. Maller et al. (eds.), *Selected Works of C.C. Heyde*, Selected Works in Probability and Statistics,
DOI 10.1007/978-1-4419-5823-5_15, © Springer Science+Business Media, LLC 2010

(2) $$\liminf_{u\to\infty} \frac{u^2 \Pr(|X_1| > u)}{\int_{|x|\leqq u} x^2 d \Pr(X_1 \leqq u)} = 0.$$

("a.s." denotes almost sure convergence; "med" is an abbreviation of median).

PROOF. Firstly, since (1) is a tail event on the $S_1, S_2, \cdots$ process it has probability one using the Hewitt-Savage zero-one law (e.g. Breiman [1, Corollary 3.50]).

Next, suppose that neither (i) nor (ii) are satisfied. From (2), there exists a constant $C>0$ such that

$$u^2 \Pr(|X_1| > u) > C \int_{|x|\leqq u} x^2 d \Pr(X_1 \leqq x)$$

for $u > U$. Thus, for $u > U$ and $\eta > 1$,

$$\begin{aligned} u^2[\Pr(|X_1| > u) - \Pr(|X_1| > \eta u)] &\leqq \int_{u<|x|\leqq \eta u} x^2 d \Pr(X_1 \leqq x) \\ &\leqq u^2 C^{-1}\eta^2 \Pr(|X_1| > \eta u), \end{aligned}$$

from which we obtain

$$\Pr(|X_1| > u) \leqq (1 + C^{-1}\eta^2) \Pr(|X_1| > \eta u).$$

Thus, for n sufficiently large we have, when $\epsilon > 1$,

$$(1 + C^{-1}\epsilon^2)^{-1} \Pr(|X_1| > b_n) \leqq \Pr(|X_1| > \epsilon b_n) \leqq \Pr(|X_1| > b_n),$$

and when $\epsilon < 1$,

$$\Pr(|X_1| > b_n) \leqq \Pr(|X_1| > \epsilon b_n) \leqq (1 + C^{-1}\epsilon^{-2})\Pr(|X| > b_n),$$

so the convergence (divergence) of $\sum \Pr(|X_1| > \epsilon b_n)$ for some $\epsilon > 0$ implies convergence (divergence) for every $\epsilon > 0$.

If $\sum \Pr(|X_1| > \epsilon b_n) = \infty$ for every $\epsilon > 0$, we have from the Borel-Cantelli Lemmas that

(3) $$\Pr(|X_n| > \epsilon b_n \text{ i.o.}) = 1$$

for every $\epsilon > 0$ ("i.o." is an abbreviation of infinitely often). Furthermore, $|S_n - S_{n-1}| = |X_n| \leqq |S_n| + |S_{n-1}|$, so that (3) implies $\Pr(|S_n| > \frac{1}{2}\epsilon b_n \text{ i.o.}) = 1$ for every $\epsilon > 0$ which contradicts the condition (1).

Finally, suppose that $\sum \Pr(|X_1| > \epsilon b_n) < \infty$, every $\epsilon > 0$. We introduce a sequence of random variables $\{Y_k, k = 1, 2, 3, \cdots\}$ defined by

$$\begin{aligned} Y_k &= X_k \qquad \text{if } |X_k| \leqq b_k \\ &= 0 \qquad \text{if } |X_k| > b_k, \end{aligned}$$

and write $T_n = \sum_{k=1}^n Y_k$. Then,

$$\sum \Pr(Y_k \neq X_k) = \sum \Pr(|X_1| > b_k) < \infty,$$

from which it follows that

$$b_n^{-1}(S_n - T_n) \xrightarrow{\text{a.s.}} 0 \tag{4}$$

as $n \to \infty$. Also, for n sufficiently large,

$$EY_n^2 \leqq Cb_n^2 \Pr(|X_1| > b_n),$$

and hence $\sum b_n^{-2} EY_n^2 < \infty$. It then follows from a theorem of Kolmogorov (e.g. Loève [**8**, p. 238]) that

$$b_n^{-1}(T_n - ET_n) \xrightarrow{\text{a.s.}} 0$$

as $n \to \infty$ and consequently, from (4), that

$$b_n^{-1}(S_n - ET_n) \xrightarrow{\text{a.s.}} 0,$$

or equivalently (e.g. Loève [**8**, p. 247]),

$$b_n^{-1}(S_n - \text{med } S_n) \xrightarrow{\text{a.s.}} 0.$$

This again provides a contradiction and completes the proof of the theorem.

In view of the important condition (ii) of Theorem 1, it is interesting to investigate the oscillation behaviour of sums when the summands do not satisfy condition (ii). Unfortunately, it turns out that centering complications prevent us from giving a complete answer in the general case. These complications can, of course, be avoided by working with symmetric random variables and then it is possible to give a full explanation of the behaviour. This is what we shall do in the next theorem.

THEOREM 2. *Suppose the X_i are symmetric and that they do not belong to the domain of partial attraction of the normal distribution. If $\{b_n, n = 1, 2, 3, \cdots\}$ is a monotone sequence of positive constants with $b_n \to \infty$ as $n \to \infty$, then for any $\epsilon > 0$, $\Pr(|S_n| > \epsilon b_n \text{ i.o.}) = 0$ or 1 according as $\sum \Pr(|X_1| > b_n)$ converges or diverges. Equivalently, $b_n^{-1} S_n \xrightarrow{\text{a.s.}} 0$ as $n \to \infty$ if and only if $\sum \Pr(|X_1| > b_n)$ converges.*

The proof of this theorem involves only minor modifications of the proof of Theorem 1 and will be omitted. The result is closely related to those of Feller [**3**].

REMARK. By way of comparison with the result of Theorem 2, it is worth noting that under the same conditions on the X_i, $b_n^{-1}S_n \xrightarrow{p} 0$ if and only if $n \Pr(|X_1| > \epsilon b_n) \to 0$ as $n \to \infty$ for every $\epsilon > 0$ ("p" denotes convergence in probability). This follows simply from the degenerate convergence criterion (e.g. Loève [**8**, p.317]) upon making use of the Lévy condition (2).

If further structural conditions are imposed on the X_i, it is often possible to investigate the matter of centering of the S_n process in sufficient detail to enable the symmetry assumption of Theorem 2 to be dispensed with while still obtaining the same behaviour. As an example of the possibilities, we have the following theorem.

THEOREM 3. *Suppose that the X_i belong to the domain of attraction of a nonnormal stable law of index $\alpha \neq 1$ which is not one-sided and that $EX_i = 0$ if $E|X_i| < \infty$. If $\{b_n, n = 1, 2, 3, \cdots\}$ is a monotone sequence of positive constants with $b_n \to \infty$ as $n \to \infty$, then for any $\epsilon > 0$, $\Pr(|S_n| > \epsilon b_n \text{ i.o.}) = 0$ or 1 according as $\sum \Pr(|X_1| > b_n)$ converges or diverges. Equivalently, $b_n^{-1}S_n \xrightarrow{\text{a.s.}} 0$ as $n \to \infty$ if and only if $\sum \Pr(|X_1| > b_n)$ converges.*

PROOF. Again we can make use of a slightly modified version of the proof of Theorem 1 and in order to complete the proof of Theorem 3 it is just necessary to show that $b_n^{-1}ET_n \to 0$ as $n \to \infty$ where T_n is defined as in the proof of Theorem 1.

Firstly we note that under the conditions of the theorem we may write for $x > 0$, using Gnedenko and Kolmogorov [**5**, Theorem 2, p. 175],

$$\Pr(|X_1| > x) = M(x)x^{-\alpha},$$

where $M(x)$ is a function which is slowly varying as $x \to \infty$. It is necessary to treat separately the cases $\alpha < 1$ and $1 < \alpha < 2$.

If $\alpha < 1$,

$$|EY_n| \leqq E|Y_n| = \int_{|x| \leqq b_n} |x| \, d \Pr(X_1 \leqq x)$$
$$\leqq \int_0^{b_n} \Pr(|X_1| > x)dx,$$

using integration by parts and, using standard properties of slowly varying functions (e.g. Feller [**4**, p. 273]), we have

$$\frac{\int_0^{b_n} \Pr(|X_1| > x)dx}{b_n \Pr(|X_1| > b_n)} \to \frac{1}{1-\alpha}$$

as $n \to \infty$ so that we can choose a constant $c_1 > 0$ such that

$$|EY_n| \leqq c_1 b_n \Pr(|X_1| > b_n) \tag{5}$$

for each n.

If, on the other hand, $1 < \alpha < 2$, we have since $EX_1 = 0$,

$$\begin{aligned} |EY_n| &= \left| \int_{|x| \leqq b_n} xd \Pr(X_1 \leqq x) \right| \\ &= \left| \int_{|x| > b_n} xd \Pr(X_1 \leqq x) \right| \\ &\leqq \int_{|x| > b_n} |x| \, d \Pr(X_1 \leqq x) \\ &= -\int_{b_n}^{\infty} xd \Pr(|X_1| > x) \\ &= b_n \Pr(|X_1| > b_n) + \int_{b_n}^{\infty} \Pr(|X_1| > x)dx, \end{aligned}$$

and

$$\frac{\int_{b_n}^{\infty} \Pr(|X_1| > x)dx}{b_n \Pr(|X_1| > b_n)} \to \frac{1}{\alpha - 1}$$

as $n \to \infty$, so that we can choose a constant $c_2 > 0$ such that

$$|EY_n| \leqq c_2 b_n \Pr(|X_1| > b_n). \tag{6}$$

Take $c = \max(c_1, c_2)$. Then, from (5) and (6) we have in general that

$$\sum_{n=1}^{\infty} b_n^{-1} |EY_n| \leqq c \sum_{n=1}^{\infty} \Pr(|X_1| > b_n) < \infty, \tag{7}$$

and the result $b_n^{-1} ET_n \to 0$ follows from an application of the Kronecker Lemma to (7). This completes the sketch of the proof.

Theorems 2 and 3 provide a convenient tool for obtaining results of the type of Chover [2]. We immediately obtain from Theorem 3, for example, the following extension of the result of [2].

COROLLARY 1. *Suppose that the X_i belong to the domain of normal attraction of a nonnormal stable law of index $\alpha \neq 1$ which is not one-sided and that $EX_i = 0$ if $E|X_i| < \infty$. Then,*

$$\Pr(\limsup_{n\to\infty} |n^{-\alpha^{-1}} S_n|^{(\log\log n)^{-1}} = e^{\alpha^{-1}}) = 1.$$

(Note that in the case of normal attraction, there is a constant $c > 0$ such that $\Pr(|X_1| > x) \sim cx^{-\alpha}$ as $x \to \infty$.)

REFERENCES

1. L. Breiman, *Probability*, Addison-Wesley, Reading, Mass., 1968.

2. J. Chover, *A law of the iterated logarithm for stable summands*, Proc. Amer. Math. Soc. **17** (1966), 441–443.

3. W. Feller, *A limit theorem for random variables with infinite moments*, Amer. J. Math. **68** (1946), 257–262.

4. ———, *An introduction to probability theory and its applications.* II, Wiley, New York, 1966.

5. B. V. Gnedenko and A. N. Kolmogorov, *Limit distributions for sums of independent random variables*, Addison-Wesley, Reading, Mass., 1954.

6. P. Hartman and A. Wintner, *On the law of the iterated logarithm*, Amer. J. Math. **63** (1941), 169–176.

7. P. Lévy, *Théorie de l'addition des variables aléatoires*, 2nd ed., Gauthier-Villars, Paris, 1954.

8. M. Loève, *Probability theory*, 3rd ed., Van Nostrand, New York, 1963.

9. V. Strassen, *A converse to the law of the iterated logarithm*, Z. Wahrscheinlichkeitstheorie **4** (1966), 265–268.

AUSTRALIAN NATIONAL UNIVERSITY

The Annals of Mathematical Statistics
1969, Vol. 40, No. 6, 2178–2179

ON EXTENDED RATE OF CONVERGENCE RESULTS FOR THE INVARIANCE PRINCIPLE

BY C. C. HEYDE
Australian National University

Let $\{X_i, i = 1, 2, 3, \cdots\}$ be a sequence of independent and identically distributed random variables for which $EX_i = 0$, $\operatorname{var} X_i = \sigma^2$, $E|X_i|^{2+a} = b < \infty$, $a > 0$. $C[0, 1]$ is the space of continuous real-valued functions on the interval $[0, 1]$ with the sup norm topology. We define a "random broken line" as follows:

$$\xi_n(t) = \textstyle\sum_{i=1}^{k} X_i/(\sigma n^{\frac{1}{2}}) \qquad \text{for } t = n^{-1}k$$

$$\xi_n(t) = \xi_n(n^{-1}k) + n[\xi_n(n^{-1}(k+1)) - \xi_n(n^{-1}k)](t - n^{-1}k)$$

$$\text{for } n^{-1}k \leqq t \leqq n^{-1}(k+1),\ 1 \leqq k \leqq n.$$

THEOREM. *Let g be any uniformly continuous functional, $g: C[0, 1] \to$ real line, such that there exists a constant $L > 0$ with the property that*

$$|\Pr\{g(\xi(t)) \leqq x + h\} - \Pr\{g(\xi(t)) \leqq x\}| \leqq L|h|$$

where $\xi(t)$ is the standard Wiener process. Then, there exists a constant $A > 0$ such that for all $n > 1$,

$$(1) \qquad |\Pr\{g(\xi_n(t)) \leqq x\} - \Pr\{g(\xi(t)) \leqq x\}| \leqq A(\log n)^{\lambda} n^{-\mu},$$

where $\lambda = (1 + \frac{1}{2}a)/(a+3)(<\frac{1}{2})$, $\mu = \min(a, 1 + \frac{1}{2}a)/2(a+3)$. This result extends that of Theorem 5 of Rosenkrantz [3] where the bound (1) is, in essence, given in the form

$$|\Pr\{g(\xi_n(t)) \leqq x\} - \Pr\{g(\xi(t)) \leqq x\}| \leqq A(\log n)^{\frac{1}{2}} n^{-\mu}$$

subject to the restriction that $a \leqq 2$. Only minor modifications of the work of Rosenkrantz and necessary to furnish this extension; we shall just indicate these and refer the reader to [3] for full details of the proof.

The condition $a \leqq 2$ was imposed by Rosenkrantz as a consequence of employing a moment inequality of von Bahr and Esseen [1] which is valid for exponent r, $1 \leqq r \leqq 2$ (note that $2r = 2 + a$). This is used to obtain the vital inequality (25) of [3]. For the case $a > 2$ ($r > 2$), we make use of a result of Dharmadhikari, Fabian and Jogdeo [2] which gives in place of (25),

$$(2) \qquad E|z_{nn}|^r \leqq C_r n^{\frac{1}{2}} \nu_n,$$

where $C_r = [8(r-1)\max(1, 2^{r-3})]^r$.

Thus, combining (2) with (25) of [3], we can replace the estimate (27) of [3]

Received 14 February 1969.

R. Maller et al. (eds.), *Selected Works of C.C. Heyde*, Selected Works in Probability and Statistics, DOI 10.1007/978-1-4419-5823-5_16, © Springer Science+Business Media, LLC 2010

by the general estimate

$$\Pr(\max_{1 \leqq k \leqq n} |z_{nk}| > \delta) \leqq \max(2, C_r) n^{-a(r)} \delta^{-r} \tag{3}$$

where $a(r) = \min(r - 1, \frac{1}{2}r)$, $r > 1$.
The details of proof of [3] are then followed exactly up to the stage of choosing ϵ_n and δ_n tending to zero at appropriate rates. Instead of the versions chosen in (37), we employ

$$\delta_n = n^{-2a(r)(2r+1)^{-1}} (\beta \log n)^{-(2r+1)^{-1}}, \qquad \epsilon_n^2 = 2\beta\delta_n \log n,$$

where $\beta > 1 + a(r)/(2r + 1)$. The result of the theorem then follows as with [3].

REFERENCES

[1] Bahr, B. von and Esseen, C. G. (1965). Inequalities for the rth absolute moment of a sum of random variables, $1 \leqq r \leqq 2$. *Ann. Math. Statist.* **36** 299–303.

[2] Dharmadhikari, S. W., Fabian, V. and Jogdeo, K. (1968). Bounds on the moments of martingales. *Ann. Math. Statist.* **39** 1719–1723.

[3] Rosenkrantz, W. A. (1967). On rates of convergence for the invariance principle. *Trans. Amer. Math. Soc.* **129** 542–552.

J. Appl. Prob. **6**, 419–429 (1969)
Printed in Israel

ON THE MAXIMUM OF SUMS OF RANDOM VARIABLES AND THE SUPREMUM FUNCTIONAL FOR STABLE PROCESSES

C. C. HEYDE*, *University of Manchester*

1. Introduction

Let X_i, $i = 1,2,3,\cdots$ be a sequence of independent and identically distributed random variables which belong to the domain of attraction of a stable law of index α. Write $S_0 = 0$, $S_n = \sum_{i=1}^{n} X_i, n \geqq 1$, and $M_n = \max_{0 \leqq k \leqq n} S_k$. In the case where the X_i are such that $\sum_1^\infty n^{-1}\Pr(S_n > 0) < \infty$, we have $\lim_{n\to\infty} M_n = M$ which is finite with probability one, while in the case where $\sum_1^\infty n^{-1}\Pr(S_n < 0) < \infty$, a limit theorem for M_n has been obtained by Heyde [9]. The techniques used in [9], however, break down in the case $\sum_1^\infty n^{-1}\Pr(S_n < 0) = \infty, \sum_1^\infty n^{-1}\Pr(S_n > 0) = \infty$ (the case of oscillation of the random walk generated by the S_n) and the only results available deal with the case $\alpha = 2$ (Erdös and Kac [5]) and the case where the X_i themselves have a symmetric stable distribution (Darling [4]). In this paper we obtain a general limit theorem for M_n in the case of oscillation. Specifically, if $\{B_n, n = 1,2,3,\cdots\}$ is a monotone sequence of constants such that $B_n^{-1}S_n$ converges in distribution to the stable law with characteristic function

$$\exp\left\{-\lambda|t|^\alpha\left(1 + i\beta\,\mathrm{sgn}\,t\tan\frac{\pi\alpha}{2}\right)\right\}, \tag{1}$$

$\lambda > 0, 0 < \alpha \leqq 2, \beta = 0$ if $\alpha = 1$, $|\beta| < 1$ if $\alpha < 1$, $|\beta| \leqq 1$ if $1 < \alpha \leqq 2$, we shall find

$$H(x) = \lim_{n\to\infty} \Pr(B_n^{-1}M_n \leqq x).$$

In connection with the parameter restrictions, we note that the stable law with characteristic function (1) is one-sided if $\alpha < 1$, $|\beta| = 1$ (e.g., Lukacs [12], page 106) so that the random walk generated by the S_n does not oscillate ([9], Lemma). The case $\alpha = 1$, $\beta \neq 0$ introduces a normalization complication and is not amenable to treatment by the methods of this paper.

It is possible to approach the problem of finding $H(x)$ in various ways. For, if $Y(t)$, $t \geqq 0$, $Y(0) = 0$, is the separable stable process with stationary independent increments which is based on (1), then also,

Received in revised form 26 August 1968.
* Now at the Australian National University.

R. Maller et al. (eds.), *Selected Works of C.C. Heyde*, Selected Works in Probability and Statistics,
DOI 10.1007/978-1-4419-5823-5_17, © Springer Science+Business Media, LLC 2010

$$H(x) = \Pr(\sup_{0 \leqq t \leqq 1} Y(t) < x).$$

The double Laplace transform of $\Pr(\sup_{0 \leqq u \leqq t} Y(u) < x)$ has, in this context, been obtained by Baxter and Donsker [1]. Their results are, however, not in a sufficiently explicit form to enable any of the properties of $H(x)$ to be deduced. In the present work, we make a more detailed investigation leading to considerably more illuminating results.

2. The supremum functional for a stable process

In this section, we shall deal with the case where the X_i themselves have a stable distribution with characteristic function (1) (and, of course, the specified parameter restrictions). This will be followed in the next section by an invariance theorem to establish the generality of the limiting distribution found in the present context.

Under the present circumstances, it is a simple matter to establish the existence of a limiting distribution for $n^{-1/\alpha}M_n$. In fact, if $Y(t)$, $t \geqq 0$, $Y(0) = 0$, is the separable stable process with stationary independent increments which is characterised by

$$E\exp\{iuY(T)\} = \exp\left\{ -T\lambda|u|^\alpha\left(1 + i\beta \operatorname{sgn} u \tan\frac{\pi\alpha}{2}\right)\right\},$$

and if we take

$$X_k = \left[Y\left(\frac{k}{2^N}\right) - Y\left(\frac{k-1}{2^N}\right)\right] 2^{N\alpha^{-1}}, \qquad k = 1, 2, \cdots, 2^N,$$

then it follows readily from Lemma 2 of Baxter and Donsker [1], that

$$\lim_{n\to\infty} \Pr(n^{-1/\alpha} M_n < x) = H(x), \tag{2}$$

where

$$H(x) = \Pr\left(\sup_{0 \leqq t \leqq 1} Y(t) < x\right) \tag{3}$$

and also

$$\Pr\left(\sup_{0 \leqq t \leqq T} Y(t) < x\right) = H(xT^{-1/\alpha}).$$

We shall proceed, using methods of Darling [4], to obtain an expression for $H(x)$.

For s real and positive, let

$$\begin{aligned}\phi_n(s) &= \Pr(S_n \leqq 0) + E(\exp\{-s Sn\};\, S_n > 0)\\ &= \Pr(X \leqq 0) + E(\exp\{-sn^{1/\alpha}X\};\, X > 0),\end{aligned}$$

since $n^{-1/\alpha} S_n$ and X_i have the same distribution. Then, using Theorem 4 of Zolotarev [18], we obtain the unilateral Laplace transform

(4) $$E(e^{-sX}; X > 0) = \frac{1}{\pi} \int_0^\infty \exp\{-(sx)^\alpha \lambda_1\} \frac{\sin \pi\rho}{x^2 + 2x\cos\pi\rho + 1} dx,$$

where

(5) $$\lambda_1 = \lambda\left(1+\beta^2 \tan^2 \frac{\pi\alpha}{2}\right)^{1/2}, \rho = \tfrac{1}{2} + \frac{1}{\pi\alpha} \arctan\left[-\beta \tan\frac{\pi\alpha}{2}\right] = \Pr(X > 0),$$

(the result $\rho = \Pr(X > 0)$ was obtained by Chung-Teh [2]), so that for $0 \leqq t < 1$,

$$\sum_1^\infty n^{-1} t^n \phi_n(s)$$
$$= -(1-\rho)\log(1-t) - \frac{\sin\pi\rho}{\pi} \int_0^\infty \log[1 - t\exp\{-(sx)^\alpha \lambda_1\} \frac{dx}{x^2+2x\cos\pi\rho+1}.$$

But, using (4),

$$\frac{\sin\pi\rho}{\pi} \int_0^\infty \frac{dx}{x^2 + 2x\cos\pi\rho + 1} = \rho,$$

and therefore,

$$\log(1-t) + \sum_1^\infty n^{-1} t^n \phi_n(s)$$
$$= -\frac{\sin\pi\rho}{\pi} \int_0^\infty \log\left[\frac{1 - t\exp\{-(sx)^\alpha \lambda_1\}}{1-t}\right] \frac{dx}{x^2 + 2x\cos\pi\rho + 1}.$$

Consequently, as $t \uparrow 1$,

$$\log(1-t) + \sum_1^\infty n^{-1} t^n \phi_n[(1-t)^{1/\alpha} s] \to -\frac{\sin\pi\rho}{\pi} \int_0^\infty \frac{\log(1 + s^\alpha x^\alpha \lambda_1)}{x^2 + 2x\cos\pi\rho + 1} dx,$$

and, using the Spitzer-Pollaczek Identity (see for example Prabhu [13], page 218, Theorem 4.1),

(6) $$\lim_{t\uparrow 1} (1-t) \sum_0^\infty E(\exp\{-s(1-t)^{1/\alpha}\} M_n) = g(s),$$

where

(7) $$g(s) = \exp\left\{-\frac{\sin\pi\rho}{\pi} \int_0^\infty \frac{\log(1 + s^\alpha x^\alpha \lambda_1)}{x^2 + 2x\cos\pi\rho + 1} dx\right\}.$$

Precisely the argument of Darling [4] then yields the relation

$$(8) \qquad A(s) = \frac{G(1-s)}{\Gamma(1-s)\Gamma(1-\alpha^{-1}+s\alpha^{-1})},$$

$|\mathscr{R}s|$ sufficiently small, between the Mellin transforms,

$$A(s) = \int_0^\infty x^{s-1} dH(x),$$

$$G(s) = \int_0^\infty x^{s-1}\, g(x)\, dx.$$

As with [4], it follows that $H(x)$ is an absolutely continuous distribution function and upon inverting in (8) we obtain the density function

$$\frac{d}{dx}H(x) = \frac{1}{2\pi i}\int_{c-i\infty}^{c+i\infty} \frac{G(1-s)x^{-s}}{\Gamma(1-s)\Gamma(1-\alpha^{-1}+s\alpha^{-1})}ds,$$

(see for example Widder [17], pages 246, 247). We have thus established the following theorem.

Theorem 1. If $0<\alpha\leqq 2$ and $\beta=0$ if $\alpha=1$, $|\beta|<1$ if $\alpha<1$, $|\beta|\leqq 1$ if $1<\alpha\leqq 2$, then

$$(9) \qquad \frac{d}{dx}H(x) = \frac{1}{2\pi i}\int_{c-i\infty}^{c+i\infty} \frac{G(1-s)x^{-s}}{\Gamma(1-s)\Gamma(1-\alpha^{-1}+s\alpha^{-1})}ds,$$

where

$$G(s) = \int_0^\infty x^{s-1} g(x) dx,$$

$g(x)$ being given by (7).

The expression in (9) does not in general simplify conveniently. We shall go on to examine an alternative method which is illuminating in its own right and has the advantage of providing more recognisable results in certain particular cases. This is based on the representation of M_n in the form

$$M_n = 0 \text{ if } R_n = 0$$

$$M_n = Z_1 + Z_2 + \cdots + Z_{R_n}, \text{ if } R_n \geqq 1,$$

where the Z_i are successive strong ascending ladder steps for the random walk generated by the S_n (that is, Z_i has the distribution of the minimum S_k with $S_k > 0$) and R_n is the number of strong ascending ladder indices in the first n steps of the random walk.

Firstly, we need to investigate the distribution of R_n. We easily see that the occurrence of a strong ascending ladder index is a recurrent event in the usual sense of Feller. This we shall denote by $\mathscr{E}$, so that R_n is the number of occurrences of $\mathscr{E}$ in n steps. R_n can be expressed as a sum of indicator random variables in the following way:

$$R_n = \sum_{k=1}^{n} \delta_k,$$

where

$$\delta_k = \begin{cases} 1 & \text{if } \mathscr{E} \text{ occurs at the } k\text{th step,} \\ 0 & \text{otherwise,} \end{cases}$$

and we can study R_n via the δ_k. If N_n is the number of positive terms in the sequence $S_1, S_2, \cdots, S_n$,

$$\begin{aligned} \Pr(\delta_k = 1) &= \Pr(S_k > 0, \quad S_k > S_1, \cdots, S_k > S_{k-1}) \\ &= \Pr(S_k > 0, \quad \sum_{2}^{k} X_i > 0, \cdots, X_k > 0) \qquad (10) \\ &= \Pr(N_k = k), \end{aligned}$$

the X_i being identically distributed. We now define sequences $\{u_n, n \geqq 0\}$, $\{f_n, n \geqq 1\}$, as follows:

$$\begin{aligned} u_0 &= 1, \quad u_n = \Pr(\delta_n = 1), \quad n \geqq 1, \\ f_n &= \Pr(\delta_1 = 0, \ \delta_2 = 0, \cdots, \delta_{n-1} = 0, \ \delta_n = 1), \qquad n \geqq 1, \end{aligned}$$

and introduce the generating functions

$$U(t) = \sum_{n=0}^{\infty} u_n t^n, \quad F(t) = \sum_{n=1}^{\infty} f_n t^n, \quad 0 \leqq t < 1.$$

These, of course, satisfy the standard identity of recurrent event theory,

$$U(t) = [1 - F(t)]^{-1}, \qquad 0 \leqq t < 1. \qquad (11)$$

Then, using a well-known result of Sparre-Andersen (see for example Spitzer [16], page 219) together with (5) and (10), we have

$$U(t) = \exp\left\{\sum_{1}^{\infty} t^k k^{-1} \Pr(X > 0)\right\} = (1 - t)^{-\rho}. \qquad (12)$$

Lemma. The recurrence time distribution of $\mathscr{E}$ belongs to the domain of attraction of the stable law of index $1 - \rho$.

Proof. If $q_n = \sum_{r=n+1}^{\infty} f_r$, we have

$$(1-t)^{-1}[1-F(t)] = \sum_{n=0}^{\infty} q_n t^n = (1-t)^{-(1-\rho)},$$

using (11) and (12), so that

$$(1-t)^{1-\rho} \sum_{n=0}^{\infty} q_n t^n = 1,$$

and applying a standard Tauberian theorem (see for example Feller [7], page 423, Theorem 5),

$$q_n \sim \frac{1}{\Gamma(1-\rho)} n^{-\rho}$$

as $n \to \infty$. The result of the lemma then follows immediately by appeal to Theorem la, page 303 of Feller [7]. Then, using the lemma in conjunction with Theorem 7 of Feller [6], we readily obtain:

Theorem 2. Let G_ρ be the distribution function of the stable law whose characteristic function is given by

$$g_\rho(t) = \exp\left\{ -|t|^\rho \left(\cos\frac{\pi\rho}{2} - i \sin\frac{\pi\rho}{2} \operatorname{sgn} t \right) \right\},$$

where $\rho = \Pr(X > 0)$. Then,

(13) $$\lim_{n\to\infty} \Pr(n^{-\rho} R_n \leqq x) = 1 - G_\rho(x^{-1/\rho}), \qquad x \geqq 0.$$

Remark. Only in the case $\rho = \frac{1}{2}$ is it possible to give a convenient alternative form for the limit distribution in (13). We have, making use of a result due to Lévy (see for example Lukacs [12], page 107),

$$\frac{d}{dx} G_{\frac{1}{2}}(x) = \tfrac{1}{2}\pi^{-\frac{1}{2}} x^{-\frac{3}{2}} \exp\{-1/4x\}, \qquad x > 0,$$

so that

(14) $$\frac{d}{dx}[1 - G_{\frac{1}{2}}(x^{-2})] = \pi^{-\frac{1}{2}} \exp\{-1/4x^2\}.$$

Thus, the limit distribution is a truncated normal.

Next, we need to examine the limit properties of the Z_i. We have (see for example Prabhu [13], page 210, Theorem 2.2)

$$1 - Ee^{-sZ} = \exp\left\{ - \sum_1^\infty n^{-1} E(\exp\{-sSn\}; S_n > 0)\right\}$$

$$= \exp\left\{ - \sum_1^\infty n^{-1} E(\exp\{-sn^{1/\alpha}X\}; X > 0)\right\}$$

$$= \exp\left\{\frac{\sin \pi\rho}{\pi} \int_0^\infty \log\left[1 - \exp\{-(sx)^\alpha \lambda_1\}\right] \frac{dx}{x^2 + 2x\cos\pi\rho + 1}\right\},$$

using (4). Furthermore, if

$$I(s) = \int_0^\infty \log[1 - \exp\{-(sx)^\alpha\lambda_1\}] \frac{dx}{x^2 + 2x\cos\pi\rho + 1},$$

then we can write $I(s) = I_1(s) + I_2(s)$, where

$$I_1(s) = \int_0^\infty \frac{\log b(sx)}{x^2 + 2x\cos\pi\rho + 1} dx, \quad I_2(s) = \int_0^\infty \frac{\log[(sx)^\alpha \lambda_1]}{x^2 + 2x\cos\pi\rho + 1} dx,$$

and $b(u) = [1 - \exp\{-u^\alpha\lambda_1\}]/(u^\alpha\lambda_1)$. Then, $I_1(s) \to 0$ as $s \to 0$ using dominated convergence. Also,

$$\int_0^\infty \frac{dx}{x^2 + 2x\cos\pi\rho + 1} = \frac{\pi\rho}{\sin\pi\rho}, \quad \int_0^\infty \frac{\log x\, dx}{x^2 + 2x\cos\pi\rho + 1} = 0,$$

using (4) and Gradshteyn and Ryzhik [8], page 533, respectively, so that

$$I_2(s) = \frac{\pi\rho}{\sin\pi\rho}(\log\lambda_1 + \alpha\log s)$$

and hence

$$s^{-\alpha\rho}(1 - Ee^{-sZ}) = s^{-\alpha\rho}\exp\left\{\frac{\sin\pi\rho}{\pi}[I_1(s) + I_2(s)]\right\}$$

$$= \lambda_1^\rho \exp\left\{\frac{\sin\pi\rho}{\pi} I_1(s)\right\} \tag{15}$$

$$\to \lambda_1^\rho = \left[\lambda\left(1 + \beta^2 \tan^2\frac{\pi\alpha}{2}\right)^{\frac{1}{2}}\right]^{\Pr(X>0)}$$

as $s \to 0$. Now it is easily checked from (5) that $\alpha\rho \leqq 1$, with equality if and only if $1 < \alpha \leqq 2$ and $\beta = 1$ when $\alpha < 2$. Thus, if $0 < \alpha \leqq 1$ or $1 < \alpha < 2$ and $-1 \leqq \beta < 1$, $n^{-1/\alpha\rho}(Z_1 + \cdots + Z_n)$ converges in distribution to the stable law with characteristic function

$$\exp\left\{-\lambda_1^\rho |t|^{\alpha\rho}\left(1 - i\,\mathrm{sgn}\,t \tan\frac{\pi\alpha\rho}{2}\right)\right\},$$

while if $1 < \alpha \leqq 2$ and $\beta = 1$ when $\alpha < 2$, we have $\alpha\rho = 1$ so that

$$s^{-1}(1 - Ee^{-sZ}) \to \left[\lambda\left(1 + \tan^2 \frac{\pi\alpha}{2}\right)^{\frac{1}{2}}\right]^{1/\alpha} = \left[\lambda \sec\left(\pi - \frac{\pi\alpha}{2}\right)\right]^{1/\alpha},$$

and therefore,

$$EZ = \left[\lambda \sec\left(\pi - \frac{\pi\alpha}{2}\right)\right]^{1/\alpha} < \infty. \tag{16}$$

These results clarify those of Rogozin [15], where it is not possible to calculate specifically the constant in (15). We are now in a position to establish the following limiting result.

Theorem 3. If $1 < \alpha \leqq 2$ and $\beta = 1$ when $\alpha < 2$, $|\beta| \leqq 1$ when $\alpha = 2$, then

$$H(x) = 1 - G_{1/\alpha}\left(-\lambda \sec \frac{\pi\alpha}{2} x^{-\alpha}\right), \qquad x \geqq 0, \tag{17}$$

where $G_{1/\alpha}$ is the distribution function of the stable law whose characteristic function is given by

$$g_{1/\alpha}(t) = \exp\left\{-|t|^{1/\alpha}\left(\cos\frac{\pi}{2\alpha} - i \sin\frac{\pi}{2\alpha} \operatorname{sgn} t\right)\right\}.$$

Proof. We may write

$$\frac{M_n}{n^{1/\alpha}} = \frac{Z_1 + \cdots + Z_{R_n}}{n^{1/\alpha}} = \frac{R_n}{n^{1/\alpha}} \cdot \frac{Z_1 + \cdots + Z_{R_n}}{R_n}. \tag{18}$$

Now, in view of (16) and using Theorem 2 of Richter [14],

$$\frac{Z_1 + \cdots + Z_{R_n}}{R_n} \xrightarrow{p} EZ = \left[\lambda \sec\left(\pi - \frac{\pi\alpha}{2}\right)\right]^{1/\alpha}, \tag{19}$$

("p" stands for convergence in probability). Also, we have from Theorem 1 that

$$\lim_{n\to\infty} \Pr(n^{-1/\alpha} R_n \leqq x) = 1 - G_{1/\alpha}(x^{-\alpha}). \tag{20}$$

Then, using a standard convergence result (see for example Cramér [3], page 254), it follows from (18), (19) and (20) that

$$H(x) = \lim_{n\to\infty} \Pr(n^{-1/\alpha} M_n \leqq x) = \lim_{n\to\infty} \Pr(n^{-1/\alpha} R_n EZ \leqq x)$$

$$= 1 - G_{1/\alpha}\left(-\lambda \sec\frac{\pi\alpha}{2} x^{-\alpha}\right).$$

This completes the proof of the theorem. In the particular case $\alpha = 2$, we see from (14) that

$$H(x) = \frac{1}{\sqrt{\pi}} \int_0^{x/\sqrt{\lambda}} e^{-\frac{1}{4}u^2} du = \frac{1}{\sqrt{(\pi\lambda)}} \int_0^x e^{-u^2/4\lambda} du. \tag{21}$$

The result (21) is well known and has a long history. Different methods of derivation can be seen in Erdös and Kac [5], Lévy [11], page 85, and Darling [4].

3. Limit theorem for maxima of partial sums

We shall establish the following invariance theorem.

Theorem 4. Let X_i, $i = 1,2,3,\cdots$ be a sequence of independent and identically distributed random variables and write $S_n = \sum_{i=1}^{n} X_i$, $n \geqq 1$, $S_0 = 0$, and $M_n = \max_{0 \leqq k \leqq n} S_k$. Suppose that there exists a monotone sequence of constants $\{B_n,\ n=1,2,3,\cdots\}$ such that $B_n^{-1} S_n$ converges in distribution to the stable law with characteristic function (1). Then,

$$\lim_{n\to\infty} \Pr(M_n \leqq B_n x) = H(x),$$

the distribution function $H(x)$ being given by (9), (17) or (21).

Proof. Let Y_i, $i = 1,2,3,\cdots$ be independent and identically distributed random variables with characteristic function (1) and write $W_n = \sum_{i=1}^{n} Y_i$. Write also, $H_n(x) = \Pr(M_n \leqq B_n x)$ and for $n_j = [jk^{-1}n]$, $j = 1,\cdots,k$, ($[u]$ here refers to the integer part of u) $H_{n,k}(x) = \Pr[\max(S_{n_1}, S_{n_2},\cdots, S_{n_k}) \leqq B_n x]$. Then, since B_n is of the form $n^{1/\alpha}L(n)$ where $L(n)$ is slowly varying as $n \to \infty$ (Ibragimov and Linnik [10], Theorem 2.1.1), we have

$$H_{n,k}(x) = \Pr\left(\operatorname{Max}(S_{n_1}, S_{n_2},\cdots, S_{n_k}) \leqq n_1^{1/\alpha}L(n_1)\left(\frac{n}{n_1}\right)^{1/\alpha} \frac{L(n)}{L(n_1)} x\right).$$

Consequently, it follows from the multidimensional central limit theorem and Equation (2) that for $x > 0$,

$$\lim_{k\to\infty}\lim_{n\to\infty} H_{n,k}(x) = \lim_{k\to\infty} \Pr\left(\max_{1\leqq j\leqq k} W_j \leqq k^{1/\alpha}x\right) = H(x). \tag{22}$$

Now, for arbitrary $\varepsilon > 0$, define

$$X_{kn} = \begin{cases} X_k & \text{if } -K\varepsilon B_n \leqq X_k \leqq \varepsilon B_n\,, \\ 0 & \text{otherwise,} \end{cases}$$

where $K = K(\varepsilon, n)$ is chosen such that $EX_{kn} = 0$. It is easily seen that $K(\varepsilon, n) \to K_0$, a positive constant, as $n \to \infty$. Write $S_{n,m} = \sum_{k=1}^{m} X_{kn}$ where $m = n_{i+1} - r$ and $n_i < r \leqq n_{i+1}$. Then, if $N = n_{i+1} - n_i$, we have

$$\begin{aligned} \Pr(|S_{n_{i+1}} - S_r| > \varepsilon B_n) &= \Pr(|S_m| > \varepsilon B_n) \\ &\leqq m[\Pr(X < -K\varepsilon B_n) + \Pr(X > \varepsilon B_n)] + \Pr(|S_{n,m}| > \varepsilon B_n) \\ &\leqq N[\Pr(X < -K\varepsilon B_n) + \Pr(X > \varepsilon B_n)] + \varepsilon^{-2} B_n^{-2} ES_{n,m}^2 \\ &\leqq N[\Pr(X < -K\varepsilon B_n) + \Pr(X > \varepsilon B_n)] + N\varepsilon^{-2}B_n^{-2}EX_{kn}^2, \end{aligned} \tag{23}$$

where we have used Chebyshev's inequality in obtaining the second last inequality. Now, $nB_n^{-2}\, EX_{kn}^2 \leqq c_1\varepsilon^{2-\alpha}$, $n\Pr(X < -K\varepsilon B_n) \leqq c_2\varepsilon^{-\alpha}$, and $n\Pr(X > \varepsilon B_n) \leqq c_3\varepsilon^{-\alpha}$ as $n \to \infty$, where c_1, c_2, c_3 are positive constants (not depending on ε) (Feller [7], pages 304, 544–546) and, furthermore, $N \sim k^{-1}n$. It follows then, from (23), that we can choose a positive constant c such that for all i, r, n,

$$\Pr(|S_{n_i+1} - S_r| > \varepsilon B_n) \leqq ck^{-1}\varepsilon^{-\alpha}. \tag{24}$$

Consequently for $x > 0$,

$$\begin{aligned}
\Pr(M_n > B_n x) &= \sum_{r=1}^{n} \Pr\Big(\max_{1 \leqq j \leqq r-1} S_j \leqq B_n x, S_r > B_n x\Big) \\
&= \sum_{i=0}^{k-1} \sum_{n_i < r \leqq n_i+1} \Pr\Big(\max_{1 \leqq j \leqq r-1} S_j \leqq B_n x, S_r > B_n x, |S_{n_i+1} - S_r| > \varepsilon B_n\Big) \\
&\quad + \sum_{i=0}^{k-1} \sum_{n_i < r \leqq n_i+1} \Pr\Big(\max_{1 \leqq j \leqq r-1} S_j \leqq B_n x, S_r > B_n x, |S_{n_i+1} - S_r| \leqq \varepsilon B_n\Big) \\
&\leqq ck^{-1}\varepsilon^{-\alpha} \sum_{r=1}^{n} \Pr\Big(\max_{1 \leqq j \leqq r-1} S_j \leqq B_n x, S_r > B_n x\Big) \\
&\quad + \sum_{i=0}^{k-1} \sum_{n_i < r \leqq n_i+1} \Pr\Big(\max_{1 \leqq j \leqq r-1} S_j \leqq B_n x, S_r > B_n x, |S_{n_i+1} - S_r| \leqq \varepsilon B_n\Big) \\
&\leqq ck^{-1}\varepsilon^{-\alpha} + 1 - H_{n,k}(x - \varepsilon),
\end{aligned} \tag{25}$$

using (24). Finally, since $H_n(x) \leqq H_{n,k}(x)$, and with the aid of (25), we deduce that

$$H_{n,k}(x - \varepsilon) - ck^{-1}\varepsilon^{-\alpha} \leqq H_n(x) \leqq H_{n,k}(x). \tag{26}$$

In view of (22), the result of the theorem is then immediate if we let $n \to \infty$ and then $k \to \infty$ in (26) since $H(x)$ is an absolutely continuous distribution function.

Addendum

The author has been informed by a referee that Theorems 3 and 4 of this paper are derivable from results of A. V. Skorokhod in "Random Processes with Independent Increments" Izdat. Nauk (Moscow) 1964 (in Russian). It has not been possible to compare methods due to the unobtainability of this book. A number of improvements to the exposition of this paper have been suggested by the referees and these the author gratefully acknowledges.

References

[1] BAXTER, G. AND DONSKER, M. D. (1957) On the distribution of the supremum functional for processes with stationary independent increments. *Trans. Amer. Math. Soc.* **85,** 73–87.

[2] CHUNG-TEH, C. (1953) Explicit formula for the stable law of distribution. *Acta Math. Sinica* **3,** 177–185.

[3] CRAMÉR, H. (1946) *Mathematical Methods in Statistics*. Princeton U. Press, Princeton.

[4] DARLING, D. A. (1956) The maximum of sums of stable random variables. *Trans. Amer. Math. Soc.* **83**, 164–169.

[5] ERDÖS, P. AND KAC, M. (1946) On certain limit theorems in the theory of probability. *Bull. Amer. Math. Soc.* **52**, 292–302.

[6] FELLER, W. (1949) Fluctuation theory of recurrent events. *Trans. Amer. Math. Soc.* **67**, 98–119.

[7] FELLER, W. (1966) *An Introduction to Probability Theory and its Applications*. Volume 2. Wiley, New York.

[8] GRADSHTEYN, I. S. AND RYZHIK, I. M. (1965) *Tables of Integrals, Series and Products*. 4th Edition. Academic Press, New York.

[9] HEYDE, C. C. (1967) A limit theorem for random walks with drift. *J. Appl. Prob.* **4**, 144–150.

[10] IBRAGIMOV, I. A. AND LINNIK, YU. V. (1965) *Independent and Stationary Related Random Variables*. Izd-vo "Nauka", Moscow. (In Russian).

[11] LÉVY, P. (1948) *Processus Stochastiques et Mouvement Brownien*. Gauthier Villars, Paris.

[12] LUKACS, E. (1960) *Characteristic Functions*. Griffin, London.

[13] PRABHU, N. U. (1965) *Stochastic Processes*. Macmillan, New York.

[14] RICHTER, W. (1965) Limit theorems for sequences of random variables with sequences of random indices. *Theor. Probability Appl.* **10**, 74–84. (English translation).

[15] ROGOZIN, B. A. (1964) On the distribution of the first jump. *Theor. Probability Appl.* **9**, 450–465. (English translation).

[16] SPITZER, F. (1964) *Principles of Random Walk*. Van Nostrand, New York.

[17] WIDDER, D. V. (1941) *The Laplace Transform*. Princeton U. Press, Princeton.

[18] ZOLOTAREV, V. M. (1957) Mellin-Stieltjes transforms in probability theory. *Theor. Probability Appl.* **2**, 433–460. (English translation).

Z. Wahrscheinlichkeitstheorie verw. Geb. 11, 181–192 (1969)

Some Properties of Metrics in a Study on Convergence to Normality

C. C. HEYDE

Received January 17, 1968

Summary. This paper is concerned with the use of the $\mathscr{L}_1$ and $\mathscr{L}_\infty$ metrics in a study of certain properties and implications of convergence rates in the central limit theorem for sums of independent and identically distributed random variables which belong to the domain of attraction of the normal distribution. Also, some general convergence rate results on the $\mathscr{L}_\infty$ metric obtained under the assumption of a finite second moment are used as a vital tool in a new proof of the classical iterated logarithm law and in extending the scope of classical methods for the proof of other similar results of a more general kind.

1. Introduction

In this paper, we consider a sequence of independent and identically distributed random variables X_i, $i = 1, 2, 3, \ldots$ which belong to the domain of attraction of the normal distribution. Since this last property implies $E|X_i| < \infty$, we shall suppose for convenience that $EX_i = 0$. Then, if $S_n = \sum_{i=1}^{n} X_i$, $n \geqq 1$, there is a monotone sequence of normalizing constants $\{B_n, n \geqq 1\}$ such that as $n \to \infty$,

$$F_n(x) = \Pr(S_n \leqq B_n x) \to \Phi(x) = \frac{1}{\sqrt{2\pi}} \int_{-\infty}^{x} e^{-(1/2)u^2}\, du,$$

and it is not difficult to see that as $n \to \infty$,

$$\int_{-\infty}^{\infty} |F_n(x) - \Phi(x)|\, dx \to 0.$$

We shall firstly establish the following theorem:

Theorem 1. *If*

$$\sum_{n=1}^{\infty} n^{-1} \int_{-\infty}^{\infty} |F_n(x) - \Phi(x)|\, dx < \infty, \tag{1}$$

then $EX_i^2 < \infty$. That is, the X_i belong to the domain of normal attraction of the normal distribution.

In connection with this theorem, we mention the results of IBRAGIMOV [7] where it is shown that if

$$\int_{-\infty}^{\infty} |F_n(x) - \Phi(x)|\, dx = O(n^{-\eta})$$

for some $\eta > 0$, then $EX_i^2 < \infty$. Also, if $EX_i^2 = \sigma^2 < \infty$, then for $0 < \delta < 1$,

$$\int_{-\infty}^{\infty} |\Pr(S_n \leqq x\sigma\sqrt{n}) - \Phi(x)|\, dx = O(n^{-\delta/2})$$

R. Maller et al. (eds.), *Selected Works of C.C. Heyde*, Selected Works in Probability and Statistics, DOI 10.1007/978-1-4419-5823-5_18, © Springer Science+Business Media, LLC 2010

if and only if, as $z \to \infty$,

$$\int_{|x|>z} x^2 \, d \Pr(X_i \leqq x) = O(z^{-\delta}).$$

We shall show in Theorem 2 that if $EX_i^2 = \sigma^2 < \infty$, then

$$\sum_1^\infty n^{-1} \int_{-\infty}^{\infty} \left| \Pr\left(S_n \leqq x \sigma \sqrt{n}\right) - \Phi(x) \right| dx < \infty$$

if and only if $EX_i^2 \log^+ |X_i| < \infty$. (Here, $\log^+ x = \max(0, \log x)$).

The other main theorem of this paper is Theorem 4.

Theorem 4. *Let* $EX_i^2 = 1$, $F(x) = \Pr(X_i \leqq x)$, *and* $F_n(x) = \Pr(S_n \leqq x \sigma_n \sqrt{n})$, *where*

$$\sigma_n^2 = \int_{|x|<\sqrt{n}} x^2 \, dF(x) - \left[\int_{|x|<\sqrt{n}} x \, dF(x) \right]^2. \tag{2}$$

Then, if $K > 0$, $C > 1$, *and* $\{n_k, k \geqq 1\}$ *is a sequence of integers with* $n_k \sim K C^{2k}$ *as* $k \to \infty$, *we have*

$$\sum_{k=1}^\infty \sup_x |F_{n_k}(x) - \Phi(x)| < \infty. \tag{3}$$

This theorem depends on refinement of the details of proof of Theorem 1 of FRIEDMAN, KATZ, and KOOPMANS [3] where it is not difficult to see that the following somewhat stronger result is, in effect, derived.

Theorem A. *If* $EX_i^2 = 1$ *and* $F_n(x) = \Pr\left(S_n \leqq x \sigma_n \sqrt{n}\right)$ *where* σ_n *is given by* (2), *then*

$$\sum_{n=1}^\infty n^{-1} \sup_x |F_n(x) - \Phi(x)| < \infty. \tag{4}$$

We remark, in passing, that conditions (3) and (4) would, of course, be equivalent if it could be shown that the L_∞ metric is ultimately monotonically convergent to zero. Theorem 4 is particularly useful from the point of view of applications since it is often convenient to study convergence along a geometric subsequence. We shall illustrate by obtaining a new proof of the law of the iterated logarithm and by showing how results of CHOVER [1] may be extended.

2. Proof of Theorem 1

Let $F(x) = \Pr(X_i \leqq x)$, and suppose that $f(t)$ and $f_n(t)$ are the characteristic functions corresponding to $F(x)$ and $F_n(x)$ respectively. Then, integrating by parts in the equation

$$f_n(t) - e^{-\frac{1}{2}t^2} = \int_{-\infty}^{\infty} e^{itx} \, d[F_n(x) - \Phi(x)],$$

we obtain

$$-\frac{f_n(t) - e^{-\frac{1}{2}t^2}}{it} = \int_{-\infty}^{\infty} e^{itx} [F_n(x) - \Phi(x)] \, dx,$$

so that

$$\left| \frac{f_n(t) - e^{-\frac{1}{2}t^2}}{t} \right| \leqq \int_{-\infty}^{\infty} |F_n(x) - \Phi(x)| \, dx,$$

and therefore using (1),

$$\sum_{1}^{\infty} n^{-1} \left| \frac{f_n(t) - e^{-\frac{1}{2}t^2}}{t} \right| < \infty . \tag{5}$$

Now, by Lemma 2.1 of [7], $f(t)$ is representable in the form

$$f(t) = \exp\{-\tfrac{1}{2}t^2 H(t)\}, \tag{6}$$

where $H(t)$ is a slowly varying function as $t \to 0$ and

$$H(t) \sim \int_{|xt| \leq 1} x^2 \, dF(x)$$

as $t \to 0$. Then, from (6),

$$f_n(t) = \left[f\left(\frac{t}{B_n}\right)\right]^n = \exp\left\{-\frac{nt^2}{2B_n^2} H\left(\frac{t}{B_n}\right)\right\},$$

so that (5) gives for $t > 0$,

$$\sum_{1}^{\infty} n^{-1} \left| \exp\left\{-\frac{t^2}{2}\left[\frac{n}{B_n^2} H\left(\frac{t}{B_n}\right) - 1\right]\right\} - 1 \right| < \infty .$$

However, as $n \to \infty$,

$$\exp\left\{-\frac{t^2}{2}\left[\frac{n}{B_n^2} H\left(\frac{t}{B_n}\right) - 1\right]\right\} - 1 = \frac{t^2}{2}\left[1 - \frac{n}{B_n^2} H\left(\frac{t}{B_n}\right)\right](1 + o(1)),$$

so that for $t > 0$,

$$\sum_{1}^{\infty} n^{-1} \left| 1 - \frac{n}{B_n^2} H\left(\frac{t}{B_n}\right) \right| < \infty .$$

Therefore,

$$\sum_{1}^{\infty} \frac{1}{B_n^2} \left| H\left(\frac{1}{B_n}\right) - H\left(\frac{2}{B_n}\right) \right|$$
$$\leqq \sum_{1}^{\infty} n^{-1} \left| 1 - \frac{n}{B_n^2} H\left(\frac{1}{B_n}\right) \right| + \sum_{1}^{\infty} n^{-1} \left| 1 - \frac{n}{B_n^2} H\left(\frac{2}{B_n}\right) \right| < \infty ,$$

or equivalently, in view of (6),

$$\sum_{1}^{\infty} \left| \log f(2B_n^{-1}) - 4 \log f(B_n^{-1}) \right| < \infty . \tag{7}$$

Now,

$$\begin{aligned} \log f(2B_n^{-1}) &- 4 \log f(B_n^{-1}) \\ &= \log[1 - \{1 - f(2B_n^{-1})\}] - 4 \log[1 - \{1 - f(B_n^{-1})\}] \\ &= -\{1 - f(2B_n^{-1})\} + 4\{1 - f(B_n^{-1})\} + A_n , \end{aligned} \tag{8}$$

where

$$A_n = \sum_{r=2}^{\infty} r^{-1} [4\{1 - f(B_n^{-1})\}^r - \{1 - f(2B_n^{-1})\}^r] .$$

Furthermore, as $n \to \infty$, we have

$$1 - f(2B_n^{-1}) \sim \tfrac{1}{2}(2B_n^{-1})^2 H(2B_n^{-1}) \sim 2n^{-1},$$
$$1 - f(B_n^{-1}) \sim \tfrac{1}{2}B_n^{-2} H(B_n^{-1}) \sim \tfrac{1}{2}n^{-1},$$

so that

$$\sum_{1}^{\infty} |A_n| < \infty .$$

13*

Consequently, from (7) and (8),

$$\sum_{1}^{\infty} |4\{1-f(B_n^{-1})\} - \{1-f(2B_n^{-1})\}| < \infty,$$

and a fortiori, taking real parts,

$$\sum_{1}^{\infty} |4\{1-\operatorname{Re} f(B_n^{-1})\} - \{1-\operatorname{Re} f(2B_n^{-1})\}| < \infty.$$

This may be rewritten as

$$\sum_{1}^{\infty} \left| \int_{-\infty}^{\infty} \{4(1-\cos B_n^{-1}x) - (1-\cos 2B_n^{-1}x)\}\, dF(x) \right| < \infty,$$

which reduces to

$$\sum_{1}^{\infty} \int_{-\infty}^{\infty} (1-\cos B_n^{-1}x)^2\, dF(x) < \infty. \tag{9}$$

Now, for $|x| \leqq \pi B_n$, we can find a positive constant C so that

$$1-\cos B_n^{-1}x \geqq C(B_n^{-1}x)^2,$$

and therefore, (9) yields

$$\sum_{1}^{\infty} B_n^{-4} \int_{|x|\leqq \pi B_n} x^4\, dF(x) < \infty. \tag{10}$$

Furthermore,

$$\begin{aligned}\sum_{n=1}^{\infty} B_n^{-4} \int_{|x|\leqq\pi B_n} x^4\, dF(x) &\geqq \sum_{n=2}^{\infty} B_n^{-4} \sum_{k=1}^{n-1} \int_{\pi B_k<|x|\leqq\pi B_{k+1}} x^4\, dF(x) \\ &\geqq \pi^4 \sum_{n=2}^{\infty} B_n^{-4} \sum_{k=1}^{n-1} B_k^4 \Pr(\pi B_k < |X_i| \leqq \pi B_{k+1}) \\ &= \pi^4 \sum_{k=1}^{\infty} B_k^4 \Pr(\pi B_k < |X_i| \leqq \pi B_{k+1}) \sum_{n=k+1}^{\infty} B_n^{-4}.\end{aligned} \tag{11}$$

Also, $B_n = \sqrt{n}\, h(n)$ where $h(n)$ is slowly varying as $n \to \infty$ ([7], Lemma 2.2) so, using a standard result on regularly varying functions (e.g. FELLER [2], II, 273), we have

$$\sum_{n=k+1}^{\infty} B_n^{-4} \sim k B_k^{-4}$$

as $k \to \infty$ and, from (10) and (11),

$$\sum_{k=1}^{\infty} k \Pr(\pi B_k < |X_i| \leqq \pi B_{k+1}) < \infty. \tag{12}$$

But, since $B_n^{-1} S_n$ converges to normality, $n \Pr(|X_i| > \varepsilon B_n) \to 0$ as $n \to \infty$ for any $\varepsilon > 0$ (e.g. LOÈVE [8], 316), and therefore, from (12),

$$\sum_{n=1}^{\infty} \Pr(|X_i| > \pi B_n) < \infty. \tag{13}$$

Now suppose that $E|X_i|^2 = \infty$, so that the slowly varying function

$$K(t) = \int_{|xt| \leq 1} x^2\, dF(x) \to \infty \quad \text{as} \quad t \to 0.$$

Under these circumstances, we shall obtain a contradiction with our assumption of convergence to normality by showing that the condition (13) implies that $B_n^{-1} S_n$ converges in probability to zero.

Using the degenerate convergence criterion (Loève [8], 317), we have $B_n^{-1} S_n \xrightarrow{p} 0$ provided that, as $n \to \infty$, and for any $\varepsilon > 0$,

$$n \Pr(|X_i| > \varepsilon B_n) \to 0, \frac{n}{B_n^2} \int_{|x| < \pi B_n} x^2\, dF(x) \to 0.$$

As we have indicated, the former is satisfied so it remains to establish the latter. We have, since $n B_n^{-2} K(B_n^{-1}) \to 1$ as $n \to \infty$,

$$\int_{|x| < \pi B_n} x^2\, dF(x) \leq \int_{|x| < \pi B_1} x^2\, dF(x) + \pi^2 \sum_{k=1}^{n} B_{k+1}^2 \Pr(\pi B_k < |X_i| \leqq \pi B_{k+1})$$

$$\leqq \int_{|x| < \pi B_1} x^2\, dF(x) + C \sum_{k=1}^{n} K(B_k^{-1})\, k \Pr(\pi B_k < |X_i| \leqq \pi B_{k+1})$$

for some positive constant C. Then, using (12) and by virtue of our assumptionthat $K(t) \to \infty$ as $t \to 0$, we have from the Kronecker Lemma (e.g. [8], 238), that

$$\frac{n}{B_n^2} \int_{|x| < \pi B_n} x^2 dF(x) \sim \frac{1}{K(B_n^{-1})} \int_{|x| < \pi B_n} x^2\, dF(x) \to 0 \quad \text{as} \quad n \to \infty.$$

This establishes the required contradiction. The result of the theorem is then immediate. (13) is only possible in the case where $K(B_n^{-1})$ approaches a positive constant as $n \to \infty$ in which case it is easily seen that $EX_i^2 < \infty$.

3. Background to Theorem 1

Theorem 2. *Let* $EX_i = 0$ *and* $EX_i^2 = \sigma^2 < \infty$. *Then, the following three conditions are equivalent:*

(i) $EX_i^2 \log^+ |X_i| < \infty$,

(ii) $\sum_1^\infty n^{-1} \int_{-\infty}^{\infty} |\Pr(S_n \leqq x \sigma \sqrt{n}) - \Phi(x)|\, dx < \infty$,

(iii) $\sum_1^\infty n^{-1} \sup_x |\Pr(S_n \leqq x \sigma \sqrt{n}) - \Phi(x)| < \infty$.

Proof. The equivalence of (i) and (iii) has been established in Heyde [5]. We shall proceed to establish the equivalence of (i) and (ii).

Suppose firstly that (ii) holds. It is easily seen that the proof of the necessity part of the theorem of [5] is still applicable and (i) follows. If, on the other hand, (i) holds, we have from Lemma 2 of Heyde [6] with $a_n^2 = (2 + \varepsilon) \log \log n$, $\varepsilon > 0$, and $B_n^2 = \sigma^{-2} \int_{|x| < \sqrt{n}} x^2\, d\Pr(X_i \leqq x)$, that

$$\sum_1^\infty n^{-1} \int_{-\infty}^{\infty} |\Pr(S_n \leqq x \sigma \sqrt{n}) - \Phi(x B_n^{-1})|\, dx < \infty \tag{14}$$

if

(a) $\sum_{1}^{\infty} n^{-1}(1 - B_n^2) < \infty$,

(b) $\sum_{1}^{\infty} n^{-1} \int_{|x| \geqq a_n} x^2 \, d\Phi(x B_n^{-1}) < \infty$,

(c) $\sum_{3}^{\infty} n^{-1} \log\log n \sup_x |\Pr(S_n \leqq x \sigma \sqrt{n}) - \Phi(x B_n^{-1})| < \infty$.

Furthermore, it is shown in the proof of the theorem of [6] that these three conditions are satisfied. Consequently, from (14), we see that (ii) holds if

$$\sum_{1}^{\infty} n^{-1} \int_{-\infty}^{\infty} |\Phi(x B_n^{-1}) - \Phi(x)| \, dx < \infty. \tag{15}$$

Now,

$$|\Phi(x B_n^{-1}) - \Phi(x)| = \frac{1}{\sqrt{2\pi}} \int_{|x|}^{|x| B_n^{-1}} e^{-\frac{1}{2}u^2} du = \frac{|x|}{\sqrt{2\pi}} \int_{1}^{B_n^{-1}} e^{-\frac{1}{2}u^2 x^2} du,$$

so that

$$\begin{aligned} \int_{-\infty}^{\infty} |\Phi(x B_n^{-1}) - \Phi(x)| \, dx &= \frac{2}{\sqrt{2\pi}} \int_0^{\infty} x \left\{ \int_1^{B_n^{-1}} e^{-\frac{1}{2}u^2 x^2} du \right\} dx \\ &= \frac{2}{\sqrt{2\pi}} \int_1^{B_n^{-1}} \left\{ \int_0^{\infty} x e^{-\frac{1}{2}u^2 x^2} dx \right\} du \\ &= \frac{2}{\sqrt{2\pi}} \int_1^{B_n^{-1}} u^{-2} du \\ &= \frac{2}{\sqrt{2\pi}} (1 - B_n) < \frac{2}{\sqrt{2\pi}} (1 - B_n^2), \end{aligned} \tag{16}$$

and (15) holds in view of (a). This completes the proof of the theorem.

Next, we shall go on to examine the effect of the choice of B_n on the magnitude of the L_1 metric $\int_{-\infty}^{\infty} |F_n(x) - \Phi(x)| \, dx$. In doing this we shall, in the light of Theorem 1, concentrate on the context of $EX_i^2 = \sigma^2 < \infty$, so that B_n may be written in the form $B_n = \sigma \sqrt{n(1 + \varepsilon_n)}$ where $\varepsilon_n \to 0$ as $n \to \infty$ ([7], Lemma 2.2).

As a starting point, we take a simple example which well illustrates the rôle of ε_n. Let the X_i be normally distributed with unit variance. Then, since $S_n/\sqrt{n}$ has the same distribution as X_i we have, as in (16),

$$\begin{aligned} \int_{-\infty}^{\infty} |\Pr(S_n \leqq x \sqrt{n(1+\varepsilon_n)}) - \Phi(x)| \, dx &= \int_{-\infty}^{\infty} |\Phi(x \sqrt{1+\varepsilon_n}) - \Phi(x)| \, dx \\ &= \frac{2}{\sqrt{2\pi}} (1 - (1 + |\varepsilon_n|)^{-1/2}) = O(|\varepsilon_n|), \end{aligned} \tag{17}$$

so that we must have $\sum_{1}^{\infty} n^{-1} |\varepsilon_n| < \infty$ in order that

$$\sum_{1}^{\infty} n^{-1} \int_{-\infty}^{\infty} |\Pr(S_n \leqq x \sqrt{n(1+\varepsilon_n)}) - \Phi(x)| \, dx < \infty.$$

With this kind of consideration in mind, we shall go on to establish the following theorem.

Theorem 3. *Let* $EX_i^2 \log^+ |X_i| < \infty$ *and write* $B_n^2 = n\, EX_i^2 (1 + \varepsilon_n)$ *where* $\varepsilon_n \to 0$ *as* $n \to \infty$. *If* $\sum_1^\infty n^{-1} |\varepsilon_n| < \infty$, *then*

$$\sum_1^\infty n^{-1} \int_{-\infty}^{\infty} |\Pr(S_n \leqq B_n x) - \Phi(x)|\, dx < \infty. \tag{18}$$

Conversely, if (18) *holds for some monotone sequence* $\{B_n\}$, *and the* X_i *belong to the domain of attraction of the normal distribution, then* $EX_i^2 < \infty$ *and*

$$B_n^2 = n\, EX_i^2 (1 + \varepsilon_n)$$

where $\varepsilon_n \to 0$ *as* $n \to \infty$. *If, in addition,*

$$\sum_1^\infty n^{-1} |\varepsilon_n| < \infty, \quad \text{then} \quad EX_i^2 \log^+ |X_i| < \infty.$$

Proof. Write $EX_i^2 = \sigma^2 (< \infty)$. We obtain from Theorem 2 that

$$\sum_1^\infty n^{-1} \int_{-\infty}^{\infty} |\Pr(S_n \leqq x \sigma \sqrt{n}) - \Phi(x)|\, dx < \infty, \tag{19}$$

while

$$\begin{aligned} &|\Pr(S_n \leqq x \sigma \sqrt{n(1+\varepsilon_n)}) - \Phi(x)| \\ &\leqq |\Pr(S_n \leqq x \sqrt{(1+\varepsilon_n)}\, \sigma \sqrt{n}) - \Phi(x \sqrt{1+\varepsilon_n})| + |\Phi(x\sqrt{1+\varepsilon_n}) - \Phi(x)|, \end{aligned} \tag{20}$$

so that the result of the first part of the theorem is immediate from (17) and (19).

Now suppose that (18) holds for some monotone sequence $\{B_n\}$, and the X_i belong to the domain of attraction of the normal distribution. It follows from Theorem 1 that $EX_i^2 = \sigma^2 < \infty$, and consequently B_n must be of the form $B_n = \sigma \sqrt{n(1+\varepsilon_n)}$ where $\varepsilon_n \to 0$ as $n \to \infty$. Further, the Eq. (20) can be rewritten to yield

$$\int_{-\infty}^{\infty} |\Pr(S_n \leqq x\sigma\sqrt{n}) - \Phi(x)|\, dx \leqq \int_{-\infty}^{\infty} |\Pr(S_n \leqq x \sigma \sqrt{n(1+\varepsilon_n)}) - \Phi(x)|\, dx + C |\varepsilon_n|,$$

where C is a positive constant, so that, if $\sum_1^\infty n^{-1} |\varepsilon_n| < \infty$, (19) holds. The result $EX_i^2 \log^+ |X_i| < \infty$ then follows from Theorem 2. This completes the proof.

4. Proof of Theorem 4

From the proof of Theorem 1 of Friedman, Katz, and Koopmans [3], we extract the information that

$$\begin{aligned} &n^{-1} |F_n(x\sigma_n^{-1}) - \Phi(x\sigma_n^{-1})| \\ &\leqq M n^{-3/2} \{a_n + |\mu_n|^3\} + n^{-1} |\Phi[\sigma_n^{-1}(x - \sqrt{n}\,\mu_n)] - \Phi(\sigma_n^{-1} x)| + c_n, \end{aligned} \tag{21}$$

where M is a positive constant,

$$\mu_n = \int_{|x| < \sqrt{n}} x\, dF(x), \quad a_n = \int_{|x| < \sqrt{n}} |x|^3\, dF(x), \quad c_n = \Pr(|X_i| > \sqrt{n}),$$

and also that $\sum_{n=1}^{\infty} n^{-3/2} a_n < \infty$, and $\sum_{n=1}^{\infty} c_n < \infty$. Now, since $EX_i = 0$ and $EX_i^2 = 1$, it follows that $\sqrt{n}\mu_n \to 0$ and $1 \geqq \sigma_n \to 1$ as $n \to \infty$. Consequently, using the mean value theorem, we obtain

$$|\Phi[\sigma_n^{-1}(x - \sqrt{n}\,\mu_n)] - \Phi(\sigma_n^{-1} x)| \leqq C\sqrt{n}\,|\mu_n|$$

for some positive constant C (independent of x). Furthermore, we note that $\mu_n = o(1) = o(a_n)$ as $n \to \infty$ and we can write instead of (21),

$$n^{-1} \sup_x |F_n(x) - \Phi(x)| \leqq A\, n^{-3/2} a_n + B n^{-1/2} b_n + c_n, \tag{22}$$

where

$$b_n = \int_{|x| \geqq \sqrt{n}} |x|\, dF(x) \geqq \Big| \int_{|x| \geqq \sqrt{n}} x\, dF(x) \Big| = |\mu_n|,$$

and A, B are positive constants.

Before proceeding, we check that the result (4) follows from (22) and to do this, in view of the abovementioned results, it is just necessary to show that $\sum_{n=1}^{\infty} n^{-1/2} b_n < \infty$. This is easily seen to be the case as $EX_i^2 < \infty$, so that

$$\begin{aligned}\sum_{n=1}^{\infty} n^{-1/2} b_n &\leqq \sum_{n=1}^{\infty} n^{-1/2} \sum_{k=n}^{\infty} (k+1)^{1/2} \Pr(k \leqq X_i^2 < k+1) \\ &= \sum_{k=1}^{\infty} (k+1)^{1/2} \Pr(k \leqq X_i^2 < k+1) \sum_{n=1}^{k} n^{-1/2} \\ &\leqq C' \sum_{k=1}^{\infty} (k+1) \Pr(k \leqq X_i^2 < k+1) < \infty,\end{aligned}$$

C' being a positive constant. We are now in a position to deduce (3) from (22) by showing that

$$\sum_{k=1}^{\infty} n_k^{-1/2} a_{n_k} < \infty, \quad \sum_{k=1}^{\infty} n_k^{1/2} b_{n_k} < \infty \quad \text{and} \quad \sum_{k=1}^{\infty} n_k c_{n_k} < \infty.$$

Since a_n is non-decreasing in n,

$$\sum_{n=1}^{\infty} n^{-3/2} a_n \geqq \sum_{k=1}^{\infty} \sum_{n=n_k}^{n_{k+1}-1} n^{-3/2} a_n \geqq \sum_{k=1}^{\infty} (n_{k+1} - n_k)(n_{k+1} - 1)^{-3/2} a_{n_k}.$$

However, since $n_k \sim K C^{2k}$ as $k \to \infty$,

$$(n_{k+1} - n_k)(n_{k+1} - 1)^{-3/2} \sim C^{-3}(C^2 - 1)\, n_k^{-1/2},$$

so that $\sum_{k=1}^{\infty} n_k^{-1/2} a_{n_k} < \infty$. Also, since b_n is non-increasing in n,

$$\sum_{n=1}^{\infty} n^{-1/2} b_n \geqq \sum_{k=1}^{\infty} \sum_{n=n_k+1}^{n_{k+1}} n^{-1/2} b_n \geqq \sum_{k=1}^{\infty} n_{k+1}^{-1/2} (n_{k+1} - n_k)\, b_{n_{k+1}}$$

and

$$n_{k+1}^{-1/2} (n_{k+1} - n_k) \sim n_{k+1}^{1/2} (1 - C^{-2}),$$

so that $\sum_{k=1}^{\infty} n_k^{1/2} b_{n_k} < \infty$. Finally, c_n is non-increasing in n, so that

$$\sum_{n=1}^{\infty} c_n \geqq \sum_{k=1}^{\infty} \sum_{n=n_k+1}^{n_{k+1}} c_n \geqq \sum_{k=1}^{\infty} (n_{k+1} - n_k)\, c_{n_{k+1}},$$

while

$$n_{k+1} - n_k \sim (1 - C^{-2})\, n_{k+1},$$

and consequently, $\sum_{k=1}^{\infty} n_k\, c_{n_k} < \infty$. The result (3) then follows from (22).

Corollary 1. *Let $EX_i^2 = 1$ and $\{\varphi(n), n \geqq 1\}$ be a monotone sequence. Then, if $K > 0$, $C > 1$, and $\{n_k, k \geqq 1\}$ is a sequence of integers with $n_k \sim KC^{2k}$ as $k \to \infty$, the following two conditions are equivalent:*

$$\text{(A)} \qquad \sum_{k=1}^{\infty} \Pr(S_{n_k} > \varphi(n_k)\sqrt{n_k}) < \infty,$$

$$\text{(B)} \qquad \sum_{k=1}^{\infty} [\varphi(n_k)]^{-1} \exp\{-\tfrac{1}{2}[\varphi(n_k)]^2 \sigma_{n_k}^{-2}\} < \infty,$$

σ_n being given by (2).

Proof. From Theorem 4, we immediately obtain the equivalence of (A) and the condition

$$\sum_{k=1}^{\infty} [1 - \Phi(\varphi(n_k)\, \sigma_{n_k}^{-1})] < \infty.$$

The result of the corollary then follows since, as $k \to \infty$,

$$\begin{aligned} 1 - \Phi(\varphi(n_k)\,\sigma_{n_k}^{-1}) &\sim \frac{1}{\sqrt{2\pi}} \frac{\sigma_{n_k}}{\varphi(n_k)} \exp\{-\tfrac{1}{2}[\varphi(n_k)]^2 \sigma_{n_k}^{-2}\} \\ &\sim \frac{1}{\sqrt{2\pi}} \frac{1}{\varphi(n_k)} \exp\{-\tfrac{1}{2}[\varphi(n_k)]^2 \sigma_{n_k}^{-2}\}, \end{aligned}$$

using a well-known asymptotic formula (e.g. FELLER [2], I, 166).

5. Applications of Theorem 4

In this section, we shall illustrate the usefullness of Theorem 4 and Corollary 1 with two applications. We start by giving a new proof of the classical iterated logarithm law based on the use of Corollary 1. By making use of the corollary, we are able to avoid the ordinary complications of truncation and the use of exponential bounds.

Theorem 5 (HARTMAN *and* WINTNER [4]). *If $EX_i = 0$ and $EX_i^2 = 1$, then*

$$\Pr\left(\limsup_{n\to\infty} \frac{S_n}{\sqrt{2n \log\log n}} = 1\right) = 1.$$

Proof. Let $b_n = (2n \log\log n)^{1/2}$ and $t_n = (2 \log\log n)^{1/2}$. Firstly we shall show that if $\delta > 0$, then $\Pr(S_n > (1 + \delta)\, b_n \text{ i.o.}) = 0$ ("i.o." stands for "infinitely often").

For $c > 1$, let n_k be the integer part of c^{2k}, $k \geqq 1$. Then, if $M_n = \max_{1 \leqq k \leqq n} S_k$, we have

$$\Pr(S_n > (1+\delta)\, b_n \text{ i.o.}) \leqq \Pr(M_{n_k} > (1+\delta)\, b_{n_{k-1}} \text{ i.o.}),$$

where

$$(1+\delta)\, b_{n_{k-1}} \sim c^{-1}(1+\delta)\, b_{n_k}.$$

Thus, taking δ' with $0 < \delta' < \delta$, we can select $c > 1$, so that $c^{-1}(1+\delta) > 1+\delta'$, and then,

$$\Pr(M_{n_k} > (1+\delta)\, b_{n_{k-1}} \text{ i.o.}) \leqq \Pr(M_{n_k} > (1+\delta')\, b_{n_k} \text{ i.o.}).$$

Consequently, the result $\Pr(S_n > (1+\delta)\, b_n \text{ i.o.}) = 0$ will follow from the Borel-Cantelli lemma if we show that

$$\sum_{k=1}^{\infty} \Pr(M_{n_k} > (1+\delta')\, b_{n_k}) < \infty. \tag{23}$$

Now, using an inequality due to KOLMOGOROV (e.g. LOÈVE [8], 248),

$$\begin{aligned}\Pr(M_{n_k} > (1+\delta')\, b_{n_k}) &\leqq 2\Pr[S_{n_k} > (1+\delta' - t_{n_k}^{-1}\sqrt{2})\, b_{n_k}] \\ &\leqq 2\Pr(S_{n_k} > (1+\delta'')\, b_{n_k})\end{aligned}$$

for $0 < \delta'' < \delta'$ and k sufficiently large. Thus, by appeal to Corollary 1, we have convergence in (23) if

$$\sum_{k=1}^{\infty} t_{n_k}^{-1} \exp\{-\tfrac{1}{2}(1+\delta'')^2 t_{n_k}^2 \sigma_{n_k}^{-2}\} < \infty.$$

This is easily seen to be the case as

$$\begin{aligned} t_{n_k}^{-1} \exp\{-\tfrac{1}{2}(1+\delta'')^2 t_{n_k}^2 \sigma_{n_k}^{-2}\} &\leqq t_{n_k}^{-1} \exp\{-\tfrac{1}{2}(1+\delta'')\, t_{n_k}^2\} \\ &= O(k^{-(1+\delta'')} (\log k)^{-1/2}) \quad \text{as} \quad k \to \infty. \end{aligned}$$

In order to complete the proof of the theorem it remains to show that

$$\Pr(S_n > (1-\delta)\, b_n \text{ i.o.}) = 1 \quad \text{for} \quad \delta > 0.$$

Take $1 > \delta > \delta' > 0$ and let

$$\begin{aligned} u_k^2 &= n_k - n_{k-1} \sim n_k(1 - c^{-2}), \\ v_k &= (2\log\log u_k^2)^{1/2} \sim (2\log\log n_k)^{1/2} = t_{n_k}, \end{aligned}$$

and

$$A_k = \{S_{n_k} - S_{n_{k-1}} > (1-\delta')\, u_k v_k\}.$$

Then,

$$\Pr(A_k) = \Pr(S_{n_k - n_{k-1}} > (1-\delta')\, u_k v_k),$$

while for $1 > \delta' > \delta'' > 0$ and k sufficiently large,

$$\begin{aligned} v_k^{-1} \exp[-\tfrac{1}{2}(1-\delta')^2 v_k^2 \sigma_{n_k - n_{k-1}}^{-2}] &\geqq v_k^{-1} \exp[-\tfrac{1}{2}(1-\delta'')^2 v_k^2] \\ &= O(k^{-(1-\delta'')^2} (\log k)^{-1/2}), \end{aligned}$$

so that from Corollary 1, $\sum_{k=1}^{\infty} \Pr(A_k) = \infty$. Therefore, since the sums $S_{n_k} - S_{n_{k-1}}$ are non-overlapping, we have from the Borel-Cantelli lemma that $\Pr(A_k \text{ i.o.}) = 1$. Furthermore, if $B_k = \{|S_{n_{k-1}}| \leqq 2b_{n_{k-1}}\}$, it follows from the first part of the proof that $\Pr(\bar{B}_k \text{ i.o.}) = 0$, where we have used a bar to denote the complementary event. We must therefore have $\Pr(A_k \cap B_k \text{ i.o.}) = 1$. Also,

$$A_k \cap B_k \subset \{S_{n_k} > (1-\delta') u_k v_k - 2 b_{n_{k-1}}\},$$

while

$$(1-\delta') u_k v_k - 2 b_{n_{k-1}} \sim [(1-\delta')(1-c^{-2})^{1/2} - 2c^{-1}] b_{n_k},$$

so that if we take c so large that

$$(1-\delta')(1-c^{-2})^{1/2} - 2c^{-1} > 1-\delta,$$

we have

$$1 - \Pr(A_k \cap B_k \text{ i.o.}) \leqq \Pr(S_{n_k} > (1-\delta) b_{n_k} \text{ i.o.}).$$

This implies, a fortiori, that $\Pr(S_n > (1-\delta) b_n \text{ i.o.}) = 1$ and thus completes the proof. We next proceed to our final application.

In the paper [9], STRASSEN obtained a deep and rather striking generalization of the iterated logarithm law. For his proof, he appealed to a result of Skorokhod which permits one to realize a sequence of independent and identically distributed random variables with finite variance in terms of random increments of a Brownian motion process. Later, CHOVER [1] developed a more classical approach to the same results but was forced to introduce an extra moment condition $E|X_i|^{2+\delta} < \infty$, some $\delta > 0$, in order to use an estimate of Esseen on convergence to normality. As we shall indicate here, Theorem 4 provides just the tool necessary to extend CHOVER's proof to obtain the results given by STRASSEN (the $\delta = 0$ case). The reader is referred to the papers [1] and [9] for notations and explainations.

The are three places in the paper [1] where the suplementary moment condition $E|X_i|^{2+\delta} < \infty$, some $\delta > 0$, is required. In each case it may be avoided by use of Theorem 4 (or Corollary 1). The applications are routine in character so we shall only carry out one of them (the last).

We need to show (see equations 30 and 32 of [1]) that for fixed ν,

$$\sum_r \Pr(|Z_r - A| < (2m)^{-1}\varepsilon) = \infty, \tag{24}$$

where

$$Z_r = (2m N_{r,\nu} \log\log n_r)^{-1/2} S_{N_{r,\nu}}, \quad A = \left[g\left(\frac{\nu+1}{m}\right) - g\left(\frac{\nu}{m}\right)\right].$$

But,

$$\Pr(|Z_r - A| < (2m)^{-1}\varepsilon) = \Pr(Z_r < A + (2m)^{-1}\varepsilon) - \Pr(Z_r \leqq A - (2m)^{-1}\varepsilon),$$

and by Theorem 4, since $N_{r,\nu} \sim m^{r-1}$,

$$\sum_r |\Pr(Z_r < A + (2m)^{-1}\varepsilon) - \Pr(Z_r \leqq A - (2m)^{-1}\varepsilon) - \Phi(a+b) + \Phi(a-b)| < \infty,$$

where $s = \sigma_{N_{r,\nu}} \to 1$ as $r \to \infty$,

$$a = A(2m \log\log n_r)^{1/2} s^{-1}, \quad \text{and} \quad b = (2m)^{-1}\varepsilon(2m\log\log n_r)^{1/2} s^{-1}.$$

Therefore, (24) holds provided that

$$\sum_r |\Phi(a+b) - \Phi(a-b)| = \infty. \tag{25}$$

If $A = 0$, this is easily seen to be the case since

$$\Phi(b) - \Phi(-b) = 2\Phi(b) - 1 \to 1$$

as $r \to \infty$. Now suppose $A \neq 0$ and that ε is small compared with A. Then, for r sufficiently large,

$$\begin{aligned}
\Phi(a+b) - \Phi(a-b) &= \Phi(|a|+b) - \Phi(|a|-b) \\
&\sim 1 - \Phi(|a|-b) \\
&\sim (4\pi m \log\log n_r)^{-1/2}(|A| - (2m)^{-1}\varepsilon)^{-1} \\
&\quad \cdot \exp\{-(|A| - (2m)^{-1}\varepsilon)^2 m s^{-2} \log\log n_r\} \\
&\geqq (4\pi m \log\log n_r)^{-1/2}(|A| - (2m)^{-1}\varepsilon)^{-1} \\
&\quad \cdot \exp\{-A^2 m \log\log n_r\} \\
&= O(r^{-A^2 m}(\log r)^{-1/2}) \qquad (26)
\end{aligned}$$

as $r \to \infty$ since $n_r \sim m^r$. Also,

$$\begin{aligned}
A^2 = \left[g\left(\frac{\nu+1}{m}\right) - g\left(\frac{\nu}{m}\right)\right]^2 &= \left[\int_{m^{-1}\nu}^{m^{-1}(\nu+1)} \dot g(t)\,dt\right]^2 \\
&\leqq \int_{m^{-1}\nu}^{m^{-1}(\nu+1)} dt \int_{m^{-1}\nu}^{m^{-1}(\nu+1)} [\dot g(t)]^2\,dt \leqq m^{-1},
\end{aligned}$$

so that from (26) we see that (25) holds and the required result is obtained.

References

1. CHOVER, J.: On Strassen's version of the log log law. Z. Wahrscheinlichkeitstheorie verw. Geb. **8**, 83–90 (1967).
2. FELLER, W.: An introduction to probability theory and its applications, Vols I, II. New York: Wiley 1957, 1966.
3. FRIEDMAN, N., M. KATZ, and L. H. KOOPMANS: Convergence rates for the central limit theorem. Proc. nat. Acad. Sci. U.S.A. **56**, 1062–1065 (1966).
4. HARTMAN, P., and A. WINTNER: On the law of the iterated logarithm. Amer. J. Math. **63**, 169–176 (1941).
5. HEYDE, C. C.: On the influence of moments on the rate of convergence to the normal distribution. Z. Wahrscheinlichkeitstheorie verw. Geb. **8**, 12–18 (1967).
6. — On the growth of a random walk. Ann. Inst. Stat. Math. **20**, 315–321 (1968).
7. IBRAGIMOV, I. A.: On the accuracy of the Gaussian approximation to the distribution function of sums of independent variables. Theor. Probab. Appl. **11**, 559–576 (1966).
8. LOÈVE, M.: Probability theory (3rd edition). New York: Van Nostrand 1963.
9. STRASSEN, V.: An invariance principle for the law of the iterated logarithm. Z. Wahrscheinlichkeitstheorie verw. Geb. **3**, 211–226 (1964).

Dr. C. C. HEYDE
Department of Statistics
Australian Natural University,
School of General Studies,
Box 4, P. O.,
Canberra, A. C. T. 2600,
Australia

The Annals of Mathematical Statistics
1970, Vol. 41, No. 2, 739–742

EXTENSION OF A RESULT OF SENETA FOR THE SUPER-CRITICAL GALTON–WATSON PROCESS

BY C. C. HEYDE

Australian National University

1. Introduction. Let $Z_0 = 1, Z_1, Z_2 \cdots$ denote a super-critical Galton–Watson process whose non-degenerate offspring distribution has probability generating function $F(s) = \sum_{j=0}^{\infty} s^j \Pr(Z_1 = j)$, $0 \leqq s \leqq 1$, where $1 < m = EZ_1 < \infty$. The Galton–Watson process evolves in such a way that the generating function $F_n(s)$ of Z_n is the nth functional iterate of $F(s)$ and, for the super-critical case in question, the probability of extinction of the process, q, is well known to be the unique real number in $[0, 1)$ satisfying $F(q) = q$. It is the main purpose of this paper to establish the following theorem which gives an ultimate form of the limit result for the case in question.

THEOREM 1. *There exists a sequence of positive constants* $\{c_n, n \geqq 1\}$ *with* $c_n \to \infty$ *and* $c_n^{-1} c_{n+1} \to m$ *as* $n \to \infty$ *such that the random variables* $c_n^{-1} Z_n$ *converge almost surely to a non-degenerate random variable* W *for which* $\Pr(W = 0) = q$ *and which has a continuous distribution on the set of positive real numbers. Let* s_0 *be any fixed number in* $(0, -\log q)$. *Then,* c_n *can be taken as* $[h_n(s_0)]^{-1}$ *where* $h_n(s)$ *is the inverse function of* $k_n(s) = -\log E\{\exp(-sZ_n)\}$.

This result constitutes an extension of the main result of Seneta [6] where convergence in distribution was established. It should be remarked that, when $EZ_1 = \infty$, it is not possible to find a sequence of positive constants $\{c_n\}$ for which $c_n^{-1} Z_n$ converges in distribution to a non-degenerate limit law ([7] Theorem 4.4).

By way of comparison with Theorem 1, we note that:

THEOREM A. (Stigum [8], Kesten and Stigum [3]). *As* $n \to \infty$, $m^{-n} Z_n$ *converges almost surely to a random variable* W_1 *for which* $\Pr(W_1 = 0) = q$ *or* 1 *and which, if* $\Pr(W_1 = 0) < 1$, *has a continuous density on the set of positive real numbers. Moreover, the following two conditions are equivalent:*

(i) $E(Z_1 \log Z_1) < \infty$.
(ii) $\Pr(W_1 = 0) = q$.

Thus, when $E(Z_1 \log Z_1) = \infty$, the norming by m^n is not appropriate and a more subtle norming is required to obtain a non-degenerate limit law. Almost sure convergence in Theorem A is based on the fact (due to Doob) that the process $\{m^{-n} Z_n\}$ is a martingale. The process $\{h_n(s_0) Z_n\}$ is, as was noted in [6], a submartingale but the submartingale convergence theorem is only applicable when $E(Z_1 \log Z_1) < \infty$.

Received March 25, 1969.

R. Maller et al. (eds.), *Selected Works of C.C. Heyde*, Selected Works in Probability and Statistics,
DOI 10.1007/978-1-4419-5823-5_19, © Springer Science+Business Media, LLC 2010

2. Proof of Theorem 1. Firstly we note the following results of [6]. $k_n(s) = -\log E\{\exp(-sZ_n)\}, s \geqq 0$, is the nth functional iterate of

$$k(s) = -\log E\{\exp(-sZ_1)\}.$$

$k_n(s)$ is continuous, strictly monotone, and strictly concave for $s \geqq 0$ and its inverse function $h_n(s)$ (the nth functional iterate of $h(s) = k^{-1}(s)$) exists for $0 \leqq s < -\log q$ and has properties which are dual to those of $k_n(s)$. Let s_0 be any fixed number in $(0, -\log q)$.

Now, for $n \geqq 1$ let $\mathscr{F}_n$ be the σ-field generated by $Z_1, \cdots, Z_n$ and consider the process $\{\exp(-h_n(s_0)Z_n)\}$. Then,

$$\begin{aligned} E[\exp(-h_{n+1}(s_0)Z_{n+1}) \mid \mathscr{F}_n] &= [E[\exp(-h_{n+1}(s_0)Z_1)]]^{Z_n} \\ &= \exp(-Z_n k(h_{n+1}(s_0))) \\ &= \exp(-h_n(s_0)Z_n), \end{aligned}$$

so that $\{\exp(-h_n(s_0)Z_n), \mathscr{F}_n\}$ is a martingale. Furthermore, $0 \leqq \exp\{-h_n(s_0)Z_n\} \leqq 1$, so the martingale convergence theorem gives the almost sure convergence of $\{\exp(-h_n(s_0)Z_n)\}$ to a finite limit. It has already been demonstrated in [6] that $h_n(s_0)Z_n$ converges in distribution to a non-degenerate limit so almost sure convergence to a non-degenerate random variable W is established.

It is not shown explicitly in [6] that $h_n(s_0)[h_{n+1}(s_0)]^{-1} \to m$ as $n \to \infty$ but it follows readily from the results given therein since

$$h_n(s_0)[h_{n+1}(s_0)]^{-1} = h_n(s_0)[h(h_n(s_0))]^{-1} \to m$$

as $n \to \infty$. Furthermore, Seneta has not shown that the limit distribution function is continuous on the set of positive real numbers. It follows simply, however, from Equation 3.1 of [6], that the characteristic function $\phi(t)$ of W satisfies the functional equation

$$\phi(mt) = F(\phi(t)) \tag{1}$$

which is just that studied by Stigum [8]. Then, following [8] and noting that $\Pr(W = 0) = q$, we define a characteristic function

$$\Psi(t) = [\phi((1-q)t) - q]/(1-q),$$

and a probability generating function

$$h(s) = [F((1-q)s + q) - q]/(1-q),$$

so that, using (1),

$$\Psi(mt) = h(\Psi(t)).$$

It can then be deduced from Lemma 2 of [8] that $\lim_{|t| \to \infty} |\Psi(t)| = 0$. This ensures that the distribution function corresponding to Ψ is continuous ([5], 27), and hence that W has a continuous distribution on the set of positive real numbers. This completes the proof of the theorem.

3. A Wald type identity. Let T be a stopping rule on the sequence $\{Z_n\}$. That is, T is an integer-valued random variable such that the event $\{T \leqq n\} \in \mathscr{F}_n$ for every $n \geqq 1$ and $P(T < \infty) = 1$. We shall establish the following theorem.

THEOREM 2. *For any s in $[0, -\log q)$, we have* $e^s E\{\exp(-h_T(s)Z_T)\} = 1$.

PROOF. We have seen in the proof of Theorem 1 that, for fixed s in $(0, -\log q)$, $\{\exp(-h_n(s)Z_n), \mathscr{F}_n\}$ is a martingale. Also, the family $\{\exp(-h_n(s)Z_n)\}$ is trivially seen to be uniformly integrable so we may apply Theorem 2.2, Chapter 7, of Doob [1] and obtain

$$E\{\exp(-h_T(s)Z_T)\} = E\{\exp(-h(s)Z_1)\}$$
$$= \exp\{-k(h(s))\} = \exp\{-s\},$$

as required.

Theorem 2 is included in this paper as it follows so simply from the proof of Theorem 1. The result will be explored elsewhere.

4. An application of Theorem 1. In this section we shall establish the consistency in a certain sense of the estimator $\sum_{j=1}^n Z_j/\sum_{j=0}^{n-1} Z_j$ for m. This estimator has been discussed by Harris [2] who has shown that it is a maximum likelihood estimator for m and that, if $EZ_1^2 < \infty$, it is consistent in the sense that

$$\lim_{n\to\infty} \Pr(|(\textstyle\sum_{j=1}^n Z_j/\sum_{j=0}^{n-1} Z_j) - m| \geqq \varepsilon \,|\, Z_n > 0) = 0$$

for every $\varepsilon > 0$. We shall strengthen this result and remove the restriction that $EZ_1^2 < \infty$.

Firstly, we need the following theorem which is of some independent interest.

THEOREM 3. *If* $c_n^{-1}Z_n \to_{\text{a.s.}} W$ *where* $c_n \to \infty, c_n^{-1}c_{n+1} \to m$ *as* $n \to \infty$, *then* $c_n^{-1}\sum_{j=0}^n Z_j \to_{\text{a.s.}} mW/(m-1)$ *as* $n \to \infty$. ("a.s." denotes almost sure convergence).

PROOF. Take $c_0 = 1$ for convenience. Since $c_n^{-1}Z_n - W \to_{\text{a.s.}} 0$ as $n \to \infty$ we have, using the Toeplitz Lemma (e.g. Loève [4] 238),

$$\{\textstyle\sum_{k=0}^n c_k[c_k^{-1}Z_k - W]/\sum_{k=0}^n c_k\} \to_{\text{a.s.}} 0,$$

which yields

$$(\textstyle\sum_{k=0}^n c_k)^{-1}\sum_{k=0}^n Z_k \to_{\text{a.s.}} W. \tag{2}$$

Also, since $c_n^{-1}c_{n+1} \to m$ as $n \to \infty$, a further application of the Toeplitz Lemma gives

$$\{\textstyle\sum_{k=0}^n c_k[c_k^{-1}c_{k+1} - m]/\sum_{k=0}^n c_k\} \to 0, \qquad \text{that is,}$$
$$\{1 + (c_{n+1} - 1)/\textstyle\sum_{k=0}^n c_k\} \to m$$

as $n \to \infty$. This yields

$$\textstyle\sum_{k=0}^n c_k \sim mc_n/(m-1) \tag{3}$$

and the desired result follows immediately from (2) and (3).

THEOREM 4. *Let $\mathscr{E}$ denote the event $\{Z_k > 0, k = 1, 2, 3, \cdots\}$. Then, for arbitrary $\varepsilon > 0$,*

$$\lim_{n\to\infty} \Pr(\max_{k\geqq n} |(\sum_{j=1}^{k} Z_j / \sum_{j=0}^{k-1} Z_j) - m| \geqq \varepsilon \mid \mathscr{E}) = 0.$$

PROOF. Define the random variables U_n, $n = 1, 2, 3, \cdots$, W^* as follows:

$$\begin{aligned} U_n &= h_n(s_0) \textstyle\sum_{j=0}^{n} Z_j && \text{if} \quad Z_n > 0, \\ &= 1 && \text{if} \quad Z_n = 0; \\ W^* &= W && \text{if} \quad W > 0, \\ &= 1 && \text{if} \quad W = 0. \end{aligned}$$

Then, it is clear from Theorem 1 and Theorem 3 that U_n converges almost surely to $mW^*/(m-1)$ as $n \to \infty$, the random variable W^* having a distribution function which is continuous at zero. We therefore have, since $\Pr(\mathscr{E}) = 1-q$,

$$\begin{aligned} &\Pr(\max_{k\geqq n} |(\textstyle\sum_{j=1}^{k} Z_j / \sum_{j=0}^{k-1} Z_j) - m| \geqq \varepsilon \mid \mathscr{E}) \\ &\quad = (1-q)^{-1} \Pr(\max_{k\geqq n} U_{k-1}^{-1} |h_{k-1}(s_0)\{[h_k(s_0)]^{-1} U_k - 1\} - mU_{k-1}| \geqq \varepsilon; \mathscr{E}) \\ &\quad \leqq (1-q)^{-1} \Pr(\max_{k\geqq n} U_{k-1}^{-1} |h_{k-1}(s_0)\{[h_k(s_0)]^{-1} U_k - 1\} - mU_{k-1}| \geqq \varepsilon). \end{aligned}$$

The result of the theorem then follows readily because $\Pr(W^* = 0) = 0$ and

$$\begin{aligned} &h_{k-1}(s_0)\{[h_k(s_0)]^{-1} U_k - 1\} - mU_{k-1} \\ &\quad = (h_{k-1}(s_0)[h_k(s_0)]^{-1} - m)U_k - h_{k-1}(s_0) + m(U_k - U_{k-1}) \to_{\text{a.s.}} 0 \end{aligned}$$

as $n \to \infty$ since $h_{k-1}(s_0)[h_k(s_0)]^{-1} \to m$, $h_{k-1}(s_0) \to 0$ and U_k converges almost surely.

REFERENCES

[1] DOOB, J. L. (1953). *Stochastic Processes.* Wiley, New York.
[2] HARRIS, T. E. (1948). Branching processes. *Ann. Math. Statist.* **19** 474–494.
[3] KESTEN, H. and STIGUM, B. P. (1966). A limit theorem for multi-dimensional Galton–Watson processes. *Ann. Math. Statist.* **37** 1211–1223.
[4] LOÈVE, M. (1963). *Probability Theory.* 3rd ed. Van Nostrand, Princeton.
[5] LUKACS, E. (1960). *Characteristic Functions.* Griffin, London.
[6] SENETA, E. (1968). On recent theorems concerning the supercritical Galton–Watson process. *Ann. Math. Statist.* **39** 2098–2102.
[7] SENETA, E. (1969). Functional equations and the Galton–Watson process. *Adv. Appl. Probability* **1** 1–42.
[8] STIGUM, B. P. (1966). A theorem on the Galton–Watson process. *Ann. Math. Statist.* **37** 695–698.

Z. Wahrscheinlichkeitstheorie verw. Geb. 16, 151–156 (1970)

On the Implication of a Certain Rate of Convergence to Normality

C. C. HEYDE

1. Introduction

Consider a sequence of independent and identically distributed random variables X_i, $i=1,2,3,\ldots$ which belong to the domain of attraction of the normal distribution. Since this last property implies $E|X_i|<\infty$, we shall suppose for convenience that $EX_i=0$. Write

$$S_n=\sum_{i=1}^{n} X_i, \qquad n\geqq 1,$$

and

$$\Phi(x)=\frac{1}{\sqrt{2\pi}}\int_{-\infty}^{x} e^{-\frac{1}{2}u^2}\,du.$$

We shall establish the following main theorem.

Theorem 1. *If there is a monotone sequence of normalizing constants $\{B_n\}$ with $B_n\to\infty$ as $n\to\infty$ such that*

$$\sum_{n=3}^{\infty} n^{-1}(\log\log n)\sup_x |Pr(S_n\leqq B_n x)-\Phi(x)|<\infty, \tag{1}$$

then $EX_i^2<\infty$.

This result extends that contained in Theorem 3.1 of Ibragimov [7] where it is shown that if

$$\sup_x |Pr(S_n\leqq B_n x)-\Phi(x)|=0(n^{-\delta})$$

for some $\delta>0$, then $EX_i^2<\infty$. It provides us with a converse to a result of Heyde [4]. Furthermore, it provides an interesting elucidation of a result of Petrov [8] concerning iterated logarithm behaviour which was proved without explicit assumptions regarding the existence of moments. Theorem 1 is proved in Section 2 while Section 3 contains a discussion of its applications.

2. Proof of Theorem 1

Suppose firstly that the X_i are symmetric. Since

$$1-\Phi(x)\sim\frac{1}{x\sqrt{2\pi}}\,e^{-\frac{1}{2}x^2}$$

as $x\to\infty$, we have

$$\sum_{n=3}^{\infty} n^{-1}(\log\log n)\left\{1-\Phi\left[((2+\varepsilon)\log\log n)^{\frac{1}{2}}\right]\right\}<\infty$$

R. Maller et al. (eds.), *Selected Works of C.C. Heyde*, Selected Works in Probability and Statistics,
DOI 10.1007/978-1-4419-5823-5_20, © Springer Science+Business Media, LLC 2010

for any $\varepsilon>0$ so that, writing $t_n=[(2+\varepsilon)\log\log n]^{\frac{1}{2}}$, it follows from (1) that

$$\sum_{n=3}^{\infty} n^{-1}(\log\log n)\, Pr(|S_n|>B_n t_n)<\infty. \tag{2}$$

But, for symmetric random variables,

$$Pr(|S_n|>B_n t_n)\geqq \tfrac{1}{2} Pr(\max_{1\leqq k\leqq n}|X_k|>B_n t_n), \tag{3}$$

(e.g. Feller [2], II, 147) and, from Bonferroni's inequalities (e.g. [2], I, 100) we have

$$\begin{aligned} n\, Pr(|X_i|>B_n t_n)\,[1-\tfrac{1}{2}(n-1)\, Pr(|X_i|>B_n t_n)]& \\ \leqq Pr(\max_{1\leqq k\leqq n}|X_k|>B_n t_n)&\leqq n\, Pr(|X_i|>B_n t_n), \end{aligned}$$

so that

$$Pr(\max_{1\leqq k\leqq n}|X_k|>B_n t_n)\sim n\, Pr(|X_i|>B_n t_n) \tag{4}$$

as $n\to\infty$ and hence, from (2), (3) and (4),

$$\sum_{n=3}^{\infty} (\log\ \log n)\, Pr(|X_i|>B_n t_n)<\infty. \tag{5}$$

Now suppose that $EX_i^2=\infty$ so that

$$K(t)=\int_{|xt|\leqq 1} x^2\, d\, Pr(X_i\leqq x)\to\infty$$

as $t\to 0$. Since the X_i belong to the domain of attraction of the normal distribution, $K(t)$ is slowly varying as $t\to 0$ (e.g. [2], II, 303) and $n B_n^{-2} K(B_n^{-1})\to 1$ as $n\to\infty$ (e.g. [2], II, 304-305).

Then, we have

$$\begin{aligned} \int_{|x|\leqq B_n t_n} x^2\, d\, Pr(X_i\leqq x) &\leqq \int_{|x|\leqq B_3 t_3} x^2\, d\, Pr(X_i\leqq x) \\ &\quad+\sum_{k=3}^{n} B_{k+1}^2 t_{k+1}^2\, Pr(B_k t_k<|X_i|\leqq B_{k+1} t_{k+1}) \\ &\leqq \int_{|x|\leqq B_3 t_3} x^2\, d\, Pr(X_i\leqq x) \\ &\quad+C\sum_{k=3}^{n} K(B_k^{-1})k\log\log k\, Pr(B_k t_k<|X_i|\leqq B_{k+1} t_{k+1}) \end{aligned} \tag{6}$$

for some positive constant C. Furthermore, it is easily shown, using (5), that

$$\sum_{k=3}^{\infty} k(\log\log k)\, Pr(B_k t_k<|X_i|\leqq B_{k+1} t_{k+1})<\infty. \tag{7}$$

Thus, using (6), (7) and by virtue of our assumption that $K(t)\to\infty$ as $t\to 0$, we have from the Kronecker lemma that

$$n B_n^{-2}\int_{|x|\leqq B_n t_n} x^2\, d\, Pr(X_i\leqq x)\sim[K(B_n^{-1})]^{-1} K(B_n^{-1} t_n^{-1})\to 0$$

as $n\to\infty$. This establishes the required contradiction since by definition,

$$[K(B_n^{-1})]^{-1}\,K(B_n^{-1}\,t_n^{-1})\geqq 1.$$

It follows that (5) is only possible in the case where $K(B_n^{-1})$ approaches a positive constant as $n\to\infty$ in which case it is easily seen that $EX_i^2<\infty$. This completes the proof for the case of symmetric X_i.

In order to deal with the general case, consider the sequence Y_i, $i=1,2,3,\ldots$ of independent, symmetrized random variables; each Y_i having the distribution of the difference between two independent X_i's. Write

$$G_n(x)=Pr(Y_1+\cdots+Y_n\leqq x\,B_n\sqrt{2}),\qquad F_n(x)=Pr(S_n\leqq B_n\,x).$$

Then, using $*$ to denote convolution,

$$G_n(x)=F_n(x\sqrt{2})*(1-F_n(-x\sqrt{2}-0)),$$

and in view of (1) we obtain

$$\begin{aligned}
&\sum_{n=3}^{\infty} n^{-1}(\log\log n)\sup_x |G_n(x)-\Phi(x)|\\
&\quad=\sum_{n=3}^{\infty} n^{-1}(\log\log n)\sup_x |F_n(x\sqrt{2})*(1-F_n(-x\sqrt{2}-0))\\
&\qquad-\Phi(x\sqrt{2})*(1-\Phi(-x\sqrt{2}))|\\
&\quad\leqq\sum_{n=3}^{\infty} n^{-1}(\log\log n)\sup_x |F_n(x\sqrt{2})*(1-F_n(-x\sqrt{2}-0))\\
&\qquad-\Phi(x\sqrt{2})*(1-F_n(-x\sqrt{2}-0))|\\
&\qquad+\sum_{n=3}^{\infty} n^{-1}(\log\log n)\sup_x |\Phi(x\sqrt{2})*(1-F_n(-x\sqrt{2}-0))\\
&\qquad-\Phi(x\sqrt{2})*(1-\Phi(-x\sqrt{2}))|<\infty.
\end{aligned}$$

Thus, our result for the symmetric case gives $EY_i^2<\infty$ which implies $EX_i^2<\infty$. This completes the proof of the theorem.

Remark 1. The following result of Davis [1] is worth noting in connection with Theorem 1. If $EX_i=0$, $EX_i^2=1$, $EX_i^2\log^+\log^+|X_i|<\infty$ and

$$\beta_n^2=\int_{|x|<\sqrt{n}} x^2\,d\,Pr(X_i\leqq x)-\Big(\int_{|x|<\sqrt{n}} x\,d\,Pr(X_i\leqq x)\Big)^2,$$

then

$$\sum_{n=3}^{\infty} n^{-1}(\log\log n)\sup_x |Pr(S_n\leqq x\,\beta_n\sqrt{n})-\Phi(x)|<\infty.$$

(Here $\log^+ x$ denotes $\max(0,\log x)$.)

Remark 2. It is conjectured that an improved version of Theorem 1 will hold with the $\log\log n$ term omitted in (1). For background to this conjecture see Heyde [5].

3. Applications of Theorem 1

Firstly we use Theorem 1 to elucidate a result of Heyde [4].

Theorem 2. *Take* $0<\delta<1$. *Let* $E|X_i|^{2+\delta}<\infty$ *and write* $B_n^2=nEX_i^2(1+\varepsilon_n)$. *Then, if* $\sum_1^\infty n^{-1+\delta/2}|\varepsilon_n|<\infty$, *we obtain*

$$\sum_{n=1}^{\infty} n^{-1+\delta/2} \sup_x |Pr(S_n \leqq B_n x)-\Phi(x)|<\infty. \tag{8}$$

Conversely, if (8) *holds for some monotone sequence* $\{B_n\}$, *and the* X_i *belong to the domain of attraction of the normal distribution, then* $EX_i^2<\infty$ *and* $B_n^2=nEX_i^2(1+\varepsilon_n)$ *where* $\varepsilon_n \to 0$ *as* $n\to\infty$. *If, in addition,* $\sum_1^\infty n^{-1+\delta/2}|\varepsilon_n|<\infty$, *then* $E|X_i|^{2+\delta}<\infty$.

Proof. Write $EX_i^2=\sigma^2(<\infty)$. If $E|X_i|^{2+\delta}<\infty$, we obtain from the theorem of [4] that

$$\sum_{n=1}^{\infty} n^{-1+\delta/2} \sup_x |Pr(S_n \leqq x\sigma\sqrt{n})-\Phi(x)|<\infty. \tag{9}$$

Furthermore,

$$\begin{aligned} &Pr(S_n \leqq x\sigma\sqrt{n(1+\varepsilon_n)})-\Phi(x)\\ &\quad = Pr(S_n \leqq (x\sqrt{1+\varepsilon_n}\,\sigma\sqrt{n})-\Phi(x\sqrt{1+\varepsilon_n})+\Phi(x\sqrt{1+\varepsilon_n})-\Phi(x), \end{aligned} \tag{10}$$

while expanding by Taylor's theorem gives

$$\Phi(x\sqrt{1+\varepsilon_n})-\Phi(x)=\frac{1}{2\sqrt{2\pi}}\varepsilon_n x e^{-\frac{1}{2}x^2}+0(\varepsilon_n^2). \tag{11}$$

Thus, upon taking sup with respect to x in (10) and using (11), we obtain

$$\sup_x |Pr(S_n \leqq x\sigma\sqrt{n(1+\varepsilon_n)})-\Phi(x)| \leqq \sup_x |Pr(S_n \leqq x\sigma\sqrt{n})-\Phi(x)|+C|\varepsilon_n|, \tag{12}$$

for some positive constant C. The result of the first part of the theorem then follows from (9) and (12).

Now suppose that (8) holds for some monotone sequence $\{B_n\}$, and the X_i belong to the domain of attraction of the normal distribution. It follows from Theorem 1 that $EX_i^2=\sigma^2<\infty$, and consequently B_n must be of the form $B_n=\sigma\sqrt{n(1+\varepsilon_n)}$ where $\varepsilon_n\to 0$ as $n\to\infty$. Further, the equation (10) can be rewritten to yield

$$\sup_x |Pr(S_n \leqq x\sigma\sqrt{n})-\Phi(x)| \leqq \sup_x |Pr(S_n \leqq x\sigma\sqrt{n(1+\varepsilon_n)})-\Phi(x)|+C'|\varepsilon_n|,$$

where C' is a positive constant, so that, if $\sum_1^\infty n^{-1+\delta/2}|\varepsilon_n|<\infty$, (9) holds. The result $E|X_i|^{2+\delta}<\infty$ then follows from the theorem of [4]. This completes the proof.

Our next application of Theorem 1 concerns its relation to questions of behaviour of iterated logarithm type. It is well-known that

$$Pr(\limsup_{n\to\infty}(2n\log\log n)^{-\frac{1}{2}}S_n=1)=1$$

if and only if $EX_i=0$, $EX_i^2=1$ (Hartman and Wintner [3] obtained the sufficiency part and Strassen [9] the necessity). However, it has been shown by Freedman (see [9]) that there exist symmetric random variables X_i with $EX_i^2=\infty$ or even $E|X_i|=\infty$ for which it is possible to choose a monotone sequence $\{b_n\}$ of positive constants with $b_n\to\infty$ as $n\to\infty$ such that

$$Pr(\limsup_{n\to\infty} b_n^{-1}S_n=1)=1. \tag{13}$$

Furthermore, Heyde [6] has shown that the behaviour (13) is only possible if the X_i belong to the domain of partial attraction of the normal distribution. Also, the construction of Freedman's is quite inexplicit and it is possible to obtain very little idea of the generality of the circumstances under which behaviour of the type (13) is possible when $EX_i^2=\infty$.

Now the following theorem has recently been obtained by Petrov [8].

Theorem. *Suppose that the independent random variables X_i, $i=1,2,3,\ldots$ have symmetric distributions. If there exists a sequence of constants $\{C_n\}$ such that*

$$C_n\uparrow\infty,\qquad C_n^{-1}C_{n+1}\to 1$$

and

$$\sup_x\left|Pr\left(\sum_{i=1}^n X_i<C_n^{\frac{1}{2}}x\right)-\Phi(x)\right|=0((\log C_n)^{-1-\delta}) \tag{14}$$

as $n\to\infty$ for some $\delta>0$, then

$$Pr\left(\limsup_{n\to\infty}(2C_n\log\log C_n)^{-\frac{1}{2}}\sum_{i=1}^n X_i=1\right)=1.$$

This theorem involves no explicit assumptions regarding the existence of moments and it is an interesting question as to whether it can be used to generate results of the type of (13) for identically distributed random variables with $EX_i^2=\infty$. The answer to this question is in the negative. This follows immediately from Theorem 1 upon noting that, in the case of identically distributed X_i,

$$\log C_n\sim\tfrac{1}{2}\log n$$

as $n\to\infty$.

References

1. Davis, J. A.: Convergence rates for the law of the iterated logarithm. Ann. math. Statistics **39**, 1479–1485 (1968).
2. Feller, W.: An introduction to probability theory and its applications, vols. I, II. New York: Wiley 1957, 1966.
3. Hartman, P., Wintner, A.: On the law of the iterated logarithm. Amer. J. Math. **63**, 169–176 (1941).

4. Heyde, C. C.: On the influence of moments on the rate of convergence to the normal distribution. Z. Wahrscheinlichkeitstheorie verw. Geb. **8**, 12–18 (1967).
5. – Some properties of metrics in a study on convergence to normality. Z. Wahrscheinlichkeitstheorie verw. Geb. **11**, 181–192 (1969).
6. – A note concerning behaviour of iterated logarithm type. Proc. Amer. math. Soc. (1969) (in press).
7. Ibragimov, I. A.: On the accuracy of the Gaussian approximation to the distribution function of sums of independent variables. Theor. Probab. Appl. **11**, 559–576 (1966).
8. Petrov, V. V.: On the law of the iterated logarithm without assumptions about the existence of moments. Proc. Nat. Acad. Sci. U.S.A. **59**, 1068–1072 (1968).
9. Strassen, V.: A converse to the law of the iterated logarithm. Z. Wahrscheinlichkeitstheorie verw. Geb. **4**, 265–268 (1966).

Dr. C. C. Heyde
Department of Statistics
Australian National University
School of General Studies
Box 4, P.O.
Canberra, A.C.T. 2600
Australia

(Received August 18, 1969)

J. Appl. Prob. **7**, 451–454 (1970)
Printed in Israel

A RATE OF CONVERGENCE RESULT FOR THE SUPER-CRITICAL GALTON-WATSON PROCESS

C. C. HEYDE, *Australian National University*

Let $Z_0 = 1, Z_1, Z_2, \cdots$ denote a super-critical Galton-Watson process whose non-degenerate offspring distribution has probability generating function $F(s) = \sum_{j=0}^{\infty} s^j P(Z_1 = j), 0 \leqq s \leqq 1$, where $1 < m = EZ_1 < \infty$. The Galton-Watson process evolves in such a way that the generating function $F_n(s)$ of Z_n is the nth functional iterate of $F(s)$. The convergence problem for Z_n, when appropriately normed, has been studied by quite a number of authors; for an ultimate form see Heyde [2]. However, no information has previously been obtained on the rate of such convergence. We shall here suppose that $EZ_1^2 < \infty$ in which case $W_n = m^{-n}Z_n$ converges almost surely to a non-degenerate random variable W as $n \to \infty$ (Harris [1], p. 13). It is our object to establish the following result on the rate of convergence of W_n to W.

Theorem. Let $\operatorname{var} Z_1 = \sigma^2 < \infty$. As $n \to \infty$, $m^{n/2}(W - W_n)$ converges in distribution. The limit law is given by the characteristic function relation

$$\lim_{n\to\infty} E[\exp\{it\,m^{n/2}(W - W_n)\}] = E[\exp\{-\tfrac{1}{2}t^2\sigma^2(m^2 - m)^{-1}W\}].$$

Proof. Firstly, we take $r > n$ and consider

$$\begin{aligned} &E[\exp\{it\,m^{n/2}(W_r - W_n)\}] \\ (1)\qquad &= \sum_{j=0}^{\infty} E[\exp\{it\,m^{n/2}(m^{-r}Z_r - m^{-n}Z_n)\} \mid Z_n = j]P(Z_n = j) \\ &= \sum_{j=0}^{\infty} \exp\{-it\,m^{-n/2}j\}\,[E[\exp\{it\,m^{n/2-r}Z_{r-n}\}]]^j P(Z_n = j) \\ &= E[\exp\{-it\,m^{-n/2}\}\,E[\exp\{itm^{n/2-r}Z_{r-n}\}]]^{Z_n}. \end{aligned}$$

Next, we let $r \to \infty$ in (1) keeping n fixed. Given $\varepsilon > 0$ we can choose N so large that $\sum_{j=N+1}^{\infty} P(Z_n = j) < \varepsilon$. Also, since W_{r-n} converges in distribution to W, $E[\exp\{it\,m^{n/2-r}Z_{r-n}]]$ converges to $E[\exp\{it\,m^{-n/2}W\}]$, uniformly in any finite t interval, so we can find R so large that

Received 2 January 1970.

R. Maller et al. (eds.), *Selected Works of C.C. Heyde*, Selected Works in Probability and Statistics, DOI 10.1007/978-1-4419-5823-5_21, © Springer Science+Business Media, LLC 2010

$$\left| E[\exp\{itm^{n/2-r}Z_{r-n}\}] - E[\exp\{itm^{-n/2}W\}] \right| < \varepsilon$$

for $r > R$. Consequently,

$$\sum_{j=0}^{\infty} \exp\{-itm^{-n/2}j\}[E[\exp\{itm^{n/2-r}Z_{r-n}\}]]^j P(Z_n = j)$$

$$= \sum_{j=0}^{\infty} \exp\{-itm^{-n/2}j\}[E[\exp\{itm^{-n/2}W\}]]^j P(Z_n = j)$$

$$+ \sum_{j=0}^{R} \exp\{-itm^{-n/2}j\}\{[E[\exp\{itm^{n/2-r}Z_{r-n}\}]]^j - [E[\exp\{itm^{-n/2}W\}]]^j\}P(Z_n=j)$$

$$+ \sum_{j=R+1}^{\infty} \exp\{-itm^{-n/2}j\}\{[E[\exp\{itm^{n/2-r}Z_{r-n}\}]]^j - [E[\exp\{itm^{-n/2}W\}]]^j\}P(Z_n=j),$$

and

$$\left| \sum_{j=R+1}^{\infty} \exp\{-itm^{-n/2}j\}\{[E[\exp\{itm^{n/2-r}Z_{r-n}\}]]^j - [E[\exp\{itm^{-n/2}W\}]]^j\}P(Z_n=j) \right|$$

$$\leqq 2 \sum_{j=R+1}^{\infty} P(Z_n = j) \leqq 2\varepsilon,$$

while for $r > R$,

$$\left| \sum_{j=0}^{R} \exp\{itm^{-n/2}j\}\{[E[\exp\{itm^{n/2-r}Z_{r-n}\}]]^j - [E[\exp\{itm^{-n/2}W\}]]^j\}P(Z_n = j) \right|$$

$$\leqq \sum_{j=0}^{R} \left| [E[\exp\{itm^{n/2-r}Z_{r-n}\}]]^j - [E[\exp\{itm^{-n/2}W\}]]^j \right| P(Z_n = j)$$

$$\leqq \varepsilon \sum_{j=0}^{R} jP(Z_n = j) \leqq \varepsilon m^n$$

since $EZ_n = m^n$. However, ε is arbitrary and n fixed so that

$$E[\exp\{itm^{n/2}(W - W_n)\}] = \lim_{r \to \infty} E[\exp\{itm^{n/2}(W_r - W_n)\}]$$

$$(2) \qquad = \sum_{j=0}^{\infty} \exp\{-itm^{-n/2}j\}[E[\exp\{itm^{-n/2}W\}]]^j P(Z_n = j)$$

$$= E[E[\exp\{itm^{-n/2}(W-1)\}]]^{Z_n}.$$

Now,

$$E[E[\exp\{itm^{-n/2}(W-1)\}]]^{Z_n}$$

$$= \int_0^{\infty} [E[\exp\{itm^{-n/2}(W-1)\}]]^{xm^n} dP(m^{-n}Z_n \leqq x)$$

$$(3) \qquad = P(Z_n = 0 + m^n \log E[\exp\{itm^{-n/2}(W-1)\}] \int_0^\infty P(m^{-n}Z_n > x)$$
$$\times [E[\exp\{itm^{-n/2}(W-1)\}]]^{xm^n} dx,$$

using integration by parts. Furthermore, under the hypotheses of the theorem, W has mean 1 and variance $\sigma^2/(m^2 - m)$ ([1], p. 13) so we may expand $E[\exp\{itm^{-n/2}(W-1)\}]$ in the form

$$(4) \qquad E[\exp\{itm^{-n/2}(W-1)\}] = 1 - \tfrac{1}{2}t^2\sigma^2 m^{-n}(m^2 - m)^{-1} + o(t^2 m^{-n}).$$

Thus,

$$(5) \qquad m^n \log E[\exp\{itm^{-n/2}(W-1)\}] \to -\tfrac{1}{2}t^2\sigma^2(m^2 - m)^{-1}$$

and

$$(6) \qquad [E[\exp\{itm^{-n/2}(W-1)\}]]^{m^n} \to \exp\{-\tfrac{1}{2}t^2\sigma^2(m^2 - m)^{-1}\}$$

as $n \to \infty$. Also, using (4), we have for fixed t and suitably large n,

$$|E[\exp\{itm^{-n/2}(W-1)\}]|^{m^n} < \exp\{-\tfrac{1}{2}t^2\sigma^2(m^2 - m)^{-1} + o(t^2)\}$$
$$< \exp\{-\tfrac{1}{4}t^2\sigma^2(m^2 - m)^{-1}\},$$

so that the integrand in the rightmost term of (3) is bounded in absolute value by $\exp\{-\frac{1}{4}t^2\sigma^2 x(m^2 - m)^{-1}\}$. It then follows from Fatou's lemma together with (5), (6) and since $P(m^{-n}Z_n > x) \to P(W > x)$ as $n \to \infty$, that

$$\lim_{n\to\infty} E[\exp\{itm^{n/2}(W - W_n)\}]$$
$$= P(W = 0) - \tfrac{1}{2}t^2\sigma^2(m^2 - m)^{-1} \int_0^\infty P(W > x)\exp\{-\tfrac{1}{2}t^2\sigma^2 x(m^2 - m)^{-1}\}dx$$
$$= \int_0^\infty \exp\{-\tfrac{1}{2}t^2\sigma^2 x(m^2 - m)^{-1}\}dP(W \leqq x)$$
$$= E[\exp\{-\tfrac{1}{2}t^2\sigma^2(m^2 - m)^{-1}W\}],$$

again using integration by parts. This completes the proof of the theorem.

Unfortunately, the distribution function corresponding to the characteristic function $E[\exp\{-\frac{1}{2}t^2\sigma^2(m^2 - m)^{-1}W\}]$ does not seem to have a useful general representation. However, in certain particular cases it may be recognised simply.

For example, in the case of a fractional linear generating function for the offspring distribution ([1], pp. 9, 17) it is easily found to have an atom at zero and a Laplace density elsewhere.

References

[1] HARRIS, T. E. (1963) *The Theory of Branching Processes.* Springer-Verlag, Berlin.

[2] HEYDE, C. C. (1970) Extension of a result of Seneta for the super-critical Galton-Watson process. *Ann. Math. Statist.* **41** (in press).

The Annals of Mathematical Statistics
1970, Vol. 41, No. 6, 2161–2165

ON THE DEPARTURE FROM NORMALITY OF A CERTAIN CLASS OF MARTINGALES

C. C. HEYDE AND B. M. BROWN

Australian National University and La Trobe University

Let $\{X_n, \mathscr{F}_n, n = 0, 1, 2, \cdots\}$ be a martingale with $X_0 = 0$ a.s., $X_n = \sum_{i=1}^n Y_i$, $n \geqq 1$, and $\mathscr{F}_n$ the σ-field generated by $X_0, X_1, \cdots, X_n$. Write

$$\sigma_n^2 = E(Y_n^2 \mid \mathscr{F}_{n-1}), \qquad s_n^2 = \sum_{i=1}^n E\sigma_i^2,$$

and suppose that there is a constant δ, with $0 < \delta \leqq 1$, such that $E|Y_n|^{2+2\delta} < \infty$, $n = 1, 2, \cdots$. It is the object of this paper to establish the following theorem on departure from normality.

THEOREM. *There exist finite constants K_1, K_2 depending only on δ, such that*

$$\begin{aligned} &\sup_x |P(X_n \leqq s_n x) - \Phi(x)| \\ (1) \qquad &\leqq K_1\{s_n^{-2-2\delta}(\textstyle\sum_{i=1}^n E|Y_i|^{2+2\delta} + E|(\sum_{i=1}^n \sigma_i^2) - s_n^2|^{1+\delta})\}^{1/(3+2\delta)} \\ &\leqq K_2\{s_n^{-2-2\delta}(\textstyle\sum_{i=1}^n E|Y_i|^{2+2\delta} + E|(\sum_{i=1}^n Y_i^2) - s_n^2|^{1+\delta})\}^{1/(3+2\delta)}, \end{aligned}$$

where

$$\Phi(x) = (2\pi)^{-\frac{1}{2}} \int_{-\infty}^x e^{-\frac{1}{2}u^2}\, du.$$

Thus, if

$$(2) \qquad \lim_{n\to\infty} s_n^{-2-2\delta} \sum_{i=1}^n E|Y_i|^{2+2\delta} = 0, \qquad \textit{and}$$

$$(3) \qquad \lim_{n\to\infty} E|(s_n^{-2}\sum_{i=1}^n Y_i^2) - 1|^{1+\delta} = 0$$

or more generally, (2) *and*

$$(4) \qquad \lim_{n\to\infty} E|(s_n^{-2}\sum_{i=1}^n \sigma_i^2) - 1|^{1+\delta} = 0,$$

then $\lim_{n\to\infty} P(X_n \leqq s_n x) = \Phi(x)$, *and a bound on the rate of convergence is given by* (1).

The interesting feature of this result is the bound on departure from normality. More general central limit results for martingales are known, under conditions related to (2), (3) and (4) (see for example Brown [1]), but rates of convergence are not available.

We note that if $Y_1, Y_2, \cdots$ are independent, or more generally if $\sigma_1^2, \sigma_2^2, \cdots$ are constants a.s., then $s_n^{-2}\sum_{i=1}^n \sigma_i^2 = 1$ a.s., (4) is trivially true, and the first bound in (1) assumes a simplified form, since its second term vanishes. The utility of the second bound in (1) is that it does not depend on the conditioning by the sequence of σ-fields $\{\mathscr{F}_n\}$.

The proof of the theorem is based on a martingale form of the Skorokhod representation theorem which we state as a lemma in the interests of clarity.

Received January 19, 1970.

R. Maller et al. (eds.), *Selected Works of C.C. Heyde*, Selected Works in Probability and Statistics,
DOI 10.1007/978-1-4419-5823-5_22, © Springer Science+Business Media, LLC 2010

LEMMA. (Strassen [3], Theorem 4.3) *Let $Z_1, Z_2, \cdots$ be random variables such that for all n, $E(Z_n^2 \mid Z_{n-1}, \cdots, Z_1)$ is defined and $E(Z_n \mid Z_{n-1}, \cdots, Z_1) = 0$ a.s. Then, there is a Brownian motion W together with a sequence of nonnegative random variables $T_1, T_2, \cdots$ such that*

$$\textstyle\sum_{i=1}^n Z_i = W(\sum_{i=1}^n T_i) \quad \text{a.s.}$$

for all n. Moreover, if $\mathscr{G}_n$ is the σ-field generated by $Z_1, Z_2, \cdots, Z_n$ and $W(t)$ for $0 \leqq t \leqq \sum_{i=1}^n T_i$, then T_n is $\mathscr{G}_n$-measurable, $E(T_n \mid \mathscr{G}_{n-1})$ is defined and

$$E(T_n \mid \mathscr{G}_{n-1}) = E(Z_n^2 \mid \mathscr{G}_{n-1}) = E(Z_n^2 \mid Z_{n-1}, \cdots, Z_1) \quad \text{a.s.}$$

If k is a real number > 1 and $E(|Z_n|^{2k} \mid Z_{n-1}, \cdots, Z_1)$ is defined, then $E(T_n^k \mid \mathscr{G}_{n-1})$ is also defined and

$$\begin{aligned} E(T_n^k \mid \mathscr{G}_{n-1}) &\leqq L_k E(|Z_n|^{2k} \mid \mathscr{G}_{n-1}) \\ &= L_k E(|Z_n|^{2k} \mid Z_{n-1}, \cdots, Z_1) \quad \text{a.s.,} \end{aligned}$$

where L_k are constants which depend only on k.

Applying the lemma to the sequence of random variables $s_n^{-1} Y_1, \cdots, s_n^{-1} Y_n$, we note that there exists a Brownian motion W and nonnegative random variables $T_{n1}, \cdots, T_{nn}$ such that

$$s_n^{-1} X_n = W(\textstyle\sum_{i=1}^n T_{ni}) \quad \text{a.s.} \tag{5}$$

for all n. Moreover, if $\mathscr{G}_{ni}$ is the σ-field generated by $Y_1, \cdots, Y_i$ and $W(t)$ for $0 \leqq t \leqq \sum_{j=1}^i T_{nj}$, $1 \leqq i \leqq n$, then T_{ni} is $\mathscr{G}_{ni}$-measurable and

$$\begin{aligned} E(T_{ni} \mid \mathscr{G}_{n,i-1}) &= s_n^{-2} E(Y_i^2 \mid \mathscr{G}_{n,i-1}) \\ &= s_n^{-2} \sigma_i^2 \quad \text{a.s.} \end{aligned} \tag{6}$$

Now define

$$Z_{nn} = \textstyle\sum_{i=1}^n T_{ni} - 1, \tag{7}$$

noting that $EZ_{nn} = 0$, from (6). Then, from (5) and (7),

$$P(X_n \leqq s_n x) = P(W(1+Z_{nn}) \leqq x),$$

so that

$$|P(X_n \leqq s_n x) - \Phi(x)| = |P(W(1+Z_{nn}) \leqq x) - P(W(1) \leqq x)|. \tag{8}$$

Next, let $\{\varepsilon_n\}$ be a sequence with $0 < \varepsilon_n < 1$ and $\lim_{n\to\infty} \varepsilon_n = 0$. We have

$$\begin{aligned} &P(W(1+Z_{nn}) \leqq x) \\ &\quad = P(W(1+Z_{nn}) \leqq x; |Z_{nn}| \leqq \varepsilon_n) + P(W(1+Z_{nn}) \leqq x; |Z_{nn}| > \varepsilon_n) \\ &\quad \leqq P(\inf_{|t| \leqq \varepsilon_n} W(1+t) \leqq x; |Z_{nn}| \leqq \varepsilon_n) + P(|Z_{nn}| > \varepsilon_n) \\ &\quad \leqq P(\inf_{|t| \leqq \varepsilon_n} W(1+t) \leqq x) + P(|Z_{nn}| > \varepsilon_n), \end{aligned} \tag{9}$$

and similarly,

$$P(W(1+Z_{nn}) > x) \leqq P(\sup_{|t| \leqq \varepsilon_n} W(1+t) > x) + P(|Z_{nn}| > \varepsilon_n),$$

or equivalently

$$P(W(1+Z_{nn}) \leqq x) \geqq P(\sup_{|t| \leqq \varepsilon_n} W(1+t) \leqq x) - P(|Z_{nn}| > \varepsilon_n). \tag{10}$$

Now,

$$\begin{aligned} P(&\inf_{|t| \leqq \varepsilon_n} W(1+t) \leqq x) \\ &= P(\inf_{|t| \leqq \varepsilon_n} [W(1+t) - W(1-\varepsilon_n)] + W(1-\varepsilon_n) \leqq x) \\ &= P(W(1-\varepsilon_n) + \inf_{0 \leqq t \leqq 2\varepsilon_n} [W(1+t) - W(1)] \leqq x) \qquad (11) \\ &= \int_{-\infty}^{0} P(W(1-\varepsilon_n) \leqq x - y) \, dP(\inf_{0 \leqq t \leqq 2\varepsilon_n} W(t) \leqq y) \\ &= (\pi\varepsilon_n)^{-\frac{1}{2}} \int_0^\infty \Phi((x+y)(1-\varepsilon_n)^{-\frac{1}{2}}) \exp(-y^2/4\varepsilon_n) \, dy \\ &= \pi^{-\frac{1}{2}} \int_0^\infty \Phi((x+\varepsilon_n^{\frac{1}{2}} z)(1-\varepsilon_n)^{-\frac{1}{2}}) e^{-z^2/4} \, dz, \end{aligned}$$

and also, using a similar procedure,

$$P(\sup_{|t| \leqq \varepsilon_n} W(1+t) \leqq x) = \pi^{-\frac{1}{2}} \int_0^\infty \Phi((x-\varepsilon_n^{\frac{1}{2}} z)(1-\varepsilon_n)^{-\frac{1}{2}}) e^{-z^2/4} \, dz. \tag{12}$$

But,

$$|\Phi((x+\varepsilon_n^{\frac{1}{2}} z)(1-\varepsilon_n)^{-\frac{1}{2}}) - \Phi(x(1-\varepsilon_n)^{-\frac{1}{2}})| \leqq (2\pi)^{-\frac{1}{2}} |z| \varepsilon_n^{\frac{1}{2}} (1-\varepsilon_n)^{-\frac{1}{2}}, \tag{13}$$

so that, using (13) in (11) and (12), we obtain

$$\begin{aligned} \pi^{-\frac{1}{2}} &\int_0^\infty [\Phi(x(1-\varepsilon_n)^{-\frac{1}{2}}) - (2\pi)^{-\frac{1}{2}} z \varepsilon_n^{\frac{1}{2}} (1-\varepsilon_n)^{-\frac{1}{2}}] e^{-z^2/4} \, dz \\ &\leqq P(\sup_{|t| \leqq \varepsilon_n} W(1+t) \leqq x) \leqq P(\inf_{|t| \leqq \varepsilon_n} W(1+t) \leqq x) \\ &\leqq \pi^{-\frac{1}{2}} \int_0^\infty [\Phi(x(1-\varepsilon_n)^{-\frac{1}{2}}) + (2\pi)^{-\frac{1}{2}} z \varepsilon_n^{\frac{1}{2}} (1-\varepsilon_n)^{-\frac{1}{2}}] e^{-z^2/4} \, dz, \end{aligned}$$

which yields, for suitably large n,

$$\begin{aligned} P(&W(1-\varepsilon_n) \leqq x) - \varepsilon_n^{\frac{1}{2}} \\ &\leqq P(\sup_{|t| \leqq \varepsilon_n} W(1+t) \leqq x) \leqq P(\inf_{|t| \leqq \varepsilon_n} W(1+t) \leqq x) \qquad (14) \\ &\leqq P(W(1-\varepsilon_n) \leqq x) + \varepsilon_n^{\frac{1}{2}}. \end{aligned}$$

Combining (9) and (10) with (14) then gives

$$|P(W(1+Z_{nn}) \leqq x) - P(W(1-\varepsilon_n) \leqq x)| \leqq \varepsilon_n^{\frac{1}{2}} + P(|Z_{nn}| > \varepsilon_n). \tag{15}$$

Also, from a Taylor series expansion for $\varepsilon_n < \frac{1}{2}$, and trivially for $\varepsilon_n \geqq \frac{1}{2}$, we have

$$|P(W(1-\varepsilon_n) \leqq x) - P(W(1) \leqq x) \leqq 2\varepsilon_n. \tag{16}$$

Consequently, (15) and (16) imply that, for n sufficiently large,

$$\begin{aligned} |P(W(1+Z_{nn}) \leqq x) - P(W(1) \leqq x)| &\leqq 2\varepsilon_n^{\frac{1}{2}} + P(|Z_{nn}| > \varepsilon_n) \qquad (17) \\ &\leqq 2\varepsilon_n^{\frac{1}{2}} + \varepsilon_n^{-1-\delta} E|Z_{nn}|^{1+\delta} \end{aligned}$$

by Markov's inequality. The sequence $\{\varepsilon_n\}$ is then chosen to provide the best possible order in (17). Take $\varepsilon_n = (E|Z_{nn}|^{1+\delta})^{2/(3+2\delta)}$, so that

(18) $$|P(W(1+Z_{nn}) \leqq x) - P(W(1) \leqq x)| \leqq 3(E|Z_{nn}|^{1+\delta})^{1/(3+2\delta)},$$

and the problem reduces to that of producing a bound on $E|Z_{nn}|^{1+\delta}$.

Write $Z_{nn} = A_{nn} + B_n + C_n$, where

$$A_{ni} = \sum_{j=1}^{i}(T_{nj} - s_n^{-2}\sigma_j^2), \qquad 1 \leqq i \leqq n,$$
$$B_n = s_n^{-2}\sum_{i=1}^{n}(\sigma_i^2 - Y_i^2), \qquad C_n = s_n^{-2}\sum_{i=1}^{n} Y_i^2 - 1,$$

and note that $\{A_{ni}, \mathscr{G}_{ni}, 1 \leqq i \leqq n\}$ and $\{s_n^2 B_n, \mathscr{F}_n, n \geqq 1\}$ are both martingale sequences. Therefore, by Burkholder's martingale extension of the Marcinkiewicz–Zygmund inequality (Theorem 9 of [2]), there exists a finite universal constant $M_{1+\delta}$ such that

(19) $$\begin{aligned} E|A_{nn}|^{1+\delta} &\leqq M_{1+\delta}E(\textstyle\sum_{i=1}^{n}(T_{ni} - s_n^{-2}\sigma_i^2)^2)^{\frac{1}{2}(1+\delta)} \\ &\leqq M_{1+\delta}E\textstyle\sum_{i=1}^{n}|T_{ni} - s_n^{-2}\sigma_i^2|^{1+\delta}, \qquad \text{since } \tfrac{1}{2}(1+\delta) \leqq 1, \\ &\leqq M_{1+\delta}2^{\delta}\textstyle\sum_{i=1}^{n}(ET_{ni}^{1+\delta} + s_n^{-2-2\delta}E\sigma_i^{2+2\delta}) \\ &\leqq M_{1+\delta}2^{\delta}s_n^{-2-2\delta}\textstyle\sum_{i=1}^{n}(L_{1+\delta}E|Y_i|^{2+2\delta} + E\sigma_i^{2+2\delta}) \end{aligned}$$

using the lemma in the last step. Similarly,

(20) $$E|B_n|^{1+\delta} \leqq M_{1+\delta}2^{\delta}s_n^{-2-2\delta}\sum_{i=1}^{n}(E\sigma_i^{2+2\delta} + E|Y_i|^{2+2\delta}),$$

so that an application of the inequality

$$E\sigma_i^{2+2\delta} = E[E(Y_i^2 \mid \mathscr{F}_{i-1})]^{1+\delta} \leqq E[E(|Y_i|^{2+2\delta} \mid \mathscr{F}_{i-1})] = E|Y_i|^{2+2\delta}$$

to (19) and (20) gives

(21) $$E|A_{nn}|^{1+\delta} \leqq G_\delta s_n^{-2-2\delta}\sum_{i=1}^{n} E|Y_i|^{2+2\delta} \quad \text{and}$$

(22) $$E|B_n|^{1+\delta} \leqq G_\delta s_n^{-2-2\delta}\sum_{i=1}^{n} E|Y_i|^{2+2\delta}$$

for all n, where G_δ is a finite constant depending only on δ. Consequently, applying the inequality $(\sum_1^2|a_i|)^{1+\delta} \leqq 2^\delta\sum_1^2|a_i|^{1+\delta}$, we have

(23) $$\begin{aligned} E|Z_{nn}|^{1+\delta} &\leqq E|A_{nn} + (B_n + C_n)|^{1+\delta} \\ &\leqq 2^\delta(E|A_{nn}|^{1+\delta} + E|B_n + C_n|^{1+\delta}) \\ &\leqq 2^\delta[G_\delta s_n^{-2-2\delta}\textstyle\sum_{i=1}^{n} E|Y_i|^{2+2\delta} \\ &\qquad + E|(s_n^{-2}\textstyle\sum_{i=1}^{n}\sigma_i^2) - 1|^{1+\delta}], \end{aligned}$$

from (21), and the first part of (1) follows from (8), (18) and (23). To obtain the second part of (1), note that

(24) $$\begin{aligned} E|(s_n^{-2}\textstyle\sum_{i=1}^{n}\sigma_i^2) - 1|^{1+\delta} &= E|B_n + C_n|^{1+\delta} \\ &\leqq 2^\delta(E|B_n|^{1+\delta} + E|C_n|^{1+\delta}) \\ &\leqq 2^\delta[G_\delta s_n^{-2-2\delta}\textstyle\sum_{i=1}^{n} E|Y_i|^{2+2\delta} \\ &\qquad + E|(s_n^{-2}\textstyle\sum_{i=1}^{n} Y_i^2) - 1|^{1+\delta}], \end{aligned}$$

from (22), so that the required result follows from (24) and the first part of (1).

REMARK. It is evident from the proof of the theorem that a sufficient condition for $\lim_{n\to\infty} P(X_n \leqq s_n x) = \Phi(x)$ for all x, is the condition $Z_{nn} \to_p 0$ as $n \to \infty$, i.e.

$$\textstyle\sum_{i=1}^n T_{ni} \to_p 1 \quad \text{as} \quad n \to \infty,$$

which in turn is equivalent to the mean convergence

$$\lim_{n\to\infty} E\left|(\textstyle\sum_{i=1}^n T_{ni}) - 1\right| = 0,$$

since $Z_{nn} \geqq -1$ a.s. and $EZ_{nn} = 0$. These conditions, though not explicitly in terms of the martingale differences $\{Y_n\}$, are in a form much simpler than other known sufficient conditions for the central limit theorem for martingales (cf. [1]).

REFERENCES

[1] BROWN, B. M. (1970). Martingale central limit theorems. To appear *Ann. Math. Statist.* **42**

[2] BURKHOLDER, D. L. (1966). Martingale transforms. *Ann. Math. Statist.* **37** 1494–1504.

[3] STRASSEN, V. (1967). Almost sure behaviour of sums of independent random variables and martingales. *Proc. Fifth Berkeley Symp. Math. Statist. Prob.* **2** 315–343. Univ. of California Press.

Z. Wahrscheinlichkeitstheorie verw. Geb. 20, 189–192 (1971)

Some Almost Sure Convergence Theorems for Branching Processes

C. C. HEYDE

1. Introduction

In this paper, we shall be concerned with a super-critical Galton-Watson process $\{Z_0=1, Z_1, Z_2, \dots\}$ whose non-degenerate offspring distribution has $1<m=EZ_1<\infty$ and $\operatorname{var} Z_1=\sigma^2<\infty$. The Galton-Watson process evolves in such a way that the probability generating function of Z_n is the n-th functional iterate of the probability generating function of Z_1.

It is well-known that, for the process in question, $m^{-n} Z_n$ converges almost surely (a.s.) to a non-degenerate random variable W as $n\to\infty$ (e.g. Harris [2], p. 13). Also, a central limit analogue holds, namely that, conditional on $Z_n>0$, $(m^2-m)^{\frac{1}{2}}\sigma^{-1} Z_n^{-\frac{1}{2}}(m^n W-Z_n)$ converges in distribution to the unit normal (Heyde [3]). It is an important question as to whether there is, in addition, an iterated logarithm analogue and it is the object of this paper to show that this is the case. Corresponding results also hold for continuous time Markov branching processes and can be obtained by minor modifications of the present arguments. In this connection, we mention that an iterated logarithm analogue for the pure birth process has been obtained by Kendall [5] but his methods are quite different. We make extensive use of results on rates of convergence to normality which were established in [4].

2. Results

Theorem. *Suppose that $EZ_1^3<\infty$. Then, on the non-extinction set $\{W>0\}$ we have almost surely,*

$$\limsup_{n\to\infty}\frac{Z_{n+r}-m^r Z_n}{(2\sigma_r^2 Z_n \log n)^{\frac{1}{2}}}=1, \qquad \liminf_{n\to\infty}\frac{Z_{n+r}-m^r Z_n}{(2\sigma_r^2 Z_n \log n)^{\frac{1}{2}}}=-1 \tag{1}$$

and

$$\begin{aligned}\limsup_{n\to\infty}\frac{m^n W-Z_n}{(2\sigma^2(m^2-m)^{-1} Z_n\log n)^{\frac{1}{2}}}&=1,\\ \liminf_{n\to\infty}\frac{m^n W-Z_n}{(2\sigma^2(m^2-m)^{-1} Z_n\log n)^{\frac{1}{2}}}&=-1,\end{aligned} \tag{2}$$

where $r\geqq 1$ is any fixed integer and

$$\sigma_r^2=\operatorname{var} Z_r=\sigma^2 m^r(m^r-1)(m^2-m)^{-1}.$$

Proof. We shall just obtain the results for lim sup; those for lim inf require only minor modifications to the arguments given. There is also no essential loss of generality in giving the proof for the case $P(Z_1=0)=0$ in which case $Z_n\to\infty$

R. Maller et al. (eds.), *Selected Works of C.C. Heyde*, Selected Works in Probability and Statistics,
DOI 10.1007/978-1-4419-5823-5_23, © Springer Science+Business Media, LLC 2010

with probability one as $n\to\infty$. We shall do this for the sake of convenience. When $P(Z_1=0)>0$, it is well known that $Z_n\to\infty$ a.s. on the set $\{W>0\}$.

Firstly we shall work on (1). We have, using Theorem 2 of [4], that

$$\sum_{n=1}^{\infty}\sup_x\left|P(\sigma_r^{-1}Z_n^{-\frac{1}{2}}(Z_{r+n}-m^r Z_n)\leqq x)-\Phi(x)\right|<\infty, \tag{3}$$

$\Phi(x)$ being the distribution function of the unit normal. Thus, for any $\delta>0$,

$$\sum_{n=2}^{\infty}P(\sigma_r^{-1}Z_n^{-\frac{1}{2}}(Z_{r+n}-m^r Z_n)>(1+\delta)(2\log n)^{\frac{1}{2}})<\infty,$$

since $\sum_{n=2}^{\infty}[1-\Phi((1+\delta)(2\log n)^{\frac{1}{2}})]<\infty$ and hence, from the Borel-Cantelli lemma,

$$\limsup_{n\to\infty}\frac{Z_{r+n}-m^r Z_n}{(2\sigma_r^2 Z_n\log n)^{\frac{1}{2}}}\leqq 1 \tag{4}$$

with probability one.

Next, we shall fix attention on the particular case $r=1$. For $0<\delta<1$, define

$$A_n=\{Z_n-mZ_{n-1}>(1-\delta)\,\sigma(2Z_{n-1}\log(n-1))^{\frac{1}{2}}\},$$

$n=2,3,\ldots$ and note that $A_n\in\mathscr{F}(Z_1,\ldots,Z_n)$, the σ-field generated by $Z_1,\ldots,Z_n$. We make use of an extension of the Borel-Cantelli lemma which gives $P(A_n\text{ i.o.})=1$ if and only if $\sum_{n=1}^{\infty}P(A_{n+1}|Z_n,\ldots,Z_1)=\infty$ with probability one (e.g. Breiman [1], Corollary 5.29, p. 96).

Now,

$$\begin{aligned}P(A_{n+1}|Z_n,\ldots,Z_1)&\\ &=P(Z_{n+1}-mZ_n>(1-\delta)\,\sigma(2Z_n\log n)^{\frac{1}{2}}|Z_n,\ldots,Z_1)\\ &=P((Z_1^{(1)}-m)+\cdots+(Z_1^{(Z_n)}-m)>(1-\delta)\,\sigma(2Z_n\log n)^{\frac{1}{2}}|Z_n),\end{aligned}$$

where the $Z_1^{(i)}$ are independent and identically distributed each with the distribution of Z_1 and are independent of $Z_1,\ldots,Z_n$. Thus, using the Berry-Esseen inequality,

$$\begin{aligned}&P((Z_1^{(1)}-m)+\cdots+(Z_1^{(Z_n)}-m)>(1-\delta)\,\sigma(2Z_n\log n)^{\frac{1}{2}}|Z_n)\\ &\quad\geqq 1-\Phi((1-\delta)(2\log n)^{\frac{1}{2}})-cZ_n^{-\frac{1}{2}}\end{aligned}$$

c being a positive constant. Also,

$$\sum_{n=2}^{\infty}[1-\Phi((1-\delta)(2\log n)^{\frac{1}{2}})]=\infty\quad\text{and}\quad\sum_{n=1}^{\infty}Z_n^{-\frac{1}{2}}<\infty\quad\text{a.s.}$$

since

$$\sum_{n=1}^{\infty}E(Z_n^{-\frac{1}{2}})\leqq\sum_{n=1}^{\infty}(EZ_1^{-1})^{+n/2}<\infty$$

using Theorem 2 of [4]. Thus,

$$\sum_{n=1}^{\infty}P(A_{n+1}|Z_n,\ldots,Z_1)=\infty$$

with probability one, from which we deduce that $P(A_n \text{ i.o.})=1$. This gives

$$\limsup_{n\to\infty} \frac{Z_{n+1}-m Z_n}{(2\sigma^2 Z_n \log n)^{\frac{1}{2}}} \geqq 1 \tag{5}$$

with probability one and the particular case $r=1$ of the lim sup part of (1) follows from (4) and (5).

To deal with the case $r>1$, we write Z_n^* for Z_{rn}, $n=0,1,2,\dots$. The process $\{Z_n^*\}$ is a super-critical Galton-Watson process with offspring distribution that of Z_r. Thus, applying the lim sup part of (1) with $r=1$ to the process $\{Z_n^*\}$, we have with probability one,

$$\limsup_{n\to\infty} \frac{Z_{n+1}^*-m^r Z_n^*}{(2\sigma_r^2 Z_n^* \log n)^{\frac{1}{2}}} = 1,$$

which gives

$$\limsup_{n\to\infty} \frac{Z_{nr+r}-m^r Z_{nr}}{(2\sigma_r^2 Z_{nr} \log n r)^{\frac{1}{2}}} = 1$$

with probability one. But,

$$\limsup_{n\to\infty} \frac{Z_{n+r}-m^r Z_n}{(2\sigma_r^2 Z_n \log n)^{\frac{1}{2}}} \geqq \limsup_{n\to\infty} \frac{Z_{nr+r}-m^r Z_{nr}}{(2\sigma_r^2 Z_{nr} \log n r)^{\frac{1}{2}}},$$

which together with (4) gives the lim sup part of (1). The lim inf part follows similarly.

In order to obtain (2) we firstly use Theorem 2 of [4] to deduce that

$$\sum_{n=1}^{\infty} \sup_x \left|P((m^2-m)^{\frac{1}{2}} \sigma^{-1} Z_n^{-\frac{1}{2}}(Z_n-m^n W) \leqq x)-\Phi(x)\right| < \infty, \tag{6}$$

from which we obtain, for any $\delta>0$,

$$\sum_{n=2}^{\infty} P(|Z_n-m^n W|>(1+\delta)(2\sigma^2(m^2-m)^{-1} Z_n \log n)^{\frac{1}{2}}) < \infty$$

since $\sum_{n=2}^{\infty} [1-\Phi((1+\delta)(2\log n)^{\frac{1}{2}})+\Phi(-(1+\delta)(2\log n)^{\frac{1}{2}})] < \infty$. Thus, from the Borel-Cantelli lemma,

$$\limsup_{n\to\infty} \frac{|Z_n-m^n W|}{(2\sigma^2(m^2-m)^{-1} Z_n \log n)^{\frac{1}{2}}} \leqq 1 \tag{7}$$

with probability one.

Next, making use of (1) we have with probability one and for any fixed integer $r \geqq 1$,

$$1=\limsup_{n\to\infty} \frac{Z_{r+n}-m^r Z_n}{(2\sigma^2 m^r(m^r-1)(m^2-m)^{-1} Z_n \log n)^{\frac{1}{2}}},$$

so that

$$\begin{aligned}\left(1-\frac{1}{m^r}\right)^{\frac{1}{2}} &= \limsup_{n\to\infty} \frac{(m^{-r} Z_{r+n}-W m^n)+(W m^n-Z_n)}{(2\sigma^2(m^2-m)^{-1} Z_n \log n)^{\frac{1}{2}}} \\ &\leqq \limsup_{n\to\infty} \frac{m^{-r} Z_{r+n}-W m^n}{(2\sigma^2(m^2-m)^{-1} Z_n \log n)^{\frac{1}{2}}} + \limsup_{n\to\infty} \frac{W m^n-Z_n}{(2\sigma^2(m^2-m)^{-1} Z_n \log n)^{\frac{1}{2}}}\end{aligned} \tag{8}$$

with probability one. But,

$$\limsup_{n\to\infty} \frac{m^{-r} Z_{r+n} - W m^n}{(2\sigma^2 (m^2-m)^{-1} Z_n \log n)^{\frac{1}{2}}} = \frac{1}{m^{r/2}} \limsup_{n\to\infty} \frac{Z_{r+n} - W m^{r+n}}{(2\sigma^2 (m^2-m)^{-1} Z_{r+n} \log(r+n))^{\frac{1}{2}}} \left(\frac{Z_{r+n} \log(r+n)}{m^r Z_n \log n} \right)^{\frac{1}{2}} \leqq \frac{1}{m^{r/2}} \tag{9}$$

with probability one in view of (7) and since $m^r Z_n^{-1} Z_{r+n} \xrightarrow{\text{a.s.}} 1$ as $n \to \infty$. Therefore, using (9) in (8), we have with probability one that

$$\limsup_{n\to\infty} \frac{W m^n - Z_n}{(2\sigma^2 (m^2-m)^{-1} Z_n \log n)^{\frac{1}{2}}} \geqq \left(1 - \frac{1}{m^r}\right)^{\frac{1}{2}} - \frac{1}{m^{r/2}}. \tag{10}$$

Now, (10) holds for any integer $r \geqq 1$ so we must have

$$\limsup_{n\to\infty} \frac{W m^n \quad Z_n}{(2\sigma^2 (m^2-m)^{-1} Z_n \log n)^{\frac{1}{2}}} \geqq 1 \tag{11}$$

with probability one. (7) and (11) then give the required lim sup part of (2). Similar reasoning gives the lim inf part.

3. Concluding Remark

The condition $EZ_1^3 < \infty$ in the theorem can at least be weakened to $EZ_1^{2+\delta} < \infty$, some $\delta > 0$. This is a simple consequence of the fact that (3) and (6) and a minor modification of the Berry-Esseen bound continue to hold under the condition $EZ_1^{2+\delta} < \infty$. See the remarks after the proof of Theorem 2 of [4] concerning geometric bounds on the rate of convergence to normality for the case $EZ_1^{2+\delta} < \infty$.

Note Added in Proof. Subsequent research has revealed that the theorem of this paper continues to hold under the condition $EZ_1^2 < \infty$. This result will appear in a paper of Heyde and Leslie in Bull. Austral. math. Soc. **5** (1971).

References

1. Breiman, L.: Probability. Reading, Mass.: Addison-Wesley 1968.
2. Harris, T. E.: The theory of branching processes. Berlin-Göttingen-Heidelberg: Springer 1963.
3. Heyde, C.C.: Some central limit analogues for super-critical Galton-Watson processes. J. appl. Probab. **8**, 52–59 (1971).
4. – Brown, B. M.: An invariance principle and some convergence rate results for branching processes. Z. Wahrscheinlichkeitstheorie verw. Geb. (1971) (to appear).
5. Kendall, D.G.: Branching processes since 1873. J. London math. Soc. **41**, 385–406 (1966).

Dr. C.C. Heyde
Department of Statistics
Australian National University
School of General Studies
Box 4, P.O. Canberra, A.C.T. 2600
Australia

(Received November 9, 1970)

SOME CENTRAL LIMIT ANALOGUES FOR SUPERCRITICAL GALTON-WATSON PROCESSES

BY

C. C. HEYDE

Reprinted from the
JOURNAL OF APPLIED PROBABILITY
Vol. 8, No. 1, pp. 52–59
March 1971

R. Maller et al. (eds.), *Selected Works of C.C. Heyde*, Selected Works in Probability and Statistics,
DOI 10.1007/978-1-4419-5823-5_24, © Springer Science+Business Media, LLC 2010

J. Appl. Prob. **8**, 52–59 (1971)
Printed in Israel

SOME CENTRAL LIMIT ANALOGUES FOR SUPER-CRITICAL GALTON-WATSON PROCESSES

C. C. HEYDE, *Australian National University*

1. Introduction

It is possible to interpret the classical central limit theorem for sums of independent random variables as a convergence rate result for the law of large numbers. For example, if X_i, $i = 1,2,3,\cdots$ are independent and identically distributed random variables with $EX_i = \mu$, $\operatorname{var} X_i = \sigma^2 < \infty$ and $S_n = \sum_{i=1}^{n} X_i$, then the central limit theorem can be written in the form

$$\lim_{n\to\infty} P(\sigma^{-1}n^{\frac{1}{2}}(n^{-1}S_n - \mu) \leqq x) = \Phi(x) = (2\pi)^{-\frac{1}{2}} \int_{-\infty}^{x} e^{-\frac{1}{2}u^2} du.$$

This provides information on the rate of convergence in the strong law $n^{-1} S_n \overset{\text{a.s.}}{\to} \mu$ as $n \to \infty$. ("a.s." denotes almost sure convergence.) It is our object in this paper to discuss analogues for the super-critical Galton-Watson process.

Let $Z_0 = 1, Z_1, Z_2, \cdots$ denote a super-critical Galton-Watson process with a non-degenerate offspring distribution having probability generating function $F(s) = \sum_{j=0}^{\infty} s^j P(Z_1 = j)$, $0 \leqq s \leqq 1$, where $1 < m = EZ_1 < \infty$. The process evolves in such a way that the generating function $F_n(s)$ of Z_n is the nth functional iterate of $F(s)$. The convergence problem for Z_n, when appropriately normed, has been studied by quite a number of authors; for an ultimate form see Heyde [2]. However, we shall herein suppose that $EZ_1 \log Z_1 < \infty$ in which case $W_n = m^{-n}Z_n$ converges almost surely to a non-degenerate random variable W as $n \to \infty$ (e.g., Stigum [8], Kesten and Stigum [4]). We shall be dealing with the case where Z_1 belongs to the domain of attraction of a stable law of index α, $1 < \alpha \leqq 2$, so that $EZ_1^a < \infty$ for any $a < \alpha$ and in particular $EZ_1 \log Z_1 < \infty$.

Now the result $m^{-n} Z_n \overset{\text{a.s.}}{\to} W$ is an analogue of the strong law of large numbers. Furthermore, the existence of central limit analogues in this context has been shown by Heyde [3] where the following theorem is established.

Theorem A. Let $\operatorname{var} Z_1 = \sigma^2 < \infty$. As $n \to \infty$, $m^{\frac{1}{2}n}(W_n - W)$ converges in distribution. The limit law is given by the characteristic function relation

$$\lim_{n\to\infty} E[\exp\{itm^{\frac{1}{2}n}(W - W_n)\}] = E[\exp\{-\tfrac{1}{2}t^2\sigma^2(m^2 - m)^{-1}W\}].$$

We shall extend this result in Theorem 2 and then go on to obtain a different central limit analogue by modifying the norming (Theorem 3).

Received in revised form 8 June 1970.

2. Results

As a preliminary to setting up the central limit analogues we shall firstly obtain the following result of independent interest.

Theorem 1. W belongs to the domain of attraction of a stable law of index α, $1 < \alpha \leqq 2$, if and only if Z_1 belongs to the same domain of attraction. More specifically, if U_i, $i = 1,2,3,\cdots$ are independent and identically distributed, each with the distribution of $W-1$ and if $Z^{(i)}$, $i = 1,2,3,\cdots$ are independent and identically distributed, each with the distribution of Z_1, then $c_n^{-1}\sum_{i=1}^{n}(Z^{(i)}-m)$ converges in distribution to a stable law Y if and only if $(m^\alpha - m)^{1/\alpha}c_n^{-1}\sum_{i=1}^{n}U_i$ converges in distribution to Y. Here $\{c_n\}$ is a monotone sequence of posi ive constants.

Next, making use of Theorem 1, we shall obtain the following extension of Theorem A.

Theorem 2. Let Z_1 belong to the domain of attraction of a stable law of index α, $1 < \alpha \leqq 2$. More specifically, suppose that if $Z^{(i)}$, $i = 1,2,3,\cdots$ are independent and identically distributed, each with the distribution of Z_1, then there is a monotone sequence $\{c_n\}$ of constants such that $c_n^{-1}\sum_{i=1}^{n}(Z^{(i)} - m)$ converges in distribution to a stable law with characteristic function $f(t)$. Then, we have

$$\lim_{n\to\infty} E[\exp\{it(m^\alpha - m)^{1/\alpha}m^n c_{m^n}^{-1}(W - W_n)\}] = E[f(t)]^W.$$

We remark that since Z_1 is non-negative, $f(t)$ must be of the form $\exp\{-c|t|^\alpha(1 - i\tan\frac{1}{2}\pi\alpha\,\mathrm{sgn}\,t)\}$ where c is a positive constant (Gnedenko and Kolmogorov [1], Theorem 2, page 175).

Now the dependence of the limit result of Theorem 2 on the distribution of W is unfortunate as comparatively little can be said about W in general. The distribution function corresponding to the characteristic function $E[f(t)]^W$ can rarely be written down in a form which is suitably explicit. It would be much more pleasing to have a limit which did not involve W.

The clue to the problem of obtaining a limit independent of W comes from noting that $E[f(t)]^W$ is the characteristic function of a product $VW^{1/\alpha}$ where V and W are independent, V is stable with characteristic function $f(t)$ and α is the index of the stable law. This indicates an appropriate change to make in the norming which is done in the following theorem.

Theorem 3. Under the same conditions on the distribution of Z_1 as in Theorem 2,

$$\lim_{n\to\infty} E[\exp\{it(m^\alpha - m)^{-1/\alpha}m^n c_{Z_n}^{-1}(W - W_n)\}\,|\,Z_n > 0] = f(t).$$

The relation between Theorems 2 and 3 is further elucidated in Section 6.

3. Proof of Theorem 1

The finite variance case of this result is well-known. In fact var $W=(m^2-m)^{-1}$ var Z_1. Furthermore, Lamperti [5] has shown that W belongs to the domain of normal attraction of a stable law of index α, $1<\alpha\leqq 2$, if and only if Z_1 belongs to the same domain of normal attraction. The result of Theorem 1 is obtained by a straightforward extension of Lamperti's methods.

Let $\phi(\lambda)$ be the Laplace transform $E[\exp\{-\lambda W\}]$ and recall that F is the probability generating function of Z_1. The connection between F and ϕ is well-known to be

$$\phi(m\lambda)=F(\phi(\lambda)), \tag{1}$$

which, together with the requirement that they be a probability generating function and a Laplace transform respectively, determines either uniquely in terms of the other (e.g., Levinson [6]). Now W, which has mean one, belongs to the domain of attraction of a stable law of index α, $1<\alpha\leqq 2$, if and only if

$$\phi(\lambda)=1-\lambda+\lambda^\alpha L(\lambda^{-1})(1+o(1)) \tag{2}$$

as $\lambda\to 0$ where L is non-negative and slowly varying at infinity. Also, Z_1 belongs to the domain of attraction of a stable law of index α, $1<\alpha\leqq 2$, if and only if

$$1-F(x)=m(1-x)-(1-x)^\alpha M((1-x)^{-1})(1+o(1)) \tag{3}$$

as $x\to 1$ where M is non-negative and slowly varying at infinity. Consequently, we must derive the form (3) from (1) and (2) and conversely the form (2) from (1) and (3).

Firstly, assume (2). Then, using (1) and writing $\phi(\lambda)=x$, we have as $\lambda\to 0$,

$$\begin{aligned}1-F(x)&=m\lambda-m^\alpha\lambda^\alpha L((m\lambda)^{-1})(1+o(1))\\&=m\lambda-m^\alpha\lambda^\alpha L(\lambda^{-1})(1+o(1)),\end{aligned} \tag{4}$$

since L is slowly varying. But

$$x=1-\lambda+\lambda^\alpha L(\lambda^{-1})(1+o(1)), \tag{5}$$

so that, using (5) in (4),

$$1-F(x)=m(1-x)-(m^\alpha-m)\lambda^\alpha L(\lambda^{-1})(1+o(1)). \tag{6}$$

However, from (5),

$$\lambda=(1-x)(1+o(1))$$

as $x\to 1$, $\lambda\to 0$ and using this result in (4),

$$1-F(x)=m(1-x)-(m^\alpha-m)(1-x)^\alpha L((1-x)^{-1})(1+o(1)),$$

which is of the form (3).

Next, let

(7) $$1 = F(x) - m(1-x) - (1-x)^{\alpha} M((1-x)^{-1})(1 + A(1-x)),$$

where $A(1-x) = o(1)$ as $x \to 1$ and write

(8) $$\phi(\lambda) = 1 - \lambda + a(\lambda)$$

where $a(\lambda) = o(\lambda)$ as $\lambda \to 0$ since W has mean one. We then obtain from (1) and (7) that

(9) $$a(m\lambda) = ma(\lambda) + \lambda^{\alpha} M(\lambda^{-1})(1 + B(\lambda))$$

for a certain $B(\lambda)$ which is a function of $a(\lambda)$ and satisfies $B(\lambda) = o(1)$ as $\lambda \to 0$.

Now consider the more general functional equation

(10) $$b(m\lambda) = mb(\lambda) + \lambda^{\alpha} M(\lambda^{-1})(1 + B(\lambda))$$

with $B(\lambda)$ as in (9). The homogeneous part of (10) has the general solution $\lambda P(\log \lambda)$ where P is any function with period $\log m$. Thus, (10) has at most one and hence exactly one solution which is $o(\lambda)$ as $\lambda \to 0$, for if there were two, their difference would be of the form $\lambda P(\log \lambda)$ which is not $o(\lambda)$ unless $P(\log \lambda) \equiv 0$. This solution is easily seen to be of the form

(11) $$b(\lambda) = (m^{\alpha} - m)^{-1} \lambda^{\alpha} M(\lambda^{-1})(1 + o(1))$$

and this is the form of $a(\lambda)$ since $a(\lambda)$ is uniquely specified. The required form (2) then follows from (8) and (11).

Now we note from the above analysis that the slowly varying functions L and M of (2) and (3) are related by

(12) $$L(\lambda^{-1}) = (m^{\alpha} - m)^{-1} M(\lambda^{-1})(1 + o(1))$$

as $\lambda \to 0$. We shall exclude from consideration the case $\alpha = 2$ and $L(\lambda^{-1}), M(\lambda^{-1})$ asymptotically constant for then the variances of W and Z_1 are finite and are related by $\operatorname{var} W = (m^2 - m)^{-1} \operatorname{var} Z_1$ and the result of the theorem is obvious. We also note that when $\alpha = 2$ and the variances of W and Z_1 are not finite, $L(\lambda^{-1})$ and $M(\lambda^{-1})$ must be unbounded above as $\lambda \to 0$. Then, since

$$E[\exp\{-\lambda Z_1\}] = F(e^{-\lambda}),$$

we have from (3) that

$$E[\exp\{-\lambda c_n^{-1} \sum_{i=1}^{n} (Z^{(i)} - m)\}] = [F(\exp\{-c_n^{-1}\lambda\})\exp\{mc_n^{-1}\lambda\}]^n$$
$$\to \exp\{c\lambda^{\alpha}\}, c > 0,$$

as $n \to \infty$ if and only if

(13) $$nc_n^{-\alpha} M(c_n) \to c$$

as $n \to \infty$. Using (2) and (12), we see that the condition (13) is also necessary and sufficient for

$$E[\exp\{-\lambda(m^\alpha - m)^{1/\alpha}c_n^{-1} \sum_{i=1}^{n} U_i\}] = [\phi(\lambda c_n^{-1}(m^\alpha - m)^{1/\alpha})\exp\{\lambda c_n^{-1}(m^\alpha - m)^{1/\alpha}\}]^n$$
$$\to \exp\{c\lambda^\alpha\}$$

as $n \to \infty$. This completes the proof of the theorem.

4. Proof of Theorem 2

This follows the pattern of that of the theorem of Heyde [3] (Theorem A). Write $b_n = c_n(m^\alpha - m)^{-1/\alpha}$. Then, exactly as in [3] we find that

$$\begin{aligned} E[\exp\{itm^n b_{m^n}^{-1}(W - W_n)\}] &= E[E[\exp\{itb_{m^n}^{-1}(W-1)\}]]^{Z_n} \\ &= \int_0^\infty (E[\exp\{itb_{m^n}^{-1}(W-1)\}])^{xm^n} dP(m^{-n}Z_n \leqq x) \\ &= 1 + m^n \log E[\exp\{itb_{m^n}^{-1}(W-1)\}] \\ &\quad \times \int_0^\infty P(m^{-n}Z_n > x)(E[\exp\{itb_{m^n}^{-1}(W-1)\}])^{xm^n} dx \end{aligned} \tag{14}$$

using integration by parts. (In the right hand side of Equation 3 of [3], the term $P(Z_n = 0)$ should be replaced by 1.) Now, we have

$$E[\exp\{itc_n^{-1} \sum_{i=1}^{n} (Z^{(i)} - m)\}] = (E[\exp\{itc_n^{-1}(Z_1 - m)\}])^n \to f(t)$$

as $n \to \infty$ and consequently, using Theorem 1,

$$(E[\exp[itb_n^{-1}(W-1)\}])^n \to f(t)$$

as $n \to \infty$. Thus,

$$(E[\exp\{itb_{m^n}^{-1}(W-1)\}])^{m^n} \to f(t) \tag{15}$$

and

$$m^n \log E[\exp\{itb_{m^n}^{-1}(W-1)\}] \to \log f(t) \tag{16}$$

as $n \to \infty$. Also,

$$|f(t)| = \exp\{-c|t|^\alpha\}$$

for some constant $c > 0$ so that for suitably large n,

$$|E[\exp\{itb_{m^n}^{-1}(W-1)\}]|^{m^n} < \exp\{-\tfrac{1}{2}c|t|^\alpha\},$$

and therefore the integrand in the rightmost term of (14) is bounded in absolute value by $\exp\{-\frac{1}{2}c|t^{\alpha}|x\}$. It then follows from the dominated convergence theorem together with (15), (16) and since $P(m^{-n}Z_n > x) \to P(W > x)$ as $n \to \infty$, that

$$\begin{aligned}\lim_{n\to\infty} E[\exp\{itm^n b_{m_n}^{-1}(W - W_n)\} &= 1 + \log f(t)\int_0^{\infty} P(W > x)[f(t)]^x dx \\ &= \int_0^{\infty}[f(t)]^x dP(W \leqq x) \\ &= E[f(t)]^W\end{aligned}$$

again using integration by parts. This completes the proof.

5. Proof of Theorem 3

Take $r > n$ and again write $b_n = c_n(m^{\alpha} - m)^{-1/\alpha}$. We have

$$E[\exp\{itm^n b_{Z_n}^{-1}(W_r - W_n)\}\,|\,Z_n > 0]$$

$$= [P(Z_n > 0)]^{-1}\sum_{j=1}^{\infty} E[\exp\{itm^n b_{Z_n}^{-1}(m^{-r}Z_r - m^{-n}Z_n)\}\,|\,Z_n = j]P(Z_n = j)$$

(17) $$= [P(Z_n > 0)]^{-1}\sum_{j=1}^{\infty}\exp\{-itjb_j^{-1}\}(E[\exp\{itb_j^{-1}W_{r-n}\}])^j P(Z_n = j)$$

$$\to [P(Z_n > 0)]^{-1}\sum_{j=1}^{\infty}\exp\{-itjb_j^{-1}\}(E[\exp\{itb_j^{-1}W\}])^j P(Z_n = j)$$

as $r \to \infty$, the justification for taking the limit inside the summation following exactly as in [3]. Now define Z_n^*, $n \geqq 1$, such that

$$P(Z_n^* = j) = P(Z_n = j\,|\,Z_n > 0) = [P(Z_n > 0)]^{-1}P(Z_n = j), \quad j = 1,2\ 3,\cdots.$$

Then, from (17),

$$\begin{aligned}E[\exp\{itm^n b_Z^{-1}(W - W_n)\}\,|\,Z_n > 0] &\\ = \lim_{r\to\infty} E[\exp\{itm^n b_{Z}^{-1}(W_r - W_n)\}\,|\,Z_n > 0] &\\ = \sum_{j=1}^{\infty}(E[\exp\{itb_j^{-1}(W - 1)\}])^j P(Z_n^* = j) &\\ = E[E[\exp\{itb_{Z_n^*}^{-1}(W - 1)\}]]^{Z_n^*}. &\end{aligned} \tag{18}$$

Now, $E[E[\exp\{itb_{Z_n^*}^{-1}(W - 1)\}]]^{Z_n^*}$ is the characteristic function of $b_{Z_n^*}^{-1}(U_1 + \cdots + U_{Z_n^*})$ where the U_i, $i = 1,2,3,\cdots$ are independent of Z_n^* and are independent and identically distributed, each with the distribution of $W - 1$. Also, as $n \to \infty$, $m^{-n}Z_n^* \overset{\text{a.s.}}{\to} W^*$ where W^* is positive and has an absolutely continuous distribution function given by

$$P(W \leqq x) = P(W=0)D(x) + P(W>0)P(W^* \leqq x)$$

where

$$D(x) = \begin{cases} 0, & x < 0 \\ 1, & x \geqq 0 \end{cases}$$

(see Stigum [8]).

Then, since $b_n^{-1}(U_1 + \cdots + U_n)$ converges in distribution to the stable law with characteristic function $f(t)$ (from Theorem 1) we can, given $\varepsilon > 0$, choose N so large that

$$\left|(E[\exp\{itb_j^{-1}U_i\}])^j - f(t)\right| < \varepsilon$$

for $j \geqq N$ uniformly in any fixed finite t interval. Thus, for any fixed δ, $0 < \delta < 1$, we can choose n so large that $[m^n\delta] \geqq N$ ($[x]$ denotes the integer part of x) and then

$$\begin{aligned} &\left|\sum_{j=1}^{\infty} [(E[\exp\{itb_j^{-1}U_i\}])^j - f(t)]P(Z_n^* = j)\right| \\ &\leqq \sum_{j=1}^{[m^n\delta]-1} \left|(E[\exp\{itb_j^{-1}U_i\}])^j - f(t)\right| P(Z_n^* = j) \\ &\quad + \sum_{j=[m^n\delta]}^{\infty} \left|(E[\exp\{itb_j^{-1}U_i\}])^j - f(t)\right| P(Z_n^* = j) \\ &\leqq 2P(Z_n^* < [m^n\delta]) + \varepsilon P(Z_n^* \geqq [m^n\delta]) \\ &\to 2P(W^* < \delta) + \varepsilon P(W^* > \delta) \end{aligned}$$

as $n \to \infty$. Thus, since W^* is absolutely continuous and δ and ε can be arbitrarily small, we have

$$\lim_{n\to\infty} \sum_{j=1}^{\infty} (E[\exp\{itb_j^{-1}U_i\}])^j P(Z_n^* = j) = f(t) \tag{19}$$

and using (19) in (18)

$$\lim_{n\to\infty} E[\exp\{itm^n b_{Z_n}^{-1}(W - W_n)\} \mid Z_n > 0] = f(t)$$

as required.

6. Concluding remarks

To illuminate the connection between Theorems 2 and 3, we note from (14) that $m^n b_{m^n}^{-1}(W - W_n)$ has the same distribution as X_n which is given by

$$X_n = \begin{cases} 0, & \text{if } Z_n = 0 \\ b_{m^n}^{-1}(U_1 + \cdots + U_{Z_n}), & \text{if } Z_n > 0, \end{cases}$$

where the U_i, $i = 1, 2, 3, \cdots$ are independent of Z_n and are independent and identically distributed each with the distribution of $W - 1$. Take Z_n^* and W^* as in the proof of Theorem 3. Then, conditional on $Z_n > 0$, $m^n b_{m^n}^{-1}(W - W_n)$ has the distribution of $b_{m^n}^{-1}(U_1 + \cdots + U_{Z_n^*})$ which converges in distribution to a product $V(W^*)^{1/\alpha}$ of independent random variables (using Theorem 2) where V has characteristic function $f(t)$. On the other hand, from the proof of Theorem 3 we see that, conditional on $Z_n > 0$, $m^n b_{Z_n}^{-1}(W - W_n)$ has the distribution of $b_{Z_n^*}^{-1}(U_1 + \cdots + U_{Z_n^*})$ which converges in distribution to V.

Finally, it is worth remarking that a central limit analogue is not obtainable for the Galton-Watson process with $EZ_1 = \infty$. Clearly it is possible to find a monotone sequence $\{c_n\}$ of positive constants such that $c_n^{-1}Z_n \xrightarrow{\text{a.s.}} 0$ as $n \to \infty$. However, there is no norming sequence $\{d_n\}$ for which $d_n^{-1}Z_n$ converges in distribution to a non-degenerate limit law ([7], Theorem 4.4).

References

[1] Gnedenko, B. V. and Kolmogorov, A. N. (1954) *Limit Distributions for Sums of Independent Random Variables.* Addison Wesley, Reading, Mass.

[2] Heyde, C. C. (1970) Extension of a result of Seneta for the super-critical Galton-Watson process. *Ann. Math. Statist.* **41**, 739–742.

[3] Heyde, C. C. (1970) A rate of convergence result for the super-critical Galton-Watson process. *J. Appl. Prob.* **7**, 451–454.

[4] Kesten, H. and Stigum, B. P. (1966) A limit theorem for multi-dimensional Galton-Watson processes. *Ann. Math. Statist.* **37**, 1211–1223.

[5] Lamperti, J. (1967) Limiting distributions for branching processes. *Proc. 5th Berkeley Symposium on Math. Statist. and Prob.* **II**, 225–241.

[6] Levinson, N. (1959) Limiting theorems for Galton-Watson branching process. *Illinois J. Math.* **3**, 554–565.

[7] Seneta, E. (1969) Functional equations and the Galton-Watson process. *Adv. Appl. Prob.* **1**, 1–42.

[8] Stigum, B. P. (1966) A theorem on the Galton-Watson process. *Ann. Math. Statist.* **37**, 695–698.

Note added in proof

A central limit analogue for the finite variance case of a continuous time Markov branching process has been obtained by K. B. Athreya (1968) Some results on multitype continuous time Markov branching processes. *Ann. Math. Statist.* **39**, 347–357.

Z. Wahrscheinlichkeitstheorie verw. Geb. 20, 271–278 (1971)

An Invariance Principle and Some Convergence Rate Results for Branching Processes

C. C. HEYDE and B. M. BROWN

1. Introduction

Let $Z_0=1, Z_1, Z_2, \ldots$ denote a super-critical Galton-Watson branching process with $1<EZ_1=m<\infty$ and $0<\operatorname{var} Z_1=\sigma^2<\infty$. It is well-known that there exists a non-degenerate random variable W such that $\lim_{n\to\infty} W_n=W$ almost surely, where $W_n=m^{-n}Z_n$ (e.g. Harris [3], p. 13). Also, a central limit analogue holds, namely that, conditional on $Z_n>0$, $(m^2-m)^{\frac{1}{2}}\sigma^{-1}Z_n^{-\frac{1}{2}}m^n(W-W_n)$ is asymptotical normal $N(0,1)$ as $n\to\infty$ (Heyde [4]). This result is called a central limit analogue by virtue of the analogy with the convergence of $\sigma^{-1}n^{\frac{1}{2}}(\mu-n^{-1}S_n)$ to $N(0,1)$ for a sum S_n of independent and identically distributed random variables with mean μ and variance σ^2.

In this paper, we explore the setting of the central limit analogue. An invariance principle in R^∞ is obtained from which it follows as a simple corollary. Also, results are obtained on the rate of convergence to normality. It is shown that, under mild restrictions, there is a geometrically decreasing bound on the L_∞ metric.

2. Results

Let the probability generating function (p.g.f.) of the offspring distribution Z_1 be $F(s)=\sum_{j=0}^{\infty} s^j P(Z_1=j)$. The p.g.f. $F_n(s)$ of Z_n is the n-th functional iterate of $F(s)$. For the super-critical process in question, the probability of eventual extinction is q, the unique real number in $[0,1)$ satisfying $q=F(q)$. Furthermore, in addition to the fact that $Z_n=0$ for n sufficiently large on the extinction set $[W=0]$, it is known that

$$Z_n\to\infty \quad \text{as} \quad n\to\infty \quad \text{on} \quad [W>0], \tag{1}$$

so that the process either dies out or becomes arbitrarily large.

The symbols P and E will, for convenience, be used to denote various different probability measures and their associated expectation operators. In addition, we use the symbol Z_n^* to denote the random variable (r.v.) Z_n under the probability measure conditional on $Z_n>0$. These conveniences should cause no ambiguities.

Let R^∞ denote the space of real sequences $x=(x_1, x_2, \ldots)$ with the metric

$$\rho(x,y)=\sup_{n\geq 1}\left\{\frac{\left|\sum_{j=1}^{n}(x_j-y_j)\right|}{1+\left|\sum_{j=1}^{n}(x_j-y_j)\right|}\right\}.$$

R. Maller et al. (eds.), *Selected Works of C.C. Heyde*, Selected Works in Probability and Statistics, DOI 10.1007/978-1-4419-5823-5_25, © Springer Science+Business Media, LLC 2010

Let $\mathscr{B}$ be the σ-field generated by the open sets of the topology given by ρ and let V be the probability measure on $(R^\infty, \mathscr{B})$ under which the co-ordinates $x_1, x_2, \ldots$ of a random element x of R^∞ become independent r.v.'s with x_j distributed as $N(0, \sigma^2 m^{-1-j})$, $j=1,2,\ldots$. Write U for the probability measure on $(R^\infty, \mathscr{B})$ which assigns probability one to the zero sequence $(0,0,\ldots)$ and define, for $j=0,1,2,\ldots$ and $n=1,2,3,\ldots$,

$$X_n = (X_{n1}, X_{n2}, \ldots)$$

$$X_{nj} = m^n (W_{n+j} - W_n)(Z_n + 1)^{-\frac{1}{2}},$$

and for $j=1,2,3,\ldots$

$$Y_{nj} = X_{nj} - X_{n,j-1}.$$

Theorem 1. *Let P_n denote the probability measure on $(R^\infty, \mathscr{B})$ defined by the distributions of $Y_n = (Y_{n1}, Y_{n2}, \ldots)$. Then,*

$$P_n \Rightarrow V P(W>0) + U P(W=0)$$

as $n \to \infty$, where $\Rightarrow$ denotes weak convergence of probability measures.

We remark that, as simple corollaries to Theorem 1, $(m^2-m)^{\frac{1}{2}} \sigma^{-1} Z_n^{-\frac{1}{2}} \cdot m^n (W - W_n)$ conditional on $Z_n > 0$ and $(m^2-m)^{\frac{1}{2}} \sigma^{-1} m^{-j/2} (m^j - 1)^{-\frac{1}{2}} Z_n^{-\frac{1}{2}} (Z_{j+n} - m^j Z_n)$ conditional on $Z_n > 0$ (fixed j) are both asymptotically $N(0,1)$. These results have previously been obtained by direct methods by Heyde [4] and by Bühler [2] respectively. They can here be deduced from Theorem 1 using the measurable mapping theorem (Billingsley [1], p. 30) with the mappings

$$h_1(x) = \limsup_{n\to\infty} \sum_{j=1}^{n} x_j$$

if this is finite, 0 otherwise and $h_2(x) = x_1 + x_2 + \cdots + x_j$ respectively.

The next theorem gives rates of convergence to normality in the limit results cited above.

Theorem 2. *Let $E Z_1^3 < \infty$. Then,*

$$\sup_x |P((m^2-m)^{\frac{1}{2}} \sigma^{-1} Z_n^{-\frac{1}{2}} m^n (W - W_n) \leqq x \mid Z_n > 0) - \Phi(x)| \\ \leqq K \sigma^{-3} (m^2-m)^{\frac{3}{2}} E(1/Z_n^*)^{\frac{1}{2}} E|W-1|^3 \tag{2}$$

and

$$\sup_x |P(\sigma_r^{-1} Z_n^{-\frac{1}{2}} (Z_{r+n} - m^r Z_n) \leqq x \mid Z_n > 0) - \Phi(x)| \\ \leqq K \sigma_r^{-3} E|Z_r - m^r|^3 E(1/Z_n^*)^{\frac{1}{2}}, \tag{3}$$

where $\Phi(x)$ is the distribution function of $N(0,1)$, $\sigma_r^2 = \operatorname{var} Z_r = \sigma^2 m^r (m^r - 1)(m^2-m)^{-1}$ and K is the universal constant in the Berry-Esseen bound ($K < 0.82$; see Zolotarev *[6]). Furthermore,*

$$E(1/Z_n^*)^{\frac{1}{2}} \leqq (E(1/Z_n^*))^{\frac{1}{2}} \leqq \{(1-F(0))^{-1} \gamma^n\}^{\frac{1}{2}} \tag{4}$$

where $\gamma < 1$ is given by

$$\gamma = \int_0^1 (F(s(1-q)+q) - q)\, s^{-1} (1-q)^{-1}\, ds,$$

with $\gamma = E Z_1^{-1}$ if $F(0) = 0$.

3. Proof of Theorem 1

Firstly we show that the finite-dimensional distributions of $\{P_n\}$ converge weakly to those of $VP(W>0)+UP(W=0)$.

Let $\mathscr{F}_k$ be the σ-field generated by $Z_0, Z_1, \ldots, Z_k$ and for fixed r write

$$\begin{aligned}\phi_n(\theta_1, \theta_2, \ldots, \theta_r) &= E\left\{\exp\left(i\sum_{j=1}^{r}\theta_j Y_{nj}\right)\right\}\\ &= E\left(\prod_{j=1}^{r} A_j\right),\end{aligned} \tag{5}$$

where

$$A_j = \exp(i\,\theta_j\, m^{-j}(Z_{n+j} - m\,Z_{n+j-1})(Z_n+1)^{-\frac{1}{2}}). \tag{6}$$

Consider the r.v.

$$\begin{aligned}R_r &= \prod_{j=1}^{r} A_j - \exp\left(-\tfrac{1}{2}\sigma^2 I(Z_n>0)\sum_{j=1}^{r}\theta_j^2 m^{-1-j}\right)\\ &= \sum_{j=1}^{r} B_j,\end{aligned} \tag{7}$$

where I is the indicator function and for $j=1, 2, \ldots, r$, using the convention that a product of the form $\prod_{k=s+1}^{s}$ is replaced by unity,

$$\begin{aligned}B_j = &\left(\prod_{k=1}^{j-1} A_k\right)\exp\left(-\tfrac{1}{2}\sigma^2 I(Z_n>0)\sum_{k=j+1}^{r}\theta_k^2 m^{-1-k}\right)\\ &\cdot(A_j - \exp(-\tfrac{1}{2}\sigma^2 I(Z_n>0)\,\theta_j^2 m^{-1-j})),\end{aligned}$$

with

$$\begin{aligned}E(B_j|\mathscr{F}_{n+j-1}) = &\left(\prod_{k=1}^{j-1} A_k\right)\exp\left(-\tfrac{1}{2}\sigma^2 I(Z_n>0)\sum_{k=j+1}^{r}\theta_k^2 m^{-1-k}\right)\\ &\cdot(E(A_j|\mathscr{F}_{n+j-1}) - \exp(-\tfrac{1}{2}\sigma^2 I(Z_n>0)\,\theta_j^2 m^{-1-j})).\end{aligned} \tag{8}$$

But, Z_{n+j} is the sum of Z_{n+j-1} independent and identically distributed r.v.'s, each with the distribution of Z_1, so that from (6),

$$E(A_j|\mathscr{F}_{n+j-1}) = \{\phi(\theta_j m^{-j}(Z_n+1)^{-\frac{1}{2}})\}^{Z_{n+j-1}}, \tag{9}$$

where $\phi(t) = E\{\exp(i\,t(Z_1-m))\}$. Therefore, $E(A_j|\mathscr{F}_{n+j-1}) \to 1$ almost surely (a.s.) as $n\to\infty$ on $[W=0]$, so that from (8),

$$\lim_{n\to\infty} E(B_j|\mathscr{F}_{n+j-1}) = 0 \quad \text{a.s. on } [W=0], \tag{10}$$

since

$$I(Z_n>0) \to I(W>0) \quad \text{a.s. as } n\to\infty. \tag{11}$$

Also, we know from the central limit theorem that for each fixed t,

$$\lim_{n\to\infty}\{\phi(t(n+1)^{-\frac{1}{2}})\}^n = \exp(-\tfrac{1}{2}\sigma^2 t^2).$$

Therefore, from (9), using (1) and since

$$\lim_{n\to\infty} Z_n^{-1} Z_{n+j-1} = m^{j-1} \quad \text{a.s. on } [W>0],$$

we have

$$\lim_{n\to\infty} E(A_j|\mathscr{F}_{n+j-1}) = \exp(-\tfrac{1}{2}\sigma^2\theta_j^2 m^{-1-j}) \quad \text{a.s. on } [W>0],$$

so that

$$\lim_{n\to\infty} E(B_j|\mathcal{F}_{n+j-1})=0 \quad \text{a.s. on } [W>0], \tag{12}$$

using (8) and (11). Clearly $|B_j|\leqq 2$ a.s. for each $j=1, 2, \ldots, r$, so that the combination of (10) and (12), with dominated convergence, gives

$$E R_r = E\sum_{j=1}^{r} B_j = E\sum_{j=1}^{r} E(B_j|\mathcal{F}_{n+j-1})\to 0$$

as $n\to\infty$. Referring now to (5) and (7), we have

$$\lim_{n\to\infty}\left\{\phi_n(\theta_1,\ldots,\theta_r)-E\exp\left(-\tfrac{1}{2}\sigma^2 I(Z_n>0)\sum_{j=1}^{r}\theta_j^2 m^{-1-j}\right)\right\}=0. \tag{13}$$

But,

$$\begin{aligned}\lim_{n\to\infty} E\exp&\left(-\tfrac{1}{2}\sigma^2 I(Z_n>0)\sum_{j=1}^{r}\theta_j^2 m^{-1-j}\right)\\ &=P(W=0)+P(W>0)\exp\left(-\tfrac{1}{2}\sigma^2\sum_{j=1}^{r}\theta_j^2 m^{-1-j}\right)\end{aligned} \tag{14}$$

from (11) and dominated convergence. (13) and (14) complete the proof of the convergence of finite-dimensional distributions.

To complete the proof of the theorem, we must show that the sequence $\{P_n\}$ is *tight*. That is, for each $\varepsilon>0$ there exists a compact set A such that $P_n(A)>1-\varepsilon$ for all $n=1, 2, \ldots$.

Consider the sets

$$C=C_r=\left\{x\colon \sup_{s\geqq k}\left|\sum_{j=k}^{s} x_j\right|\leqq m^{-k/3} \text{ for all } k\geqq r\right\}$$

and

$$D=D_M=\left\{x\colon \sup_{s\geqq 1}\left|\sum_{j=1}^{s} x_j\right|\leqq M\right\}.$$

The set $C\cap D$ has compact closure, for any sequence belonging to it can readily be shown to possess a limit point ([1], p. 217). Tightness will be established if we can show that, given $\varepsilon>0$, $P_n(C\cap D)>1-\varepsilon$ for all n, when M and r are suitably large.

We first note that $C=\bigcap_{k\geqq r} A_k$, where

$$A_k=\left\{x\colon \sup_{s\geqq k}\left|\sum_{j=k}^{s} x_j\right|\leqq m^{-k/3}\right\}.$$

Kolmogorov's inequality for martingales, when applied to the martingale $\left\{\sum_{j=k}^{r} Y_{nj},\ r=k, k+1, \ldots\right\}$ yields the inequality

$$P\left(\max_{k\leqq r\leqq s}\left|\sum_{j=k}^{r} Y_{nj}\right|>C\right)\leqq C^{-2}\sum_{j=k}^{s} E\,Y_{nj}^2. \tag{15}$$

But,

$$\begin{aligned}E(Y_{nj}^2|\mathcal{F}_{n+j-1})&=E\left(m^{-2j}(Z_n+1)^{-1}(Z_{n+j}-m Z_{n+j-1})^2|\mathcal{F}_{n+j-1}\right)\\ &=m^{-2j}(Z_n+1)^{-1}\sigma^2 Z_{n+j-1},\end{aligned}$$

and

$$\begin{aligned} E Y_{nj}^2 &= E\big(E(Y_{nj}^2 | \mathscr{F}_{n+j-1})\big) \\ &= m^{-2j} \sigma^2 \sum_{k \geqq 0} E\big((Z_n+1)^{-1} Z_{n+j-1} | Z_n = k\big) P(Z_n = k) \\ &= m^{-2j} \sigma^2 \sum_{k \geqq 0} k(k+1)^{-1} E Z_{j-1} P(Z_n = k) \\ &\leqq \sigma^2 m^{-1-j}. \end{aligned}$$

Therefore, from (15),

$$P\left(\max_{k \leqq r \leqq s} \left|\sum_{j=k}^{r} Y_{nj}\right| > C\right) \leqq \sigma^2 m^{-k} C^{-2} (m-1)^{-1}. \tag{16}$$

By applying the inequality (16) to A_k' and D' (the prime denoting complementation), we obtain

$$P_n(A_k') \leqq \sigma^2 (m-1)^{-1} m^{-k/3}$$

and

$$P_n(D') \leqq \sigma^2 m^{-1} (m-1)^{-1} M^{-2},$$

for all $n = 1, 2, \ldots$. It now follows that

$$\begin{aligned} P_n(C) &= 1 - P_n\Big(\bigcup_{k \geqq r} A_k'\Big) \\ &\geqq 1 - \sum_{k \geqq r} P_n(A_k') \\ &\geqq 1 - \sigma^2 (m-1)^{-1} (1 - m^{-\frac{1}{3}})^{-1} m^{-r/3} \end{aligned} \tag{17}$$

and

$$P_n(D) \geqq 1 - \sigma^2 (m-1)^{-1} m^{-1} M^{-2}, \tag{18}$$

for all $n = 1, 2, \ldots$. Thus, from (17) and (18), $P_n(C \cap D) > 1 - \varepsilon$ for all n if r and M are chosen suitably large, so that tightness is established. The proof of the theorem is thus complete.

It is worth remarking that the result of Theorem 1 can be generalized to cover the case where Z_1 belongs to the domain of attraction of a stable law of index α, $1 < \alpha \leqq 2$. This provides a complete extension of the result of Theorem 3 of Heyde [4]. Theorem 1 has been given in the less general form in the interests of a unified exposition in this paper. In order to obtain the general form we would use

$$X_{nj} = m^n b_{Z_n+1}^{-1} (W_{n+j} - W_n),$$

in the notation of [4].

4. Proof of Theorem 2

We first need the following lemma.

Lemma. *Let Y_i, $i = 1, 2, 3, \ldots$ be independent and identically distributed r.v.'s with $E Y_1 = 0$, $\operatorname{var} Y_1 = \alpha^2$ and $E|Y_1|^3 < \infty$. Let N be a positive integer-valued r.v. which is independent of the $\{Y_i\}$. Then,*

$$\sup_x \left|P(\alpha^{-1} N^{-\frac{1}{2}} (Y_1 + \cdots + Y_N) \leqq x) - \Phi(x)\right| \leqq C E(N^{-\frac{1}{2}}),$$

where $C = K \alpha^{-3} E|Y_1|^3$, K being the universal constant of the Berry-Esseen bound.

Proof. Using the Berry-Esseen bound,

$$-CE(N^{-\frac{1}{2}}) = -C\sum_{j=1}^{\infty} j^{-\frac{1}{2}} P(N=j)$$
$$\leq \sum_{j=1}^{\infty} \{P(\alpha^{-1} j^{-\frac{1}{2}}(Y_1+\cdots+Y_j)\leq x) - \Phi(x)\} P(N=j) \tag{19}$$
$$\leq CE(N^{-\frac{1}{2}}).$$

But, since N is independent of the $\{Y_i\}$,

$$\sum_{j=1}^{\infty} \{P(\alpha^{-1} j^{-\frac{1}{2}}(Y_1+\cdots+Y_j)\leq x) - \Phi(x)\} P(N=j)$$
$$= \sum_{j=1}^{\infty} \{P(\alpha^{-1} j^{-\frac{1}{2}}(Y_1+\cdots+Y_j)\leq x \mid N=j) - \Phi(x)\} P(N=j) \tag{20}$$
$$= P(\alpha^{-1} N^{-\frac{1}{2}}(Y_1+\cdots+Y_N)\leq x) - \Phi(x),$$

and the required result follows from (19) and (20).

We now proceed with the proof of the theorem. Firstly we establish (2) and (3). Heyde [4] has shown that, conditional on $Z_n>0$, $m^n Z_n^{-\frac{1}{2}}(W-W_n)$ has the same distribution as $(Z_n^*)^{-\frac{1}{2}}(U_1+\cdots+U_{Z_n^*})$, where the U_i are independent of Z_n^* and are independent and identically distributed, each with the distribution of $W-1$. Furthermore, $\operatorname{var} W=\sigma^2(m^2-m)^{-1}$, and hence

$$P\big((m^2-m)^{\frac{1}{2}} \sigma^{-1} Z_n^{-\frac{1}{2}} m^n (W-W_n)\leq x \mid Z_n>0\big)$$
$$= P\big((\operatorname{var} U_1)^{-\frac{1}{2}} (Z_n^*)^{-\frac{1}{2}} (U_1+\cdots+U_{Z_n^*})\leq x\big),$$

so that (2) follows immediately from the lemma. (3) is obtained similarly by noting that, conditional on $Z_n>0$, $Z_n^{-\frac{1}{2}}(Z_{r+n}-m^r Z_n)$ has the same distribution as $(Z_n^*)^{-\frac{1}{2}}(V_1+\cdots+V_{Z_n^*})$, where the V_i are independent of Z_n^* and are independent and identically distributed, each with the distribution of Z_r-m^r.

Next, in order to obtain (4) we note that

$$E(Z_n^*)^{-\frac{1}{2}} \leq (E(Z_n^*)^{-1})^{\frac{1}{2}}, \tag{21}$$

by a standard moment inequality, so that it suffices to study $E(Z_n^*)^{-1}$, which is given by

$$E(Z_n^*)^{-1} = \int_0^1 (F_n(s)-F_n(0))\, s^{-1}(1-F_n(0))^{-1}\, ds. \tag{22}$$

Now define a p.g.f.

$$H(s)=(1-q)^{-1}(F(s(1-q)+q)-q), \qquad 0\leq s\leq 1.$$

Then, if $H_n(s)$ is the n-th functional iterate of $H(s)$, it is easily seen that

$$H_n(s)=(1-q)^{-1}(F_n(s(1-q)+q)-q),$$

and $H_n(s)$ is the p.g.f. of a super-critical Galton-Watson process $\{Y_n\}$ (say) with $P(Y_n=0)=0$ for each n. The growth properties of $\{Y_n\}$ are closely related to those of $\{Z_n\}$; in particular $E Y_n = E Z_n = m^n$. Also, we note that

$$U(s)=F_n(s(1-q)+q)-(1-q)\, F_n(s)$$

satisfies

$$U(0)=F_n(q)-(1-q)\, F_n(0)=q-(1-q)\, F_n(0)$$

and

$$U'(s)=(1-q)\big(F_n'(s(1-q)+q)-F_n'(s)\big)\geqq 0$$

since $s(1-q)+q\geqq s$ $(0\leqq s\leqq 1)$ and $F_n'(t)$ is monotone increasing in t. Thus for $0\leqq s\leqq 1$,

$$F_n(s(1-q)+q)-(1-q)\,F_n(s)\geqq q-(1-q)\,F_n(0),$$

that is,

$$H_n(s)\geqq F_n(s)-F_n(0).$$

Consequently, from (22),

$$\begin{aligned} E(Z_n^*)^{-1} &\leqq (1-F_n(0))^{-1}\int_0^1 s^{-1}H_n(s)\,ds \\ &=(1-F_n(0))^{-1}E(Y_n^{-1}). \end{aligned} \tag{23}$$

Now let $\mathscr{F}_n$ be the σ-field generated by $Y_1,\ldots,Y_n$, and note that $Y_n\geqq 1$ for each n. Then,

$$\begin{aligned} E(Y_n^{-1}) &= E\big(Y_{n-1}^{-1}\,E(Y_{n-1}\,Y_n^{-1}\,|\,\mathscr{F}_{n-1})\big) \\ &= E\big(Y_{n-1}^{-1}\,E(Y_{n-1}(Y_1^{(1)}+\cdots+Y_1^{(Y_{n-1})})^{-1}\,|\,\mathscr{F}_{n-1})\big), \end{aligned} \tag{24}$$

where the $Y_1^{(i)}$ are independent of Y_{n-1} and are independent and identically distributed, each with the distribution of Y_1. Furthermore, using the harmonic mean-arithmetic mean inequality,

$$Y_{n-1}(Y_1^{(1)}+\cdots+Y_1^{(Y_{n-1})})^{-1}\leqq Y_{n-1}^{-1}\big((Y_1^{(1)})^{-1}+\cdots+(Y_1^{(Y_{n-1})})^{-1}\big),$$

so that

$$E\big(Y_{n-1}(Y_1^{(1)}+\cdots+Y_1^{(Y_{n-1})})^{-1}\,|\,\mathscr{F}_{n-1}\big)\leqq E(Y_1^{-1}).$$

Hence, from (24),

$$E(Y_n^{-1})\leqq E(Y_1^{-1})\,E(Y_{n-1}^{-1}),$$

so that

$$E(Y_n^{-1})\leqq \big(E(Y_1^{-1})\big)^n. \tag{25}$$

(4) then follows from (21), (23) and (25) by setting $\gamma=E(Y_1^{-1})$, which is in turn given by

$$\gamma=E(Y_1^{-1})=\int_0^1 s^{-1}(1-q)^{-1}\big(F(s(1-q)+q)-q\big)\,ds,$$

with $\gamma=E(Z_1^{-1})$ when $P(Z_1=0)=0$, for then $q=0$. This completes the proof of Theorem 2.

5. Remarks on Theorem 2

The bound (4) of Theorem 2 appears to be rather crude and with reference to it we note that, by Jensen's inequality,

$$E(Y_1^{-1})\geqq (EY_1)^{-1}=m^{-1}.$$

How, if $E(W^{-1}|W>0)<\infty$, it is plausible that under suitable conditions, $E(Z_n^*)^{-1}\sim m^{-n}E(W^{-1}|W>0)$, since $\lim_{n\to\infty} m^n Z_n^{-1}=W^{-1}$ a.s. on $[W>0]$. However, it emerges that $E(W^{-1}|W>0)$ is often not finite and there are wide ranging possibilities for the asymptotic behaviour of $E(Z_n^*)^{-1}$. For example, take

$$F(s)=s\big(m-(m-1)\,s^k\big)^{-1/k},\qquad m>1,$$

where k is a positive integer. Then,

$$F_n(s)=s\big(m^n-(m^n-1)\,s^k\big)^{-1/k},$$

so that

$$E(Z_n^*)^{-1} = E(Z_n^{-1}) = \int_0^1 s^{-1} F_n(s)\, ds$$

$$= \int_0^1 (m^n - (m^n - 1) s^k)^{-1/k}\, ds,$$

and, putting $y = 1 - s^k(1 - m^{-n})$, we obtain

$$E(Z_n^{-1}) = \frac{1}{k(m^n - 1)^{1/k}} \int_{m^{-n}}^{1} \frac{dy}{y^{1/k}(1-y)^{(k-1)/k}}.$$

Thus, if $k=1$,

$$E(Z_n^{-1}) = (m^n - 1)^{-1} n \log m,$$

and if $k>1$,

$$E(Z_n^{-1}) \sim k^{-1} m^{-n/k} B(1-k^{-1}, k^{-1})$$

$$\sim k^{-1} m^{-n/k} \pi (\sin(\pi/k))^{-1}$$

as $n \to \infty$ since

$$B(1-k^{-1}, k^{-1}) = \Gamma(k^{-1}) \Gamma(1-k^{-1}) = \pi (\sin(\pi/k))^{-1}.$$

In conclusion, we make two observations. Firstly, Theorem 2 may be extended to deal with the case $EZ_1^{2+\delta} < \infty$, $0 < \delta \leqq 1$, by using results of Ibragimov [5]. In this case we obtain bounds of the form $c(E(Z_n^*)^{-1})^{\delta/2}$ in (2) and (3), c denoting a positive constant. We have not framed Theorem 2 in this more general context because an explicit form for c is not available.

Secondly, an improved rate of convergence to normality can be obtained by conditioning on the value of Z_n as well as on $[Z_n > 0]$. In this case it is apparent that, after minor modifications in the foregoing proof, the bound in (2) and (3) becomes $O((Z_n^*)^{-\frac{1}{2}})$, which is $O(m^{-\frac{1}{2}n})$ as $n \to \infty$.

References

1. Billingsley, P.: Convergence of probability measures. New York: Wiley 1968.
2. Bühler, W.J.: Ein zentraler Grenzwertsatz für Verzweigungsprozesse. Z. Wahrscheinlichkeitstheorie verw. Geb. **11**, 139–141 (1969).
3. Harris, T.E.: The theory of branching processes. Berlin-Göttingen-Heidelberg: Springer 1963.
4. Heyde, C.C.: Some central limit analogues for super-critical Galton-Watson processes. J. appl. Probab. **8**, 52–59 (1971).
5. Ibragimov, I.A.: On the accuracy of the Gaussian approximation to the distribution function of sums of independent variables. Theor. Probab. Appl. **11**, 559–579 (1966).
6. Zolotarev, V.M.: A sharpening of the inequality of Berry-Esseen. Z. Wahrscheinlichkeitstheorie verw. Geb. **8**, 332–342 (1967).

Dr. C.C. Heyde
Department of Statistics
Australian National University
School of General Studies
Box 4, P.O.
Canberra, A.C.T. 2600
Australia

Dr. B.M. Brown
Department of Mathematics
La Trobe University
Bundoora
Victoria 3083
Australia

(Received November 9, 1970/March 23, 1971)

Reprinted from

BULLETIN Volume 5 (1971)

of the Australian Mathematical Society

Number 2

Bull. Austral. Math. Soc. **Vol. 5 (1971) No. 2, pp. 145-288**

THE UNIVERSITY OF QUEENSLAND PRESS
ST LUCIA, QUEENSLAND 4067

THE AUSTRALIAN MATHEMATICAL SOCIETY

Membership and correspondence

Applications for membership, notices of change of address or title or position, and other correspondence, except as noted below, should be sent to:

The Secretary, Australian Mathematical Society,
Secondary Teachers' College,
Swanston Street, Parkville, Victoria 3052, Australia.

Subscriptions and correspondence related to accounts should be addressed to:

The Treasurer, Australian Mathematical Society,
Department of Mathematics,
University of Queensland,
St Lucia, Queensland 4067, Australia.

Correspondence about the distribution of the Society's JOURNAL and BULLETIN should be addressed to:

The Business Manager, Australian Mathematical Society,
Department of Mathematics,
University of Queensland,
St Lucia, Queensland 4067, Australia.

The annual subscription is $50 for Institutions and $16 in 1971 for Ordinary Members. Junior Members (undergraduate and junior graduate students) may ask to have their subscription reduced to $6 in 1971. Members of other mathematical societies with which the Australian Mathematical Society has entered into reciprocity agreements are admitted to membership and its privileges at half the current rates, if they are not normally resident in Australia.

Members of the Society receive copies of the JOURNAL free, and may subscribe to the BULLETIN at the price of $4 per volume ($8 p.a.). Members may elect to receive the BULLETIN instead of the JOURNAL.

The JOURNAL of the Australian Mathematical Society

The JOURNAL began publication in 1959. From 1972, two volumes of four parts will be published annually, at a price of $25 per volume. Subscription orders and orders for back numbers should be sent to the Business Manager (address above). Manuscripts for publication should be sent to the Editor:

Professor B. Mond,
Department of Mathematics,
La Trobe University,
Bundoora, Victoria 3083, Australia,

or to one of the Associate Editors listed in a recent part of the JOURNAL.

(Continued inside back cover)

The BULLETIN of the Australian Mathematical Society

The BULLETIN began publication in 1969. Two volumes of three numbers each are published annually, at a subscription of $25 in 1972. Subscription orders and orders for back numbers should be sent to the Business Manager (address inside front cover). Manuscripts for publication should be sent to the Editor:

Professor B. H. Neumann,
Department of Mathematics,
Institute of Advanced Studies,
Australian National University,
POB 4, Canberra, ACT 2600, Australia,

or to the Associate Editor, Dr L. G. Kovács, at the same address. Authors should keep a complete copy of their manuscripts, because manuscripts are not normally returned to them.

Authors receive 50 free offprints of their papers and may order more at cost. Correspondence about proofs and offprint orders should be addressed to one of the Assistant Editors:

Associate Professor A. McMullen, Department of Mathematics, Royal Military College, Duntroon, ACT 2600, Australia

or Dr S. Yamamuro, Department of Mathematics, Institute of Advanced Studies, Australian National University, POB 4, Canberra, ACT 2600, Australia.

Editorial Policy

The BULLETIN of the Australian Mathematical Society aims at quick publication of original research in all branches of Mathematics. Only manuscripts that are suitable for quick publication will be accepted, and only if space permits. Manuscripts must be in their final, publishable form when submitted; refereeing will be kept to a minimum, and no revision of manuscripts will be allowed.

The BULLETIN is printed by a photo-offset process from a carefully prepared typescript. The type founts and special symbols available are, accordingly, restricted.

To facilitate the work of reviewing journals and the Mathematical Offprint Service (MOS) of the American Mathematical Society, the BULLETIN publishes an abstract at the head of each paper that is not itself an abstract, and also a tentative MOS classification in the top right-hand corner of the first page of each paper. Authors are invited to assist by supplying abstracts of their papers and supplying or checking MOS classification numbers. Copies of the MOS classification scheme are available to authors on request from MOS, American Mathematical Society, POB 6248, Providence, R.I. 02904, USA. The scheme and instructions for its use can be also found as an appendix to Mathematical Reviews, Index to Volume 39 (1970).

Copyright

The BULLETIN of the Australian Mathematical Society is published for the Society by

The University of Queensland Press,
St Lucia, Queensland 4067,
Australia,

and printed by

Watson Ferguson & Co.,
Brisbane, Queensland 4101,
Australia.

BULLETIN of the Australian Mathematical Society Vol. **5 (1971)** No. **2**

BULL. AUSTRAL. MATH. SOC. MOS 60J80
VOL. 5 (1971), 145-155. (60J15)

Improved classical limit analogues for Galton-Watson processes with or without immigration

C.C. Heyde and J.R. Leslie

It has recently emerged that the central limit theorem and iterated logarithm law for random walk processes have natural counterparts for Galton-Watson processes with or without immigration. Much of the work on these counterparts has previously involved the imposition of supplementary moment conditions. In this paper we show how to dispense with these supplementary conditions and in so doing make the analogy with the random walk results complete.

1. Introduction

Let $Z_0 = 1, Z_1, Z_2, \ldots$ denote a super-critical Galton-Watson process with $1 < EZ_1 = m$ and $0 < \operatorname{var} Z_1 = \sigma^2 < \infty$. It is well-known that there exists a non-degenerate random variable W such that $\lim_{n\to\infty} W_n = W$ almost surely, where $W_n = m^{-n}Z_n$ (for example, Harris [2], p. 13). Furthermore, some central limit analogues have been established in this context by Heyde [5] and Bühler [1]. These results are that

$\left(m^2-m\right)^{\frac{1}{2}}\sigma^{-1}Z_n^{-\frac{1}{2}}m^n\left(W-W_n\right)$ conditional on $Z_n > 0$ and

$\left(m^2-m\right)^{\frac{1}{2}}\sigma^{-1}m^{-\frac{1}{2}j}\left(m^j-1\right)^{-\frac{1}{2}}Z_n^{-\frac{1}{2}}\left(Z_{n+j}-m^jZ_n\right)$ conditional on $Z_n > 0$ (fixed j)

are both asymptotically normal $N(0, 1)$. See [5] for an explanation of

Received 23 February, 1971.

why these results can be regarded as central limit analogues.

Under the further restriction that $EZ_1^3 < \infty$, rates of convergence in the central limit analogues cited above have been given in [7]. These have been used in [6] to obtain almost sure convergence results for the Galton-Watson process which are analogues of the law of the iterated logarithm for random walks. In this paper we shall show how the restriction that $EZ_1^3 < \infty$ may be removed. In Section 2 we shall obtain convergence rates in the central limit analogues and also iterated logarithm analogues under the basic condition that $EZ_1^2 < \infty$.

In Section 3 of this paper we shall deal with the Galton-Watson process with immigration. The development of the corresponding limit results in this case has followed the pattern described previously for the case without immigration. Heyde and Seneta [8] have obtained central limit analogues under $EZ_1^2 < \infty$ and rate results and iterated logarithm analogues under $EZ_1^3 < \infty$. We shall again show how to dispense with the moment restriction and will obtain the rate results and iterated logarithm analogues under $EZ_1^2 < \infty$.

2. The process without immigration

We shall establish the following theorems. The reader is referred to the papers [5], [6] and [7] for background details.

THEOREM 1. *Let* $1 < m = EZ_1$ *and* $0 < \operatorname{var} Z_1 = \sigma^2 < \infty$. *Then*

$$\sup_x \left| P\left((m^2-m)^{\frac{1}{2}} \sigma^{-1} u_{Z_n}^{-1} Z_n^{-\frac{1}{2}} m^n (W-W_n) \le x \mid Z_n > 0 \right) - \Phi(x) \right| \le c_n$$

and

$$\sup_x \left| P\left(\sigma_r^{-1} v_{Z_n}^{-1} Z_n^{-\frac{1}{2}} \left(Z_{n+r} - m^r Z_n \right) \le x \mid Z_n > 0 \right) - \Phi(x) \right| \le d_n ,$$

where $\{c_n\}$, $\{d_n\}$ *are certain sequences of positive constants satisfying* $\sum_{n=1}^{\infty} c_n < \infty$ *and* $\sum_{n=1}^{\infty} d_n < \infty$. *Here*

$$\sigma_r^2 = \operatorname{var} Z_r = \sigma^2 m^r (m^r - 1)(m^2-m)^{-1} ,$$

r any fixed integer,

$$u_n = \int_{|x|<\sqrt{n}} x^2 dP\left(\sigma^{-1}\left(m^2-m\right)^{\frac{1}{2}}(W-1) \le x\right)$$

$$v_n = \int_{|x|<\sqrt{n}} x^2 dP\left(\sigma_r^{-1}\left(Z_r-m^r\right) \le x\right)$$

and $\Phi(x)$ *is the distribution function of* $N(0, 1)$.

Explicit forms for c_n and d_n can be found by applying the lemma below. We note also that $u_n \uparrow 1$ and $v_n \uparrow 1$ as $n \to \infty$.

THEOREM 2. *Suppose that* $1 < m = EZ_1$ *and* $0 < \operatorname{var} Z_1 = \sigma^2 < \infty$. *Then, on the non-extinction set* $\{W > 0\}$ *we have almost surely*

$$\limsup_{n\to\infty} \frac{Z_{n+r}-m^r Z_n}{\left(2\sigma_r^2 Z_n \log n\right)^{\frac{1}{2}}} = 1 , \quad \liminf_{n\to\infty} \frac{Z_{n+r}-m^r Z_n}{\left(2\sigma_r^2 Z_n \log n\right)^{\frac{1}{2}}} = -1 ,$$

and

$$\limsup_{n\to\infty} \frac{m^n W-Z_n}{\left(2\sigma^2\left(m^2-m\right)^{-1} Z_n \log n\right)^{\frac{1}{2}}} = 1 , \quad \liminf_{n\to\infty} \frac{m^n W-Z_n}{\left(2\sigma^2\left(m^2-m\right)^{-1} Z_n \log n\right)^{\frac{1}{2}}} = -1 ,$$

where r *is any fixed positive integer.*

Theorems 1 and 2 extend the scope of results given in Heyde and Brown [7] and in Heyde [6] respectively under the additional condition that $EZ_1^3 < \infty$. The form of the bounds obtained in Theorem 1 is however, of necessity, much more complicated in the general case. An explanation of this is not difficult to deduce from results given in [3] and [4]. Our Theorem 2 preserves exactly the form of the Theorem of [6] under the more general conditions.

In order to establish the above results we need the following key lemma. The result of the lemma is given in two parts; the first is needed in the present section and the second to obtain corresponding results for the process with immigration in Section 3.

LEMMA. *Let* ξ_i , $i = 1, 2, 3, \ldots$ *be independent and identically distributed random variables with* $E(\xi_1) = 0$ *and* $\operatorname{var}\xi_1 = \alpha^2 < \infty$. *Let* N_n *be a positive integer valued random variable which is independent of the* $\{\xi_i\}$. *Then,*

$$(1)\quad \sup_x \left| P\left(\alpha^{-1} d_{N_n}^{-1} N_n^{-\frac{1}{2}} \left(\xi_1 + \ldots + \xi_{N_n}\right) \le x\right) - \Phi(x)\right|$$
$$\le AE\left(N_n^{-\frac{1}{2}} a_{N_n}\right) + BE\left(N_n^{\frac{1}{2}} b_{N_n}\right) + E\left(N_n c_{N_n}\right)$$

where A, B *are positive constants and*

$$a_n = \int_{|x|<\sqrt{n}} |x|^3 dP\left(\alpha^{-1}\xi_1 \le x\right), \quad b_n = \int_{|x|\ge\sqrt{n}} |x| dP\left(\alpha^{-1}\xi_1 \le x\right),$$

$$c_n = P\left(\alpha^{-1}|\xi_1| > \sqrt{n}\right), \quad d_n^2 = \int_{|x|<\sqrt{n}} x^2 dP\left(\alpha^{-1}\xi_1 \le x\right).$$

If η_n *with* $E|\eta_n| < \infty$ *is a random variable which is independent of the* $\{\xi_i\}$ *and of* N_n, *then for any sequence* $\{\varepsilon_n\}$ *of positive constants with* $\varepsilon_n \to 0$ *as* $n \to \infty$,

$$(2)\quad \sup_x \left| P\left(\alpha^{-1} d_{N_n}^{-1} N_n^{-\frac{1}{2}} \left(\xi_1 + \ldots + \xi_{N_n} + \eta_n\right) \le x\right) - \Phi(x)\right|$$
$$\le AE\left(N_n^{-\frac{1}{2}} a_{N_n}\right) + BE\left(N_n^{\frac{1}{2}} b_{N_n}\right) + E\left(N_n c_{N_n}\right) + \alpha^{-1}\varepsilon_n^{-1} E|\eta_n| E\left(N_n^{-\frac{1}{2}} d_{N_n}^{-1}\right) + \frac{1}{2}\varepsilon_n .$$

Proof. Let

$$e_n^2 = \int_{|x|<\sqrt{n}} x^2 dP\left(\alpha^{-1}\xi_1 \le x\right) - \left(\int_{|x|<\sqrt{n}} x dP\left(\alpha^{-1}\xi_1 \le x\right)\right)^2 .$$

We have

$$(3)\quad \sup_x \left| P\left(\alpha^{-1} d_{N_n}^{-1} N_n^{-\frac{1}{2}} \left(\xi_1 + \ldots + \xi_{N_n}\right) \le x \mid N_n = k\right) - \Phi(x)\right|$$
$$\le \sup_x \left| P\left(\alpha^{-1} e_{N_n}^{-1} N_n^{-\frac{1}{2}} \left(\xi_1 + \ldots + \xi_{N_n}\right) \le x \mid N_n = k\right) - \Phi(x)\right|$$
$$+ \sup_x \left| \Phi(x) - \Phi\left(e_k d_k^{-1} x\right)\right| .$$

Also, using the mean value theorem,

$$(4)\quad \sup_x \left|\Phi(x) - \Phi\left(e_k d_k^{-1} x\right)\right| \le c\left(1 - e_k d_k^{-1}\right)$$

$$= c d_k^{-1}(e_k + d_k)^{-1}\left(\int_{|x|<\sqrt{k}} x dP\left(\alpha^{-1}\xi_1 \le x\right)\right)^2$$

$$= c d_k^{-1}(e_k + d_k)^{-1}\left(\int_{|x|\ge\sqrt{k}} x dP\left(\alpha^{-1}\xi_1 \le x\right)\right)^2$$

$$\le c_1 b_k^2 \le c_1 b_k ,$$

c, c_1 being positive constants. Furthermore, from (22) of Heyde [4] we find that

$$(5)\quad \sup_x \left|P\left(\alpha^{-1} e_{N_n}^{-1} N_n^{-\frac{1}{2}}\left(\xi_1 + \ldots + \xi_{N_n}\right) \le x \mid N_n = k\right) - \Phi(x)\right|$$

$$\le A k^{-\frac{1}{2}} a_k + B_1 k^{\frac{1}{2}} b_k + k c_k ,$$

so that using (3), (4) and (5),

$$(6)\quad \sup_x \left|P\left(\alpha^{-1} d_{N_n}^{-1} N_n^{-\frac{1}{2}}\left(\xi_1 + \ldots + \xi_{N_n}\right) \le x \mid N_n = k\right) - \Phi(x)\right|$$

$$\le A k^{-\frac{1}{2}} a_k + B k^{\frac{1}{2}} b_k + k c_k .$$

The result (1) follows readily from (6) using the argument of the lemma in §4 of Heyde and Brown [7]. (2) is obtained using exactly the method of Lemma 2.1 of Heyde and Seneta [8] with the aid of (1) instead of the Berry-Esseen bound.

Proof of Theorem 1. Suppose that Z_n^* has the distribution of Z_n conditional on $Z_n > 0$. We firstly note that (see [5], [7]), conditional on $Z_n > 0$, $m^n Z_n^{-\frac{1}{2}}(W - W_n)$ has the same distribution as $(Z_n^*)^{-\frac{1}{2}}\left(U_1 + \ldots + U_{Z_n^*}\right)$, where the U_i are independent of Z_n^* and are independent and identically distributed, each with the distribution of $W - 1$. Also, conditional on $Z_n > 0$, $Z_n^{-\frac{1}{2}}\left(Z_{n+r} - m^r Z_n\right)$ has the same distribution as $(Z_n^*)^{-\frac{1}{2}}\left(V_1 + \ldots + V_{Z_n^*}\right)$ where the V_i are independent of

Z_n^* and are independent and identically distributed, each with the distribution of $Z_r - m^r$. We can thus apply the lemma in both cases and obtain bounds which we call c_n, d_n respectively. It remains to show that $\sum_{n=1}^{\infty} c_n < \infty$, $\sum_{n=1}^{\infty} d_n < \infty$. We shall indicate the proof for $\sum_{n=1}^{\infty} c_n$; that for $\sum_{n=1}^{\infty} d_n$ follows similarly.

What we have to demonstrate is that $\sum_{n=1}^{\infty} E\left((Z_n^*)^{-\frac{1}{2}} a_{Z_n^*}\right) < \infty$, $\sum_{n=1}^{\infty} E\left((Z_n^*)^{\frac{1}{2}} b_{Z_n^*}\right) < \infty$ and $\sum_{n=1}^{\infty} E\left(Z_n^* c_{Z_n^*}\right) < \infty$ where a_n, b_n, c_n are defined in the lemma with ξ_i having the distribution of $W - 1$. The proofs of the convergence of these three series are identical in form. They depend on results of [4] where it is, in essence, shown in the proof of Theorem 4 that under the conditions of the theorem and if $\{n_k, k = 1, 2, 3, \ldots\}$ is a sequence of integers with $n_k \sim Kc^{2k}$ as $k \to \infty$ $(K > 0, c > 1)$, then

$$\sum_{k=1}^{\infty} n_k^{-\frac{1}{2}} a_{n_k} \le K_1 < \infty, \quad \sum_{k=1}^{\infty} n_k^{\frac{1}{2}} b_{n_k} \le K_2 < \infty, \quad \sum_{k=1}^{\infty} n_k c_{n_k} \le K_3 < \infty$$

for certain K_1, K_2, K_3 independent of K.

For $u > 0$, let

$$a_u = \int_{|x|<\sqrt{u}} |x|^3 dP\left(\sigma^{-1}(m^2-m)^{\frac{1}{2}}(W-1) \le x\right).$$

We have

$$E\left((Z_n^*)^{-\frac{1}{2}}a_{Z_n^*}\right) = \int_0^\infty (xm^n)^{-\frac{1}{2}}a_{xm^n}dP\left(m^{-n}Z_n^* \le x\right)$$

$$= \sum_{k=0}^{\infty}\int_{k\le x<k+1}(xm^n)^{-\frac{1}{2}}a_{xm^n}dP\left(m^{-n}Z_n^* \le x\right)$$

$$\le \sum_{k=0}^{\infty} k^{-\frac{1}{2}}m^{-\frac{1}{2}n}a_{(k+1)m^n}P\left(k \le m^{-n}Z_n^* < k+1\right)$$

$$\le c_1\sum_{k=0}^{\infty}[(k+1)m^n+1]^{-\frac{1}{2}}a_{[(k+1)m^n+1]}P\left(k \le m^{-n}Z_n^*\right) ,$$

where c_1 is a suitable positive constant and $[x]$ denotes the integer part of x. Then using Chebyshev's inequality, $P\left(m^{-n}Z_n^* \ge k\right) \le ck^{-2}$, and

$$\sum_{n=1}^{\infty} E\left((Z_n^*)^{-\frac{1}{2}}a_{Z_n^*}\right) \le c_1\sum_{n=1}^{\infty}\sum_{k=1}^{\infty}[km^n+1]^{-\frac{1}{2}}a_{[km^n+1]}P\left(k-1 \le m^{-n}Z_n^*\right)$$

$$\le c_1\sum_{n=1}^{\infty}[m^n+1]^{-\frac{1}{2}}a_{[m^n+1]}$$

$$+ c_1c\sum_{k=2}^{\infty}(k-1)^{-2}\sum_{n=1}^{\infty}[km^n+1]^{-\frac{1}{2}}a_{[km^n+1]}$$

$$\le c_2 + c_3\sum_{k=2}^{\infty}(k-1)^{-2} < \infty ,$$

as required. That $\sum_{n=1}^{\infty} E\left((Z_n^*)^{\frac{1}{2}}b_{Z_n^*}\right) < \infty$ and $\sum_{n=1}^{\infty} E\left(Z_n^*c_{Z_n^*}\right) < \infty$ follow in the same fashion.

Proof of Theorem 2. This follows the same lines as the proof of the theorem of [6]. We just make use of Theorem 1 and the inequality (6) instead of the results based on the Berry-Esseen inequality employed in [6]. We then obtain

$$\limsup_{n\to\infty}\frac{Z_{n+r}-m^rZ_n}{\left(2\sigma_r^2v_{Z_n}^2 Z_n \log n\right)^{\frac{1}{2}}} = 1 , \quad \liminf_{n\to\infty}\frac{Z_{n+r}-m^rZ_n}{\left(2\sigma_r^2v_{Z_n}^2 Z_n \log n\right)^{\frac{1}{2}}} = -1$$

almost surely on $\{W > 0\}$. The required results for $\left(Z_{n+r}-m^rZ_n\right)\left(2\sigma_r^2Z_n \log n\right)^{-\frac{1}{2}}$ then follow since $v_n \uparrow 1$ as $n \to \infty$. We have, for example, since $Z_n \to \infty$ almost surely on $\{W > 0\}$,

$$\limsup_{n\to\infty} \frac{Z_{n+r}-m^r Z_n}{\left(2\sigma_r^2 Z_n \log n\right)^{\frac{1}{2}}} \le \limsup_{n\to\infty} \frac{Z_{n+r}-m^r Z_n}{\left(2\sigma_r^2 v_{Z_n}^2 Z_n \log n\right)^{\frac{1}{2}}}$$

$$\le \limsup_{n\to\infty} \frac{Z_{n+r}-m^r Z_n}{\left(2\sigma_r^2 Z_n \log n\right)^{\frac{1}{2}}} \limsup_{n\to\infty} \frac{1}{v_{Z_n}}$$

$$= \limsup_{n\to\infty} \frac{Z_{n+r}-m^r Z_n}{\left(2\sigma_r^2 Z_n \log n\right)^{\frac{1}{2}}}$$

on $\{W > 0\}$. The remainder of the proof goes through exactly as in [6]. We show

$$\limsup_{n\to\infty} \frac{|Z_n-m^n W|}{\left(2\sigma^2 (m^2-m)^{-1} u_{Z_n}^2 Z_n \log n\right)^{\frac{1}{2}}} \le 1 ,$$

from which it follows that

$$\limsup_{n\to\infty} \frac{|Z_n-m^n W|}{\left(2\sigma^2 (m^2-m)^{-1} Z_n \log n\right)^{\frac{1}{2}}} \le 1 ,$$

and the remainder of the argument of [6] can be repeated word for word.

3. The process with immigration

Let $X_0 = 1, X_1, X_2, \ldots$ denote a Galton-Watson process with immigration whose offspring distribution has the distribution of Z_1 with $1 < EZ_1 = m$ and $0 < \mathrm{var} Z_1 = \sigma^2 < \infty$. We shall also suppose that the immigration distribution has a finite mean. We refer the reader to Seneta [9] for a detailed description of the process. Under the present conditions, the theorem of [9] ensures that $m^{-n}X_n$ converges almost surely to a proper random variable V with finite mean EV and such that $P(V = 0) = 0$. We shall here obtain the following theorems which extend results presented in Heyde and Seneta [8].

THEOREM 3. *Let* $1 < m = EZ_1$ *and* $0 < \mathrm{var} Z_1 = \sigma^2 < \infty$. *Then*

$$\sup_x \left| P\left((m^2-m)^{\frac{1}{2}} \sigma^{-1} u_{X_n}^{-1} X_n^{-\frac{1}{2}} \left(m^n V - X_n \right) \le x \mid X_n > 0 \right) - \Phi(x) \right| \le \alpha_n$$

and

$$\sup_x \left| P\left(\sigma_r^{-1} v_{X_n}^{-1} X_n^{-\frac{1}{2}} \left(X_{n+r} - m^r X_n \right) \le x \mid X_n > 0 \right) - \Phi(x) \right| \le \beta_n ,$$

where $\{\alpha_n\}$, $\{\beta_n\}$ *are certain sequences of positive constants satisfying* $\sum_{n=1}^{\infty} \alpha_n < \infty$ *and* $\sum_{n=1}^{\infty} \beta_n < \infty$. *Here* σ_r, u_n, v_n *are as defined in the statement of Theorem* 1.

Explicit forms for α_n and β_n can be found by applying the lemma.

THEOREM 4. *Suppose that* $1 < m = EZ_1$ *and* $0 < \mathrm{var}Z_1 = \sigma^2 < \infty$. *Then, with probability one,*

$$\limsup_{n\to\infty} \frac{X_{n+r} - m^r X_n}{\left(2\sigma_r^2 X_n \log n \right)^{\frac{1}{2}}} = 1 , \quad \liminf_{n\to\infty} \frac{X_{n+r} - m^r X_n}{\left(2\sigma_r^2 X_n \log n \right)^{\frac{1}{2}}} = -1 ,$$

$$\limsup_{n\to\infty} \frac{m^n V - X_n}{\left(2\sigma^2 (m^2-m)^{-1} X_n \log n \right)^{\frac{1}{2}}} = 1 , \quad \liminf_{n\to\infty} \frac{m^n V - X_n}{\left(2\sigma^2 (m^2-m)^{-1} X_n \log n \right)^{\frac{1}{2}}} = -1 ,$$

where r *is any fixed positive integer.*

Theorems 3 and 4 extend the scope of results given in Theorems 2 and 3 of [8] under the additional condition that $EZ_1^3 < \infty$. The form of the bounds obtained in the present Theorem 3 is however, of necessity, much more complicated in the more general case. The present Theorem 4 preserves exactly the form of Theorem 3 of [8] under the more general conditions.

The proofs of Theorems 2 and 3 follow the same lines as those of Theorems 1 and 2, again using the lemma and (6). We make use of the representations

$$m^n V - X_n = \left(W^{(1)} - 1 \right) + \ldots + \left(W^{(X_n)} - 1 \right) + I^{(n)} \quad \text{a.s.}$$

and

$$X_{n+r} - m^r X_n = \left(Z_r^{(1)} - m^r\right) + \ldots + \left(Z_r^{(X_n)} - m^r\right) + Y_{r,n}$$

of [8] and the points noted in the proof of Theorem 3 of [8]. The only real point of difference in the proofs involves showing that we can choose a sequence $\{\varepsilon_n\}$ with $\varepsilon_n \to 0$ as $n \to \infty$ such that $\sum_{n=1}^{\infty} \varepsilon_n < \infty$ and $\sum_{n=1}^{\infty} \varepsilon_n^{-1} E\left(X_n^{-\frac{1}{2}} w_{X_n}^{-1} \mid X_n > 0\right) < \infty$ where w_n is either u_n or v_n.

We know that $u_n \uparrow 1$, $v_n \uparrow 1$ so that $w_n \uparrow 1$ as $n \to \infty$. Thus, conditional on $X_n > 0$, $w_{X_n} \geq w_1$ and hence

$$\sum_{n=1}^{\infty} \varepsilon_n^{-1} E\left(X_n^{-\frac{1}{2}} w_{X_n}^{-1} \mid X_n > 0\right) \leq w_1^{-1} \sum_{n=1}^{\infty} \varepsilon_n^{-1} E\left(X_n^{-\frac{1}{2}} \mid X_n > 0\right)$$

$$\leq w_1^{-1} \sum_{n=1}^{\infty} \varepsilon_n^{-1} \left(E\left(X_n^{-1} \mid X_n > 0\right)\right)^{\frac{1}{2}} .$$

Now, using Lemma 2.3 of [8] we have $\varepsilon_n = \theta^n$ with $0 < \gamma < \theta < 1$,

$$\sum_{n=1}^{\infty} \varepsilon_n^{-1} \left(E\left(X_n^{-1} \mid X_n > 0\right)\right)^{\frac{1}{2}} \leq \sum_{n=1}^{\infty} \theta^{-n} \gamma^n < \infty .$$

Thus, with this choice of ε_n, $\sum_{n=1}^{\infty} \varepsilon_n < \infty$ and

$\sum_{n=1}^{\infty} \varepsilon_n^{-1} E\left(X_n^{-\frac{1}{2}} w_{X_n}^{-1} \mid X_n > 0\right) < \infty$, as required.

References

[1] Wolfgang J. Bühler, "Ein zentraler Grenzwertsatz für Verzweigungsprozesse", *Z. Wahrscheinlichkeitstheorie verw. Gebiete* 11 (1969), 139-141.

[2] Theodore E. Harris, *The theory of branching processes* (Die Grundlehren der Mathematischen Wissenschaften in Einzeldarstellungen, Band 119. Springer-Verlag, Berlin, Göttingen, Heidelberg, 1963; Prentice-Hall, Englewood Cliffs, New Jersey, 1963).

[3] C.C. Heyde, "On the influence of moments on the rate of convergence to the normal distribution", *Z. Wahrscheinlichkeitstheorie verw. Gebiete* 8 (1967), 12-18.

[4] C.C. Heyde, "Some properties of metrics in a study on convergence to normality", *Z. Wahrscheinlichkeitstheorie verw. Gebiete* 11 (1969), 181-192.

[5] C.C. Heyde, "Some central limit analogues for super-critical Galton-Watson processes", *J. Appl. Probability* 8 (1971), 52-59.

[6] C.C. Heyde, "Some almost sure convergence theorems for branching processes", *Z. Wahrscheinlichkeitstheorie verw. Gebiete* (to appear).

[7] C.C. Heyde and B.M. Brown, "An invariance principle and some convergence rate results for branching processes", *Z. Wahrscheinlichkeitstheorie verw. Gebiete* (to appear).

[8] C.C. Heyde and E. Seneta, "Analogues of classical limit theorems for the supercritical Galton-Watson process with immigration", *Math. Biosci.* (to appear).

[9] E. Seneta, "A note on the supercritical Galton-Watson process with immigration", *Math. Biosci.* 6 (1970), 305-312.

Department of Statistics,
School of General Studies,
Australian National University,
Canberra, ACT.

Analogues of Classical Limit Theorems for the Supercritical Galton–Watson Process with Immigration

C. C. HEYDE AND E. SENETA
Australian National University
Canberra, Australia

Communicated by Richard Bellman

ABSTRACT

In [6] it was shown that under standard conditions $X_n/m^n \to V > 0$ with probability one, where $\{X_n\}$ is a Galton–Watson process with immigration, and offspring mean $m > 1$. The authors obtain convergence-rate results under additional conditions for this asymptotic behavior in the form of analogues of the central limit theorem and the law of the iterated logarithm, and similar results for the estimator X_{n+r}/X_n, which for fixed $r \geqslant 1$ approaches m^r with probability one.

INTRODUCTION

The classical central limit theorem and the law of the iterated logarithm may be regarded as convergence-rate results for the strong law of large numbers. Recently, one of the authors has adopted this interpretation with regard to the well-known relation $Z_n/m^n \overset{\text{a.s.}}{\to} W$ for the ordinary Galton–Watson process $\{Z_n\}$ with offspring mean $m > 1$, and proved analogous convergence-rate theorems for this strong convergence result [1, 2].

A similar strong convergence result is known for the supercritical Galton–Watson process with immigration $\{X_n\}$; this has been derived simply in an earlier article in this journal [6], to which we refer the reader for a detailed description, and whose notation and assumptions we adopt. (The result may also be eventually deduced from a general limit theorem on an ordinary decomposable multitype Galton–Watson process, namely, Theorem 2.1 of Kesten and Stigum [4].) Thus, our basic assumptions are that the process is initiated by a single ancestor; and that the offspring and immigration distributions $\{f_j\}$, $\{b_j\}$, generated respectively by the probability generating functions (pgf's) $F(s)$, $B(s)$, $s \in [0, 1]$ satisfy:

$$1 < F'(1-) \equiv m < \infty; \; f_j \neq 1, \; j \geqslant 0; \; 0 < \lambda \equiv B'(1-) < \infty.$$

Mathematical Biosciences **11** (1971), 249–259

R. Maller et al. (eds.), *Selected Works of C.C. Heyde*, Selected Works in Probability and Statistics, DOI 10.1007/978-1-4419-5823-5_27, © Springer Science+Business Media, LLC 2010

The theorem of [6] then implies that if $\Sigma\, j(\log j)f_j < \infty$, $\{X_n/m^n\}$ converges almost surely to a proper random variable V with finite mean EV, and such that $P(V = 0) = 0$. One of the aims of this article is to indicate that for the process with immigration also analogous convergence-rate results hold, under further conditions, to be introduced as required.

We note also that the foregoing convergence result implies that for any fixed integer $r \geqslant 1$, $X_{n+r}/X_n \to m^r$ with probability one. It will be seen from the representation (2), which follows, that this may be regarded as an analogue of the strong law of large numbers for the estimator X_{n+r}/X_n of m^r. We give a parallel set of convergence-rate results for this relationship also, which may have some statistical utility.

The overall treatment is relatively brief, because much of the approach may be taken over directly from that for the ordinary supercritical Galton–Watson case; in fact, the present situation is intrinsically simpler in that $X_n \overset{\text{a.s.}}{\to} \infty$, a property not shared in general by the ordinary process. Indeed, the main difficulty is to obtain for the present process an analogue of the Berry–Esseen bound on the deviation from normality, which is needed to prove the analogue of the iterated logarithm law. This is done via Lemma 3.1, which may be of some independent interest, and two further lemmas.

DESCRIPTION OF THE RESULTS

The keys to the limit results are representations for $X_n - m^nV$ and $X_{n+r} - m^rX_n$ as sums of independent random variables.

In the first case, we have

$$X_n - m^nV = \lim_{r\to\infty}\left(X_n - m^n\frac{X_{n+r}}{m^{n+r}}\right) \quad \text{a.s.} \tag{1}$$

Furthermore,

$$X_{n+r} = Z_r^{(1)} + \cdots + Z_r^{(X_n)} + Y_{r,n} \tag{2}$$

where

$$Y_{r,n} = U_{n+r}^{(n+1)} + \cdots + U_{n+r}^{(n+r)}.$$

Here $Z_r^{(i)}$, $i = 1, 2, \ldots, X_n$ each have the distribution of Z_r, where $\{Z_n\}$ is an ordinary Galton–Watson process generated by $F(s)$; and $U_{n+r}^{(i)}$ is the number of direct descendants in the $(n + r)$th generation from immigration at the ith. Further, the $Z_r^{(i)}$, $i = 1, 2, \ldots, X_n$ and $U_{n+r}^{(i)}$, $i = n + 1, \ldots, n + r$ are all independent, and $Y_{r,n}$ has the same distribution for each n. Moreover, it has been shown in [6] that, for fixed n, $\lim_{r\to\infty} m^{-r}Y_{r,n} = I^{(n)}$ a.s. where $I^{(n)}$ has the distribution of $I = V - W$. From (1) and (2)

$$X_n - m^nV = \lim_{r\to\infty}\left[\left(1 - \frac{Z_r^{(1)}}{m^r}\right) + \cdots + \left(1 - \frac{Z_r^{(X_n)}}{m^r}\right) - \frac{Y_{r,n}}{m^r}\right]$$
$$= (1 - W^{(1)}) + \cdots + (1 - W^{(X_n)}) - I^{(n)}$$
$$= T_1 + \cdots + T_{X_n} - I^{(n)} \quad \text{a.s.,} \tag{3}$$

(say) where the components on the right-hand side are all independent and are each distributed independently of X_n.

The representation for $X_{n+r} - m^rX_n$ follows simply from (2):

$$X_{n+r} - m^rX_n = (Z_r^{(1)} - m^r) + \cdots + (Z_r^{(X_n)} - m^r) + Y_{r,n} \tag{4}$$

where again the components on the right-hand side are all independent, and are independent of X_n; and $Y_{r,n}$ has the same distribution for each n.

Making use of (3) and (4) together with an adaptation of the techniques of [2] and [3] yields the following results.

THEOREM 1. *Let* $\operatorname{var} Z_1 = \sigma^2 < \infty$. *Then*

$$\lim_{n\to\infty} P((m^2 - m)^{1/2}\sigma^{-1}X_n^{-1/2}(m^nV - X_n) \leqslant x|X_n > 0) = \Phi(x)$$

and

$$\lim_{n\to\infty} P(\sigma_r^{-1}X_n^{-1/2}(X_{n+r} - m^rX_n) \leqslant x|X_n > 0) = \Phi(x)$$

where $\Phi(x)$ *is the distribution function of* $N(0, 1)$ *and*

$$\sigma_r^2 = \operatorname{var} Z_r = \sigma^2 m^r(m^r - 1)(m^2 - m)^{-1}.$$

THEOREM 2. *Let* $EZ_1^3 < \infty$. *Then there exists a* δ, $0 < \delta < 1$, *such that*

$$\sup_x |P((m^2 - m)^{1/2}\sigma^{-1}X_n^{-1/2}(m^nV - X_n) \leqslant x|X_n > 0) - \Phi(x)| = 0(\delta^n)$$

and

$$\sup_x |P(\sigma_r^{-1}X_n^{-1/2}(X_{n+r} - m^rX_n) \leqslant x|X_n > 0) - \Phi(x)| = 0(\delta^n)$$

as $n \to \infty$.

(Explicit forms for the remainders may be found from Lemmas 2 and 3 of the next section.)

THEOREM 3. *Let* $EZ_1^3 < \infty$. *Then with probability one*

$$\limsup_{n\to\infty} \frac{m^nV - X_n}{(2\sigma^2(m^2 - m)^{-1}X_n \log n)^{1/2}} = 1,$$

$$\liminf_{n\to\infty} \frac{m^nV - X_n}{(2\sigma^2(m^2 - m)^{-1}X_n \log n)^{1/2}} = -1,$$

$$\limsup_{n\to\infty} \frac{X_{n+r} - m^rX_n}{(2\sigma_r^2X_n \log n)^{1/2}} = 1, \quad \liminf_{n\to\infty} \frac{X_{n+r} - m^rX_n}{(2\sigma_r^2X_n \log n)^{1/2}} = -1.$$

PROOFS

Proof of Theorem 1. Theorem 1 is very easily obtained from the representations (3) and (4). Since $X_n \xrightarrow{\text{a.s.}} \infty$ as $n \to \infty$, we certainly have $X_n^{-1/2} I^{(n)}$ and $X_n^{-1/2} Y_{r,n}$ converging in probability to zero on $\{X_n > 0\}$ as $n \to \infty$. Then, noting that X_n is independent of the summands in each case, an appeal to the classical central limit theorem yields the result, since from ordinary Galton–Watson theory the remaining (identically and independently distributed) summands have zero mean. ∎

In order to prove Theorem 2 we need first to extend the lemma in Section 4 of [3] so that the troublesome components $I^{(n)}$ and $Y_{r,n}$ in the representations (3) and (4) can be accounted for.

LEMMA 1. *Let ξ_i, $i = 1, 2, \ldots$, be independent and identically distributed random variables with $E\xi_1 = 0$, var $\xi_1 = \alpha^2$, and $E|\xi_1|^3 < \infty$. Let η_n, with $E|\eta_n| < \infty$, be a random variable independent of $\{\xi_i\}$ and with the same distribution for each n; and N_n be a positive integer-valued random variable that is independent of the $\{\xi_i\}$ and η_n. Then for any sequence $\{\varepsilon_n\}$ of positive constants with $\varepsilon_n \to 0$ as $n \to \infty$*

$$\sup_x |P(\alpha^{-1} N_n^{-1/2}(\xi_1 + \cdots + \xi_{N_n} + \eta_n) \leqslant x) - \Phi(x)|$$
$$\leqslant CE(N_n^{-1/2}) + \alpha^{-1}\varepsilon_n^{-1} E|\eta_n| E(N_n^{-1/2}) + \frac{\varepsilon_n}{2}$$

where $C = K\alpha^{-3} E|\xi_1|^3$, K being the universal constant in the Berry–Esseen bound.

Proof. Using the Berry–Esseen bound and since η_n is independent of $\{\xi_i\}$,

$$-CE(N_n^{-1/2})$$
$$= -C \sum_{j=1}^{\infty} j^{-1/2} P(N_n = j)$$
$$= -C \sum_{j=1}^{\infty} j^{-1/2} \left[\int_{-\infty}^{\infty} dP(\alpha^{-1} j^{-1/2} \eta_n \leqslant y)\right] P(N_n = j)$$
$$\leqslant \sum_{j=1}^{\infty} \left[\int_{-\infty}^{\infty} \{P(\alpha^{-1} j^{-1/2}(\xi_1 + \cdots + \xi_j) \leqslant x - y) - \Phi(x - y)\} \, dP(\alpha^{-1} j^{-1/2} \eta_n \leqslant y)\right] P(N_n = j)$$
$$\leqslant CE(N_n^{-1/2}).$$

But,

$$\sum_{j=1}^{\infty} \left[\int_{-\infty}^{\infty} \{P(\alpha^{-1} j^{-1/2}(\xi_1 + \cdots + \xi_j) \leqslant x - y) - \Phi(x - y)\} \, dP(\alpha^{-1} j^{-1/2} \eta_n \leqslant y)\right] P(N_n = j)$$

$$= \sum_{j=1}^{\infty} [P(\alpha^{-1}j^{-1/2}(\xi_1 + \cdots + \xi_j + \eta_n) \leqslant x) - P(\zeta + \alpha^{-1}j^{-1/2}\eta_n \leqslant x)]P(N_n = j)$$

$$= \sum_{j=1}^{\infty} [P(\alpha^{-1}j^{-1/2}(\xi_1 + \cdots + \xi_j + \eta_n) \leqslant x | N_n = j) - P(\zeta + \alpha^{-1}j^{-1/2}\eta_n \leqslant x | N_n = j)]P(N_n = j)$$

$$= P(\alpha^{-1}N_n^{-1/2}(\xi_1 + \cdots + \xi_{N_n} + \eta_n) \leqslant x) - P(\zeta + \alpha^{-1}N_n^{-1/2}\eta_n \leqslant x)$$

where ζ has an $N(0, 1)$ distribution and is independent of η_n, so that

$$\sup_x |P(\alpha^{-1}N_n^{-1/2}(\xi_1 + \cdots + \xi_{N_n} + \eta_n) \leqslant x) - P(\zeta + \alpha^{-1}N_n^{-1/2}\eta_n \leqslant x)| \leqslant CE(N_n^{-1/2}). \tag{5}$$

To this point, the argument is a straightforward generalization of the lemma of [3].

Now, for any $\varepsilon_n > 0$,

$$\begin{aligned} P(\zeta + \alpha^{-1}N_n^{-1/2}\eta_n \leqslant x) &= P(\zeta + \alpha^{-1}N_n^{-1/2}\eta_n \leqslant x, \alpha^{-1}N_n^{-1/2}|\eta_n| \leqslant \varepsilon_n) \\ &\quad + P(\zeta + \alpha^{-1}N_n^{-1/2}\eta_n \leqslant x, \alpha^{-1}N_n^{-1/2}|\eta_n| > \varepsilon_n) \\ &\leqslant P(\zeta \leqslant x + \varepsilon_n) + P(\alpha^{-1}N_n^{-1/2}|\eta_n| > \varepsilon_n), \end{aligned} \tag{6a}$$

and similarly

$$P(\zeta + \alpha^{-1}N_n^{-1/2}\eta_n > x) \leqslant P(\zeta > x - \varepsilon_n) + P(\alpha^{-1}N_n^{-1/2}|\eta_n| > \varepsilon_n),$$

or equivalently,

$$P(\zeta + \alpha^{-1}N_n^{-1/2}\eta_n \leqslant x) \geqslant P(\zeta \leqslant x - \varepsilon_n) - P(\alpha^{-1}N_n^{-1/2}|\eta_n| > \varepsilon_n). \tag{6b}$$

Also, it is readily checked that

$$\sup_x |P(\zeta \leqslant x - \varepsilon_n) - P(\zeta \leqslant x)| < \frac{\varepsilon_n}{2} \tag{7a}$$

from, for example, the mean value theorem, whence

$$\sup_x |P(\zeta \leqslant x) - P(\zeta \leqslant x + \varepsilon_n)| < \frac{\varepsilon_n}{2}. \tag{7b}$$

Using the double inequality for $P(\zeta + \alpha^{-1}N_n^{-1/2}\eta_n \leqslant x)$ provided by (6a) and (6b) in conjunction with (7a) and (7b)

$$\begin{aligned} \sup_x |P(\zeta + \alpha^{-1}N_n^{-1/2}\eta_n \leqslant x) - \Phi(x)| &\leqslant P(\alpha^{-1}N_n^{-1/2}|\eta_n| > \varepsilon_n) + \frac{\varepsilon_n}{2} \\ &\leqslant \alpha^{-1}\varepsilon_n^{-1}E(N_n^{-1/2}|\eta_n|) + \frac{\varepsilon_n}{2} \end{aligned}$$

by Markov's inequality;

$$= \alpha^{-1}\varepsilon_n^{-1}E(N_n^{-1/2})E|\eta_n| + \frac{\varepsilon_n}{2} \tag{8}$$

by the independence of η_n and N_n. The result of the lemma then follows from (5) and (8). ∎

LEMMA 2. *Let $EZ_1^3 < \infty$. Then for any sequence $\{\varepsilon_n\}$ of positive constants such that $\varepsilon_n \to 0$ as $n \to \infty$*

$$\sup_x |P((m^2 - m)^{1/2}\sigma^{-1}X_n^{-1/2}(m^nV - X_n) \leqslant x|X_n > 0) - \Phi(x)|$$
$$\leqslant K\sigma^{-3}(m^2 - m)^{3/2}E|W - 1|^3E(X_n^{-1/2}|X_n > 0)$$
$$+\sigma^{-1}(m^2 - m)^{1/2}\varepsilon_n^{-1}E(I)E(X_n^{-1/2}|X_n > 0) + \frac{\varepsilon_n}{2} \tag{9}$$

and

$$\sup_x |P(\sigma_r^{-1}X_n^{-1/2}(X_{n+r} - m^rX_n) \leqslant x|X_n > 0) - \Phi(x)|$$
$$\leqslant K\sigma_r^{-3}E|Z_r - m^r|^3E(X_n^{-1/2}|X_n > 0)$$
$$+\sigma_r^{-1}\varepsilon_n^{-1}E(Y_{r,1})E(X_n^{-1/2}|X_n > 0) + \frac{\varepsilon_n}{2} \tag{10}$$

where $K(<0.82)$ is the universal constant in the Berry–Esseen bound.

Proof. The inequalities (9) and (10) follow respectively from the representations (3) and (4), and application of Lemma 1, having noted that all expectations occurring on the right-hand sides of (9) and (10) are finite on account of the totality of conditions being assumed to be satisfied. ∎

LEMMA 3. *Let γ be any positive number satisfying*

$$1 > \gamma^2 > \max(B(q), m^{-1}).$$

Then

$$E(X_n^{-1}|X_n > 0) = 0(\gamma^n) \tag{11}$$

as $n \to \infty$. (An explicit, of somewhat inconvenient, form for the right-hand side may be deduced from the body of the proof.)

Proof. Let us denote by $P_n(x)$ the pgf of X_n, and write

$$P_n(x) = F_n(x)\tilde{P}_n(x),$$

$x \in [0, 1]$, where from [6], $\tilde{P}_n(x) \equiv \Pi_{j=0}^{n-1} B\{F_j(x)\}$, $n \geqslant 1$, denotes the pgf of the contribution to X_n resulting from immigration since generation 1. It is easily checked that

$$D_n \equiv E[X_n^{-1}|X_n > 0] = \int_0^1 \frac{P_n(x) - P_n(0)}{x\{1 - P_n(0)\}}\,dx,$$

so that we proceed to the estimation of the right-hand integral. *We assume initially that* $q > 0$, where q is the unique root in $[0, 1)$ of the equation $F(x) = x$, and decompose the integral into components.

$$D_n = \int_0^q \frac{P_n(x) - P_n(0)}{x\{1 - P_n(0)\}}\,dx + \int_q^1 \frac{P_n(x) - P_n(0)}{x\{1 - P_n(0)\}}\,dx$$

$$= D_n(1) \qquad\qquad + D_n(2), \quad \text{say}.$$

To estimate the first integral we note that

$$\begin{aligned} D_n(1) &= \{1 - P_n(0)\}^{-1} \int_0^q \left(\sum_{j=1}^{\infty} P(X_n = j) x^{j-1} \right) dx \\ &= \{1 - P_n(0)\}^{-1} \sum_{j=1}^{\infty} j^{-1} P(X_n = j) q^j \\ &\leqslant \{1 - P_n(0)\}^{-1} P_n(q) \\ &= \{1 - P_n(0)\}^{-1} q(B(q))^n \\ &\leqslant C_1(B(q))^n \end{aligned} \tag{12}$$

(since $P_n(0) \to 0$ as $n \to \infty$), where C_1 is a positive constant.

To estimate $D_n(2)$ we first note that

$$D_n(2) \leqslant C_2(1 - q)^{-1} \int_q^1 P_n(x)\,dx$$

where C_2 is a constant. If we now make the change of variable $x = q + (1 - q)y$, we get

$$\begin{aligned} D_n(2) &\leqslant C_2 \int_0^1 P_n(q + (1 - q)y)\,dy \\ &\leqslant C_2 \int_0^1 \tilde{P}_n(q + (1 - q)y)\,dy. \end{aligned}$$

Now, for $j \geqslant 0$

$$B\{F_j(q + (1 - q)y)\} = B^*\{F_j^*(y)\}$$

where

$$B^*(y) = B(q + (1 - q)y) \quad \text{and} \quad F_j^*(y) = \frac{F_j(q + (1 - q)y) - q}{1 - q},$$

$F_j^*(y)$ being easily seen to be the jth functional iterate of $F_1^*(y)$, which is a pgf with mean m, and $F_1^*(0) = 0$. Thus, for $n \geqslant 1$

$$\tilde{P}_n(q+(1-q)y) \equiv \prod_{j=0}^{n-1} B^*\{F_j^*(y)\} = P_n^*(y),$$

say, where $P_n^*(y)$ may itself be construed as the pgf of the nth generation of a *supercritical* Galton–Watson process with immigration, with offspring pgf $F_1^*(y)$ and immigration pgf $B^*(y)$, and no initial ancestor. Thus

$$P_n^*(y) = P_k^*(y)P_{n-k}^*(F_k^*(y)) \tag{13}$$

for any integer k, $1 \leqslant k \leqslant n$.

Let us note at this stage, for further use, that for any fixed $y_0 \in (0, 1)$,

$$P_k^*(y_0) \leqslant \{B^*(y_0)\}^k \tag{14}$$

since in the present situation $F_1^*(y_0) < y_0$. Also, if we let $H_n(y)$, $y \in [0, 1]$, denote the inverse function of $F_n^*(y)$, $y \in [0, 1]$, then it is easily checked that $H_n(y)$ is the nth functional iterate of $H_1(y)$, and that $H_n(y_0) \uparrow 1$ as $n \to \infty$. In fact, since by the mean value theorem

$$1 - H_n(y_0) = H_1'(\xi_n)\{1 - H_{n-1}(y_0)\}$$

where $H_{n-1}(y_0) < \xi_n < 1$, and since $H_1'(y) \downarrow m^{-1}$ as $y \uparrow 1$, it follows that for any fixed $\varepsilon > 0$ satisfying $m^{-1} + \varepsilon < 1$, there exists a fixed $r_0 \equiv r_0(y_0, \varepsilon)$ such that

$$1 - H_n(y_0) \leqslant (m^{-1} + \varepsilon)^n\{(m^{-1} + \varepsilon)^{-r_0}(1 - H_{r_0}(y_0))\}$$

for $n \geqslant r_0$.

Then

$$D_n(2) \leqslant C_2 \int_0^1 P_n^*(y)\, dy \tag{15}$$

$$= C_2\left\{\int_0^{H_k(y_0)} P_n^*(y)\, dy + \int_{H_k(y_0)}^1 P_n^*(y)\, dy\right\}$$

$$\leqslant C_2\{H_k(y_0)P_k^*(H_k(y_0))P_{n-k}^*(y_0) + 1 - H_k(y_0)\}$$

providing $1 \leqslant k \leqslant n$, by dominating by the maximum value of the integrand in each case, and in the first part using (13);

$$\leqslant C_2\{P_{n-k}^*(y_0) + 1 - H_k(y_0)\}$$

$$\leqslant C_2\{(B^*(y_0))^{n-k} + 1 - H_k(y_0)\}$$

on account of (14).

If we now choose $k = [n/2]$ and choose n so large that $[n/2] \geqslant r_0$,

$$\leqslant C_2\{(B^*(y_0))^{n/2} + \rho(\varepsilon, y_0)(m^{-1} + \varepsilon)^{n/2}\} \tag{16}$$

where $\rho > 0$ does not depend on n.

Since initially both y_0 and ε could have been chosen as close to zero as desired and $B^*(y_0) = B(q + (1 - q)y_0)$, the assertion of the lemma follows from (12) and (16).

It remains only to estimate $D_n \equiv D_n(2)$ when $q = 0$. We have then that $P_n(x) = F_n(x)\tilde{P}_n(x) = F_n(x)P_n^*(x)$, with $P_n(0) = F_n(0) = 0$, $F_n(x) < x$. Thus

$$D_n = \int_0^1 \frac{P_n(x)}{x}\,dx \leqslant \int_0^1 P_n^*(x)\,dx \tag{17}$$

whence we may proceed as before from (15). ■

NOTES. (a) The technique for estimating the rate of geometric convergence was suggested by a method of A. V. Nagaev [5] in a similar situation in connection with the ordinary supercritical Galton–Watson process. (There appear to be some imprecise steps in his development.)

(b) The rate of convergence in (17) may also be estimated with the aid of the estimate in [3] for a similar ordinary supercritical Galton–Watson process.

(c) Other choices of k, for example, $k = [\delta n]$, for a δ fixed in $(0, 1)$, would give similar results. The optimal choice of such k, giving optimal geometric rate at least within the context of the present proof, would appear to depend on the absolute sizes of the two values $B(q)$, $1/m$.

PROOF OF THEOREM 2. We first note that, by a standard moment inequality,

$$E(X_n^{-1/2}|X_n > 0) \leqslant E(X_n^{-1}|X_n > 0)^{1/2}.$$

The assertions of the theorem then follow from Lemmas 2 and 3 by putting $\varepsilon_n = \gamma^{n/4}$, in which case $\delta = \gamma^{1/4}$. ■

PROOF OF THEOREM 3. This follows exactly the pattern of that of [2]. There are two main points of difference for which some modification is required:

(i) that $\Sigma_{n=1}^{\infty} P(A_{n+1}|X_n, \ldots, X_1) = \infty$ a.s. when

$$A_n = \{X_n - mX_{n-1} > (1 - \delta)\sigma(2X_{n-1}\log(n-1))^{1/2}\}$$

for any δ, $0 < \delta < 1$;

(ii) that for any integer $r > 1$, $\{X_{rn}\}$, $n = 0, 1, 2, \ldots$, also describes a supercritical Galton–Watson process with immigration; and we shall confine ourselves to establishing the validity of these points.

(i) Using (4) we have, if $X_n > 0$,

$$P(A_{n+1}|X_n, \ldots, X_1) = P((Z_1^{(1)} - m) + \cdots + (Z_1^{(X_n)} - m) + Y > (1 - \delta)\sigma(2X_n \log n)^{1/2}|X_n)$$

where Y has the immigration distribution, and the $Z_1^{(i)}$ are independent and identically distributed, each with the distribution of Z_1, and are independent of $X_1, \ldots, X_n$ and Y; and if $X_n = 0$

$$P(A_{n+1}|X_n, \ldots, X_1) = P(Y > 0).$$

In the former case

$$P((Z_1^{(1)} - m) + \cdots + (Z_1^{(X_n)} - m) + Y > (1 - \delta)\sigma(2X_n \log n)^{1/2}|X_n)$$
$$\geqslant P((Z_1^{(1)} - m) + \cdots + (Z_1^{(X_n)} - m) > (1 - \delta)\sigma(2X_n \log n)^{1/2}|X_n)$$

since $Y \geqslant 0$

$$\geqslant 1 - \Phi((1 - \delta)(2 \log n)^{1/2}) - CX_n^{-1/2}$$

using the Berry–Esseen bound. Now, with probability one, $X_n > 0$ for n sufficiently large, since $X_n \to \infty$ a.s. Hence the last inequality may be used to show divergence with probability one of $\Sigma_1^\infty P(A_{n+1}|X_n, \ldots, X_1)$ as in [2].

(ii) Consideration of the structure of the process reveals that indeed $\{X_{nr}\}$, $n \geqslant 0$, may be considered as a branching process with immigration, the offspring distribution being $F_r(x)$ and the immigration distribution $\Pi_{i=0}^{r-1} B\{F_i(x)\}$. This last may also be easily confirmed from the fact that X_{nr} has the pgf $P_{nr}(x)$ and from the fundamental recursion for the Galton–Watson process with immigration $\{X_n\}$

$$P_{nr}(x) = B(x)P_{nr-1}(F(x))$$

whence, iterating back,

$$P_{nr}(x) = B(x)B(F(x)) \cdots B(F_{r-1}(x))P_{(n-1)r}(F_r(x)). \blacksquare$$

SUPPLEMENTARY REMARKS

The results given above have been proved under conditions that are convenient, but frequently more restrictive than necessary, and we now briefly note how these can be relaxed.

Reference to [7] in conjunction with [6] reveals that Theorem 1 remains valid if the condition $0 < \lambda < \infty$ on the immigration distribution is replaced by the milder assumption that $\Sigma_{j=1}^\infty b_j \log j < \infty$ with $b_0 < 1$.

In Theorems 2 and 3 the condition $EZ_1^3 < \infty$ may be replaced by the weaker one that $EZ_1^{2+\kappa} < \infty$ where $0 < \kappa \leqslant 1$, through the application of generalized forms of Berry–Esseen bounds. In passing we also note the Lemma 3 remains valid without any auxiliary moment conditions at all (even $m = \infty$ is permissible). Finally, it may be possible to relax the condition $0 < \lambda < \infty$, which with the other conditions implies $EI < \infty$ and $EY_{r,1} < \infty$ in (9) and (10), and replace EI and $EY_{r,1}$ by something correspondingly weaker. The reader wishing to pursue this interaction may begin by consulting Lemma 2.2 of [9] in conjunction with the representation (9) of [6].

Finally, we note that it is not difficult to verify the validity of the second part of Theorem 1 even when $m = 1$, where $\sigma_r^2 = \operatorname{var} Z_r = r\sigma^2$, since $P(X_n = k | X_n > 0) \to 0, k = 1, 2, \ldots,$ as $n \to \infty$ (see [8]).

Note added in proof: Subsequent research by C. C. Heyde and J. R. Leslie has revealed that versions of Theorems 2 and 3 hold subject to only the condition $EZ_1^2 < \infty$ (to appear in *Bull. Austral. Math. Soc.*).

REFERENCES

1 C. C. Heyde, Some central limit analogues for supercritical Galton–Watson processes *J. Appl. Prob.* (to appear in 1971).

2 C. C. Heyde, Some almost sure convergence theorems for branching processes (submitted for publication).

3 C. C. Heyde and B. M. Brown, An invariance principle and some convergence rate results for branching processes (submitted for publication).

4 H. Kesten and B. P. Stigum, Limit theorems for decomposable multidimensional Galton–Watson processes. *J. Math. Anal. Applic.* **17**(1967), 309–338.

5 A. V. Nagaev, On the estimation of the mean number of direct descendants of a particle in a branching random process, *Teor. Veroyat. Primenen.* **12**(1967), 363–369 (in Russian).

6 E. Seneta, A note on the supercritical Galton–Watson process with immigration, *Math. Biosci.*, **6**(1970), 305–312.

7 E. Seneta, On the supercritical Galton–Watson process with immigration, *Math. Biosci.*, **7**(1970), 9–14.

8 E. Seneta, An explicit-limit theorem for the critical Galton–Watson process with immigration, *J. Roy. Statist. Soc.* **32**(1970), 149–152.

9 W. L. Smith, Remarks on renewal theory when quality of renewals varies, *Inst. Statist. Mimeo Ser.* **548**(1967) (Univ. of North Carolina).

The Annals of Mathematical Statistics
1972, Vol. 43, No. 6, 2058-2066

ON LIMIT THEOREMS FOR QUADRATIC FUNCTIONS OF DISCRETE TIME SERIES

BY E. J. HANNAN AND C. C. HEYDE
Australian National University

In this paper it is shown how martingale theorems can be used to appreciably widen the scope of classical inferential results concerning autocorrelations in time series analysis. The object of study is a process which is basically the second-order stationary purely non-deterministic process and contains, in particular, the mixed autoregressive and moving average process. We obtain a strong law and a central limit theorem for the autocorrelations of this process under very general conditions. These results show in particular that, subject to mild regularity conditions, the classical theory of inference for the process in question goes through if the best linear predictor is the best predictor (both in the least squares sense).

1. Introduction. A great deal of time series analysis is based upon quadratic functions of the data. In particular, many inferential results relate to theorems concerning the autocorrelations

$$r(j) = \frac{\sum_{n=1}^{N-j} \{x(n) - \bar{x}\}\{x(n+j) - \bar{x}\}}{\sum_{n=1}^{n} \{x(n) - \bar{x}\}^2}, \qquad j \geqq 0, \tag{1}$$

$$r(-j) = r(j),$$

$x(1), x(2), \cdots, x(N)$ being a sample of N consecutive observations on some process $\{x(n)\}$. It is well known that, under certain conditions on the process $\{x(n)\}$, a strong law of large numbers and a central limit theorem hold for $r(j)$ (see, for example, Hannan [6], Chapter IV, VI). In this paper it is our object to show, using limit theorems for martingales, that the scope of the classical inferential theory can be appreciably widened in a natural way.

We shall be concerned with a process of the form

$$x(n) - \mu = \sum_{j=0}^{\infty} \alpha(j)\varepsilon(n-j), \qquad \sum_{j=0}^{\infty} \alpha^2(j) < \infty, \qquad \alpha(0) = 1; \tag{2}$$
$$E\varepsilon(n) = 0, \qquad E\{\varepsilon(m)\varepsilon(n)\} = 0, \qquad m \neq n.$$

If $x(n) - \mu$ is a second-order stationary, purely non-deterministic, process ([6], Chapter III) then it may be represented in this form with the $\varepsilon(n)$ as the linear prediction errors, having variance $\sigma^2 > 0$. As is well known, there will be many representations of such a stationary process in the form (2) but for only one of these will the $\varepsilon(n)$ be the prediction errors. However, our results extend beyond the stationary case so that we do not assume stationarity for the $\{\varepsilon(n)\}$ but only that (2) holds for $n \geqq 0$, together with other conditions to be discussed shortly. A process of the kind (2) arises from a wide variety of contexts; for example from a mixed autoregressive and moving averge process ([6], Chapter I) and as

Received August 7, 1971; revised April 7, 1972.

R. Maller et al. (eds.), *Selected Works of C.C. Heyde*, Selected Works in Probability and Statistics, DOI 10.1007/978-1-4419-5823-5_28, © Springer Science+Business Media, LLC 2010

the response of a physically realizable filter to an uncorrelated sequence (Gikhman and Skorokhod [5], Chapter 5).

Now the classical theory of inference for the process (2) usually requires that the $\varepsilon(n)$ be independent and identically distributed with zero mean and variance σ^2 (which we paraphrase as i.i.d. $(0, \sigma^2)$). The essential feature of this paper is that, subject to some reasonable additional conditions, the classical theory goes through if the independence assumption is replaced by the weaker condition

$$(3) \qquad E(\varepsilon(n) \mid \mathscr{F}_{n-1}) = 0 \quad \text{a.s.}, \qquad \text{all} \quad n,$$

where $\mathscr{F}_n$ is the σ-field generated by the $\varepsilon(m)$, $m \leqq n$. This requirement has a simple and natural interpretation in the case where $\{x(n)\}$ is stationary and thus purely non-deterministic and the $\varepsilon(n)$ are the linear prediction errors, for then $\mathscr{F}_n$ is also the σ-field generated by the $x(m)$, $m \leqq n$ so that, because of (3),

$$(4) \qquad \varepsilon(n) = x(n) - E(x(n) \mid \mathscr{F}_{n-1}).$$

To see this, write $\mathscr{G}_n$ for the σ-field generated by $x(m)$, $m \leqq n$. Clearly $\mathscr{F}_n \supseteqq \mathscr{G}_n$ and, when the ε's are the prediction errors

$$\varepsilon(n) = x(n) - E(x(n) \mid \mathscr{G}_{n-1})$$

which is $\mathscr{G}_n$ measurable. Thus $\mathscr{G}_n \equiv \mathscr{F}_n$. Then, $E(x(n) \mid \mathscr{F}_{n-1})$ is the best linear predictor and the best linear predictor is the best predictor (both in the least squares sense). Conversely, if this is so, (4) must hold and hence (3). Thus (3) is equivalent to the condition that the best predictor is the best linear predictor, both in the least squares sense. In the stationary case our additional conditions are, for example, the regularity condition (7), below, together with the requirement that $E(\varepsilon^2(n) \mid \mathscr{F}_{n-1}) = \sigma^2$ a.s. Our results give that, subject to the mild regularity condition (7), the classical theory of inference for (2) goes through when the $\varepsilon(n)$ are the prediction errors provided the best linear predictor is the best predictor and the prediction variance, given the past, is a constant.

2. Strong law for autocorrelations. Here we consider the process (2) where $\varepsilon(n)$ satisfy the condition (3). We shall not require stationarity of $\{\varepsilon(n)\}$ but instead the condition

$$(5) \qquad \lim_{N\to\infty} N^{-1} \textstyle\sum_{n=1}^{N} E(\varepsilon^2(n) \mid \mathscr{F}_{n-1}) = \sigma^2 > 0 \quad \text{a.s.}$$

and the condition that there exists a random variable X with $EX^2 < \infty$ such that

$$(6) \qquad P(|\varepsilon(n)| > u) \leqq cP(|X|) > u)$$

for some $0 < c < \infty$ and all n, all $u \geqq 0$. If $x(n)$ is stationary we modify (5) to

$$(5)' \qquad E(\varepsilon^2(n) \mid \mathscr{F}_{n-1}) = \sigma^2 > 0 \quad \text{a.s.}$$

and (6) is redundant. We shall, in Section 3, make use of the condition

$$(7) \qquad \textstyle\sum_{j=1}^{\infty} j^{\frac{1}{2}} \alpha(j)^2 < \infty .$$

Define

$$c(j) = c(-j) = N^{-1} \textstyle\sum_{n=1}^{N-j} \{x(n) - \bar{x}\}\{x(n+j) - \bar{x}\}, \qquad j \geqq 0,$$

where $\bar{x}$ is the sample mean of the $x(n)$, $n = 1, \cdots, N$. If $x(n)$ is stationary its spectral density is

$$f(\lambda) = \frac{\sigma^2}{2\pi} |\sum_0^\infty \alpha(j)e^{ij\lambda}|^2$$

and the autocovariances $\gamma(j)$ satisfy

$$\gamma(j) = \int_{-\pi}^{\pi} e^{ij\lambda} f(\lambda)\, d\lambda = \sigma^2 \sum_{u=0}^\infty \alpha(u)\alpha(u+j) .$$

However we may define $f(\lambda)$ and $\gamma(j)$ by these formulae whether or not $x(n)$ is stationary. We now have the following theorem

THEOREM 1. *If* (3), (6) *and* $\sum |\alpha(k)| < \infty$ *hold*, $\bar{x}$ *converges* a.s. *to* μ *and if* (5) *holds also, then* $c(j)$ *converges in probabilty to* $\gamma(j)$. *If* $x(n)$ *is stationary*, $\bar{x}$ *converges* a.s. *to* μ *and if* (3), (5)′ *hold* $c(j)$ *converges* a.s. *to* $\gamma(j)$.

PROOF. If $\varepsilon(n)$ satisfies (3), (6) and $\sum |\alpha(k)| < \infty$, then it is easily seen that $\bar{x}$ has a variance which is $O(N^{-1})$ as $N \to \infty$. It consequently follows from the proof given in Doob [4], Theorem X 6.2, that $\bar{x}$ converges a.s. to μ. It is then clear that $c(j)$ has the same a.s. behaviour as

$$c^*(j) = N^{-1} \sum_{n=1}^{N-j} (x(n) - \mu)(x(n+j) - \mu)$$
$$= \sum_{u=0}^\infty \sum_{v=0}^\infty \alpha(u)\alpha(v) N^{-1} \sum_{n=1}^{N-j} \varepsilon(n-u)\varepsilon(n+j-v) .$$

Now

$$E|N^{-1} \sum_{n=1}^{N-j} \varepsilon(n-u)\varepsilon(n+j-v)| < K < \infty$$

by virtue of (6) and if $\sum |\alpha(k)| < \infty$ then

$$\lim_{p\to\infty} E|\sum_{u=p+1}^\infty \sum_{v=0}^\infty \alpha(u)\alpha(v) N^{-1} \sum_{n=1}^{N-j} \varepsilon(n-u)\varepsilon(n+j-v)| = 0 . \tag{8}$$

The same is true if in the left term in (8) the first two sums are over $0 \leqq u \leqq p$, $p < v < \infty$. On the other hand,

$$N^{-1} \sum_{n=1}^N E\{\varepsilon(n)\varepsilon(n-k) \,|\, \mathscr{F}_{n-1}\} = 0 \quad \text{a.s.}, \qquad k > 0 ,$$

and it follows from (6) and a law of large numbers for martingales due to Heyde (Theorem 1 of Heyde and Seneta [8]) that $N^{-1} \sum_{n=1}^N \varepsilon(n)\varepsilon(n-k)$, $k > 0$, and hence $N^{-1} \sum_{n=1}^{N-j} \varepsilon(n-u)\varepsilon(n+j-v)$, $u \neq v - j$ converges in probability to zero. (In order to obtain the uniform bound on the distribution of $\varepsilon(n)\varepsilon(n-k)$ required to justify the application of Theorem 1 of [8] we note that

$$\begin{aligned} P(|\varepsilon(n)\varepsilon(n-k)| > u) &\leqq P(\varepsilon^2(n) + \varepsilon^2(n-k) > 2u) \\ &\leqq P(\varepsilon^2(n) > u) + P(\varepsilon^2(n-k) > u) \\ &\leqq 2cP(X^2 > u) \end{aligned} \tag{9}$$

using (6).) Furthermore, by the same theorem together with (5), $N^{-1} \sum_{n=1}^{N-j} \varepsilon^2(n)$ converges in probability to σ^2. Thus

$$\sum_{u=0}^p \sum_{v=0}^p \alpha(u)\alpha(v) N^{-1} \sum_{n=1}^{N-j} \varepsilon(n-u)\varepsilon(n+j-v) \to \sigma^2 \sum{}' \alpha(u)\alpha(u+j)$$

in probability as $N \to \infty$ where $\sum'$ is a sum over $0 \leqq u$, $u + j \leqq p$. It follows from this together with (8) and Markov's inequality that $c^*(j)$, and hence $c(j)$, converges in probability to $\gamma(j)$.

If $x(n)$ is stationary with a.c. spectrum it is well known that $\bar{x}$ converges a.s. to μ. In place of $c^*(j)$ consider

$$\tilde{c}(j) = c^*(j) - \sum_{u=0}^{\infty} \alpha(u)\alpha(u+j)N^{-1} \sum_{n=1}^{N-j} \varepsilon^2(n-u) .$$

Assuming $x(n)$ stationary and (3), (5)′ we may show that the mean of $\tilde{c}(j)$ is zero and its variance converges to zero. The proof of the first is obvious and we prove the second, for simplicity, in case $j = 0$. The variance is

$$\sum\sum_{p\neq q,=0}^{\infty} \alpha(p)\alpha(q) \sum\sum_{r\neq s,=0}^{\infty} \alpha(r)\alpha(s) \times [N^{-2} \sum\sum_{m,n=1}^{N} E\{\varepsilon(m-p)\varepsilon(m-q)\varepsilon(n-r)\varepsilon(n-s)\}] .$$

We evaluate the expectation using (3) and (5)′. The only contribution comes when $m - p = n - r$ and $s = q - p + r$ or $m - p = n - s$ and $s = p - q + r$. Both sets of identifications give the same result and we take the first. After evaluating the expectation let us add back

$$N^{-1}\sigma^4 \sum\sum_{|p-r|<N,=0}^{\infty} \alpha(p)^2\alpha(r)^2 \left(1 - \frac{|p-r|}{N}\right)$$

which clearly converges to zero as N increases. Then we obtain

$$N^{-1}\sigma^4 \sum_{j=-N+1}^{N-1} \left(1 - \frac{|j|}{N}\right) \{\sum_0^{\infty} \alpha(p)\alpha(p+|j|)\}^2 .$$

This is N^{-1} by the Cesàro sum of the Fourier series, evaluated at the origin, of the convolution of $f(\lambda)$ with itself. It thus converges to zero. Thus $\tilde{c}(0)$, and in the same way $\tilde{c}(j)$, converges in probability to zero. However

$$\sum_{u=0}^{\infty} \alpha(u)\alpha(u+j)N^{-1} \sum_{n=1}^{N-j} \varepsilon^2(n-u)$$

converges in probability to $\gamma(j)$ by the same kind of argument as was used earlier in the proof (the convergence of $\sum |\alpha(u)|$ not now being needed). Thus $c^*(j)$ and hence $c(j)$ converges in probability to $\gamma(j)$ and since, by the ergodic theorem, $c(j)$ converges almost surely it must converge almost surely to $\gamma(j)$. This completes the proof.

3. Central limit theorems for autocorrelations. From Theorem 1 we note that, depending on the conditions, $r(j) = c(j)/c(0)$ converges either in probability or almost surely to $\rho(j) = \gamma(j)/\gamma(0)$. Here we offer two theorems, the first of which is for the non-stationary case.

THEOREM 2. *Suppose* (7) *holds,* $\sum |\alpha(j)| < \infty$ *and* $\{\varepsilon(n)\}$ *is a stochastic sequence, with* $E\varepsilon^2(n) = \sigma^2$, *all* n, *satisfying* (3), (5), *and* (6) *with bounding random variable* X *having finite fourth moment. Suppose also that* $E\{\varepsilon^2(n)\varepsilon(n-r)\varepsilon(n-s)\} = \sigma^4\tau_{rs}$ *is finite and uniformly bounded for every* n, $r \geqq 1$, $s \geqq 1$, *and*

(10) $$n^{-1} \sum_{t=1}^{n} \varepsilon(t-r)\varepsilon(t-s)E(\varepsilon^2(t) \mid \mathscr{F}_{t-1}) \to_{\text{a.s.}} \sigma^4\tau_{rs}$$

as $n \to \infty$ *for any* $r \geqq 1$, $s \geqq 1$. *Then, the joint distribution of* $N^{\frac{1}{2}}(r(j) - \rho(j))$,

$1 \leqq j \leqq s$, *converges to the s-variate normal distribution with zero mean and non-singular covariance matrix* $W = [w_{ij}]$ *where*

$$w_{ij} = \sum_{r=1}^{\infty} \sum_{s=1}^{\infty} \tau_{rs}\{\rho(r+i) + \rho(r-i) - 2\rho(r)\rho(i)\} \\ \times \{\rho(r+j) + \rho(r-j) - 2\rho(r)\rho(j)\}.$$

PROOF. This theorem may be established by adapting the proof of the theorem of Anderson and Walker [1], where the case $\{\varepsilon(n)\}$ i.i.d. $(0, \sigma^2)$ is considered. The only parts which cannot immediately be adapted involve the replacement of their condition $\sum j\alpha^2(j) < \infty$ by our (7), their application of a central limit theorem of Diananda in the proof of their Lemma 1 and their demontstration that $N^{-1} \sum_{n=1}^{N} \{x(n) - \mu\}^2$ converges in probability to $E\{x(1) - \mu\}^2$ as $N \to \infty$. This last result, however, is covered by our Theorem 1.

The replacement of $\sum j\alpha^2(j) < \infty$ by our (7) is easy to justify. The only point at issue concerns the proof of their Lemma 3 where it is necessary to note that the bound on the expectation of their (2.23) can be suitably sharpened for $|i| > n$. Full details are given in the proof that we give for Theorem 3 below.

It then remains to consider the central limit part. What is required is just that, for any sequence of constants $c_1, \cdots, c_m$, $n^{-\frac{1}{2}} \sum_{r=1}^{m} c_r \sum_{t=1}^{n} \varepsilon(t)\varepsilon(t+r)$ converges in distribution to a certain normal law. In order to obtain this under our conditions we first note that it suffices to establish the convergence result for $n^{-\frac{1}{2}} \sum_{r=1}^{m} c_r \sum_{t=1}^{n} \varepsilon(t)\varepsilon(t-r)$ which differs from the former in a fixed finite number of terms (and hence the difference goes in probability to zero as $n \to \infty$). We shall obtain this last result with the aid of a central limit theorem for martingales due to Brown [3].

Define $X_t = \sum_{r=1}^{m} c_r \varepsilon(t)\varepsilon(t-r)$, noting that $\{S_n = \sum_{t=1}^{n} X_t, \mathscr{F}_n, n \geqq 1\}$ is a martingale. Let

$$V_n^2 = \sum_{t=1}^{n} E(X_t^2 \mid \mathscr{F}_{t-1}) = \sum_{t=1}^{n} \sum_{r=1}^{m} \sum_{s=1}^{m} c_r c_s \varepsilon(t-r)\varepsilon(t-s) E(\varepsilon^2(t) \mid \mathscr{F}_{t-1})$$

and

$$s_n^2 = EV_n^2 = \sum_{t=1}^{n} EX_t^2 .$$

In order to apply Theorem 2 of [3] we need to show that

$$\text{(i)} \quad s_n^{-2} V_n^2 \to_p 1 \qquad \text{and} \qquad \text{(ii)} \quad s_n^{-2} \sum_{t=1}^{n} E(X_t^2 I(|X_t| \geqq \varepsilon s_n)) \to 0$$

for any $\varepsilon > 0$, $I(\cdot)$ being the indicator function.

The condition (10) clearly ensures that (i) holds upon noting that

$$s_n^2 = n\sigma^4 \sum_{r=1}^{m} \sum_{s=1}^{m} c_r c_s \tau_{rs} . \tag{11}$$

To obtain (ii) we first note that for any $u \geqq 0$,

$$\begin{aligned} P(|X_n| > u) &\leqq P(\textstyle\sum_{r=1}^{m} |c_r \varepsilon(n)\varepsilon(n-r)| > u) \\ &\leqq P(\textstyle\bigcup_{r=1}^{m} \{|c_r \varepsilon(n)\varepsilon(n-r)| > u/m\}) \\ &\leqq \textstyle\sum_{r=1}^{m} P(|\varepsilon(n)\varepsilon(n-r)| > u/m|c_r|) \qquad \text{(using (9))} \\ &\leqq 2c \textstyle\sum_{r=1}^{m} P(X^2 < u/m|c_r|) \\ &\leqq 2cmP(X^2 < u/mc^*) \end{aligned} \tag{12}$$

where $c^* = \max_{1 \leq r \leq m} |c_r|$. Thus, using integration by parts and (12),

$$EX_n^2 I(|X_n| \geqq \varepsilon s_n) \leqq 2 \int_{\varepsilon s_n}^{\infty} xP(|X_n| > x)\, dx$$
$$\leqq 2cm \int_{\varepsilon s_n}^{\infty} xP(X^2 > x/mc^*)\, dx \to 0$$

as $n \to \infty$ and (ii) follows. Brown's theorem then gives that $s_n^{-1} \sum_{t=1}^{n} X_t$ converges in distribution to $N(0, 1)$. That is, $n^{-\frac{1}{2}} \sum_{r=1}^{m} c_r \sum_{t=1}^{n} \varepsilon(t)\varepsilon(t-r)$ converges in distribution to $N(0, \sigma^4 \sum_{r=1}^{m} \sum_{s=1}^{m} c_r c_s \tau_{rs})$. The proof of Theorem 2 is then completed along the lines of Anderson and Walker [1]. Our second theorem relates to the stationary case.

THEOREM 3. *Let $x(n)$ be stationary and satisfy* (3), (5)′ *and* (7). *Then the joint distribution of $N^{\frac{1}{2}}(r(j) - \rho(j))$, $1 \leqq j \leqq s$, converges to the s-variate multivariate normal distribution with zero mean vector and covariance matrix $W = [w_{ij}]$ where*

$$\begin{aligned} w_{ij} &= \sum_{r=1}^{\infty} \{\rho(r+i) + \rho(r-i) - 2\rho(r)\rho(i)\} \\ &\qquad \times \{\rho(r+j) + \rho(r-j) - 2\rho(r)\rho(j)\} \\ &= \sum_{r=-\infty}^{\infty} \{\rho(r)\rho(r+i-j) + \rho(r)\rho(r+i+j) + 2\rho^2(r)\rho(i)\rho(j) \\ &\qquad - 2\rho(r)\rho(i)\rho(r+j) - 2\rho(r)\rho(j)\rho(r+i)\}\,. \end{aligned}$$

This proof also follows [1] and improves the theorem of that paper in three ways. Firstly the independence of the $\varepsilon(n)$ is replaced by (3), (5)′. Secondly $\sum j\alpha^2(j) < \infty$ is replaced by $\sum j^{\frac{1}{2}}\alpha^2(j) < \infty$. Thirdly $\sum |\alpha(j)| < \infty$ is avoided. Because we use the conditions (3), (5)′ we must confine ourselves to a one-sided sum, $\sum_0^{\infty} \alpha(j)\varepsilon(n-j)$ (i.e., to the purely non-deterministic case) while [1] treats the case of a two-sided moving average. The second and third relaxations of the conditions of the theorem in [1] may be made under the other conditions of that theorem.

PROOF. Let us take the case where μ is known to be zero since mean correction makes no difference to the truth of the theorem. Instead of $N^{\frac{1}{2}}\{r((j) - \rho(j)\}$ we may consider the limit distribution of $N^{\frac{1}{2}}\{c^*(j) - \rho(j)c^*(0)\}/c^*(0)$ and since $c^*(0)$ converges a.s. to $\gamma(0)$, by Theorem 1, we thus consider

$$N^{\frac{1}{2}}\{c^*(j) - \rho(j)c^*(0)\}\,. \tag{13}$$

We first show that we may omit all terms involving an $\varepsilon^2(n)$. These terms are (see [1] page 1301)

$$N^{-\frac{1}{2}}\{\textstyle\sum_0^{\infty} \alpha(k)\alpha(k+j)T_{N,k}^{(j)} - \rho(j) \sum_0^{\infty} \alpha^2(k)T_{N,k}^{(0)}\} \tag{14}$$

where

$$T_{N,k}^{(j)} = \sum_{1-k}^{N-j-k} \varepsilon^2(n) - \sum_1^N \varepsilon^2(n)$$

and

$$E|T_{N,k}^{(j)}| = \sigma^2 \min(2k+j, 2N-j)\,.$$

(In [1] a bound by $(2k+j)\sigma^2$ is given for $E|T_{N,k}^{(j)}|$ and this accounts for the replacement of our $\sum j^{\frac{1}{2}}\alpha^2(j) < \infty$ by $\sum j\alpha^2(j) < \infty$ in [1].) Thus, taking the first term in (14), for example,

$$\begin{aligned} N^{-\frac{1}{2}}E\{|\textstyle\sum_0^\infty \alpha(k)\alpha(k+j)T_{N,k}^{(j)}|\} &\leqq N^{-\frac{1}{2}}\sigma^2 \textstyle\sum_0^\infty |\alpha(k)\alpha(k+j)| \min(2k+j, 2N-j) \\ &\leqq \sigma^2[\{N^{-\frac{1}{2}} \textstyle\sum_0^\infty \alpha^2(k) \min(2k+j, 2N-j)\}^{\frac{1}{2}} \\ &\qquad \times \{N^{-\frac{1}{2}} \textstyle\sum_0^\infty \alpha^2(k+j) \min(2k+j, 2N-j)\}^{\frac{1}{2}}] . \end{aligned}$$

However,

$$\begin{aligned} N^{-\frac{1}{2}} \textstyle\sum_0^\infty \alpha^2(k) \min(2k+j, 2N-j) \\ \leqq 2 \textstyle\sum_0^{N-j} \alpha^2(k)k^{\frac{1}{2}}\{k/N\}^{\frac{1}{2}} + jN^{-\frac{1}{2}} \textstyle\sum_0^{N-j} \alpha^2(k) + 2 \textstyle\sum_{N-j}^\infty \alpha^2(k)k^{\frac{1}{2}} , \end{aligned}$$

which converges to zero. The same is true of the second factor and the second term in (14) and thus (14) converges in probability to zero.

Let us put $x(n) = x_1(n) + x_2(n)$ where

$$x_2(n) = \textstyle\sum_{K+1}^\infty \alpha(j)\varepsilon(n-j) .$$

We also put

$$c_{ij}(k) = N^{-1} \textstyle\sum_1^N x_i(n)x_j(n+k) ; \qquad i, j = 1, 2 .$$

If $\gamma_{ij}(k) = Ec_{ij}(k)$, then

$$\gamma_{ij}(k) = \textstyle\int_{-\pi}^{\pi} e^{ik\lambda} f_{ij}(\lambda)\, d\lambda$$

and

$$f_{11}(\lambda) = \frac{\sigma^2}{2\pi} |\textstyle\sum_0^K \alpha(j)e^{ij\lambda}|^2 , \qquad f_{22}(\lambda) = \frac{\sigma^2}{2\pi} |\textstyle\sum_{K+1}^\infty \alpha(j)e^{ij\lambda}|^2 ,$$

$$f_{12}(\lambda) = \overline{f_{21}(\lambda)} = \frac{\sigma^2}{2\pi} \textstyle\sum_0^K \alpha(j)e^{ij\lambda} \textstyle\sum_{K+1}^\infty \alpha(j)e^{-ij\lambda} .$$

All of these functions are square integrable over $[-\pi, \pi]$. In fact to see that, for example, $f(\lambda)$ is square integrable under (7) note that this is equivalent to

$$\textstyle\sum_{k=0}^\infty \{\textstyle\sum_{j=1}^\infty \alpha(j)\alpha(j+k)\}^2 < \infty$$

and

$$\begin{aligned} \textstyle\sum_{k=0}^\infty \{\textstyle\sum_{j=1}^\infty j^{\frac{1}{4}}\alpha(j) \cdot j^{-\frac{1}{4}}\alpha(j+k)\}^2 &\leqq \textstyle\sum_{k=0}^\infty \{\textstyle\sum_{j=1}^\infty j^{\frac{1}{2}}\alpha^2(j)\}\{\textstyle\sum_{j=1}^\infty j^{-\frac{1}{2}}\alpha^2(j+k)\} \\ &\leqq \{\textstyle\sum_{j=1}^\infty j^{\frac{1}{2}}\alpha^2(j)\} \textstyle\sum_{k=0}^\infty \textstyle\sum_{j=1}^\infty j^{-\frac{1}{2}}\alpha^2(j+k) \\ &\leqq \{\textstyle\sum_{j=1}^\infty j^{\frac{1}{2}}\alpha^2(j)\} \textstyle\sum_{k=1}^\infty \alpha^2(k) \textstyle\sum_{j=1}^k j^{-\frac{1}{2}} < \infty \end{aligned}$$

under (7).

Now (13) becomes

$$(15) \qquad N^{\frac{1}{2}}[\{c_{11}(j) + c_{22}(j) + c_{12}(j) + c_{21}(j)\} - \rho(j)\{c_{11}(0) + c_{22}(0) + 2c_{12}(0)\}] .$$

We call $c'_{ij}(k)$ the expression $c_{ij}(k)$ with all terms involving an $\varepsilon^2(n)$ omitted. We wish to show that the contribution to the primed form of (15) from the $c'_{ij}(k)$ for i, j not both equal to unity has a variance which, for all sufficiently large N, may be made arbitrarily small by taking K large. To this end we put

$$N^{\frac{1}{2}}\{c_{ij}(k) - Ec_{ij}(k)\} = N^{\frac{1}{2}}c'_{ij}(k) + N^{\frac{1}{2}}\{c''_{ij}(k) - Ec''_{ij}(k)\} ,$$

wherein $c''_{ij}(k)$ contains all terms in $c_{ij}(k)$ involving an $\varepsilon^2(n)$. If we evaluate the variance of this as if the $\varepsilon(n)$ and $x(n)$ were Gaussian we shall not affect the variance of $N^{\frac{1}{2}}c'_{ij}(k)$. Since on this Gaussian assumption the two terms on the right are uncorrelated, we obtain the variance of the left-hand term as an upper bound to the variance of $N^{\frac{1}{2}}c'_{ij}(k)$. The variance of the left-hand term is, on the Gaussian assumption, ([6] page 210)

$$\sum_{-N+1}^{N-1} (1 - |u|/N)\{\gamma_{ii}(u)\gamma_{jj}(u) + \gamma_{ij}(u+k)\gamma_{ji}(u-k)\}$$

which converges to

$$(16) \qquad 2\pi \int_{-\pi}^{\pi} \{f_{ii}(\lambda)f_{jj}(\lambda) + |f_{ij}(\lambda)|^2 e^{2ik\lambda}\}\, d\lambda$$

because of the square integrability of the $f_{ij}(\lambda)$ and Parseval's theorem. However, as $K \to \infty$,

$$\int_{-\pi}^{\pi} f_{22}^2(\lambda)\, d\lambda\,, \qquad \int_{-\pi}^{\pi} f_{22}(\lambda)\, d\lambda$$

converge to zero because the Fourier series of a function in $L_p(-\pi, \pi)$, $1 < p < \infty$, converges in the L_p norm to the function ([9] page 50). (In our case the function is $\sum_{K+1}^{\infty} \alpha(j) \exp ij\lambda \in L_4$.) Thus taking K sufficiently large we may, if i, j are not both unity, make (16) arbitrarily small and hence the variance of $N^{\frac{1}{2}}c'_{ij}(k)$, for all sufficiently large N, arbitrarily small.

By what is sometimes called Bernstein's lemma ([6] page 242) the theorem will now result if it is shown that the $N^{\frac{1}{2}}\{c'_{11}(j) - \rho(j)c'_{11}(0)\}$ are jointly asymptotically normal with a covariance matrix which converges, as K is increased, to W. The proof of the asymptotic normality is the same as that given in the course of proving Theorem 2. Putting

$$\rho'(j) = \sum_{k \leq K} \alpha(k)\alpha(k+j) / \sum_{k \leq K} \alpha^2(k)\,,$$

the covariance of the $N^{\frac{1}{2}}\{c'_{11}(j) - \rho'(j)c'_{11}(0)\}$ converges to W' where W' is obtained from W by replacing $f(\lambda)$ by $f_{11}(\lambda)$. Since $N^{\frac{1}{2}}(\rho(j) - \rho'(j))c'_{11}(0)$ evidently converges in probability to zero and W' converges to W as $K \to \infty$ because of the theorem, quoted above, on L_p convergence, the theorem is proved.

4. Some applications and extensions. The most obvious application of the results of Sections 2 and 3 is to the autoregression

$$(17) \qquad \sum_{k=0}^{q} \beta(k)\{x(n-k) - \mu\} = \varepsilon(n)\,, \qquad \beta(0) = 1\,,$$

wherein we assume that

$$\sum_{k=0}^{q} \beta(k)z^k \neq 0\,, \qquad |z| \leqq 1\,,$$

so that $x(n)$ can be represented in the form (2), and (7) is satisfied (see, for example, [6], Chapter I). The $\beta(j)$ and σ^2 $(=E\varepsilon^2(n))$ are estimated through

$$\sum_{j=0}^{q} \hat{\beta}(j)c(k-j) = \delta_{0,k}\hat{\sigma}^2\,, \qquad k = 0, 1, \cdots, q\,,$$

([6], Chapter VI) and, remembering that $\beta(0) = 1$, we see that $\hat{\beta}(1), \cdots, \hat{\beta}(q)$ are functions only of the $r(j)$, $j = 1, 2, \cdots, q$ and thus Theorem 3 may be applied. Similar considerations apply to the mixed autoregressive and moving average process obtained when (17) is altered only by the replacement of the right-hand side by

$$\sum_{j=0}^{r} \delta(j)\varepsilon(n-j) .$$

It seems that all of the classical inferential theory for the $\beta(j)$, $\delta(j)$ (see [2]; [6], Chapter VI) will continue to apply under the appropriate conditions of the present paper.

There are other problems that yield to the same treatment as we have applied to the $r(j)$ in the present paper. For example, if $x(n)$ is generated by (2) and is stationary with finite fourth moment and (3) holds, then

$$\lim_{N\to\infty} \sup_{|\lambda|\leq\pi} |N^{-1} \sum_{n=1}^{N} (x(n) - \mu)e^{in\lambda}| = 0 \quad \text{a.s.}$$

The proof of this proposition, which is important in connection with the estimation of the frequencey of a sinusoidal signal received together with noise, follows the same lines as that given in Hannan [7].

Acknowledgment. We are indebted to E. Seneta for some useful conversations during the initial stage of this work.

REFERENCES

[1] Anderson, T. W. and Walker, A. M. (1964). On the asymptotic distribution of the autocorrelations of a sample from a linear stochastic process. *Ann. Math. Statist.* **35** 1296–1303.

[2] Box, G. E. P. and Jenkins, G. M. (1970). *Time Series Analysis Forecasting and Control.* Holden-Day, San Francisco.

[3] Brown, B. M. (1971). Martingale central limit theorems. *Ann. Math. Statist.* **42** 59–66.

[4] Doob, J. L. (1953). *Stochastic Processes.* Wiley, New York.

[5] Gikhman, I. I. and Skorokhod, A. V. (1969) *Introduction to the Theory of Random Processes.* Saunders, Philadelphia.

[6] Hannan, E. J. (1970). *Multiple Time Series.* Wiley, New York.

[7] Hannan, E. J. (1971). Non linear time series regression. *J. Appl. Probability* **8** 767–780.

[8] Heyde, C. C. and Seneta, E. (1972). Estimation theory for growth and immigration rates in a muliplicative process. To appear in *J. Appl. Probability.*

[9] Katznelson, Y. (1968). *Harmonic Analysis.* Wiley, New York.

Department of Statistics
Australian National University
Box 4, Post Office, Canberra A.C.T. 2600
Australia

Reprinted from *The Australian Journal of Statistics*, Vol. 14, No. 1, pages 1–9, April, 1972

MARTINGALES: A CASE FOR A PLACE IN THE STATISTICIAN'S REPERTOIRE

C. C. Heyde
Australian National University

R. Maller et al. (eds.), *Selected Works of C.C. Heyde*, Selected Works in Probability and Statistics,
DOI 10.1007/978-1-4419-5823-5_29,

Austral. J. Statist., **14 (1)**, 1972, 1–9

MARTINGALES: A CASE FOR A PLACE IN THE STATISTICIAN'S REPERTOIRE[1]

C. C. HEYDE
Australian National University

1. Introduction

It is usually the case that martingale methods do not form a valued part of the statistician's repertoire of techniques. The reasons are partly historical and partly environmental, and while modern texts on probability theory deal in great detail with sums of independent random variables, the essentially broader subject of martingales is often relegated to a small section or the back pages. The object of this note is to give a brief exposition of what martingales are and what they can do. It is my personal feeling that the rôle of martingale theory in probability and statistics will increase rapidly in importance within the next few years.

2. The Basic Ideas

In order to give a feel for the subject, we start with the basic concepts in the case of a countable state space.

Let Ω be a sample space consisting of countably many points $\omega_1, \omega_2, \ldots$ to which are assigned probabilities $p_1, p_2, \ldots$ respectively, with $p_j \geq 0$ and $\Sigma_j p_j = 1$. If X is a random variable its *expectation* is defined as

$$EX = \Sigma_j X(\omega_j) p_j$$

(provided $\Sigma_j \mid X(\omega_j) \mid p_j < \infty$) and the *conditional expectation* of X relative to A is defined correspondingly, when $P(A) > 0$, as

$$E(X \mid A) = \sum_{\{j: \omega_j \varepsilon A\}} X(\omega_j) p_j / P(A).$$

If $P(A) = 0$, the conditional expectation can be defined arbitrarily.

Now let $\{A_n, n \geq 1\}$ be a partition of Ω, that is, a countable class of disjoint sets with union Ω. This partition generates, and is in turn determined by, a σ-field, $\mathscr{F}$, namely the class of all unions of sets of the partition. If X is a random variable with an expectation, define $E(X \mid \mathscr{F})$, the *conditional expectation* of X relative to $\mathscr{F}$, as the random variable with constant value $E(X \mid A_n)$ on each set A_n. The definition specifies $E(X \mid \mathscr{F})$ almost surely (a.s.), that is, everywhere except on partition sets (if any) of probability zero. As an example, if $Y_1, \ldots, Y_k$ are random variables, they induce a partition each of of whose sets is determined by a condition of the form

$$\{Y_1 = a_1, \ldots, Y_k = a_k\}.$$

Let $\mathscr{F}_k$ be the σ-field generated by this partition. In this case $E(X \mid \mathscr{F}_k)$ (which is commonly written as $E(X \mid Y_1, \ldots, Y_k)$) is the random variable with value $E(X \mid Y_1 = a_1, \ldots Y_k = a_k)$ on the set $\{Y_1 = a_1, \ldots, Y_k = a_k\}$.

[1] Manuscript received February 22, 1972.

Let $\mathscr{F}_1 \subset \mathscr{F}_2 \subset \ldots$ be an increasing sequence of σ-fields generated by partitions of the sample space as just described. It is useful to interpret $\mathscr{F}_n$ as representing the class of all relevant past events up to and including time n. The monotoneity relation, then, corresponds to the idea that the past up to time $n+1$ includes more events than the past up to time n. Let $X_1, X_2, \ldots$ be a sequence of random variables such that X_n is determined by the history up to and including time n (i.e. each event of the form $\{X_n=a\} \in \mathscr{F}_n$; X_n is then called $\mathscr{F}_n$ measurable). Then, the sequence $\{X_n, n \geq 1\}$ is defined as a *martingale* (MG) relative to $\{\mathscr{F}_n, n \geq 1\}$ if each X_n has finite expectation and if for $m<n$,

$$E(X_n \mid \mathscr{F}_m)=X_m \text{ a.s.} \tag{1}$$

This is a relation between functions on the sample space and is to hold almost surely (a.s.), that is, everywhere except perhaps on a subset of the sample space of probability zero. Note that it is sufficient to replace (1) by

$$E(X_{n+1} \mid \mathscr{F}_n)=X_n \text{ a.s., each } n \geq 1,$$

for then

$$\begin{aligned} E(X_n \mid \mathscr{F}_m) &= E\{E(X_n \mid \mathscr{F}_{n-1}) \mid \mathscr{F}_m\} \\ &= E(X_{n-1} \mid \mathscr{F}_m) \\ &= \ldots = E(X_m \mid \mathscr{F}_m) = X_m \text{ a.s.} \end{aligned}$$

using continued reduction.

Now the basic definition of a MG is inspired by notions of fairness in gambling. The idea of a fair game is that the knowledge of the past should not enable the gambler to improve on his fortunes. As a simple example of the above idea let $Z_i = \pm 1$ each with probability $\frac{1}{2}$ and suppose that $Z_1, Z_2, \ldots$ are independent (i.e. basically a simple coin-tossing situation). Suppose the gambler's strategy is that he bets $b_n(Z_1, \ldots, Z_n)$ on the $(n+1)$st trial and let

$$S_n = S_n(b_1, \ldots, b_n; \; Z_1, \ldots, Z_n)$$

denote his fortune after n trials. $\{S_n\}$ satisfies the recurrence relation

$$S_{n+1} = S_n + Z_{n+1} b_n(Z_1, \ldots, Z_n).$$

Then, if $\mathscr{F}_n$ denotes the σ-field generated by $Z_1, \ldots, Z_n$, we observe that $\{S_n, n \geq 1\}$ is a MG with respect to $\{\mathscr{F}_n, n \geq 1\}$. We have for $n \geq 1$,

$$\begin{aligned} E(S_{n+1} \mid \mathscr{F}_n) &= E(S_n + Z_{n+1} b_n(Z_1, \ldots, Z_n) \mid \mathscr{F}_n) \\ &= S_n + b_n(Z_1, \ldots, Z_n) E(Z_{n+1}) \\ &= S_n \end{aligned}$$

as required. Of course if $b_n(Z_1, \ldots, Z_n) \equiv 1$, then S_n is just a sum of independent and identically distributed random variables.

Now suppose that the gambler can decide to skip individual trials. Introduce a decision function ε_n, which is $\mathscr{F}_{n-1}$ measurable, such that ε_n is 0 or 1. If ε_n is 0 the gambler skips the nth trial and if $\varepsilon_n = 1$ he bets. Then, denoting his fortune after n trials by T_n, we have

$$T_{n+1} = T_n + \varepsilon_{n+1} Z_{n+1} b_n(Z_1, \ldots, Z_n)$$

so that

$$\begin{aligned} E(T_{n+1} \mid \mathscr{F}_n) &= T_n + \varepsilon_{n+1} b_n(Z_1, \ldots, Z_n) E(Z_{n+1}) \\ &= T_n \end{aligned}$$

and the game remains " fair " under this system of optional sampling.

Martingale theory forms a natural setting for many such gambling problems. It can be used to establish intuitively striking propositions such as *the persistent gambler goes broke with probability one* (Breiman [2], pp. 101, 102).

3. A General Definition

In many problems the assumption of the countability of the sample space Ω is not satisfied. We go on to sketch a general framework for a definition of a MG.

Let $\{\Omega, \mathscr{F}, P\}$ be a measure space: Ω is a set, $\mathscr{F}$ is a σ-field of subsets of Ω and P a probability measure defined on $\mathscr{F}$. Let $\mathscr{F}_1 \subset \mathscr{F}_2 \subset \ldots$ be an increasing sequence of σ-fields of $\mathscr{F}$ sets. Let $\{X_n, n \geq 1\}$ be a sequence of functions on Ω satisfying

(i) X_n is measurable relative to $\mathscr{F}_n$,

(ii) $E \mid X_n \mid < \infty$,

(iii) If $m < n$, then $E(X_n \mid \mathscr{F}_m) = X_m$ a.s.

Then, the sequence of random variables is said to be a MG relative to $\{\mathscr{F}_n, n \geq 1\}$.

In many applications, the σ-field $\mathscr{F}_n$ is the past as determined by $X_1, \ldots, X_m$; that is $\mathscr{F}_n$ is generated by $X_1, \ldots, X_n$. In this case the basic martingale condition (iii) is commonly written

$$E(X_n \mid X_m, \ldots, X_1) = X_m \text{ a.s.}$$

The more flexible setting of the definition is chosen since, for example, one has often to replace the conditioning variables $X_m, \ldots, X_1$ by some functions of them.

4. The Martingale Convergence Theorem

A key reason for the importance of the concept of a MG is provided by the following powerful theorem.

The MG convergence theorem. Let $\{X_n\}$ be a MG such that $\limsup_{n \to \infty} E \mid X_n \mid < \infty$. Then, there exists a random variable X such that $\lim_{n \to \infty} X_n = X$ a.s. and $E \mid X \mid \leq \liminf_{n \to \infty} E \mid X_n \mid$.

This is an existence theorem; it tells us nothing about the limit random variable save that it has a finite first absolute moment. The theorem seems rather unexpected a priori and it is an extremely powerful tool which has led to a number of interesting results for which it seems essentially a unique method of approach. Of course one is still faced with finding the limit law, but that can usually be done by other methods.

A proof of the MG convergence theorem is particularly easy in the special case where $\limsup_{n \to \infty} EX_n^2 < \infty$. See for example Feller [6], p. 236. A general proof can be found in many of the standard probability texts.

As a simple example of the power of the theorem, consider its application to show that if $S_n = \sum_{i=1}^{n} X_i$ is a sum of independent random variables with S_n converging in distribution as $n \to \infty$, then S_n converges a.s. This result is a straightforward consequence of the MG convergence theorem when it is noted that

$$Z_n = e^{itS_n}/E(e^{itS_n}) = e^{itS_n}/\prod_{i=1}^{n} E(e^{itX_i})$$

is a MG. The result is rather difficult to prove by other methods.

Another interesting application of the MG convergence theorem concerns the likelihood ratio. Suppose $X_1, X_2, \ldots$ is a stochastic process such that the joint density of $(X_1, \ldots, X_n)$ is either p_n or q_n; the problem is to decide which. Define Y_n by

$$Y_n = \frac{q_n(X_1, \ldots, X_n)}{p_n(X_1, \ldots, X_n)}.$$

Y_n is likely to be small or large according as the true density is p_n or q_n and the asymptotic behaviour of $\{Y_n\}$ is clearly of importance.

For simplicity we assume that the densities p_n are strictly positive and continuous (although it certainly suffices to assume $q_n(x_1, \ldots, x_n)=0$ whenever $p_n(x_1, \ldots, x_n)=0$; see Doob [5], p. 348, or Chow, Robbins and Siegmund [4], p. 12). Then, if the p_n are the true densities, the conditional density of X_{n+1} given $X_1, \ldots, X_n$ is $p_{n+1}(X_1, \ldots, X_n, y)/p_n(X_1, \ldots, X_n)$ and hence

$$\begin{aligned} E(Y_{n+1} \mid X_n, \ldots, X_1) &= \int_{-\infty}^{\infty} \frac{q_{n+1}(X_1, \ldots, X_n, y)}{p_{n+1}(X_1, \ldots, X_n, y)} \cdot \frac{p_{n+1}(X_1, \ldots, X_n, y)}{p_n(X_1, \ldots, X_n)} dy \\ &= \frac{q_n(X_1, \ldots, X_n)}{p_n(X_1, \ldots, X_n)} = Y_n \end{aligned}$$

so that $\{Y_n, n \geq 1\}$ is a MG with respect to the sequence of σ-fields generated by $X_1, \ldots, X_n$, $n \geq 1$. Clearly $Y_n \geq 0$ and $EY_n = 1$, so that the MG convergence theorem ensures the existence of a random variable Y, with $EY \leq 1$, given by $\lim_{n\to\infty} Y_n = Y$ a.s.

Under suitable hypotheses, detailed information can be obtained on Y. For example, suppose the p_n's and q_n's correspond to independent X_j's with a common distribution. Then, it is not difficult to show that $Y=0$ a.s. ([5], p. 349; [4], p. 19). This is a result of precisely the type desired.

5. Stopping

Martingale theory is closely related to problems of *stopping* and in particular *optimal stopping*. We shall just sketch these concepts here; a comprehensive discussion of the material in this section is given by Chow, Robbins and Siegmund [4].*

Let $\{\mathscr{F}_n, n \geq 1\}$ be an increasing sequence of σ-fields. A stopping time is by definition a random variable t such that

$$P(t < \infty) = 1$$

and

$$\{t = n\} \in \mathscr{F}_n \text{ for each } n.$$

* Reviewed in this issue.—Editor.

If $\{X_n, n \geq 1\}$ is a MG with respect to $\{\mathscr{F}_n, n \geq 1\}$ we may for concreteness interpret the MG as the values assumed by the gambler's fortune as he plays a sequence of "fair" games. A stopping time is a strategy by which the gambler decides to stop playing and X_t is his terminal fortune. The requirement $\{t=n\} \varepsilon \mathscr{F}_n$ is just that the gambler's decision to stop at time n depends only on his past experience. One interesting and important problem is whether

$$EX_t = EX_1,$$

i.e., whether the property of "fairness" is preserved under the stopping time t.

This theory provides a simple method of deriving the Wald identity of sequential analysis. Let $Y_1, Y_2, \ldots$ be independent and identically distributed, $S_n = \sum_{i=1}^{n} Y_i$, and assume that for some real $\lambda \neq 0$, $\varphi(\lambda) = Ee^{\lambda Y_1}$ exists and that $\varphi(\lambda) \geq 1$. Then, if N is a stopping time for the sums $S_1, S_2, \ldots$ such that $|S_n|$ is uniformly bounded for $n \leq N$ and $EN < \infty$, we have

$$E[\{\varphi(\lambda)\}^{-N} e^{\lambda S_N}] = 1.$$

This identity is derived by noting that

$$X_n = [\varphi(\lambda)]^{-n} e^{\lambda S_n}$$

is a MG and showing that

$$EX_N = EX_1 = 1$$

(e.g. Breiman [2], p. 100).

Now let $\{X_n, n \geq 1\}$ be any stochastic sequence with X_n measurable with respect to $\mathscr{F}_n$ and let t be any stopping time. Suppose also that $E|X_n| < \infty$ for each n. Define $V = \sup EX_t$ where the supremum is taken over all stopping times for which EX_t exists. Optimal stopping theory is concerned with the problems of existence and nature of an optimal stopping variable (i.e. one for which $EX_t = V$).

As an example, we mention the problem of testing a simple hypothesis against a simple alternative with independent and identically distributed observations of constant cost. It is well known that a Bayes solution to the problem is a Wald sequential probability ratio test. Now, the problem of minimizing the risk in this context can be re-stated as a problem on optimal stopping ([4], pp. 46–49) and the stopping variable appropriate to the sequential probability ratio test turns out to be optimal ([4], pp. 71, 105). Some variants are also given in [4].

6. Limit Theorems

Zero mean random walks and sums of independent random variables centred at their expectations are MG's. The standard theory for sums of independent random variables, however, continues to proceed in its own right. Limit results such as the laws of large numbers, central limit theorem and law of the iterated logarithm have been major achievements and have had a substantial impact. Now, for the most part, the standard theory for sums of independent random variables can be put into a more general MG setting. From the point of view of applications there is considerable advantage in doing this. We shall sketch the basic ideas.

Let $\{S_n, n\geq 1\}$ be a zero mean MG with respect to $\{\mathscr{F}_n, n\geq 1\}$ and write $S_n=\sum_{j=1}^{n} X_j$. The first thing to note is that many results on sums of independent random variables in fact require only orthogonality of the increments $(E(X_iX_j)=0,\ i\neq j)$ and that this property holds for MG's whose increments have finite variance. As a simple example we see that the so-called Chebyshev inequality continues to hold for MG's. By Markov's inequality we have for any $\varepsilon>0$,

$$P(|S_n|\geq\varepsilon)\leq\varepsilon^{-2}ES_n^2,$$

while

$$ES_n^2=\sum_{j=1}^{n} EX_j^2+2\sum_{i>j} EX_iX_j=\sum_{j=1}^{n} EX_j^2$$

since for $i>j$,

$$E(X_iX_j)=E(X_jE(X_i \mid \mathscr{F}_{i-1}))=E(X_j\{E(S_i \mid \mathscr{F}_{i-1})-S_{i-1}\})=0.$$

This inequality for MG's gives immediately a *weak law of large numbers*. In fact, for MG's whose increments have finite variance,

$$P(|S_n|\geq n\varepsilon)\leq n^{-2}\varepsilon^{-2}\sum_{j=1}^{n} EX_j^2,$$

so that $n^{-1}S_n\to 0$ in probability if $n^{-2}\sum_{j=1}^{n} EX_j^2\to 0$.

In the general case of the weak law of large numbers, where the finite variance assumption is dropped, there is a well-known set of necessary and sufficient conditions for convergence in probability of $n^{1-}S_n$, $S_n=\sum_{j=1}^{n} X_j$, denoting a sum of independent random variables. In fact (Loève [11], p. 278) $n^{-1}S_n\to 0$ in probability if and only if

(i) $\sum_{j=1}^{n} P(|X_j|\geq n)\to 0$

(ii) $n^{-1}\sum_{j=1}^{n} EX_{jn}\to 0$

(iii) $n^{-2}\sum_{j=1}^{n} [EX_{jn}^2-(EX_{jn})^2]\to 0$

where X_{jn} is defined by $X_{jn}=X_j$ or 0 according as $|X_j|<n$ or $|X_j|\geq n$. An inspection of the proof given by Loève shows that the " if " part (which is after all the most important part) depends only on truncation and the use of the so-called Chebyshev inequality. A minor variant obtained by, in (ii), replacing EX_{jn} by $E(X_{jn} \mid X_{j-1},\ldots,X_1)$ and $\to 0$ by $\to 0$ in probability and in (iii) replacing $(EX_{jn})^2$ by $E(E(X_{jn} \mid X_{j-1},\ldots X_1))^2$, holds in the MG context. This of course reduces to the Loève condition in the special case of independence. The " only if " part of Loève's proof involves the use of characteristic functions and only this rests on the assumption of independence.

In the case of the *strong law of large numbers*, the most basic tool is probably the Kolmogorov criterion that if the X_i are independent with zero mean and $EX_i^2<\infty$ each i, then $\Sigma b_n^{-2}EX_n^2<\infty$, $b_n\uparrow\infty$ entails $\lim_{n\to\infty} b_n^{-1}S_n=0$ a.s. (e.g. Loève [11], p. 238). That this result continues to hold in the case when $\{S_n\}$ is a MG is a simple consequence of the MG convergence theorem (e.g. Feller [6], p. 238).

In the case of the *central limit theorem* there is the following result of Brown [3]. Let $\{S_n,\mathscr{F}_n\}$ denote a zero mean MG whose increments have finite variance. Write

$$S_n=\sum_{i=1}^n X_i,\quad V_n^2=\sum_{j=1}^n E(X_j^2\mid\mathscr{F}_{j-1})\text{ and }s_n^2=EV_n^2=ES_n^2.$$

If ("p" denotes convergence in probability)

$$s_n^{-2}V_n^2\overset{p}{\to}1$$

and

$$s_n^{-2}\sum_{j=1}^n E(X_j^2I(|X_j|\geq\varepsilon s_n))\to 0$$

as $n\to\infty$ for any $\varepsilon>0$ ($I(.)$ denotes the indicator function) then

$$\lim_{n\to\infty}P(s_n^{-1}S_n\leq x)=(2\pi)^{-\frac{1}{2}}\int_{-\infty}^{x}e^{-\frac{1}{2}u^2}du.$$

This result reduces to the standard Lindeberg-Feller result (e.g. [11], p. 280) in the case of independent random variables as the condition $s_n^{-2}V_n^2\overset{p}{\to}1$ is then trivially satisfied. The proof is a little more delicate than that of the Lindeberg-Feller result but uses similar ideas (whose use in the MG context dates back to Lévy; see for example [5], p. 383).

The following *law of the iterated logarithm* has been obtained by Stout [12]. Let $\{S_n\}$ denote a stationary and ergodic zero mean MG whose increments have finite variance σ^2. Then

$$n^{-1}S_n=\sigma\zeta(n)(2\log\log n/n)^{\frac{1}{2}}$$

where $\zeta(n)$ has its set of limit points confined to $[-1,1]$ and $\limsup_{n\to\infty}\zeta(n)=+1$ a.s., $\liminf_{n\to\infty}\zeta(n)=-1$ a.s. The law of the iterated logarithm clearly provides detailed information on the rate of convergence in the strong law of large numbers ($\lim_{n\to\infty} n^{-1}S_n=0$ a.s.). The result contains the classical law of the iterated logarithm for sums of independent and identically distributed random variables (due to Hartman and Wintner).

7. Some Applications of the Limit Theory

In this section we shall mention some applications of the results of Section 6. The examples chosen involve problems amenable to treatment using MG limit theorems but not using the corresponding results for sums of independent random variables. Basically, the MG limit theory of Section 6 seems relevant to any context where conditional expectations, given the past, have a simple form. This is largely a consequence of results such as if $\{Z_n, n\geq 1\}$ is a sequence of random variables with finite means, then

$$\left\{\sum_{i=1}^{n} [Z_i - E(Z_i \mid Z_{i-1}, \ldots, Z_1)]\right\}$$

is a MG relative to the sequence of σ-fields generated by Z_i, $i\leq n$.

Our first application deals with a time series context. Consider a stationary autoregression $\{x(n)\}$ of the form

$$\sum_{k=0}^{q} \beta(k)\{x(n-k)-\mu\}=\varepsilon(n)\ ;\ \ \beta(0)=1, \tag{2}$$

$$E\varepsilon(n)=0,\ E\{\varepsilon(m)\varepsilon(n)\}=0,\ m\neq n,$$

where we assume that

$$\sum_{k=0}^{q} \beta(k)z^k \neq 0,\ \mid z\mid\leq 1.$$

We also suppose that $\{\varepsilon(n)\}$ is a stationary stochastic sequence with $E\varepsilon^2(1)=\sigma^2$ satisfying

$$E(\varepsilon(n) \mid \mathscr{F}_{n-1})=0 \text{ a.s., all } n, \tag{3}$$

where $\mathscr{F}_n$ is the σ-field generated by $\varepsilon(m)$, $m\leq n$ and

$$E(\varepsilon^2(n) \mid \mathscr{F}_{n-1})=\sigma^2 \text{ a.s., all } n.$$

In this context, $\mathscr{F}_n$ is also the σ-field generated by $x(m)$, $m\leq n$ and

$$\varepsilon(n)=x(n)-E(x(n) \mid \mathscr{F}_{n-1})$$

so that (3) (i.e. the assumption that any $\left\{\sum_{n=-r}^{s} \varepsilon(n)\right\}$ is a MG) is equivalent to assuming that $E(x(n) \mid \mathscr{F}_{n-1})$ is the best linear predictor and the best linear predictor is the best predictor (both in the least squares sense). We are also assuming that $E(\varepsilon^2(n) \mid \mathscr{F}_{n-1})$, the prediction variance given the past, is a.s.-constant. Both these assumptions, and particularly the former, seem natural and intuitively satisfying.

Now the classical theory of inference for the process (2) requires that the $\varepsilon(n)$ be independent and identically distributed with zero mean and variance σ^2 (e.g. Hannan [7], Chapter VI). However, it has been shown by Hannan and Heyde [8], with proofs resting primarily on a law of large numbers and a central limit theorem for MG's, that the classical theory goes through unchanged under the present conditions. In fact, the results of [8] are considerably stronger and deal with the general purely non-deterministic process

$$x(n)-\mu=\sum_{j=0}^{\infty} \alpha(j)\varepsilon(n-j)\ ;\ \ \sum_{j=0}^{\infty} \alpha^2(j)<\infty\ ;\ \ \alpha(0)=1\ ;$$

$$E\varepsilon(n)=0,\ E\{\varepsilon(m)\varepsilon(n)\}=0,\ m\neq n,$$

but we have specialized here for clarity of exposition. We shall sketch the relevant results.

Let $x(1), x(2), \ldots, x(N)$ be a sample of N consecutive observations on the process $\{x(n)\}$ and define the autocovariances

$$c(j)=c(-j)=N^{-1}\sum_{n=1}^{N-j} (x(n)-\bar{x})(x(n+j)-\bar{x}),\ j\geq 0,$$

where $\bar{x}=N^{-1}\sum_{j=1}^{N} x(n)$. The classical results, namely that $\beta(j)$ and σ^2 can be estimated through

$$\sum_{j=0}^{q} \hat{\beta}(j)c(k-j)=\delta_{ok}\hat{\sigma}^2, \quad k=0, 1, \ldots, q$$

where δ_{ok} is the Kronecker delta, while μ can be estimated by $\bar{x}$, remain true in the present context. So does the Anderson and Walker [1] central limit theorem for the autocorrelations $r(j)=c(j)/c(O)$. Furthermore, Heyde [9] has used an iterated logarithm result for MG's to give detailed (almost sure) information on the error involved in the estimation of $\beta(j)$, σ^2 and μ. MG methods form a natural tool in this context since the conditional expectations take a simple and relevant form. Note that

$$\left\{\sum_{n=k+1}^{N} (x(n-k)-\mu)\varepsilon(n),\ N>k\geq 1\right\}, \left\{\sum_{n=1}^{N} [(x(n)-\mu)\varepsilon(n)-\sigma^2],\ N\geq 1\right\},$$

and $\left\{\sum_{n=1}^{N} \varepsilon(n),\ N\geq 1\right\}$ are all zero mean MG's with respect to $\{\mathscr{F}_N\}$.

Another setting where MG results of the style of Section 6 provide a rather natural tool is in the problem of estimating growth and immigration parameters in a subcritical branching process with immigration. Heyde and Seneta [10] have proved consistency and asymptotic normality of their recommended estimators using a law of large numbers and a central limit theorem for MG's. Again the setting is such that conditional expectations have a very simple form. In particular, if $\{X_n\}$ denotes a subcritical Galton-Watson branching process with immigration with mean of offspring distribution $m(<1)$ and mean of immigration distribution λ, then

$$E(X_n \mid X_{n-1}, \ldots, X_1)=mX_{n-1}+\lambda.$$

References

[1] Anderson, T. W., and Walker, A. M. (1964). "On the asymptotic distribution of the autocorrelations of a sample from a linear stochastic process." *Ann. Math. Statist.*, 35, 1296–1303.
[2] Breiman, L. (1968). *Probability*. Addison-Wesley, Reading, Mass.
[3] Brown, B. M. (1971). "Martingale central limit theorems." *Ann. Math. Statist.*, 42, 59–66.
[4] Chow, Y. S., Robbins, H., and Siegmund, D. (1971). *Great Expectations: The Theory of Optimal Stopping*. Houghton-Mifflin, Boston.
[5] Doob, J. L. (1953). *Stochastic Processes*. Wiley, New York.
[6] Feller, W. (1966). *An Introduction to Probability Theory and its Applications*. Vol. II. Wiley, New York.
[7] Hannan, E. J. (1970). *Multiple Time Series*. Wiley, New York.
[8] Hannan, E. J., and Heyde, C. C. (1971). "On limit theorems for quadratic functions of discrete time series." Mimeographed (to appear in *Ann. Math. Statist.*).
[9] Heyde, C. C. (1971). "An iterated logarithm result for martingales and its application in estimation theory for autoregressive processes." Mimeographed (to appear in *J. Applied Prob.*).
[10] Heyde, C. C., and Seneta, E. (1971). "Estimation theory for growth and immigration rates in a multiplicative process." Mimeographed (to appear in *J. Applied Prob.*).
[11] Loève, M. (1963). *Probability Theory*. Van Nostrand, Princeton.
[12] Stout, W. F. (1970). "The Hartman-Wintner law of the iterated logarithm for martingales." *Ann. Math. Statist.*, 41, 2158–2160.

AUSTRALASIAN MEDICAL PUBLISHING CO LTD.
71-79 ARUNDEL ST., GLEBE, N.S.W., 2037

Z. Wahrscheinlichkeitstheorie verw. Geb. 21, 255–268 (1972)

On the Influence of Moments on Approximations by Portion of a Chebyshev Series in Central Limit Convergence

C. C. HEYDE and J. R. LESLIE

1. Introduction

Let X_i, $i=1, 2, 3, \ldots$ be a sequence of independent and identically distributed random variables with $EX_i=0$ and $\operatorname{var} X_i=1$. Write $F(x)$ for the distribution function and $f(t)$ for the characteristic function of X_i and put $S_n=\sum_{i=1}^{n} X_i$. Then,

$$F_n(x)=P(S_n \leqq x\sqrt{n}) \rightarrow \Phi(x)=\frac{1}{\sqrt{2\pi}} \int_{-\infty}^{x} e^{-\frac{1}{2}u^2} du$$

as $n \rightarrow \infty$. We shall herein be concerned with the influence of moments of X_i on the rate of convergence to zero of

$$A_{kn}=\sup_x |F_n(x)-G_{kn}(x)|$$

where

$$G_{kn}(x)=\Phi(x)+\frac{1}{\sqrt{2\pi}} e^{-\frac{1}{2}x^2} \sum_{s=1}^{k} Q_s(x)\, n^{-\frac{1}{2}s}$$

is a given portion of the Chebyshev series corresponding to the X_i (see for example Gnedenko and Kolmogorov [2], Section 38), the $Q_j(x)$ being polynomials of degree $3j-1$ whose coefficients depend on the first $(j+2)$ moments of X_i. Now, Cramér (see [2], Section 45) has shown that for distributions satisfying the condition (C) (that is, $\limsup_{t\rightarrow\infty} |f(t)|<1$) and if $E|X_i|^{k+2}<\infty$ $(k \geqq 1)$, then $A_{kn}=o(n^{-k/2})$) as $n \rightarrow \infty$. Furthermore, Ibragimov [4] has produced necessary and sufficient conditions, under (C), for (i) $A_{kn}=o(n^{-k/2})$ $(k \geqq 1)$ and (ii) $A_{kn}=O(n^{-(k+\delta)/2})$, $0<\delta \leqq 1$, $k \geqq 1$, but these conditions are not in general moment conditions. We shall provide, also under (C), some necessary and sufficient conditions in terms of moments on the rate of convergence of A_{kn} to zero.

In order to avoid presupposing the existence of moments of higher order than the second, we shall follow the formulation of [4] and prescribe an arbitrary numerical sequence $\beta_1=0$, $\beta_2=1$, $\beta_3, \beta_4, \ldots$. On the basis of this sequence we form polynomials $Q_k(x)$ in such a way that their coefficients are expressed in terms of $\beta_1, \ldots, \beta_{k+2}$ in the same way as the coefficients of the classical polynomials $Q_k(x)$ are expressed in terms of the cumulants $\kappa_1, \ldots, \kappa_{k+2}$ of X_i. That is,

$$Q_k(x)=-\sum \frac{1}{j_1! \cdots j_k!} \left(\frac{\beta_3}{3!}\right)^{j_1} \cdots \left(\frac{\beta_{k+2}}{(k+2)!}\right)^{j_k} H_{3j_1+\cdots+(k+2)j_k-1},$$

R. Maller et al. (eds.), *Selected Works of C.C. Heyde*, Selected Works in Probability and Statistics, DOI 10.1007/978-1-4419-5823-5_30, © Springer Science+Business Media, LLC 2010

where the summation is over all non-negative solutions of $j_1+2j_2+\cdots+kj_k=k$ and $H_m(x)$ is the Hermite-Chebyshev polynomial

$$H_m(x)=(-1)^m e^{\frac{1}{2}x^2}\frac{d^m}{dx^m}e^{-\frac{1}{2}x^2}$$

(Petrov [8]). $Q_k(x)$ will henceforth be interpreted in this way.

Let $\alpha_1=0$, $\alpha_2=1$, $\alpha_3, \alpha_4, \ldots$ be the "moment" sequence corresponding to the "cumulant" sequence $\beta_1=0$, $\beta_2=1$, $\beta_3, \beta_4, \ldots$. We shall establish the following results.

Theorem 1. *In order that*

$$\sum_{n=1}^{\infty} n^{-1+(k+\delta)/2} \sup_x |F_n(x)-G_{kn}(x)|<\infty \tag{1}$$

where k is a non-negative integer and $0<\delta<1$, *it is necessary and for* $k=0$ *or for distributions satisfying* (*C*) *also sufficient that*

$$E|X_i|^{k+2+\delta}<\infty \quad \text{and} \quad \alpha_j=EX_i^j, \quad j=1,2,\ldots,k+2. \tag{2}$$

Theorem 2. *In order that the relation* (1) *hold, where* $0<\delta<1$, *it is necessary and for distributions satisfying* (*C*) *or for* $k=0$ *also sufficient that*

$$f(t)=\exp\left\{\sum_{s=2}^{k+2}\frac{(i\,t)^s}{s!}\beta_s+|t|^{k+2}\gamma(t)\right\}, \tag{3}$$

where for $A>0$,

$$\int_0^A \frac{|\gamma(t)|}{t^{1+\delta}}\,dt<\infty. \tag{4}$$

Unfortunately it has not been possible to treat the case $\delta=0$ in general and then not without certain presuppositions on the existence of moments. In this case we find the following result.

Theorem 3. *Suppose* $E|X_i|^{k+2}<\infty$ *where k is a non-negative even integer and* $\alpha_j=EX_i^j$, $j=1,2,\ldots,k+2$. *Then, for* (1) *to hold with* $\delta=0$ *it is necessary and for* $k=0$ *or for distributions satisfying* (*C*) *also sufficient that* $E|X_i|^{k+2}\log(1+|X_i|)<\infty$.

These theorems extend the work of Heyde [3] where the results for the case $k=0$ were obtained.

2. Preliminary Lemmas

Lemma 1. *Suppose* $E|X_i|^r<\infty$ *for some integer* $r\geqq 2$. *Then* $f(t)$ *is representable in the form*

$$f(t)=\exp\left\{\sum_{s=2}^{r}\frac{(i\,t)^s}{s!}\kappa_s+|t|^r\gamma(t)\right\} \tag{5}$$

where $\gamma(t)=o(1)$ *as* $t\to 0$ *and* κ_s *denotes the s-th cumulant of* X_i. *Furthermore, there exists an* $\varepsilon>0$ *such that for* $0<t<\varepsilon$, $|\gamma(t)|>0$ *or* $|\gamma(t)|\equiv 0$. *If* $|\gamma(t)|\equiv 0$ *for* $0<t<\varepsilon$, X_i *has a normal distribution.*

Proof. The representation in the form (5) with $\gamma(t)=o(1)$ as $t\to 0$ follows simply from a Taylor expansion of $\log f(t)$ (e.g. [2], p. 64).

Next, suppose $\gamma(t)=0$ for all $t\in\{t_k\}$ where $\{t_k\}$ is a sequence of non-zero real numbers converging to zero. Then,

$$\exp\left\{\sum_{s=2}^{r}\frac{(i\,t)^s}{s!}\kappa_s+|t|^r\gamma(t)\right\}=\exp\left\{\sum_{s=2}^{r}\frac{(i\,t)^s}{s!}\kappa_s\right\} \tag{6}$$

for all $t\in\{t_k\}$ and applying Theorem 4.2.1 of Linnik [5], we have that (6) holds for all real t. However, this is impossible unless $r=2$ since the left hand side of (6) represents a characteristic function and the right hand side does not, in view of Marcinkiewicz's Theorem (e.g. Lukacs [7], p. 147), unless $r=2$. Thus, if $r>2$ we must be able to choose $\varepsilon>0$ so that $\gamma(t)$ has no zeros in $(0,\varepsilon)$. If $r=2$, on the other hand, either (6) holds for all t in which case $\gamma(t)\equiv 0$ and X_i has a normal distribution or zero is not a limit point of a sequence of zeros of $\gamma(t)$ and hence we can choose an interval $(0,\varepsilon)$ containing no zeros of $\gamma(t)$.

Lemma 2. *Suppose $E|X_i|^r<\infty$ for some integer $r\geqq 2$. Then, $f(t)$ is representable in the form*

$$f(t)=\sum_{s=0}^{r}\frac{(i\,t)^s}{s!}\mu_s+|t|^r\beta(t) \tag{7}$$

where $\beta(t)=o(1)$ as $t\to 0$ and μ_s denotes the s-th moment of X_i. Furthermore, for any $A>0$ and $0\leqq\delta<1$, the conditions $\int_0^A|\beta(t)|\,t^{-(1+\delta)}\,dt<\infty$ and $\int_0^A|\gamma(t)|\,t^{-(1+\delta)}\,dt<\infty$ are equivalent, $\gamma(t)$ being given by (5) and these conditions are in turn equivalent to $E|X_i|^{r+\delta}<\infty$ if $\delta>0$, $E|X_i|^r\log(1+|X_i|)<\infty$ if r is even and $\delta=0$.

Proof. Firstly, the representation of $f(t)$ in the form (7) with $\beta(t)=o(1)$ as $t\to 0$ follows simply from a Taylor expansion of $f(t)$. Also, from Lemma 3 of [4] we find that as $t\to 0$,

$$\log f(t)-\sum_{s=2}^{r}\frac{(i\,t)^s}{s!}\kappa_s=f(t)-\sum_{s=0}^{r}\frac{(i\,t)^s}{s!}\mu_s+\frac{\Lambda_{r+1}}{(r+1)!}(i\,t)^{r+1}+o(t^{r+1}),$$

where Λ_{k+1} is a constant, so that

$$|t|^r\gamma(t)=|t|^r\beta(t)+\frac{\Lambda_{r+1}}{(r+1)!}(i\,t)^{r+1}+o(t^{r+1}) \tag{8}$$

and we readily deduce the equivalence of

$$\int_0^A|\beta(t)|\,t^{-(1+\delta)}\,dt<\infty \quad\text{and}\quad \int_0^A|\gamma(t)|\,t^{-(1+\delta)}\,dt<\infty, \qquad 0\leqq\delta<1.$$

Now suppose that

$$\int_0^A|\beta(t)|\,t^{-(1+\delta)}\,dt<\infty.$$

This implies

$$\int_0^A|\mathrm{Re}\,\beta(t)|\,t^{-(1+\delta)}\,dt<\infty$$

18*

where Re denotes the real part and hence that

$$\left|\int_0^A \operatorname{Re}\beta(t)\, t^{-(1+\delta)}\, dt\right| < \infty .$$

It is with this last condition that we shall work. We have

$$\begin{aligned}
\left|\int_0^A \frac{\operatorname{Re}\beta(t)}{t^{1+\delta}}\, dt\right| &= \left|\int_0^A \frac{1}{t^{r+\delta+1}}\left(\int_{-\infty}^{\infty} \operatorname{Re}\left(e^{itx} - \sum_{s=0}^{r} \frac{(i\,t\,x)^s}{s!}\right) dF(x)\right) dt\right| \\
&= \left|\int_{-\infty}^{\infty} dF(x) \int_0^A \frac{\operatorname{Re}\left(e^{itx} - \sum_{s=0}^{r} \frac{(i\,t\,x)^s}{s!}\right)}{t^{r+\delta+1}}\, dt\right| \qquad (9) \\
&= \left|\int_{-\infty}^{\infty} dF(x) \int_0^A \frac{\left(\cos t\,x - \sum_{s=0}^{R} (-1)^s \frac{(t\,x)^{2s}}{(2s)!}\right)}{t^{r+\delta+1}}\, dt\right|
\end{aligned}$$

where $R=[r/2]$, the integer part of $r/2$. But, after two integrations by parts,

$$\begin{aligned}
&\int_0^A \frac{\left(\cos t\,x - \sum_{s=0}^{R} (-1)^s \frac{(t\,x)^{2s}}{(2s)!}\right)}{t^{r+\delta+1}}\, dt \\
&= \frac{-\left(\cos A\,x - \sum_{s=0}^{R} (-1)^s \frac{(A\,x)^{2s}}{(2s)!}\right)}{(r+\delta)\, A^{r+\delta}} - \frac{x\left(\sum_{s=0}^{R-1} (-1)^s \frac{(A\,x)^{2s+1}}{(2s+1)!} - \sin A\,x\right)}{(r+\delta)(r+\delta-1)\, A^{r+\delta-1}} \\
&\quad + \frac{x^2}{(r+\delta)(r+\delta-1)} \int_0^A \frac{\left(\sum_{s=0}^{R-1} (-1)^s \frac{(t\,x)^{2s}}{(2s)!} - \cos t\,x\right)}{t^{r+\delta-1}}\, dt,
\end{aligned}$$

so that, recalling that $E|X_i|^r < \infty$, we must have using (9),

$$\left|\int_{-\infty}^{\infty} x^2\, dF(x) \int_0^A \frac{\left(\cos t\,x - \sum_{s=0}^{R-1} (-1)^s \frac{(t\,x)^{2s}}{(2s)!}\right)}{t^{r+\delta-1}}\, dt\right| < \infty .$$

Continuing this reduction, we find ultimately that

$$\left|\int_{-\infty}^{\infty} x^{2R}\, dF(x) \int_0^A \frac{(\cos t\,x - 1)}{t^{r+\delta+1-2R}}\, dt\right| < \infty$$

which transforms to give

$$\int_{-\infty}^{\infty} |x|^{r+\delta} \left\{\int_0^{A|x|} \frac{1-\cos u}{u^{r+\delta+1-2R}}\, du\right\} dF(x) < \infty .$$

Thus, if r is even, $R=r/2$ and

$$\int_{-\infty}^{\infty} |x|^{r+\delta} \left\{\int_0^{A|x|} \frac{1-\cos u}{u^{1+\delta}}\, du\right\} dF(x) < \infty , \qquad (10)$$

while if r is odd, $R=(r-1)/2$ and

$$\int_{-\infty}^{\infty}|x|^{r+\delta}\left\{\int_{0}^{A|x|}\frac{1-\cos u}{u^{2+\delta}}\,du\right\}dF(x)<\infty. \tag{11}$$

(10) and (11) are clearly equivalent to $E|X_i|^{r+\delta}<\infty$ if $\delta>0$. On the other hand, when $\delta=0$, we have for $|x|>1$

$$\int_{0}^{A|x|}\frac{1-\cos u}{u}\,du=\int_{0}^{A}\frac{1-\cos u}{u}\,du+\int_{A}^{A|x|}\frac{du}{u}-\int_{A}^{A|x|}\frac{\cos u}{u}\,du\sim\log|x|$$

as $|x|\to\infty$ so that (10) is equivalent to the condition $E|X_i|^r\log(1+|X_i|)<\infty$ when $\delta=0$. We have thus shown that $\int_0^A|\beta(t)|\,t^{-(1+\delta)}\,dt<\infty$ implies $E|X_i|^{r+\delta}<\infty$ if $0<\delta<1$, $E|X_i|^r\log(1+|X_i|)<\infty$ if r is even and $\delta=0$.

Finally, suppose that $E|X_i|^{r+\delta}<\infty$, $\delta>0$, or that $E|X_i|^r\log(1+|X_i|)<\infty$ if r is even and $\delta=0$. Then,

$$\begin{aligned}\int_0^A\frac{|\beta(t)|}{t^{1+\delta}}\,dt&=\int_0^A\frac{\left|f(t)-\sum_{s=0}^{r}\frac{(it)^s}{s!}\mu_s\right|}{t^{r+1+\delta}}\,dt\\&=\int_0^A\frac{1}{t^{r+1+\delta}}\left|\int_{-\infty}^{\infty}\left(e^{itx}-\sum_{s=0}^{r}\frac{(itx)^s}{s!}\right)dF(x)\right|dt\\&\leqq\int_{-\infty}^{\infty}\left\{\int_0^A\left|\frac{\left(e^{itx}-\sum_{s=0}^{r}\frac{(itx)^s}{s!}\right)}{t^{r+1+\delta}}\right|dt\right\}dF(x)\\&=\int_{-\infty}^{\infty}|x|^{r+\delta}\left\{\int_0^{A|x|}\frac{\left|e^{iu}-\sum_{s=0}^{r}\frac{(iu)^s}{s!}\right|}{u^{r+1+\delta}}\,du\right\}dF(x).\end{aligned}\tag{12}$$

But, using the inequalities

$$\left|e^{ix}-\sum_{s=0}^{r}\frac{(ix)^s}{s!}\right|\leqq\frac{|x|^{r+1}}{(r+1)!}\qquad\text{for } |x|\leqq1,$$

$$\left|e^{ix}-\sum_{s=0}^{r}\frac{(ix)^s}{s!}\right|\leqq(1+e)|x|^r\qquad\text{for } |x|\geqq1,$$

(for the first of these see e.g. Lemma 1 of [4] while the second is obtained by taking the modulus of each of the terms and bounding this) we have

$$\begin{aligned}I(|x|)&=\int_0^{A|x|}\frac{\left|e^{iu}-\sum_{s=0}^{r}\frac{(iu)^s}{s!}\right|}{u^{r+1+\delta}}\,du\\&\leqq\begin{cases}\dfrac{1}{(r+1)!}\displaystyle\int_0^1\frac{du}{u^{\delta}} & \text{for } |x|\leqq A^{-1}\\[2ex] \dfrac{1}{(r+1)!}\displaystyle\int_0^1\frac{du}{u^{\delta}}+(1+e)\int_1^{A|x|}\frac{du}{u^{1+\delta}} & \text{for } |x|>A^{-1}.\end{cases}\end{aligned}$$

Thus, $I(|x|) \leqq c_1$ if $\delta > 0$, $I(|x|) \leqq c_2 \log(|x|+1)$ if $\delta = 0$, where c_1 and c_2 are positive constants and using these results in (12) we have

$$\int_0^A |\beta(t)|\, t^{-(1+\delta)}\, dt < \infty$$

if $\delta > 0$ or if r is even and $\delta = 0$. This completes the proof of the lemma.

Lemma 3. *Suppose* (1) *holds and that* $E|X_i|^r < \infty$ *for some integer* $2 \leqq r \leqq k+2$. *Then,* $\alpha_j = EX_i^j$, $j = 1, 2, \ldots, r$.

Proof. The result of the lemma is true by specification for $r = 2$ and we develop a proof by induction.

Suppose that $E|X_i|^s < \infty$, some $s \geqq 2$ and $\alpha_j = EX_i^j$, $j = 1, 2, \ldots, s$. Then, if $E|X_i|^{s+1} < \infty$, let $Q_j^*(x)$, $1 \leqq j \leqq s-1$ be the classical Chebyshev polynomials expressed in terms of the cumulants κ_j, $j = 1, 2, \ldots, s+1$ of X_i and write

$$G_{s-1,n}^*(x) = \Phi(x) + \frac{1}{\sqrt{2\pi}} e^{-x^2/2} \sum_{j=1}^{s-1} Q_j^*(x) \frac{1}{n^{j/2}}.$$

We have

$$\sup_x |G_{s-1,n}(x) - G_{s-1,n}^*(x)| \leqq \sup_x |F_n(x) - G_{s-1,n}(x)| + \sup_x |F_n(x) - G_{s-1,n}^*(x)|, \tag{13}$$

and from Theorem 1 of [4],

$$n^{(s-1)/2} \sup_x |F_n(x) - G_{s-1,n}^*(x)| = o(1) \tag{14}$$

as $n \to \infty$. Also, from (1),

$$\liminf_{n\to\infty} n^{(k+\delta)/2} \sup_x |F_n(x) - G_{kn}(x)| = 0,$$

so that

$$\begin{aligned} &\liminf_{n\to\infty} n^{(s-1)/2} \sup_x |F_n(x) - G_{s-1,n}(x)| \\ &\quad \leqq \liminf_{n\to\infty} n^{(s-1)/2} \sup_x |F_n(x) - G_{kn}(x)| + \liminf_{n\to\infty} n^{(s-1)/2} \sup_x |G_{kn}(x) - G_{s-1,n}(x)| \quad (15) \\ &\quad = 0 \end{aligned}$$

since

$$G_{kn}(x) - G_{s-1,n}(x) = \frac{1}{\sqrt{2\pi}} e^{-x^2/2} \sum_{j=s}^{k} Q_j(x)\, n^{-j/2}.$$

Consequently, using (14) and (15) in (13),

$$\liminf_{n\to\infty} n^{(s-1)/2} \sup_x |G_{s-1,n}(x) - G_{s-1,n}^*(x)| = 0. \tag{16}$$

But,

$$G_{s-1,n}(x) - G_{s-1,n}^*(x) = \frac{1}{\sqrt{2\pi}} e^{-x^2/2} [Q_{s-1}(x) - Q_{s-1}^*(x)] \frac{1}{n^{(s-1)/2}}$$

since $\alpha_j = EX_i^j$, $j = 1, 2, \ldots, s$ implies $Q_j(x) = Q_j^*(x)$, $1 \leqq j \leqq s-2$, and hence (16) can only hold if $Q_{s-1}(x) = Q_{s-1}^*(x)$. This gives $\kappa_{s+1} = \beta_{s+1}$ upon identifying coefficients and hence $\alpha_{s+1} = EX_i^{s+1}$ as required.

3. Proof of Theorems

We start by proving Theorems 1 and 2 simultaneously in the following three steps. Firstly we note that the equivalence of (2) and (3), (4) follows immediately from Lemmas 1 and 2. Next we shall show that (3), (4) ensures (1) under (*C*) or if $k=0$ and lastly that (1) ensures (2).

(3), (4) $\Rightarrow$ (1) *under* (*C*) *or if* $k=0$.

Firstly we note that $G_{kn}(x)$ has a bounded first derivative; let $|G'_{kn}(x)|\leqq B$. Then, using a bound due to Esseen ([2], Section 39), we have for any $T>0$,

$$|F_n(x)-G_{kn}(x)|\leqq\frac{1}{\pi}\int_{-T}^{T}\left|\frac{f_n(t)-g_{kn}(t)}{t}\right|dt+c\frac{B}{T} \tag{17}$$

where $f_n(t)$ and $g_{kn}(t)$ are the characteristic functions corresponding to $F_n(x)$ and $G_{kn}(x)$ respectively and c is a positive constant.

Now, using (3) we have

$$\begin{aligned} f_n(t)&=\left[f\left(\frac{t}{\sqrt{n}}\right)\right]^n\\ &=\exp\left\{\sum_{s=2}^{k+2}\frac{(i\,t)^s}{s!}\,\frac{\beta_s}{n^{(s-2)/2}}+n\left|\frac{t}{\sqrt{n}}\right|^{k+2}\gamma\left(\frac{t}{\sqrt{n}}\right)\right\},\end{aligned}$$

while from Lemma 5 of [4],

$$g_{kn}(t)=\exp\left\{\sum_{s=2}^{k+2}\frac{(i\,t)^s}{s!}\,\frac{\beta_s}{n^{(s-2)/2}}+D(t)\right\},$$

where $D(t)=O(|n^{-\frac{1}{2}}\,t|^{k+1})$ as $t\to 0$ so that

$$\begin{aligned}|f_n(t)-g_{kn}(t)|\leqq&\left|\exp\left\{\sum_{s=2}^{k+2}\frac{(i\,t)^s}{s!}\,\frac{\beta_s}{n^{(s-2)/2}}\right\}\right|\left|\exp\left(n\left|\frac{t}{\sqrt{n}}\right|^{k+2}\gamma\left(\frac{t}{\sqrt{n}}\right)\right)-1\right|\\ &+\left|\exp\left\{\sum_{s=2}^{k+2}\frac{(i\,t)^s}{s!}\,\frac{\beta_s}{n^{(s-2)/2}}\right\}-g_{kn}(t)\right|.\end{aligned} \tag{18}$$

Now in view of Lemma 1, we may choose a, $0<a<1$, so small that $\max_{|t|<a}|\gamma(t)|\leqq\frac{1}{4}$. Then for $|t|<a\sqrt{n}$, using the inequality $|e^x-1|\leqq|x|\,e^{|x|}$, the first term on the right hand side of (18) is bounded by

$$\begin{aligned}&\exp\left\{\sum_{s=1}^{[\frac{1}{2}(k+2)]}(-1)^s\frac{t^{2s}}{(2s)!}\,\frac{\beta_{2s}}{n^{s-1}}\right\}\frac{|t|^{k+2}}{n^{k/2}}\left|\gamma\left(\frac{t}{\sqrt{n}}\right)\right|\exp\left\{\frac{|t|^{k+2}}{n^{k/2}}\left|\gamma\left(\frac{t}{\sqrt{n}}\right)\right|\right\}\\ &\leqq\exp\left\{n\left(\sum_{s=1}^{[\frac{1}{2}(k+2)]}(-1)^s\frac{t^{2s}}{(2s)!}\,\frac{\beta_{2s}}{n^s}+\frac{1}{4}\left|\frac{t}{\sqrt{n}}\right|^{k+2}\right)\right\}n\left|\frac{t}{\sqrt{n}}\right|^{k+2}\left|\gamma\left(\frac{t}{\sqrt{n}}\right)\right|\\ &\leqq\exp(-t^2/8)\,n\left|\frac{t}{\sqrt{n}}\right|^{k+2}\left|\gamma\left(\frac{t}{\sqrt{n}}\right)\right|\end{aligned} \tag{19}$$

for n sufficiently large since $\beta_2=1$. Also, from Lemma 5 of [4], the second term on the right hand side of (18) is bounded by

$$\frac{c}{n^{(k+1)/2}}\left(|t|^{3(k+1)}+|t|^{k+1}\right)e^{-t^2/4} \tag{20}$$

for $|t|<b\sqrt{n}$ when b is sufficiently small, c being a positive constant. Then, choosing $\alpha=\min(a,b)$ and using (18), (19) and (20), we have

$$\begin{aligned}\sum_{n=1}^{\infty} n^{-1+(k+\delta)/2}\int_{-\alpha\sqrt{n}}^{\alpha\sqrt{n}}\left|\frac{f_n(t)-g_{kn}(t)}{t}\right|dt \\ \leqq c\sum_{n=1}^{\infty} n^{-1-(1-\delta)/2}\int_{-\alpha\sqrt{n}}^{\alpha\sqrt{n}}\left(|t|^{3k+2}+|t|^k\right)e^{-t^2/4}\,dt \\ +\sum_{n=1}^{\infty} n^{-1+\delta/2}\int_{-\alpha\sqrt{n}}^{\alpha\sqrt{n}} e^{-t^2/8}\,|t|^{k+1}\left|\gamma\left(\frac{t}{\sqrt{n}}\right)\right|dt \\ \leqq A+\sum_{n=1}^{\infty} n^{(k+\delta)/2}\int_{-\alpha}^{\alpha}|u|^{k+1}\,|\gamma(u)|\,e^{-nu^2/8}\,du,\end{aligned} \tag{21}$$

A being a finite constant. Furthermore, using a standard Abelian theorem (e.g. Feller [1], Vol. II, p. 423), we have

$$\lim_{t\uparrow 1}(1-t)^{(k+2+\delta)/2}\sum_{n=1}^{\infty} n^{(k+\delta)/2}\,t^n=\Gamma((k+2+\delta)/2)$$

so that for $u\neq 0$ it is possible to choose a constant $K_1>0$ such that

$$\sum_{n=1}^{\infty} n^{(k+\delta)/2}\,e^{-nu^2/8}\leqq K_1\left(1-e^{-u^2/8}\right)^{-(k+2+\delta)/2}$$

Also, $|u|\leqq\alpha<1$, so that

$$1-e^{-u^2/8}>\tfrac{1}{8}u^2\left(1-\tfrac{1}{16}\alpha^2\right)$$

and hence

$$\sum_{n=1}^{\infty} n^{(k+\delta)/2}\,e^{-nu^2/8}\leqq K_2\,|u|^{-(k+2+\delta)}$$

for some $K_2>0$. Consequently, noting that $|\gamma(t)|$ is symmetric in t, we have

$$\begin{aligned}\sum_{n=1}^{\infty} n^{(k+\delta)/2}\int_{-\alpha}^{\alpha}|u|^{k+1}\,|\gamma(u)|\,e^{-nu^2/8}\,du \\ =2\int_0^{\alpha}u^{k+1}\,|\gamma(u)|\left\{\sum_{n=1}^{\infty} n^{(k+\delta)/2}\,e^{-nu^2/8}\right\}du \\ \leqq 2K_2\int_0^{\alpha}\frac{|\gamma(u)|}{u^{1+\delta}}\,du<\infty\end{aligned} \tag{22}$$

in view of (4).

Next write $T_\alpha=\{t\colon \alpha\sqrt{n}\leqq|t|\leqq\alpha n^{(k+1)/2}\}$, noting that T_α is empty if $k=0$. Then, using condition (C), $\max_{|t|\geqq\alpha}|f(t)|=\theta<1$, so that for t in T_α,

$$|f_n(t)|=\left|f\left(\frac{t}{\sqrt{n}}\right)\right|^n\leqq\theta^n$$

and hence

$$\int_{T_\alpha}\left|\frac{f_n(t)}{t}\right|dt\leqq 2\theta^n\int_{\alpha\sqrt{n}}^{\alpha n^{(k+1)/2}}\frac{dt}{t}\sim\theta^n(k+1)\log n$$

as $n\to\infty$. Consequently,

$$\sum_{n=1}^{\infty}n^{-1+(k+\delta)/2}\int_{T_\alpha}\left|\frac{f_n(t)}{t}\right|dt<\infty. \tag{23}$$

Furthermore, by the rules for forming the polynomials $Q_j(x)$,

$$g_{kn}(t)=e^{-\frac{1}{2}t^2}\left[1+\sum_{j=1}^{k}P_j(i\,t)\,n^{-\frac{1}{2}j}\right]$$

where P_j is a polynomial of degree $3j$ determined from the formal identity

$$\exp\left\{\sum_{j=3}^{\infty}\frac{\beta_j}{j!}\,\frac{(i\,t)^j}{n^{(j-2)/2}}\right\}=1+\sum_{j=1}^{\infty}P_j(i\,t)\,n^{-j/2}.$$

Thus,

$$|g_{kn}(t)|\leqq e^{-\frac{1}{2}t^2}\left[1+\sum_{j=1}^{3k}a_{nj}|t|^j\right]$$

where the a_{nj} are polynomials in $n^{-\frac{1}{2}}$ which tend to zero as $n\to\infty$. Consequently, we certainly have

$$\int_{T_\alpha}\left|\frac{g_{kn}(t)}{t}\right|dt\leqq K\,e^{-\alpha^2 n/2}\,n^{3k(k+1)/2}$$

for K a positive constant which gives

$$\sum_{n=1}^{\infty}n^{-1+(k+\delta)/2}\int_{T_\alpha}\left|\frac{g_{kn}(t)}{t}\right|dt<\infty, \tag{24}$$

and from (23) and (24) we obtain

$$\int_{T_\alpha}\left|\frac{f_n(t)-g_{kn}(t)}{t}\right|dt\leqq\int_{T_\alpha}\left|\frac{f_n(t)}{t}\right|dt+\int_{T_\alpha}\left|\frac{g_{kn}(t)}{t}\right|dt<\infty. \tag{25}$$

The required result (1) then follows, using (17) with $T=\alpha n^{\frac{1}{2}(k+1)}$, in view of (21), (22) and (25).

(1) $\Rightarrow$ (2)

Firstly we symmetrize the X_i's. Consider the sequence Y_i, $i=1,2,3,\ldots$ of independent symmetrized random variables; each Y_i having the distribution of the difference between two independent X_i's. Clearly the Y_i have characteristic function $|f(t)|^2$ and the distribution function of $Z_n = n^{-\frac{1}{2}} \sum_{i=1}^{n} Y_i$ is $F_n(x) * (1-F_n(-x-0)) = F_n^*(x)$. Write $G_{kn}^*(x)$ for the convolution $G_{kn}(x) * (1-G_{kn}(-x))$. Then,

$$
\begin{aligned}
&\sum_{n=1}^{\infty} n^{-1+(k+\delta)/2} \sup_x |F_n^*(x) - G_{kn}^*(x)| \\
&\quad = \sum_{n=1}^{\infty} n^{-1+(k+\delta)/2} \sup_x \left|F_n(x) * (1-F_n(-x-0)) - G_{kn}(x) * (1-G_{kn}(-x))\right| \\
&\quad \leq \sum_{n=1}^{\infty} n^{-1+(k+\delta)/2} \sup_x \left|F_n(x) * (1-F_n(-x-0)) - G_{kn}(x) * (1-F_n(-x-0))\right| \\
&\quad + \sum_{n=1}^{\infty} n^{-1+(k+\delta)/2} \sup_x \left|G_{kn}(x) * (1-F_n(-x-0)) - G_{kn}(x) * (1-G_{kn}(-x))\right| \\
&\quad < \infty
\end{aligned}
\tag{26}
$$

in view of (1).

Next we shall show that (26) implies $E|Y_i|^{k+2} < \infty$ and hence $E|X_i|^{k+2} < \infty$. We note that

$$\sum_{n=1}^{\infty} n^{-1+(k+\delta)/2} \{1 - G_{kn}^*(x_n)\} < \infty$$

where $x_n = \{(k+\delta+1)\log n\}^{\frac{1}{2}}$, so that from (26),

$$\sum_{n=1}^{\infty} n^{-1+(k+\delta)/2} P\left(\left|\sum_{i=1}^{n} Y_i\right| > n^{\frac{1}{2}} x_n\right) < \infty. \tag{27}$$

But, for symmetric random variables,

$$P\left(\left|\sum_{i=1}^{n} Y_i\right| > n^{\frac{1}{2}} x_n\right) \geq \tfrac{1}{2} P\left(\max_{1 \leq k \leq n} |Y_k| > n^{\frac{1}{2}} x_n\right) \tag{28}$$

(e.g. [1], Vol. II, p. 147) and from Bonferroni's inequalities (e.g. [1], Vol. I, p. 100) we have

$$
\begin{aligned}
n P(|Y_i| > n^{\frac{1}{2}} x_n)\{1 - \tfrac{1}{2}(n-1) P(|Y_i| > n^{\frac{1}{2}} x_n)\} \\
\leq P\left(\max_{1 \leq k \leq n} |Y_k| > n^{\frac{1}{2}} x_n\right) \leq n P(|Y_i| > n^{\frac{1}{2}} x_n),
\end{aligned}
$$

while $n P(|Y_i| > n^{\frac{1}{2}} x_n) \to 0$ as $n \to \infty$ since $E Y_i^2 = 2 E X_i^2 < \infty$. Consequently,

$$P\left(\max_{1 \leq k \leq n} |Y_k| > n^{\frac{1}{2}} x_n\right) \sim n P(|Y_i| > n^{\frac{1}{2}} x_n) \tag{29}$$

as $n \to \infty$ and hence, from (27), (28) and (29),

$$\sum_{n=1}^{\infty} n^{(k+\delta)/2} P(|Y_i| > n^{\frac{1}{2}} x_n) < \infty. \tag{30}$$

But,

$$\begin{aligned}
E|Y_i|^{k+2} &= -\int_0^\infty x^{k+2}\, dP(|Y_i|>x) \\
&\leqq (k+2)\int_0^\infty x^{k+1} P(|Y_i|>x)\, dx \\
&= (k+2)\sum_{n=0}^\infty \int_{x_n n^{\frac{1}{2}}}^{x_{n+1}(n+1)^{\frac{1}{2}}} x^{k+1} P(|Y_i|>x)\, dx \\
&\leqq \sum_{n=0}^\infty P(|Y_i|>n^{\frac{1}{2}} x_n)\{((n+1)\, x_{n+1}^2)^{(k+2)/2} - (n\, x_n^2)^{(k+2)/2}\} \\
&\leqq c\sum_{n=0}^\infty n^{k/2}(\log n)^{k/2} P(|Y_i|>n^{\frac{1}{2}} x_n) < \infty
\end{aligned}$$

in view of (30), c being a positive constant. It then follows that $E|X_i|^{k+2}<\infty$ and an appeal to Lemma 3 gives $\alpha_j = EX_i^j$, $j=1,2,\ldots,k+2$.

Now we note that the characteristic functions of $F_n^*(x)$ and $G_{kn}^*(x)$ are $|f_n(t)|^2$ and $|g_{kn}(t)|^2$ respectively. Then, integrating by parts in the equation

$$|f_n(t)|^2 - |g_{kn}(t)|^2 = \int_{-\infty}^{\infty} e^{itx}\, d\{F_n^*(x) - G_{kn}^*(x)\},$$

we obtain

$$-\frac{|f_n(t)|^2 - |g_{kn}(t)|^2}{it} = \int_{-\infty}^{\infty} e^{itx}\{F_n^*(x) - G_{kn}^*(x)\}\, dx.$$

Also,

$$it\, e^{-t^2/2} = \int_{-\infty}^{\infty} e^{itx} \frac{x}{\sqrt{2\pi}}\, e^{-x^2/2}\, dx,$$

and we obtain from Parseval's identity

$$\int_{-\infty}^{\infty} \{|f_n(t)|^2 - |g_{kn}(t)|^2\}\, e^{-t^2/2}\, dt = \sqrt{2\pi}\int_{-\infty}^{\infty} \{F_n^*(x) - G_{kn}^*(x)\}\, x\, e^{-x^2/2}\, dx.$$

Thus from (26)

$$\begin{aligned}
&\sum_{n=1}^\infty n^{-1+(k+\delta)/2}\left|\int_{-\infty}^{\infty}\{|f_n(t)|^2 - |g_{kn}(t)|^2\}\, e^{-t^2/2}\, dt\right| \\
&= \sqrt{2\pi}\sum_{n=1}^\infty n^{-1+(k+\delta)/2}\left|\int_{-\infty}^{\infty}\{F_n^*(x) - G_{kn}^*(x)\}\, x\, e^{-x^2/2}\, dx\right| \qquad (31)\\
&\leqq 2\sqrt{2\pi}\sum_{n=1}^\infty n^{-1+(k+\delta)/2}\sup_x |F_n^*(x) - G_{kn}^*(x)| < \infty.
\end{aligned}$$

Furthermore, we note that for $0<\alpha<\frac{1}{4}$,

$$\begin{aligned}
&\sum_{n=1}^\infty n^{-1+(k+\delta)/2}\left|\int_{n^\alpha}^{\infty} e^{-t^2/2}\{|f_n(t)|^2 - |g_{kn}(t)|^2\}\, dt\right| \\
&\leqq 2\sum_{n=1}^\infty n^{-1+(k+\delta)/2}\int_{n^\alpha}^{\infty} e^{-t^2/2}\, dt < \infty,
\end{aligned}$$

so that in view of (31)

$$\sum_{n=1}^{\infty} n^{-1+(k+\delta)/2} \left| \int_0^{n^\alpha} e^{-t^2/2} \{|f_n(t)|^2 - |g_{kn}(t)|^2\} \, dt \right| < \infty. \tag{32}$$

Now from Lemma 5 of [4], letting $R = [\frac{1}{2}(k+2)]$, we have for $t < c'\sqrt{n}$, c' some suitably small positive constant,

$$\left| |g_{kn}(t)|^2 - \exp\left(\sum_{s=1}^{R} (-1)^s \frac{t^{2s} 2\kappa_{2s}}{(2s)!\, n^{s-1}} \right) \right| \leq \frac{c}{n^{(k+1)/2}} (|t|^{3(k+1)} + |t|^{k+1}) \, e^{-t^2/4}$$

for some $c > 0$ since $\alpha_j = EX_i^j$, $j = 1, 2, \ldots, k+2$ ensures $\beta_j = \kappa_j$, $j = 1, 2, \ldots, k+2$. Thus,

$$\begin{aligned} &\sum_{n=1}^{\infty} n^{-1+(k+\delta)/2} \left| \int_0^{n^\alpha} \left\{ \exp\left\{ \sum_{s=1}^{R} (-1)^s \frac{t^{2s}}{(2s)!} \frac{2\kappa_{2s}}{n^{s-1}} \right\} - |g_{kn}(t)|^2 \right\} e^{-t^2/2} \, dt \right| \\ &\quad \leq c \sum_{n=1}^{\infty} n^{-1-(1-\delta)/2} \int_0^{n^\alpha} e^{-3t^2/4} (t^{3(k+1)} + t^{k+1}) \, dt < \infty. \end{aligned} \tag{33}$$

Also, from Lemma 1,

$$|f_n(t)|^2 = \left| f\left(\frac{t}{\sqrt{n}}\right) \right|^{2n} = \exp\left\{ \sum_{s=1}^{R} (-1)^s \frac{t^{2s}}{(2s)!} \frac{2\kappa_{2s}}{n^{s-1}} + 2 \frac{|t|^{k+2}}{n^{k/2}} \operatorname{Re} \gamma\left(\frac{t}{\sqrt{n}}\right) \right\},$$

and using this result in conjunction with (32) and (33),

$$\begin{aligned} &\sum_{n=1}^{\infty} n^{-1+(k+\delta)/2} \\ &\quad \cdot \left| \int_0^{n^\alpha} e^{-t^2/2} \exp\left\{ \sum_{s=1}^{R} (-1)^s \frac{t^{2s}}{(2s)!} \frac{2\kappa_{2s}}{n^{s-1}} \right\} \left[1 - \exp\left(2 \frac{t^{k+2}}{n^{k/2}} \operatorname{Re} \gamma\left(\frac{t}{\sqrt{n}}\right) \right) \right] dt \right| < \infty. \end{aligned}$$

Now Lemma 1 tells us that for n large enough, $\operatorname{Re} \gamma\left(\frac{t}{\sqrt{n}}\right)$ will be of constant sign for $0 < t \leq n^\alpha$ and hence

$$\begin{aligned} &\sum_{n=1}^{\infty} n^{-1+(k+\delta)/2} \\ &\quad \cdot \int_0^{n^\alpha} e^{-t^2/2} \exp\left\{ \sum_{s=1}^{R} (-1)^s \frac{t^{2s}}{(2s)!} \frac{2\kappa_{2s}}{n^{s-1}} \right\} \left| 1 - \exp\left(2 \frac{t^{k+2}}{n^{k/2}} \operatorname{Re} \gamma\left(\frac{t}{\sqrt{n}}\right) \right) \right| dt < \infty \end{aligned}$$

so that

$$\begin{aligned} &\sum_{n=1}^{\infty} n^{-1+(k+\delta)/2} \\ &\quad \cdot \int_0^1 \exp\left\{ -3t^2/2 + \sum_{s=2}^{R} (-1)^s \frac{t^{2s}}{(2s)!} \frac{2\kappa_{2s}}{n^{s-1}} \right\} \left| 1 - \exp\left(2 \frac{t^{k+2}}{n^{k/2}} \operatorname{Re} \gamma\left(\frac{t}{\sqrt{n}}\right) \right) \right| dt < \infty \end{aligned}$$

which implies

$$\sum_{n=1}^{\infty} n^{-1+(k+\delta)/2} \int_0^1 \left| 1 - \exp\left(2 \frac{t^{k+2}}{n^{k/2}} \operatorname{Re} \gamma\left(\frac{t}{\sqrt{n}}\right) \right) \right| dt < \infty.$$

However, as $n \to \infty$,

$$\left|1-\exp\left(2\frac{t^{k+2}}{n^{k/2}}\operatorname{Re}\gamma\left(\frac{t}{\sqrt{n}}\right)\right)\right| = \frac{2|t|^{k+2}}{n^{k/2}}\left|\operatorname{Re}\gamma\left(\frac{t}{\sqrt{n}}\right)\right|(1+o(1)),$$

so that

$$\sum_{n=1}^{\infty} n^{-1+\delta/2}\int_0^1 t^{k+2}\left|\operatorname{Re}\gamma\left(\frac{t}{\sqrt{n}}\right)\right| dt < \infty,$$

and, upon making the transformation $u = t/\sqrt{n}$, this yields

$$\sum_{n=1}^{\infty} n^{(k+1+\delta)/2}\int_0^{1/\sqrt{n}} u^{k+2}|\operatorname{Re}\gamma(u)|\,du < \infty. \tag{34}$$

Now $\int_0^{1/\sqrt{x}} u^{k+2}|\operatorname{Re}\gamma(u)|\,du$ is monotone decreasing as x increases so for $X > 1$,

$$\begin{aligned}
&\int_1^X x^{(k+1+\delta)/2}\left\{\int_1^{1/\sqrt{x}} u^{k+2}|\operatorname{Re}\gamma(u)|\,du\right\}dx \\
&\leqq \sum_{n=1}^{[X]}\int_n^{n+1} x^{(k+1+\delta)/2}\left\{\int_1^{1/\sqrt{x}} u^{k+2}|\operatorname{Re}\gamma(u)|\,du\right\}dx \\
&\leqq \sum_{n=1}^{[X]}(n+1)^{(k+1+\delta)/2}\int_1^{1/\sqrt{n}} u^{k+2}|\operatorname{Re}\gamma(u)|\,du \\
&\leqq c\sum_{n=1}^{[X]} n^{(k+1+\delta)/2}\int_1^{1/\sqrt{n}} u^{k+2}|\operatorname{Re}\gamma(u)|\,du,
\end{aligned}$$

c being a positive constant such that $(n+1)^{(k+1+\delta)/2} < c\,n^{(k+1+\delta)/2}$ for all positive integral n. Consequently, using (34), we have

$$\int_1^{\infty} x^{(k+1+\delta)/2}\left\{\int_0^{1/\sqrt{x}} u^{k+2}|\operatorname{Re}\gamma(u)|\,du\right\}dx < \infty. \tag{35}$$

Now in view of (35) we must have

$$\int_{\omega}^{2\omega} x^{(k+1+\delta)/2}\left\{\int_0^{1/\sqrt{x}} u^{k+2}|\operatorname{Re}\gamma(u)|\,du\right\}dx \to 0$$

as $\omega \to \infty$ and

$$\begin{aligned}
&\int_{\omega}^{2\omega} x^{(k+1+\delta)/2}\left\{\int_0^{1/\sqrt{x}} u^{k+2}|\operatorname{Re}\gamma(u)|\,du\right\}dx \\
&\geqq \omega^{(k+3+\delta)/2}\int_0^{1/\sqrt{2\omega}} u^{k+2}|\operatorname{Re}\gamma(u)|\,du \geqq 0,
\end{aligned}$$

so that, putting $v = 1/\sqrt{2\omega}$, we conclude that

$$v^{-(k+3+\delta)}\int_0^{v} u^{k+2}|\operatorname{Re}\gamma(u)|\,du \to 0 \tag{36}$$

as $v \to 0$. Then, upon making the transformation $v = 1/\sqrt{x}$, (35) becomes

$$\int_0^1 \left\{ \int_0^v u^{k+2} |\mathrm{Re}\,\gamma(u)|\, du \right\} v^{-(k+4+\delta)}\, dv < \infty$$

and, in view of (36), the fact that $\int_0^1 |\mathrm{Re}\,\gamma(t)|\, t^{-(1+\delta)}\, dt < \infty$ holds follows immediately from an integration by parts. Lemma 2 then enables us to conclude that $E|Y_i|^{k+2+\delta} < \infty$ from which we deduce that $E|X_i|^{k+2+\delta} < \infty$. This completes the proof that (1) $\Rightarrow$ (2) and hence the proof of Theorems 1 and 2.

For the proof of Theorem 3, we note firstly that the above proof that (3), (4) $\Rightarrow$ (1) under (C) or if $k=0$ together with Lemmas 1 and 2 show that $E|X_i|^{k+2} \log(1+|X_i|) < \infty$ implies (1) with $\delta = 0$ under (C) or if $k=0$. On the other hand, the above proof that (1) $\Rightarrow$ (2) shows that (1) with $\delta = 0$ implies $\int_0^1 |\mathrm{Re}\,\gamma(t)|\, t^{-1}\, dt < \infty$. Thus, from Lemmas 1 and 2 we have $E|Y_i|^{k+2} \log(1+|Y_i|) < \infty$ from which we deduce the required result that $E|X_i|^{k+2} \log(1+|X_i|) < \infty$.

References

1. Feller, W.: An introduction to probability theory and its applications, vols. I, II. New York: Wiley 1957, 1966.
2. Gnedenko, B.V., Kolmogorov, A.N.: Limit distributions for sums of independent random variables. Translated and annotated by K.L. Chung. Cambridge (Mass.): Addison Wesley 1954.
3. Heyde, C.C.: On the influence of moments on the rate of convergence to the normal distribution. Z. Wahrscheinlichkeitstheorie verw. Geb. **8**, 12–18 (1967).
4. Ibragimov, I.A.: On the Chebyshev-Cramér asymptotic expansions. Theor. Probab. Appl. **12**, 455–469 (1967).
5. Linnik, Yu.V.: Decompositions of probability distributions. Translation ed. by S.J. Taylor. London: Oliver & Boyd 1964.
6. Loève, M.: Probability theory (3rd ed.). New York: Van Nostrand 1963.
7. Lukacs, E.: Characteristic functions. London: Griffin 1960.
8. Petrov, V.V.: On certain polynomials encountered in probability theory. Vestnik Leningrad Univ. **19**, 150–153 (1962) [in Russian].

Dr. C.C. Heyde
Mr. J.R. Leslie
Department of Statistics, SGS
Australian National University
Box 4, P.O.
Canberra, A.C.T. 2600
Australia

(Received July 15, 1970)

J. Appl. Prob. **9**, 235–256 (1972)
Printed in Israel

ESTIMATION THEORY FOR GROWTH AND IMMIGRATION RATES IN A MULTIPLICATIVE PROCESS

C. C. HEYDE AND
E. SENETA, *Australian National University*

Abstract

This paper deals with the simple Galton-Watson process with immigration, $\{X_n\}$ with offspring probability generating function (p.g.f.) $F(s)$ and immigration p.g.f. $B(s)$, under the basic assumption that the process is subcritical $(0 < m \equiv F'(1-) < 1)$, and that $0 < \lambda \equiv B'(1-) < \infty$, $0 < B(0) < 1$, together with various other moment assumptions as needed. Estimation theory for the rates m and λ on the basis of a single terminated realization of the process $\{X_n\}$ is developed, in that (strongly) consistent estimators for both m and λ are obtained, together with associated central limit theorems in relation to m and $\mu \equiv \lambda(1-m)^{-1}$ Following this, historical antecedents are analysed, and some examples of application of the estimation theory are discussed, with particular reference to the continuous-time branching process with immigration. The paper also contains a strong law for martingales; and discusses relation of the above theory to that of a first order autoregressive process.

ESTIMATION THEORY; GROWTH AND IMMIGRATION RATES; MULTIPLICATIVE PROCESSES; STRONG CONSISTENCY; STRONG LAW; MARTINGALES; CENTRAL LIMIT THEOREM; FIRST ORDER AUTOREGRESSION; PARTICLE FLUCTUATION

1. Introduction

The fundamental process with which we shall be concerned is the simple Galton-Watson process with immigration, $\{X_n\}$, with offspring probability generating function (p.g.f.) $F(s)$, and immigration p.g.f. $B(s)$. In the present paper we confine ourselves to the subcritical case $(0 < m \equiv F'(1-) < 1)$, assuming throughout also that $\lambda \equiv B'(1-) < \infty$, $0 < B(0) < 1$, and that the respective variances σ^2 and b^2 of the offspring and immigration distributions are finite. We do not make one of the usual assumptions that $X_0 = 1$ or $X_0 = 0$, but allow a general distribution, which, however, is assumed to satisfy $E(X_0^2) < \infty$ throughout. (Additional moment assumptions are made as needed.)

Our main aim in this paper is to develop an estimation theory for the rates m and λ, on the basis of a single realization of the process $\{X_n\}$ (observed up to some fixed time).

The process, under the present assumptions, is not stationary in general, so

Received 25 June 1971.

R. Maller et al. (eds.), *Selected Works of C.C. Heyde*, Selected Works in Probability and Statistics,
DOI 10.1007/978-1-4419-5823-5_31, © Springer Science+Business Media, LLC 2010

it is not possible to avail oneself directly of the large body of estimation theory known for such processes. However, it is known (Heathcote (1966); Seneta (1969), Section 5.2) that the process approaches a limiting-stationary distribution as $n \to \infty$, so it is not unexpected that our results resemble known ones for stationary regimes.

In the same vein, the present paper considers techniques which (in the usual manner of time-series analysis) attempt to exploit the totality of information contained in a realization of the process. Another approach, based more on recurrence properties of individual states, is possible and in some ways more general and convenient. It does, however, seem to have the disadvantage of not utilizing all the available information, and so will be discussed in a sequel to this paper. This sequel will also contain estimation theory for λ in the critical case $m = 1$; and for the individual variances σ^2 and b^2.

Following the development of the general estimation theory in the present paper, we analyse its historical antecedents, and discuss some examples of its application, with particular reference to the continuous-time branching process with immigration.

In conclusion to this introduction, we note that in the supercritical ($1 < m < \infty$) case for both the processes with and without immigration, results for the estimation of m are already available (Heyde (1971); Heyde and Seneta (1971)).

2. Statement of main results and discussion

Theorem A. Let $C_n = \sum_{i=1}^{n} X_i/n \ (\equiv \bar{X})$ and $D_n = \sum_{i=1}^{n} \hat{Y}_{i+1}\hat{Y}_i / \sum_{i=1}^{n} \hat{Y}_i^2$, where $\hat{Y}_i = X_i - C_n$. Then $C_n \xrightarrow{P} \mu \equiv \lambda(1-m)^{-1}$, and $D_n \xrightarrow{P} m$ as $n \to \infty$, so that $C_n(1-D_n)$ and D_n are consistent estimators of λ and m respectively. If the offspring, immigration and initial distributions all have finite third moments, then the above convergences in probability can be replaced by almost sure convergence. That is, the respective estimators are strongly consistent.

It will also be shown in the sequel that, in the above theorem, it is possible to use the statistic

$$D_n^* = 1 - \frac{1}{2}\,\frac{\sum_{i=1}^{n}(X_{i+1} - X_i)^2}{\sum_{i=1}^{n}(X_i - \bar{X})^2} \tag{2.1}$$

in place of D_n, D_n^* being possibly more convenient computationally and aesthetically.

The other main result is the following theorem, which provides an indication of the convergence rates of C_n and D_n to their respective limits, in the form of central-limit type results. It thus makes possible statistical inferences on the offspring mean, m, and the mean, μ, of the limiting-stationary distribution, on the basis of a single long realization of $\{X_n\}$.

For purposes of this theorem, and the sequel in general, we introduce the notation c^2 for the frequently occurring quantity $b^2 + \sigma^2\mu$.

Theorem B.

$$\lim_{n\to\infty} P(\alpha^{-1}\sqrt{n}(C_n - \mu) \leqq x) = \Phi(x),$$

where $\alpha^2 = (1-m)^{-2}c^2$, and $\Phi(x) = (2\pi)^{-1/2}\int_{-\infty}^{x} e^{-u^2/2}du$.

If, in addition to the basic assumptions of Section 1, all of the offspring, immigration, and initial distributions have finite fourth moments[1], then

$$\lim_{n\to\infty} P(\beta^{-1}\sqrt{n}(D_n - m) \leqq x) = \Phi(x)$$

when $\beta = B(1 - m^2)c^{-2}$. Here

$$B = (1-m^2)^{-1}c^4 + \sigma^3(1-m)^{-3}\{E(I-\lambda)^3 + \mu E(Z_1-m)^3 + 3m\sigma^2(1 - m^2)^{-1}c^2\},$$

where I and Z_1 are random variables, respectively having the distributions of the immigration and offspring mechanisms.

In this theorem also, D_n may be replaced by D_n^*. The nature of the above results and of the statistics entering into them may readily be understood by comparison with another, simpler (time-series) context. To this end we first note that for our process $\{X_n\}$

$$E(X_{n+1} | X_n) = mX_n + \lambda, \quad n = 0, 1, 2, \cdots$$

so that if we introduce the 'limit-mean corrected' process $\{Y_n\}$ where $Y_n = X_n - \mu$, $\mu \equiv \lambda(1-m)^{-1}$ being the mean of the limiting-stationary distribution of $\{X_n\}$, we have

$$E(Y_{n+1} | Y_n) = mY_n, \tag{2.2}$$

so that

$$Y_{n+1} = mY_n + \delta_n, \quad n = 0, 1, 2, \cdots \tag{2.3}$$

where $E(\delta_n) = 0$ and (the residual) δ_n is uncorrelated with Y_n.

The relation (2.3) is similar to the defining relation

$$x_{n+1} = \rho x_n + \varepsilon_n, \quad n = 0, \pm 1, \pm 2, \cdots (|\rho| < 1) \tag{2.4}$$

of the second-order stationary process which is called a first-order autoregression, $\{\varepsilon_n\}$ being a sequence of independent and identically distributed normal variables

[1] It is likely that this condition can be weakened to finiteness of third moments.

with zero mean (i.e., 'white noise'). (For this process ρ is the first autocorrelation.) Thus in our process m (roughly) corresponds to ρ, and it is easily checked that our estimators C_n and D_n are analogous in form to the large sample maximum likelihood estimators of the corresponding quantities in (2.4), C_n being the sample mean, and D_n effectively the first circular serial correlation coefficient, while D_n^* corresponds to the correlation statistic r_A related to Von Neumann's ratio (see Hannan (1960), pp. 84–85).

Whereas our process $\{X_n\}$ is far, in general, from satisfying the conditions of the first-order autoregression as regards stationarity, and normality and independence of residuals, our Theorems A and B may be regarded as supporting the thesis that theory developed on the basis of (2.4) is often robust for large samples under deviations from assumption, a problem of some interest in time-series analysis. (Of course the process $\{X_n\}$ *may* be in its stationary regime, in which case (2.3) implies that m is its first autocorrelation; however the distribution of residuals would still appear to be far from normal. It is relevant to note however that in its stationary regime the process $\{X_n\}$ has in fact its *entire* correlogram identical with that of the first-order autoregression.)

3. Proof of Theorem A

We carry out the proof in a sequence of steps, which are each of somewhat broader significance. The first step is a law of large numbers for martingales describing the effect of averaging the successive differences between each of a sequence of random variables and its best predictor on the basis of its past. This theorem partially extends a result given by Loève ((1963), p. 387).

Theorem 1. Let $\{\mathscr{F}_n,\ n = 0,1,2,\cdots\}$ be an increasing sequence of σ-fields, and $\{U_n,\ n = 0,1,2,\cdots\}$ a sequence of random variables, with U_n measurable with respect to $\mathscr{F}_n$; and U a random variable such that $E(|U|) < \infty$ and $P(|U_n| > x) \leqq P(|U| > x)$ for each $x \geqq 0$ and $n \geqq 0$. Then

$$n^{-1} \sum_{k=1}^{n} [U_k - E(U_k | \mathscr{F}_{k-1})] \xrightarrow{P} 0$$

as $n \to \infty$. If $E|U|\log^+|U| < \infty$, then the convergence in probability may be strengthened to almost sure (a.s.) convergence.

Proof. Firstly we form the truncated random variables $U_n' = U_n I(|U_n| \leqq n)$ and note that the sequence $\{U_n' - E(U_n' | \mathscr{F}_{n-1}), \mathscr{F}_n, n \geqq 1\}$ is a martingale difference sequence. We have

$$\begin{aligned}\sum_{n=1}^{\infty} \frac{1}{n^2} E(U_n' - E(U_n' \mid \mathscr{F}_{n-1}))^2 &\leqq \sum_{n=1}^{\infty} \frac{1}{n^2} E(U_n')^2 \\ &\leqq 2 \sum_{n=1}^{\infty} \frac{1}{n^2} \int_{0<x\leqq n} xP(|U_n| > x)dx \\ &\leqq 2 \sum_{n=1}^{\infty} \frac{1}{n^2} \int_{0<x\leqq n} xP(|U| > x)dx \\ &= 2 \sum_{n=1}^{\infty} \frac{1}{n^2} \sum_{k=1}^{n} \int_{k-1<x\leqq k} xP(|U| > x)dx \\ &\leqq 2 \sum_{n=1}^{\infty} \frac{1}{n^2} \sum_{k=1}^{n} kP(|U| > k-1) \\ &= 2 \sum_{k=1}^{\infty} kP(|U| > k-1) \sum_{n=k}^{\infty} \frac{1}{n^2} \\ &\leqq 4 \sum_{k=1}^{\infty} P(|U| > k-1) < \infty\end{aligned}$$

since $E(|U|) < \infty$. Thus the application of a standard martingale convergence theorem (e.g., Neveu (1965), p. 147) gives

$$n^{-1} \sum_{k=1}^{n} [U_k' - E(U_k' \mid \mathscr{F}_{k-1})] \xrightarrow{\text{a.s.}} 0. \tag{3.1}$$

Next, we note

$$\sum_{k=1}^{\infty} P(U_k \neq U_k') = \sum_{k=1}^{\infty} P(|U_k| > k) \leqq \sum_{k=1}^{\infty} P(|U| > k) < \infty,$$

so that the sequences $\{U_k\}$ and $\{U_k'\}$ are tail-equivalent, and hence from (3.1)

$$n^{-1} \sum_{k=1}^{n} [U_k - E(U_k' \mid \mathscr{F}_{k-1})] \xrightarrow{\text{a.s.}} 0. \tag{3.2}$$

Now,

$$n^{-1} E \left| \sum_{k=1}^{n} E(U_k I(|U_k| > k) \mid \mathscr{F}_{k-1}) \right| \leqq n^{-1} \sum_{k=1}^{n} E(|U_k| I(|U_k| > k)) \to 0$$

as $n \to \infty$ since

$$E(|U_n| I(|U_n| > n)) \leqq \int_n^{\infty} P(|U_n| > x)dx \leqq \int_n^{\infty} P(|U| > x)dx \to 0$$

as $n \to \infty$ so that using Markov's inequality,

$$n^{-1}\sum_{k=1}^{n} E(U_k I(|U_k| > k)|\mathscr{F}_{k-1}) = n^{-1}\sum_{k=1}^{n} [E(U_k|\mathscr{F}_{k-1}) - E(U'_k|\mathscr{F}_{k-1})] \xrightarrow{p} 0$$

and hence

$$n^{-1}\sum_{k=1}^{n} [U_k - E(U_k|\mathscr{F}_{k-1})] \xrightarrow{p} 0.$$

Furthermore, if $E|U|\log^+|U| < \infty$, then

$$\sum_{n=1}^{\infty} n^{-1}E(|U_n|I(|U_n| > n)|\mathscr{F}_{n-1}) < \infty \text{ a.s.}$$

since

$$\begin{aligned}
&\sum_{n=1}^{\infty} n^{-1}E(E(|U_n|I(|U_n| > n)|\mathscr{F}_{n-1}))\\
&= \sum_{n=1}^{\infty} n^{-1}E(|U_n|I(|U_n| > n))\\
&\leqq \sum_{n=1}^{\infty} n^{-1}\int_n^{\infty} P(|U_n| > x)dx\\
&\leqq \sum_{n=1}^{\infty} n^{-1}\int_n^{\infty} P(|U| > x)dx\\
&= \sum_{n=1}^{\infty} n^{-1}\sum_{k=n}^{\infty}\int_{k<x\leqq k+1} P(|U| > x)dx\\
&\leqq \sum_{k=1}^{\infty} P(|U| > k)\sum_{n=1}^{k} n^{-1} \leqq \sum_{k=1}^{\infty}(1+\log k)P(|U| > k) < \infty.
\end{aligned}$$

Thus, using the Kronecker lemma

$$n^{-1}\sum_{k=1}^{n} E(|U_k|I(|U_k| > k)|\mathscr{F}_{k-1}) \xrightarrow{\text{a.s.}} 0$$

and hence

$$n^{-1}\sum_{k=1}^{n} E(U_k I(|U_k| > k)|\mathscr{F}_{k-1}) = n^{-1}\sum_{k=1}^{n}[E(U_k|\mathscr{F}_{k-1}) - E(U'_k|\mathscr{F}_{k-1})] \xrightarrow{\text{a.s.}} 0$$

as required. This completes the proof of the theorem.

The following result is the one which is responsible for the consistency of the estimators.

Theorem 2. Let $Y_i = X_i - \mu$, $i = 0, 1, 2, \cdots$ as before. Then

(i) $\sum_{i=1}^{n} Y_i/n \xrightarrow{P} 0;$

(ii) $\sum_{i=1}^{n} Y_i^2/n \xrightarrow{P} (1-m^2)^{-1}c^2;$

(iii) $\sum_{i=1}^{n} Y_{i+1}Y_i/n \xrightarrow{P} m(1-m^2)^{-1}c^2.$

If the offspring, immigration and initial distributions have finite third moments, then the above convergences can be replaced by almost sure convergence. (Note that if $\{X_n\}$ is stationary, $c^2 = E(\text{var}(Y_{n+1} \mid Y_n))$ and $(1-m^2)^{-1}c^2 = \text{var}\, Y_n$.)

Proof. All of (i)–(iii) follow easily from three applications of Theorem 1, once the appropriate 'dominating' random variable has been found, and it is this point which takes up most of the proof.

Firstly, we note that the state space, S, of our process $\{X_n\}$, considered as a denumerable Markov chain, consists (a) of an irreducible positive recurrent set, J^*, of states including the state 0, plus, possibly (b) 'ephemeral' states from which transition into the positive recurrent set, J^*, occurs in one step with probability one (see Seneta (1969), Section 5.2; this reference excludes the case $F(s) = \alpha + \beta s$, independent consideration of which reveals that $S = J^*$).

Thus it follows that $X_n = 0$ infinitely often with probability one, and the time between successive returns to zero has finite mean.

Next, for our process $\{X_n\}$, let $T = \min[n : X_n = 0]$ and consider $X = \sup_{0 \leq k \leq T} X_k$. We clearly have $E(T) < \infty$ in view of the finiteness of the mean time between successive returns to zero and we go on to show that $E(X^2) < \infty$. We do this by noting that

$$X^2 = \sup_{0 \leqq k \leqq T} X_k^2 \leqq \sum_{k=0}^{T} X_k^2$$

and showing that $E(\sum_{k=0}^{T} X_k^2) < \infty$.

Now for $r = 1$ or 2,

$$\begin{aligned} E\left(\sum_{k=1}^{T} X_k^r\right) &= \sum_{n=1}^{\infty}\sum_{k=1}^{n}\int_{\{T=n\}} X_k^r dP \\ &= \sum_{k=1}^{\infty}\sum_{n=k}^{\infty}\int_{\{T=n\}} X_k^r dP \\ &= \sum_{k=1}^{\infty}\int_{\{T\geqq k\}} X_k^r dP \\ &= \sum_{k=1}^{\infty}\int_{\{T\geqq k\}} E(X_k^r \mid \mathscr{F}_{k-1}) dP \end{aligned} \tag{3.4}$$

since $\{T \geqq k\} \in \mathscr{F}_{k-1}$, the σ-field generated by $X_0, \cdots, X_{k-1}$.

First we look at the case $r = 1$. We have for $k \geqq 0$,

$$
\begin{aligned}
c_{k+1} &= \int_{\{T \geqq k+1\}} E(X_{k+1} \mid \mathscr{F}_k)dP \\
&= \int_{\{T \geqq k, X_k > 0\}} (mX_k + \lambda)dP \\
&= m \int_{\{T \geqq k, X_k > 0\}} X_k dP + \lambda P(T \geqq k+1) \qquad (3.5) \\
&= m \int_{\{T \geqq k\}} X_k dP + \lambda P(T \geqq k+1) \\
&= m c_k + \lambda P(T \geqq k+1)
\end{aligned}
$$

and $c_0 = E(X_0) < \infty$. From the recurrence relation (3.5) we find that

$$
c_k = m^k E(X_0) + \lambda \sum_{j=1}^{k} m^{k-j} P(T \geqq j), \quad k \geqq 1,
$$

and hence, using (3.4),

$$
\begin{aligned}
E\left\{\sum_{k=1}^{T} X_k\right\} = \sum_{k=1}^{\infty} c_k &= \frac{mEX_0}{1-m} + \lambda \sum_{k=1}^{\infty} \sum_{j=1}^{k} m^{k-j} P(T \geqq j) \qquad (3.6) \\
&= \frac{mEX_0}{1-m} + \frac{\lambda}{1-m} \sum_{j=1}^{\infty} P(T \geqq j) < \infty
\end{aligned}
$$

since $E(T) < \infty$.

We now adopt a similar approach for the case $r = 2$. First note that

$$
\begin{aligned}
E(X_{k+1}^2 \mid \mathscr{F}_k) &= E((X_{k+1} - E(X_{k+1} \mid \mathscr{F}_k))^2 \mid \mathscr{F}_k) + (E(X_{k+1} \mid \mathscr{F}_k))^2 \qquad (3.7) \\
&= X_k \sigma^2 + b^2 + (mX_k + \lambda)^2
\end{aligned}
$$

since $X_{k+1} - E(X_{k+1} \mid \mathscr{F}_k)$, conditional on $\mathscr{F}_k$, is the sum of $X_k + 1$ independent random variables, X_k of them having the distribution of $Z_1 - m$ and the other the distribution of $I - \lambda$. Then, for $k \geqq 0$ and using (3.7),

$$
\begin{aligned}
d_{k+1} &= \int_{\{T \geqq k+1\}} E(X_{k+1}^2 \mid \mathscr{F}_k)dP \\
&= \int_{\{T \geqq k, X_k > 0\}} [X_k \sigma^2 + b^2 + (mX_k + \lambda)^2]dP \qquad (3.8) \\
&= \int_{\{T \geqq k\}} [X_k \sigma^2 + b^2 + (mX_k + \lambda)^2]dP - (b^2 + \lambda^2) P(T = k) \\
&= m^2 d_k + (\sigma^2 + 2m\lambda)c_k + (b^2 + \lambda^2) P(T \geqq k+1)
\end{aligned}
$$

and $d_0 = E(X_0^2)$. From the recurrence relation (3.8) we find that

$$d_{k+1} = m^{2(k+1)}EX_0^2 + (\sigma^2 + 2m\lambda)\sum_{j=0}^{k} m^{2(k-j)}c_j + (b^2+\lambda^2)\sum_{j=0}^{k} m^{2(k-j)}P(T \geqq j+1),$$

$k \geqq 0$, and hence, using (3.4), (3.6) and $E(T) < \infty$,

$$E\left(\sum_{k=1}^{T} X_k^2\right) = \sum_{k=1}^{\infty} d_k$$

$$= \frac{m^2E(X_0^2)}{1-m^2} + (\sigma^2 + 2m\lambda)\sum_{k=0}^{\infty}\sum_{j=0}^{k} m^{2(k-j)}c_j$$

$$+ (b^2+\lambda^2)\sum_{k=0}^{\infty}\sum_{j=0}^{k} m^{2(k-j)}P(T \geqq j+1)$$

$$= \frac{m^2E(X_0^2)}{1-m^2} + \frac{\sigma^2+2m\lambda}{1-m^2}\sum_{j=0}^{\infty} c_j + \frac{b^2+\lambda^2}{1-m^2}\sum_{j=0}^{\infty} P(T \geqq j+1) < \infty.$$

We thus find $E(X^2) < \infty$ as required. This completes the preliminaries and we now go on to establish the results of the theorem via three applications of Theorem 1.

Firstly, note that for any $x \geqq 0$ and $n \geqq 0$, $P(X_n > x) \leqq P(X > x)$; and of course, $E(X) < \infty$. Thus we may apply Theorem 1 with $\mathscr{F}_k$ the σ-field generated by $X_0, X_1, \cdots, X_k$ to obtain

$$n^{-1}\sum_{k=1}^{n}[Y_k - E(Y_k \mid \mathscr{F}_{k-1})] \equiv n^{-1}\sum_{k=1}^{n}[X_k - E(X_k \mid \mathscr{F}_{k-1})] \tag{3.9}$$
$$\xrightarrow{\text{P}} 0.$$

Now, from (2.2),

$$n^{-1}\sum_{k=1}^{n}[Y_k - E(Y_k \mid \mathscr{F}_{k-1})] = n^{-1}\sum_{k=1}^{n}[Y_k - mY_{k-1}]$$

$$= n^{-1}\sum_{k=1}^{n}[(Y_k - mY_k) + m(Y_k - Y_{k-1})]$$

$$= \frac{(1-m)}{n}\sum_{k=1}^{n} Y_k + m\left(\frac{Y_n - Y_0}{n}\right).$$

Thus $\sum_{k=1}^{n} Y_k/n \xrightarrow{\text{P}} 0$ since $(Y_n - Y_0)/n \xrightarrow{\text{P}} 0$. The last remark follows from an application of Markov's inequality since $E\left|Y_n - Y_0\right|/n \to 0$ as $n \to \infty$.

Next, note that $P(X_n^2 > x) \leqq P(X^2 > x)$ for all $n \geqq 0, x \geqq 0$, and $E(X^2) < \infty$. Thus by Theorem 1

$$n^{-1}\sum_{k=1}^{n}[X_k^2 - E(X_k^2 \mid \mathscr{F}_{k-1})] \xrightarrow{\text{P}} 0, \tag{3.10}$$

where $\mathscr{F}_{k-1}$ is the σ-field generated by $X_0, X_1, X_2, \cdots, X_{k-1}$, as before. Now

$$[X_k^2 - E(X_k^2 \mid \mathscr{F}_{k-1})]$$
$$= [X_k^2 - \operatorname{var}(X_k \mid \mathscr{F}_{k-1}) - \{E(X_k \mid \mathscr{F}_{k-1})\}^2]$$
$$= [(Y_k + \mu)^2 - \operatorname{var}(Y_k \mid \mathscr{F}_{k-1}) - \{E(Y_k \mid \mathscr{F}_{k-1}) + \mu\}^2]$$
$$= [Y_k^2 + \mu^2 + 2\mu Y_k - \sigma^2 Y_{k-1} - c^2 - m^2 Y_{k-1}^2 - \mu^2 - 2\mu m Y_{k-1}].$$

Thus from (3.10) and part (i) of the present Theorem

$$n^{-1} \sum_{k=1}^{n} (Y_k^2 - m^2 Y_{k-1}^2) - c^2 \xrightarrow{P} 0$$

i.e.,

$$\frac{(1 - m^2)}{n} \sum_{k=1}^{n} Y_k^2 + \frac{m^2}{n}(Y_n^2 - Y_0^2) \xrightarrow{P} c^2$$

and hence the result (ii) of the Theorem follows (since $Y_n^2/n \equiv (X_n^2 - 2X_n\mu + \mu^2)/n \xrightarrow{P} 0$, since $E(X_n^2/n) \leqq E(X^2/n) \to 0$).

To prove (iii), we first note that for any $x \geqq 0$, $n \geqq 0$

$$P(|Y_n Y_{n+1}| > x) \leqq P(Z^2 > x)$$

where Z has the distribution of $\max(X, \mu)$. Also, $EZ^2 < \infty$ since $EX^2 < \infty$. Thus by Theorem 1 we have

$$\text{(3.11)} \qquad n^{-1} \sum_{k=1}^{n} [Y_k Y_{k+1} - E(Y_k Y_{k+1} \mid \mathscr{F}_k)] \xrightarrow{P} 0.$$

Now

$$E(Y_k Y_{k+1} \mid \mathscr{F}_k) = Y_k E(Y_{k+1} \mid Y_k)$$

by Markovicity; and by (2.2)

$$= m Y_k^2,$$

whence

$$n^{-1} \sum_{k=1}^{n} [Y_k Y_{k+1} - m Y_k^2] \xrightarrow{P} 0$$

and so by part (ii) of the theorem,

$$n^{-1} \sum_{k=1}^{n} Y_k Y_{k+1} \xrightarrow{P} \frac{m}{1 - m^2} c^2$$

as required.

This completes the proof of the convergence in probability results. Almost sure convergence under the additional conditions follows straightforwardly along similar lines. When $E(X_0^3) < \infty$, $E(Z_1^3) < \infty$, $E(I^3) < \infty$, it follows that $E(X^3) < \infty$ by an argument similar to that used to show $E(X) < \infty$, $E(X^2) < \infty$. Then, the a.s. convergence form of Theorem 1 can be used in three applications. The only other point that requires noting is that $Y_n^2/n \xrightarrow{a.s.} 0$ as $n \to \infty$ and this follows from the Borel-Cantelli lemma since $Y_n^2/n = (X_n^2 - 2X_n\mu + \mu^2)/n$ and for any $\varepsilon > 0$

$$\sum_{n=1}^{\infty} P(X_n^i > n\varepsilon) \leqq \sum_{n=1}^{\infty} P(X^i > n\varepsilon) < \infty$$

since $E(X^i) < \infty$, $i = 1,2$.

Remark. It is not difficult to see that the above proof of part (i) holds without the assumptions of finite variances for the initial, offspring, and immigration distributions.

The proof of Theorem A now follows easily. That $C_n \xrightarrow{P} \mu$, is just a restatement of Theorem 2, (i). Let us turn to D_n:

$$n^{-1} \sum_{i=1}^{n} \hat{Y}_{i+1}\hat{Y}_i = n^{-1} \sum_{i=1}^{n} (Y_{i+1} + \mu - C_n)(Y_i + \mu - C_n)$$

$$= n^{-1} \sum_{i=1}^{n} Y_{i+1}Y_i + R_n,$$

where $R_n \xrightarrow{P} 0$ as $n \to \infty$ from Theorem 2, (i), therefore

$$n^{-1} \sum_{i=1}^{n} \hat{Y}_{i+1}\hat{Y}_i \xrightarrow{P} m(1 - m^2)^{-1}c^2 \tag{3.12}$$

from part (iii) of the same theorem. Similarly

$$n^{-1} \sum_{i=1}^{n} \hat{Y}_i^2 \xrightarrow{P} (1 - m^2)^{-1}c^2 \tag{3.13}$$

from Theorem 2, parts (i) and (ii). The result of Theorem A follows immediately.

Remarks. (1) As a 'bonus' result, the above provides a consistent estimator of the variance of the limiting-stationary distribution of $\{X_n\}$ *viz.* $(1-m^2)^{-1}c^2$. This is clearly the 'obvious' one

$$E_n = \sum_{i=1}^{n} \hat{Y}_i^2/n. \tag{3.14}$$

(2) The estimator

$$D_n^* = 1 - \frac{1}{2}\frac{\Sigma_{i=1}^n (X_{i+1} - X_i)^2}{\Sigma_{i=1}^n (X_i - \bar{X})^2}$$

$$= 1 - \frac{1}{2}\frac{\Sigma_{i=1}^n (Y_{i+1} - Y_i)^2}{\Sigma_{i=1}^n \hat{Y}_i^2}$$

$$= 1 - \frac{1}{2n}\left(\sum_{i=1}^n Y_{i+1}^2 + \sum_{i=1}^n Y_i^2 - 2\sum_{i=1}^n Y_{i+1}Y_i\right) \Big/ E_n$$

$$= \left(\sum_{i=1}^n Y_{i+1}Y_i/n\right) \Big/ E_n + \bar{R}_n,$$

where $\bar{R}_n \xrightarrow{P} 0$ as $n \to \infty$, by Theorem 2 and (3.13) so that

$$D_n^* \xrightarrow{P} m.$$

Thus D_n^* is also consistent for m.

4. Proof of Theorem B

We proceed in the manner of Section 3, first proving the following result.

Theorem 3.

$$\lim_{n\to\infty} P\left(\alpha^{-1}\sqrt{n}\left(\sum_{i=1}^n Y_i/n\right) \leqq x\right) = \Phi(x).$$

If, in addition to the basic assumptions of Section 1, all of the offspring, immigration and initial distributions have finite fourth moments, then

$$\lim_{n\to\infty} P\left(\frac{\Sigma_{i=1}^n Y_i(Y_{i+1} - mY_i)}{B\sqrt{n}} \leqq x\right) = \Phi(x),$$

where B is a constant defined in the statement of Theorem B.

Proof. The results are obtained by making use of a central limit theorem for martingales due to Brown (1971).

First write $W_i = Y_i - E(Y_i \mid \mathscr{F}_{i-1})$, $i = 1, 2, \cdots$ and $S_n = \Sigma_{i=1}^n W_i$, noting that $\{S_n, \mathscr{F}_n, n = 1, 2, \cdots\}$ is a zero-mean martingale. Let

$$V_n^2 = \sum_{j=1}^n E(W_j^2 \mid \mathscr{F}_{j-1}),\ s_n^2 = E(V_n^2) = E(S_n^2).$$

Then

$$V_n^2 = \sum_{j=1}^n (\sigma^2 Y_{j-1} + c^2) = \sigma^2 \sum_{j=0}^{n-1} Y_j + nc^2$$

and

$$
\begin{aligned}
s_n^2 &= \sigma^2 \sum_{j=0}^{n-1} \{m^j E(Y_0)\} + nc^2 \\
&= \sigma^2 E(Y_0) \sum_{j=0}^{n-1} m^j + nc^2 \\
&\sim nc^2 \qquad \text{as } n \to \infty .
\end{aligned} \tag{4.1}
$$

Moreover, since $V_n^2 = \sigma^2 \sum_{j=0}^{n-1} Y_j + nc^2$, it follows from Theorem 2(i) that $V_n^2/s_n^2 \xrightarrow{p} 1$ as $n \to \infty$.

Now, let $U(x) = x(1+x)^{-1}$; then

$$
0 \leqq \frac{1}{n} \sum_{j=1}^{n} W_j^2 U(|W_j|\varepsilon^{-1}s_n^{-1}) \leqq \frac{1}{\varepsilon n s_n} \sum_{j=1}^{n} |W_j|^3 . \tag{4.2}
$$

Now, using Theorem 2 we see that $n^{-1} \sum_{j=1}^{n} W_j^2$ converges in probability to a finite constant. Also, using the Borel-Cantelli lemma, $n^{-1} W_n^2 \xrightarrow{a.s.} 0$ and hence $n^{-\frac{1}{2}} \sup_{1 \leqq k \leqq n} |W_k| \xrightarrow{a.s.} 0$. It then follows that

$$
n^{-3/2} \sum_{j=1}^{n} |W_j|^3 \leqq \left(n^{-1} \sum_{j=1}^{n} W_j^2\right)(n^{-\frac{1}{2}} \sup_{1 \leqq k \leqq n} |W_k|) \xrightarrow{p} 0
$$

and hence by (4.1), (4.2) and an application of Theorem 1,

$$
n^{-1} \sum_{j=1}^{n} E(W_j^2 U(|W_j|\varepsilon^{-1}s_n^{-1}) | \mathscr{F}_{j-1}) \xrightarrow{p} 0 .
$$

It therefore follows from Lemma 2 of Brown (1971), and the fact that $s_n^{-2} V_n^2 \xrightarrow{p} 1$, that Theorem 2 of the same reference is applicable and hence

$$
\lim_{n \to \infty} P(s_n^{-1} S_n \leqq x) = \Phi(x) .
$$

The result of the first part of the theoren is then immediate from (4.1) and the fact that

$$
S_n = (1-m) \sum_{k=1}^{n} Y_k + m(Y_n - Y_0) .
$$

The proof of the second part of the theorem follows the same pattern as the first. Write $\zeta_i = Y_i(Y_{i+1} - mY_i)$, $i = 0, 1, 2, \cdots$ and $T_n = \sum_{i=1}^{n} \zeta_i$, noting that $\{T_n, \mathscr{F}_{n+1}, n = 1, 2, \cdots\}$ is a zero-mean martingale.

Let

$$
V_n^{*2} = \sum_{j=1}^{n} E(\zeta_j^2 | \mathscr{F}_j), s_n^{*2} = E(V_n^{*2}) = E(T_n^2) .
$$

Then

$$
\begin{aligned}
V_n^{*2} &= \sum_{j=1}^{n} Y_j^2 E\{(Y_{j+1} - mY_j)^2 \mid \mathscr{F}_j\} \\
&= \sum_{j=1}^{n} Y_j^2 \operatorname{var}(Y_{j+1} \mid \mathscr{F}_j) \\
&= \sum_{j=1}^{n} Y_j^2 [\sigma^2 Y_j + c^2]
\end{aligned}
$$

i.e.,

$$V_n^{*2} = c^2 \sum_{j=1}^{n} Y_j^2 + \sigma^2 \sum_{j=1}^{n} Y_j^3. \tag{4.3}$$

Now, a straightforward but tedious calculation shows that

$$n^{-1} s_n^{*2} \to \gamma\sigma^2 + c^4(1 - m^2)^{-1}, \tag{4.4}$$

where

$$\gamma = \frac{1}{1 - m^3} \left[E(I - \lambda)^3 + \mu E(Z_1 - m)^3 + \frac{3mc^2}{1 - m^2} \right],$$

(and is $E(Y_n^3)$ when Y_n has its 'limiting-stationary' distribution).

Furthermore, when $E(X_0^4) < \infty$, $E(Z_1^4) < \infty$, $E(I^4) < \infty$, as is the case at present, it follows using the methods developed in the proof of Theorem 2, that $E(X^4) < \infty$. Thus we can certainly apply Theorem 1 to obtain

$$n^{-1} \sum_{j=1}^{n} [Y_j^3 - E(Y_j^3 \mid \mathscr{F}_{j-1})] \xrightarrow{\text{p}} 0$$

from which, and more straightforward but tedious analysis,

$$n^{-1} \sum_{j=1}^{n} Y_j^3 \xrightarrow{\text{p}} \gamma. \tag{4.5}$$

It therefore follows from (ii) of Theorem 2, (4.3), (4.4) and (4.5) that $s_n^{*-2} V_n^{*2} \xrightarrow{\text{p}} 1$ as $n \to \infty$.

Next, again let $U(x) = x/\{1 + x\}$ so that

$$0 \leqq \frac{1}{n} \sum_{j=1}^{n} \zeta_j^2 U(|\zeta_j| \varepsilon^{-1} s_n^{*-1}) \leqq \frac{1}{\varepsilon n s_n^*} \sum_{j=1}^{n} |\zeta_j|^3. \tag{4.6}$$

Since $E(X^4) < \infty$ we can apply Theorem 1 to obtain

$$n^{-1} \sum_{j=1}^{n} [\zeta_j^2 - E(\zeta_j^2 \mid \mathscr{F}_j)] \xrightarrow{\text{p}} 0$$

which gives, in view of (4.4),

$$n^{-1} \sum_{j=1}^{n} \zeta_j^2 \xrightarrow{\text{P}} \gamma\sigma^2 + c^4(1-m^2)^{-1}.$$

Thus, as in the proof of the first part of the theorem, we have from the Borel-Cantelli lemma that $n^{-1}\zeta_n^2 \xrightarrow{\text{a.s.}} 0$ so that $n^{-\frac{1}{2}}\sup_{1\le k\le n}|\zeta_k| \xrightarrow{\text{a.s.}} 0$ and

$$n^{-\frac{3}{2}} \sum_{j=1}^{n} |\zeta_j|^3 \le \left(n^{-1} \sum_{j=1}^{n} \zeta_j^2\right)(n^{-\frac{1}{2}} \sup_{1\le j\le n} |\zeta_j|) \xrightarrow{\text{P}} 0.$$

Consequently, from (4.4), (4.6) and an application of Theorem 1,

$$n^{-1} \sum_{j=1}^{n} E(\zeta_j^2\, U(|\zeta_j|\varepsilon^{-1}s_n^{*-1})|\mathscr{F}_{j-1}) \xrightarrow{\text{P}} 0.$$

We have thus again completed the formalities necessary to apply Theorem 2 of Brown (1971). This gives

$$\lim_{n\to\infty} P(s_n^{*-1}\, T_n \le x) = \Phi(x)$$

which provides the desired result.

Remarks. (1) Theorem 3 is given in the above setting in the interests of clarity of exposition; much more actually holds. In fact, invariance principles for suitably defined random functions in $C[0,1]$ hold for both parts of the theorem. This is a consequence of the fact that the conditions of Theorem 3 of Brown (1971) are satisfied in both parts of the theorem. The interested reader is referred to Brown's paper for details of the setting.

(2) The result of the first part of our Theorem 3 has been obtained by Pakes (1971), who uses classical results on sums of independent random variables.

The proof of Theorem B is now immediate. The first part is just the first part of Theorem 3. For the second part note that

$$n^{-1} \sum_{i=1}^{n} \hat{Y}_i(\hat{Y}_{i+1} - m\hat{Y}_i) = n^{-1} \sum_{i=1}^{n} \hat{Y}_i\hat{Y}_{i+1} - mn^{-1} \sum_{i=1}^{n} \hat{Y}_i^2$$

so that, as in the argument leading to (3.12) and (3.13),

$$= n^{-1} \sum_{i=1}^{n} Y_iY_{i+1} - mn^{-1} \sum_{i=1}^{n} Y_i^2 + \bar{\bar{R}}_n$$

$$= n^{-1} \sum_{i=1}^{n} Y_i(Y_{i+1} - mY_i) + \bar{\bar{R}}_n,$$

where $\sqrt{n}\,\bar{\bar{R}}_n \xrightarrow{\text{P}} 0$ as $n \to \infty$ in view of part (i) of Theorem 2 and the first part of Theorem 3, and the second part follows from this, (3.13), and the second part of Theorem 3.

Remarks. (1) The values α^2/n and β^2/n give an indication of the asymptotic decline of the variances of the estimators C_n and D_n respectively; this may be of importance in practical considerations, and we shall return to it in the next section.

(2) It can be seen without difficulty from the kind of discussion given at the conclusion of Section 3, and above, that D_n^* may be inserted in place of D_n in Theorem B.

5. Analysis of historical development

The simple Galton-Watson process with immigration, where $F(s) = 1 - m + ms$ and $B(s) = \exp\{\lambda(1-s)\}$, in its stationary state, was developed as early as 1915 by Smoluchowski (for an account of this early work see Chapter III of the article of Chandrasekhar (1943)). It was used as a model for the fluctuation in the number of particles contained in a geometrically well-defined small element of volume v, in a much larger volume of solution containing particles exhibiting random motion, the system being in equilibrium, and observations on v being made at constant intervals of time apart. The number $P = 1-m$ was called the 'probability after-effect factor', and, of course (as also λ) depends on the precise circumstances of the problem.

Moreover, it was noted from the above physical situation, and can be easily deduced analytically, that the stationary distribution is also Poisson, with p.g.f. $\exp\{\mu(1-s)\}$, $\mu P = \lambda$. The quantity of interest was in fact the number P, which was to be estimated from the experimental data. The method used was to note first that (since the regime is stationary, and the stationary distribution is Poisson)

$$E((X_{n+1} - X_n)^2) = 2\mu P. \tag{5.1}$$

Then from a long series of observations on the sequence $\{X_n\}$, $C_n = \bar{X}$ was found, and μ (the variance as well as the mean of the stationary distribution) was estimated by $\sum_{i=1}^n \hat{Y}_i^2/n$, and $E((X_{n+1} - X_n)^2)$ by $\sum_{i=1}^n (X_{i+1} - X_i)^2/n$, which led to the estimate $\hat{P}_1$ of P:

$$\hat{P}_1 \equiv \frac{1}{2}\,\frac{\sum_{i=1}^n (X_{i+1} - X_i)^2}{\sum_{i=1}^n (X_i - \bar{X})^2} = 1 - D_n^*, \tag{5.2}$$

so that in fact $m = 1 - P$ *is estimated by our estimator* D_n^*.

Once P has been estimated statistically, its value may be compared with that obtained analytically on the basis of the parameters of the system. Experimental verification on Smoluchowski's theory using colloidal solutions was carried out by Westgren ((1916), (1918)) in particular, on this basis; the agreement was found to be very good. Another early experiment performed on the basis of Smoluchowski's theory was that of Fürth ((1918), (1919)) (whose data we shall use in the next section). The physical context was different; systematic counts were made of the number of pedestrians in a block every five seconds (a short enough time interval for any pedestrian to be observed at least twice with a large probability).

It is clear that the estimation of P may also in fact be *used to estimate parameters of the system*, and this fact is used in Westgren's work to determine Avogadro's constant (and could be used to estimate the 'average speed of a pedestrian' in Fürth's context, as Chandrasekhar points out).

Also, it is necessary to mention that, whereas recourse to the very early memoirs reveals that in fact the estimator $\hat{P}_1$ was used to estimate P, Chandrasekhar's account of the Smoluchowski theory suggests that in fact μ in (5.1) was actually estimated by $\bar{X}$, which in turn leads to the estimator of P:

$$\hat{P}_2 = \frac{1}{2}\,\frac{\sum_{i=1}^{n}(X_{i+1} - X_i)^2}{n\bar{X}},$$

so that m would be estimated by

$$(5.3) \qquad 1 - \hat{P}_2 = 1 - \frac{1}{2}\,\frac{\sum_{i=1}^{n}(X_{i+1} - X_i)^2}{n\bar{X}},$$

which is, of course, also consistent for m, in this special situation, since $\bar{X}$ is also consistent for the *variance* μ of the stationary-distribution. (Fürth in fact used the estimator (5.3), as an analysis of his data reveals.)

These last two paragraphs are particularly relevant to a more recent application of Smoluchowski's theory, by Rothschild (1953) who (under the influence of Chandrasekhar's account) uses the estimator (5.3) in the estimation of sperm speeds from counts of successive numbers in cinemicrographs. Rothschild's paper brings out several other features not noticeable in the earlier literature, for firstly he notes that previous authors paid little attention to the precision of the estimation method, and presents an unproved formula for the variance, obtained from moments, up to the fourth, of $\{X_n\}$. According to an account in Bartlett ((1955), p. 248) this translates as

$$\operatorname{var}(1 - \hat{P}_2) \sim \{(1-m)^2(3+m)(1+m)^{-1} + m(1-m)\mu^{-1}\}/n,$$

which is of the same order of decline as our formula β^2/n, although we are not aware as to the nature of the approximations used in the derivation. Secondly, Rothschild's account emphasizes the fact that the process is actually occurring in *continuous* time, although (even with a movie camera, as in his experiments) observations are *intermittent*; and estimation is therefore for the discrete skeleton of the process. One consequence of this is that the choice of interval between observations is important for the branching assumptions to be more closely satisfied. A second is that in such a situation it is necessary to translate the estimates from the discrete to continuous time context, which requires a knowledge of the relationship between the process and its discrete skeleton. We take up this point in the next paragraph.

A maximum-likelihood approach to the problem of estimating m and μ in the simple Smoluchowski model (but dropping the assumption of stationary regime) was made by Bartlett and his student Patankar (see Bartlett (1955), p. 246–247). The discussion begins with the more fundamental continuous-time Markov branching process with immigration (underlying the Smochulowski model), which has p.g.f.

$$E\{s^{X(t)} \mid X(0) = k\} = [1 + e^{-\nu t}(s-1)]^k \exp\{\mu(1 - e^{-\nu t})(s-1)\}.$$

Observations on this process are made at unit time intervals (w.l.o.g.), so that in fact one observes the process $\{X_n\}$ with p.g.f.

$$E\{s^{X_n} \mid X_0 = k\} = [1 + m^n(s-1)]^k \exp\{\mu(1 - m^n)(s-1)\},$$

where $m = e^{-\nu}$. It is easy to check that $\{X_n\}$ corresponds to a Galton-Watson process with immigration having offspring and immigration p.g.f.'s

$$F(s) = (1-m) + ms,$$

$$B(s) = \exp\{\lambda(s-1)\},$$

where $\lambda = \mu(1-m)$, as before, so that the parameters μ and ν of the continuous-time process are in fact validly estimated from the techniques mentioned before. The method of maximum likelihood then yields the estimator $\hat{\mu} = \sum_{i=1}^{n} X_i/n$, but the other maximum-likelihood estimator, for m, does not emerge in useful form.

In general where the offspring and immigration distributions are suitably parametrized as in the above simple model, an approach to the estimation problem is possible using the method of maximum likelihood. Such an approach has theoretical advantages, insofar as one is able to make use of the general theory of maximum likelihood estimation in the Markov chain context (e.g., Billingsley (1961)). However, the few other examples investigated indicate that difficulties similar to the above case arise in the estimation of m.

In conclusion to this historical discussion, we mention that several other physical models making use of the Galton-Watson process with immigration are mentioned in Chapter 5 of Seneta (1969).

6. Some complements

(a) *The discrete skeleton of a Markov branching process with immigration.* We shall show that if a regular homogeneous continuous-time Markov branching process with immigration, $\{X(t)\}$, is observed at equally spaced instants of time, we obtain a Galton-Watson process with immigration. We shall consider unit time intervals for simplicity of exposition.

Let $\Phi(t;s)$ be the p.g.f. of a regular Markov branching process initiated by a single individual (ignoring immigration) i.e., the p.g.f. of the direct descendants, at time t. Let $\delta_{k0} + q_k\Delta t + o(\Delta t)$ be the probability of k immigrants in time interval Δt, so that $g(1) = \sum_{k=0}^{\infty} q_k = 0$, where $g(s) = \sum_{k=0}^{\infty} q_k s^k$. Let us consider, further, a process with immigration $\{X(t)\}$ initiated by a fixed number, $i \geqq 0$, of individuals (extension to an initial distribution is trivial). Then

$$P_i(t;s) \equiv E[s^{X(t)} | X(0) = i] = \Phi^i(t;s)A(t;s),$$

where Sevast'yanov (1957) has shown that

$$A(t;s) = E[s^{X(t)} | X(0) = 0] = \exp\left\{\int_0^t g[\Phi(u;s)]du\right\}.$$

Now, it is well known that the discrete skeleton of an ordinary Markov branching process (without immigration) is a Galton-Watson process, so that $\Phi(n;s) = \Phi_n(1;s)$, where $\Phi_n(1;s)$ is the nth functional iterate of $\Phi(1;s)$. We know in fact that, more generally, for any $u, v > 0$, the characteristic branching property implies that $\Phi(u;\Phi(v;s)) = \Phi(u+v;s)$.

Thus, if we put $X(n) = X_n$, we have

$$E\{s^{X_{n+1}} | X_0 = i\} = P_i(n+1;s) = \Phi^i(n;\Phi(1;s))A(n+1;s) \tag{6.1}$$

since

$$\Phi(n+1;s) = \Phi_{n+1}(1;s) = \Phi_n(1;\Phi(1;s)) = \Phi(n;\Phi(1;s)).$$

Now, consider the following argument:

$$\begin{aligned} A(1;s)A(n;\Phi(1;s)) &= \exp\left\{\int_0^1 g[\Phi(u;s)]du\right\}\exp\left\{\int_0^n g[\Phi(u;\Phi(1;s))]du\right\} \\ &= \exp\left\{\int_0^1 g[\Phi(u;s)]du\right\}\exp\left\{\int_0^n g[\Phi(u+1;s)]du\right\} \\ &= \exp\left\{\int_0^1 g[\Phi(u;s)]du\right\}\exp\left\{\int_1^{n+1} g[\Phi(u;s)]du\right\} \\ &= \exp\left\{\int_0^{n+1} g[\Phi(u;s)]du\right\}, \end{aligned}$$

i.e.,

$$A(1;s)A(n;\Phi(1;s)) = A(n+1;s),$$

so that (6.1) yields

$$P_i(n+1;s) = A(1;s)P_i(n;\Phi(1;s))$$

which is the evolutionary equation of a Galton-Watson process with immigration, as required, with offspring p.g.f. $F(s) \equiv \Phi(1;s)$, as expected, and immigration p.g.f. of rather complex form

$$B(s) \equiv A(1;s) = \exp\left\{\int_0^1 g[\Phi(u;s)]du\right\}. \tag{6.2}$$

Henceforth we assume the existence of first and second derivatives at $s = 1-$ of the continuous-time immigration and offspring generating functions, $g(s)$ and $k(s)$. Let us put in particular $k'(1-) = -\nu$ for the mean offspring rate, assuming $\nu > 0$ for a subcritical process; and put $\tau \equiv g'(1-)$ for the immigration rate.

It follows that using our general results we are able to estimate from the process $\{X_n\}$ the mean μ of the limiting-stationary distribution (the same for the continuous-time process and its discrete skeleton), and the decay parameter ν where $e^{-\nu} = F'(1-) = m$.

To estimate the immigration intensity, we note that, differentiating (6.2) with respect to s gives

$$B'(s) = \left\{\int_0^1 \frac{d(g[\Phi(u;s)]}{ds}du\right\}B(s),$$

differentiation under the integral at $s \in (0,1)$ being permissible. Let $s \to 1-$; we have by dominated convergence

$$\lambda \equiv B'(1-) = g'(1-)\int_0^1 \Phi'(u;1)du = \tau\nu^{-1}(1-e^{-\nu}) \tag{6.3}$$

since $\Phi'(u;1) = [d\Phi(u;s)/ds]_{s=1} = e^{-\nu\mu}$. Thus

$$\tau\nu^{-1} = \lambda(1-e^{-\nu})^{-1} = \lambda(1-m)^{-1} = \mu,$$

so that $\tau = \nu\mu$ and a consistent estimator for τ is obtained by multiplying two consistent estimators, of ν and μ respectively.

Clearly this discussion subsumes that at the conclusion of Section 5.

Clearly also, Theorem B is usable in drawing inferences about μ, and about ν, so long as the constants α, β can be determined. This can in fact be done with the aid of the generators $k(s)$ and $g(s)$. Thus the moments corresponding to $F(s) \equiv \Phi(1;s)$ may be obtained from a knowledge of $k(s)$ as indicated in Harris ((1963), p. 103). Those corresponding to $B(s) = A(1;s)$ can be obtained in an analogous manner involving both $k(s)$ and $g(s)$; or from differentiation of (6.2) using $g(s)$ and the known moments of $\Phi(u;s)$. (The deduction of (6.3) is an example.)

(b) *A numerical example.* Below we give numerical values of estimators obtained for the Smoluchowski model, using Fürth's data of 505 readings (observation stopped with 506th reading zero) (Fürth (1918). Tabelle 1). For this specific model, there are two estimators for μ and three for m on the basis of the foregoing discussion. This leads to six possible estimators for λ, from $\lambda \equiv \mu(1-m)$. The λ values are given in the body of the table.

(Westgren's (1916) data are much more extensive; for example, Versuchsreihe C, pp. 12–13, contains 1583 readings. However, Versuchsreihe C, as also his other tables, contains numerous breaks, and the longest run, occurring in Versuchsreihe A is itself of about 480 readings, with one number missing. His analysis more or less ignores the fact that he is therefore dealing in each experimental circumstance with several realizations, and his estimation procedure is not purely statistical, 'true' values being partially used. Neither difficulty arises in Fürth's circumstances.)

$$n = 505, \quad \sum_{i=1}^{n} (X_{i+1} - X_i)^2/n = 1.006$$

μ \ m	D_n 0.666	D_n^* 0.666	$1-\hat{P}_2$ 0.684
$\bar{X}$ 1.592	0.532	0.532	0.503
$\sum_{i=1}^{n}(X_i-\bar{X})^2/n$ 1.505	0.503	0.503	0.476

References

BARTLETT, M. S. (1955) *An Introduction to Stochastic Processes.* Cambridge University Press, Cambridge.

BILLINGSLEY, P. (1961) *Statistical Inference for Markov Processes.* University of Chicago Press, Chicago.

BROWN, B. M. (1971) Martingale central limit theorems. *Ann. Math. Statist.* **42,** 59–66.

CHANDRASEKHAR, S. (1943) Stochastic problems in physics and astronomy. *Rev. Mod. Phys.* **15,** 1–89. (Also in *Selected Papers on Noise and Stochastic Processes*, ed. N. Wax, Dover, New York, 1954.)

FÜRTH, R. (1918) Statistik und Wahrscheinlichkeitsnachwirkung. *Z. Physik.* **19,** 421–426; *ibid.* **20** (1919) 21.

HANNAN, E. J. (1960) *Time Series Analysis.* Methuen, London.

HARRIS, T. E. (1963) *The Theory of Branching Processes.* Springer, Berlin.

HEATHCOTE, C. R. (1966) Corrections and comments on the paper "A branching process allowing immigration". *J. R. Statist. Soc.* B **28,** 213–217.

HEYDE, C. C. (1971) Some central limit analogues for supercritical Galton-Watson processes. *J. Appl. Prob.* **8**, 52–59.

HEYDE, C. C. AND SENETA, E. (1971) Analogues of classical limit theorems for the supercritical Galton-Watson process with immigration. *Math. Biosciences* **11**, 249–259.

LOÈVE, M. (1963) *Probability Theory.* 3rd Ed. Van Nostrand, New York.

NEVEU, J. (1965) *Mathematical Foundations of the Calculus of Probability.* Holden-Day, San Francisco.

PAKES, A. G. (1971) Branching processes with immigration. *J. Appl. Prob.* **8**, 32–42.

ROTHSCHILD, Lord (1953) A new method of measuring the activity of spermatozoa. *J. Exp. Biol.* **30**, 178–199.

SENETA, E. (1969) Functional equations and the Galton-Watson process. *Adv. Appl. Prob.* **1**, 1–42.

SEVAST'YANOV, B. A. (1957) Limit theorems for branching random processes of special form (in Russian). *Teor. Veroyat. Primenen.* **2**, 339–348.

WESTGREN, A. (1916) Die Veränderungsgeschwindigkeit der lokalen Teilchenkonzentration in kolloiden Systemen. (Erste Mitteilung). *Ark. Mat. Astronom. Fys.* Band 11, No. 14, 24 pp. (Zweite Mitteilung: 1918; Band 13, No. 14, 18 pp.).

AN ITERATED LOGATITHM RESULT FOR MARTINGALES AND ITS APPLICATION IN ESTIMATION THEORY FOR AUTOREGRESSIVE PROCESSES

BY

C. C. HEYDE

Reprinted from the

JOURNAL OF APPLIED PROBABILITY

Vol. 10, No. 1, pp. 146–157

March 1973

R. Maller et al. (eds.), *Selected Works of C.C. Heyde*, Selected Works in Probability and Statistics,
DOI 10.1007/978-1-4419-5823-5_32, © Springer Science+Business Media, LLC 2010

J. Appl. Prob. 10, 146–157 (1973)
Printed in Israel

AN ITERATED LOGARITHM RESULT FOR MARTINGALES AND ITS APPLICATION IN ESTIMATION THEORY FOR AUTOREGRESSIVE PROCESSES

C. C. HEYDE, *The Australian National University*

Abstract

The paper begins with an iterated logarithm law of classical Hartman-Wintner form for stationary martingales. This is then used to obtain iterated logarithm results giving information on rates of convergence of estimators of the parameters in a stationary autoregressive process. In the case of an autoregression of small order, detailed rate results for each autocorrelation and for the estimators of all parameters can be obtained. A rate result for the convergence of the sample mean is given in the general case.

ITERATED LOGARITHM LAW; MARTINGALES; TIME SERIES ESTIMATION; AUTOREGRESSIVE PROCESS

1. Results and discussion

In this paper we begin by obtaining the following iterated logarithm law for martingales.

Theorem 1. Let $\{Y_n, n \geqq 0\}$ be a stationary stochastic sequence such that $E(Y_n \mid \mathscr{G}_{n-1}) = 0$ a.s., all $n \geqq 1$, and $EY_0^2 = 1$. Here $\mathscr{G}_n$ is the σ-field generated by Z_i, $i \leqq n$, where $\{\cdots, Z_{-1}, Z_0, Z_1, \cdots\}$ is a stationary stochastic sequence such that $Y_k = \phi(Z_k, Z_{k-1}, \cdots)$, $k \geqq 0$, for some measurable ϕ. We suppose also that $n^{-1}\sum_{i=1}^n E(Y_i^2 \mid \mathscr{G}_{i-1})$ converges almost surely (a.s.) to 1 as $n \to \infty$. Then,

$$\limsup_{n\to\infty} \sum_{i=1}^{n} Y_i/(2n \log\log n)^{\frac{1}{2}} = 1 \text{ a.s.},$$

$$\liminf_{n\to\infty} \sum_{i=1}^{n} Y_i/(2n \log\log n)^{\frac{1}{2}} = -1 \text{ a.s.}$$

This result is an extension of that of Stout [3] who has taken $Z_i = Y_i$ and assumed that $\{Y_i\}$ is a stationary *ergodic* martingale difference sequence ($n^{-1}\sum_{i=1}^n E(Y_i^2 \mid \mathscr{G}_{i-1}) \overset{\text{a.s.}}{\to} 1$ then follows as a consequence of the ergodic theorem). Our result seems more useful for applications; in particular the essentially uncheckable ergodicity assumption is replaced by something rather more concrete.

Received 13 December 1971.

The point of introducing the Z's will be clear from some of our applications where a conditioning on the σ-fields generated by the Y's themselves is not at all convenient.

We go on to apply Theorem 1 in a time series context. We consider a stationary autoregression $\{x(n)\}$ of the form

$$\sum_{k=0}^{q} \beta(k)\{x(n-k)-\mu\} = \varepsilon(n); \qquad \beta(0)=1; \tag{1}$$
$$E\varepsilon(n)=0,\ E\{\varepsilon(m)\varepsilon(n)\}=0, \qquad m \neq n,$$

where we assume that

$$\sum_{k=0}^{q} \beta(k)z^k \neq 0, \qquad |z| \leqq 1. \tag{2}$$

We shall also suppose that $\{\varepsilon(n)\}$ is a stationary stochastic sequence with $E\varepsilon^2(1)=\sigma^2$ satisfying

$$E(\varepsilon(n)\,|\,\mathscr{F}_{n-1}) = 0 \text{ a.s., all } n, \tag{3}$$

where $\mathscr{F}_n$ is the σ-field generated by $\varepsilon(m)$, $m \leqq n$ and

$$E(\varepsilon^2(n)\,|\,\mathscr{F}_{n-1}) = \sigma^2 \text{ a.s., all } n. \tag{4}$$

In this context, $\mathscr{F}_n$ is also the σ-field generated by $x(m)$, $m \leqq n$, and

$$\varepsilon(n) = x(n) - E(x(n)\,|\,\mathscr{F}_{n-1})$$

so that (3) is equivalent to assuming that $E(x(n)\,|\,\mathscr{F}_{n-1})$ is the best linear predictor and the best linear predictor is the best predictor (both in the least squares sense). We are also assuming that $E(\varepsilon^2(n)\,|\,\mathscr{F}_{n-1})$, the prediction variance, given the past, is a.s. constant.

Let $x(1), x(2),\cdots,x(N)$ be a sample of N consecutive observations on the process $\{x(n)\}$ and define the autocovariances

$$c(j) = N^{-1}\sum_{n=1}^{N-j}(x(n)-\bar{x})(x(n+j)-\bar{x}), \qquad j \geqq 0,$$
$$c(-j) = c(j),$$

where $\bar{x} = N^{-1}\sum_{j=1}^{N} x(j)$. It follows from more general results given in Hannan and Heyde [2] that the $\beta(j)$ and σ^2 can be estimated through

$$\sum_{j=0}^{q} \hat{\beta}(j)c(k-j) = \delta_{0k}\hat{\sigma}^2, \qquad k = 0,1,\cdots,q,$$

where δ_{0k} is the Kronecker delta, while μ can be estimated by $\bar{x}$ (see also Hannan [1], Chapter VI). In fact,

(5) $$\lim_{N\to\infty} \sum_{j=0}^{q} \beta(j)c(k-j) = \delta_{0k}\sigma^2 \text{ a.s.}, \qquad k = 0,1,\cdots,q,$$

and if $r(j) = c(j)/c(0)$ are the autocorrelations,

(6) $$\lim_{N\to\infty} \sum_{j=0}^{q} \beta(j)r(k-j) = 0 \text{ a.s.}, \qquad k = 1,2,\cdots,q.$$

Also,

(7) $$\lim_{N\to\infty} \bar{x} = \mu \text{ a.s.}$$

The results of [2] extend the classical theory in which the $\varepsilon(n)$ are assumed to be independent and identically distributed (i.i.d.).

We shall here obtain iterated logarithm results which illuminate the Relations (5), (6) and (7) in the sense of providing detailed information on the rate of a.s. convergence. The results are new even under the additional assumption that the $\varepsilon(n)$ are i.i.d. First we deal with (6) in the following theorem.

Theorem 2. Suppose $E\varepsilon^4(n) < \infty$. For any k, $1 \leqq k \leqq q$, we have

$$\limsup_{N\to\infty} N^{\frac{1}{2}}(2\log\log N)^{-\frac{1}{2}} \sum_{j=0}^{q} \beta(j)r(k-j) = c^{-\frac{1}{2}} \text{ a.s.},$$

$$\liminf_{N\to\infty} N^{\frac{1}{2}}(2\log\log N)^{-\frac{1}{2}} \sum_{j=0}^{q} \beta(j)r(k-j) = -c^{-\frac{1}{2}} \text{ a.s.},$$

where $c = \sum_{j=0}^{\infty} \alpha^2(j)$, the α's being defined by $\sum_{j=0}^{q} \beta(j)\alpha(n-j) = \delta_{0n}$, $\alpha(k) = 0$, $k < 0$. Thus, for $1 \leqq k \leqq q$,

$$\sum_{j=0}^{q} \beta(j)r(k-j) = \zeta_k(N)c^{-\frac{1}{2}}N^{-\frac{1}{2}}(2\log\log N)^{\frac{1}{2}},$$

where $\zeta_k(N)$ has its set of limit points confined to $[-1,1]$ with $\limsup_{N\to\infty} \zeta_k(N) = 1$ a.s. and $\liminf_{N\to\infty} \zeta_k(N) = -1$ a.s.

Remark 1. It should be noted that, using Theorem 2 of [2], $N^{\frac{1}{2}} \sum_{j=0}^{q} \beta(j)r(k-j)$ converges in distribution as $N \to \infty$ to a normal law with zero mean and specified variance $v(k)$, $1 \leqq k \leqq q$.

Remark 2. In the case of a first order autoregression ($q = 1$), Theorem 2 gives

$$r(1) = \rho + \zeta_1(N)(1-\rho^2)^{\frac{1}{2}}N^{-\frac{1}{2}}(2\log\log N)^{\frac{1}{2}},$$

where $\beta(1) = -\rho$ since in this case $\alpha(n) = \rho^n$, $n \geqq 0$. For $q \geqq 2$, Theorem 2 does not generally give such precise results on the individual autocorrelations. For example, for $q = 2$ we find

$$r(1) = -\frac{\beta(1)}{1+\beta(2)} + \zeta_1(N)\frac{c^{-\frac{1}{2}}}{1+\beta(2)} N^{-\frac{1}{2}}(2\log\log N)^{\frac{1}{2}}.$$

However, the best we can do for $r(2)$ is

$$r(2) = \frac{\beta^2(1)}{1+\beta(2)} - \beta(2) + \eta(N)\, c^{-\frac{1}{2}} N^{-\frac{1}{2}}(2\log\log N)^{\frac{1}{2}},$$

where

$$\limsup_{N\to\infty} \eta(N) \leqq 1 + \left|\frac{\beta(1)}{1+\beta(2)}\right| \text{ a.s.}$$

and

$$\liminf_{N\to\infty} \eta(N) \geqq -1 - \left|\frac{\beta(1)}{1+\beta(2)}\right| \text{ a.s.}$$

Remark 3. Since the $\hat{\beta}$'s are linear combinations of the r's we can, at least for small q, also use Theorem 2 to find convenient rate results for the convergence of $\hat{\beta}(j)$ to $\beta(j)$, $1 \leqq j \leqq q$.

In Theorem 2 we have dealt with the System (6). As a by-product of the proof we shall obtain for $1 \leqq k \leqq q$,

$$\limsup_{N\to\infty} N^{\frac{1}{2}}(2\log\log N)^{-\frac{1}{2}} \sum_{j=0}^{q} \beta(j)c(k-j) = \sigma^2 c^{\frac{1}{2}} \text{ a.s.},$$

$$\liminf_{N\to\infty} N^{\frac{1}{2}}(2\log\log N)^{-\frac{1}{2}} \sum_{j=0}^{q} \beta(j)c(k-j) = -\sigma^2 c^{\frac{1}{2}} \text{ a.s.},$$

(see (23)) which deals with the System (5) save for the case $k = 0$. This case we treat in the next theorem. Here we find it convenient, in order to avoid unpleasant extra assumptions, to specialize to the case where the $\varepsilon(n)$ are independent.

Theorem 3. Suppose that in addition to our basic assumptions the $\varepsilon(n)$ are independent with $E\varepsilon^4(n) = \mu_4 < \infty$. Then,

$$\sum_{j=0}^{q} \beta(j)c(j) = \sigma^2 + [\mu_4 + \sigma^4(c-2)]^{\frac{1}{2}}\eta(N)N^{-\frac{1}{2}}(2\log\log N)^{\frac{1}{2}},$$

where $\eta(N)$ has its set of limit points confined to $[-1, 1]$ with $\limsup_{N\to\infty} \eta(N) = 1$ a.s. and $\liminf_{N\to\infty} \eta(N) = -1$ a.s.

Finally, we deal with (7) in the following theorem.

Theorem 4. We have

$$\bar{x} = \mu + \sigma\left(\sum_{k=0}^{q} \beta(k)\right)^{-1} \delta(N)\, N^{-\frac{1}{2}}(2\log\log N)^{\frac{1}{2}},$$

where $\delta(N)$ has its set of limit points confined to $[-1, 1]$ with $\limsup_{N\to\infty}\delta(N) = 1$ a.s. and $\liminf_{N\to\infty} \delta(N) = -1$ a.s.

2. Proofs

We require first the following lemma which provides a strong law for stationary martingales.

Lemma. Let $\{U_n, n \geqq 0\}$ be a stationary sequence of random variables and $\{\mathscr{G}_n, n \geqq 0\}$ an increasing sequence of σ-fields such that U_n is measurable with respect to $\mathscr{G}_n$ and $\{E(U_n | \mathscr{G}_{n-1}), n \geqq 1\}$ is a stationary sequence. If $E|U_n| < \infty$, then

$$n^{-1} \sum_{k=1}^{n} [U_k - E(U_k | \mathscr{G}_{k-1})] \overset{\text{a.s.}}{\to} 0$$

as $n \to \infty$.

Proof. First we show that

$$n^{-1} \sum_{k=1}^{n} [U_k - E(U_k | \mathscr{G}_{k-1})] \overset{p}{\to} 0 \tag{8}$$

as $n \to \infty$ ("p" denotes convergence in probability). Let $U_{kn} = U_k$ or 0 according as $|U_k| \leqq n$ or $|U_k| > n$. Then

$$\begin{aligned} &P\left(n^{-1} \sum_{k=1}^{n} [U_k - E(U_k | \mathscr{G}_{k-1})] \neq n^{-1} \sum_{k=1}^{n} [U_{kn} - E(U_k | \mathscr{G}_{k-1})]\right) \\ &\qquad \leqq \sum_{k=1}^{n} P(U_k \neq U_{kn}) = n\, P(|U_k| > n) \to 0 \end{aligned} \tag{9}$$

as $n \to \infty$ since $E|U_k| < \infty$. Furthermore,

$$\begin{aligned} E&\left|n^{-1} \sum_{k=1}^{n} \{E(U_k | \mathscr{G}_{k-1}) - E(U_{kn} | \mathscr{G}_{k-1})\}\right| \\ &\leqq n^{-1} \sum_{k=1}^{n} E\{E(|U_k| I(|U_k| > n) | \mathscr{G}_{k-1})\} \\ &= E(|U_k| I(|U_k| > n)) \to 0 \end{aligned}$$

as $n \to \infty$ since $E|U_k| < \infty$ (I denotes the indicator function). We then conclude, using Markov's inequality, that

$$n^{-1} \sum_{k=1}^{n} \{E(U_k | \mathscr{G}_{k-1}) - E(U_{kn} | \mathscr{G}_{k-1})\} \overset{p}{\to} 0. \tag{10}$$

Next, using Chebyshev's inequality and the orthogonal increments property of the martingale,

$$P\left(n^{-1}\left|\sum_{k=1}^{n}[U_{kn}-E(U_{kn}|\mathscr{G}_{k-1})]\right|>\varepsilon\right)$$

$$\begin{aligned}(11)\qquad &\leqq n^{-2}\varepsilon^{-2}\sum_{k=1}^{n}E[U_{kn}-E(U_{kn}|\mathscr{G}_{k-1})]^2\\ &\leqq n^{-1}\varepsilon^{-2}E(U_{kn})^2\\ &\leqq 2n^{-1}\varepsilon^{-2}\int_0^n xP(|U_k|>x)\,dx\to 0\end{aligned}$$

as $n\to\infty$ since $x\,P(|U_k|>x)\to 0$ as $x\to\infty$ by virtue of $E|U_k|<\infty$. Thus, from (9), (10) and (11) we obtain (8).

We now make two applications of the ergodic theorem and obtain

$$n^{-1}\sum_{k=1}^{n}U_k\overset{\text{a.s.}}{\to}E(U_1|\mathscr{G}),$$

$$n^{-1}\sum_{k=1}^{n}E(U_k|\mathscr{G}_{k-1})\overset{\text{a.s.}}{\to}E(E(U_1|\mathscr{G}_0)|\mathscr{H}),$$

where $\mathscr{G}$, $\mathscr{H}$ are the respective σ-fields of invariant events. Thus,

$$(12)\qquad n^{-1}\sum_{k=1}^{n}[U_k-E(U_k|\mathscr{G}_{k-1})]\overset{\text{a.s.}}{\to}E(U_1|\mathscr{G})-E(E(U_1|\mathscr{G}_0)|\mathscr{H}),$$

and the result of the lemma follows upon identifying in (8) and (12).

Proof of Theorem 1. We retain the notation of [3] but replace $\mathscr{F}_i$ by our $\mathscr{G}_i(\supseteq\mathscr{F}_i)$. All that we need to do is to show that $n^{-1}\sum_{i=1}^{n}E[(Z_i')^2|\mathscr{G}_{i-1}]\overset{\text{a.s.}}{\to}1$ continues to hold and then the remainder of the proof follows as in [3].

Now suppose that $Y_0(\omega)$ is a random variable on $(\Omega,\mathscr{F},P)$ and let T be the measure preserving transformation of Ω onto itself such that $Y_n(\omega)=Y_0(T^n\omega)$. Then,

$$\begin{aligned}Y_0''(T^n\omega)&=Y_0(T^n\omega)I(|Y_0(T^n\omega)|\leqq M)+MI(Y_0(T^n\omega)>M)-MI(Y_0(T^n\omega)<-M)\\ &=Y_n(\omega)I(|Y_n(\omega)|\leqq M)+MI(Y_n(\omega)>M)-MI(Y_n(\omega)<-M)\\ &=Y_n''(\omega),\end{aligned}$$

and from the ergodic theorem

$$\begin{aligned}\lim_{n\to\infty}n^{-1}\sum_{i=1}^{n}(Y_0''(T^i\omega))^2&=\lim_{n\to\infty}n^{-1}\sum_{i=1}^{n}(Y_i''(\omega))^2\\ &=E[(Y_0'')^2|\mathscr{H}]\ \text{a.s.},\end{aligned}$$

where $\mathscr{H}$ is the σ-field of invariant events determined by T, $\mathscr{F}$. Also, in view of our stationarity assumptions we can apply the lemma to give

$$n^{-1} \sum_{i=1}^{n} \{(Y_i'')^2 - E[(Y_i'')^2 | \mathcal{G}_{i-1}]\} \xrightarrow{\text{a.s.}} 0,$$

and hence

$$\text{(13)} \qquad n^{-1} \sum_{i=1}^{n} E[(Y_i'')^2 | \mathcal{G}_{i-1}] \xrightarrow{\text{a.s.}} E[(Y_0'')^2 | \mathcal{H}].$$

Next, we note that for $b_i \geqq M$, $|Y_i| \geqq |Y_i'| \geqq |Y_i''|$ so that

$$E[Y_i^2 | \mathcal{G}_{i-1}] \geqq E[(Y_i')^2 | \mathcal{G}_{i-1}] \geqq E[(Y_i'')^2 | \mathcal{G}_{i-1}] \text{ a.s.}$$

and hence, making use of (13),

$$\begin{aligned} 1 = \lim_{n\to\infty} n^{-1} \sum_{i=1}^{n} E[Y_i^2 | \mathcal{G}_{i-1}] &\geqq \limsup_{n\to\infty} n^{-1} \sum_{i=1}^{n} E[(Y_i')^2 | \mathcal{G}_{i-1}] \\ &\geqq \liminf_{n\to\infty} n^{-1} \sum_{i=1}^{n} E[(Y_i')^2 | \mathcal{G}_{i-1}] \\ &\geqq \lim_{n\to\infty} n^{-1} \sum_{i=1}^{n} E[(Y_i'')^2 | \mathcal{G}_{i-1}] \\ &= E[(Y_0'')^2 | \mathcal{H}] \text{ a.s.} \end{aligned} \qquad \text{(14)}$$

Also, as $M \uparrow \infty$, $|Y_0''| \uparrow |Y_0|$ so that $E[(Y_0'')^2 | \mathcal{H}] \xrightarrow{\text{a.s.}} E[Y_0^2 | \mathcal{H}]$.

But, an application of the lemma gives

$$n^{-1} \sum_{i=1}^{n} [Y_i^2 - E(Y_i^2 | \mathcal{G}_{i-1})] \xrightarrow{\text{a.s.}} 0,$$

while $n^{-1} \sum_{i=1}^{n} E(Y_i^2 | \mathcal{G}_{i-1}) \xrightarrow{\text{a.s.}} 1$ by hypothesis and hence

$$\lim_{n\to\infty} n^{-1} \sum_{i=1}^{n} Y_i^2 = E[Y_0^2 | \mathcal{H}] = 1 \text{ a.s.}$$

Thus, letting $M \to \infty$ in (14) gives

$$\text{(15)} \qquad n^{-1} \sum_{i=1}^{n} E[(Y_i')^2 | \mathcal{G}_{i-1}] \xrightarrow{\text{a.s.}} 1.$$

Now take $\{b_n\}$ as chosen in [3] and observe that

$$E(Y_i' | \mathcal{G}_{i-1}) = -E(Y_i - Y_i' | \mathcal{G}_{i-1}),$$

while $|Y_i'| \leqq b_i$ so that

$$E[E(Y_i' | \mathcal{G}_{i-1})]^2 \leqq b_i E|Y_i - Y_i'|.$$

Then, $\sum i^{-1}[E(Y_i' | \mathcal{G}_{i-1})]^2 < \infty$ a.s. since

$$\sum i^{-1}E[E(Y_i'\,|\,\mathscr{G}_{i-1})]^2 \leqq \sum b_i\, i^{-1}E\left|\,Y_i - Y_i'\,\right| = \sum K_i(i\log_2 i)^{-\frac{1}{2}}E\left|\,Y_i - Y_i'\right| < \infty,$$

it having been shown in [3] that $\sum(i\log_2 i)^{-\frac{1}{2}}E\left|\,Y - Y_i'\right| < \infty$ while $K_i \to 0$ as $i \to \infty$. Thus, using the Kronecker lemma,

$$n^{-1}\sum_{i=1}^{n}[E(Y_i'\,|\,\mathscr{G}_{i-1})]^2 \overset{\text{a.s.}}{\to} 0$$

and hence, making use of (15),

$$n^{-1}\sum_{i=1}^{n}E[(Z_i')^2\,|\,\mathscr{G}_{i-1}] = n^{-1}\sum_{i=1}^{n}\{E[(Y_i')^2|\,\mathscr{G}_{i-1}] - [E(Y_i'\,|\,\mathscr{G}_{i-1})]^2\}\overset{\text{a.s.}}{\to} 1.$$

The remainder of the proof is then completed as in [3].

We shall next prove Theorem 4 as it is convenient to use it in the proof of Theorems 2 and 3.

Proof of Theorem 4. First note that an application of Theorem 1 gives

$$\limsup_{N\to\infty}(2N\log\log N)^{-\frac{1}{2}}\sum_{n=1}^{N}\varepsilon(n) = \sigma \text{ a.s.}, \tag{16}$$

with a corresponding result for lim inf.

Now,

$$\begin{aligned}\sum_{n=1}^{N}\varepsilon(n) &= \sum_{n=1}^{N}\sum_{k=0}^{q}\beta(k)\{x(n-k)-\mu\}\\ &= \sum_{k=0}^{q}\beta(k)\left\{N\bar{x} - N\mu + \sum_{n=1}^{k}x(n-k) - \sum_{n=N+1}^{N+k}x(n-k)\right\}.\end{aligned} \tag{17}$$

Also, $E(\sum_{n=1}^{k}x(n-k))^2 < \infty$ and the $R_N = \sum_{n=N+1}^{N+k}x(n-k)$ are identically distributed (as R, say) with $ER^2 < \infty$. Thus,

$$\sum_N P\left(\left|\sum_{n=1}^{k}x(n-k)\right| \geqq \varepsilon(N\log\log N)^{\frac{1}{2}}\right)$$

$$\leqq \sum_N P\left(\left|\sum_{n=1}^{k}x(n-k)\right| \geqq \varepsilon N^{\frac{1}{2}}\right) < \infty,$$

and

$$\sum_N P(|R_N| \geqq \varepsilon(N\log\log N)^{\frac{1}{2}}) \leqq \sum_N P(|R| \geqq \varepsilon N^{\frac{1}{2}}) < \infty$$

for any $\varepsilon > 0$. Hence, from the Borel-Cantelli lemma,

$$\left(\sum_{n=1}^{k}x(n-k) - R_N\right)/(N\log\log N)^{\frac{1}{2}} \overset{\text{a.s.}}{\to} 0$$

as $N \to \infty$ and the result of the theorem follows from (16) and (17).

Proof of Theorem 2. Write $y(n) = x(n) - \mu$ and observe that for $k \geqq 1$,

$$
\begin{aligned}
& y(n)y(n-k) - E(y(n)y(n-k) \mid \mathscr{F}_{n-1}) \\
(18) \qquad & = y(n)y(n-k) + y(n-k) \sum_{j=1}^{q} \beta(j)y(n-j) \\
& = y(n-k) \sum_{j=0}^{q} \beta(j)y(n-j) = y(n-k)\varepsilon(n),
\end{aligned}
$$

so that $\{\sum_{n=k+1}^{N} y(n-k)\varepsilon(n),\ \mathscr{F}_N,\ N > k\}$ is a zero mean martingale with stationary increments. Also, in view of (2), $y(n)$ is representable in the form

$$y(n) = \sum_{j=0}^{\infty} \alpha(j)\varepsilon(n-j),$$

where $\sum_{j=0}^{q} \beta(j)\alpha(n-j) = \delta_{0n}$; $\alpha(k) = 0$, $k < 0$ (e.g., Hannan [1], Chapter 1) and hence

$$
\begin{aligned}
E\{y^2(n-k)\varepsilon^2(n)\} & = E\{y^2(n-k)E(\varepsilon^2(n) \mid \mathscr{F}_{n-1})\} \\
& = \sigma^2 E\, y^2(n-k) \\
& = \sigma^4 \sum_{j=0}^{\infty} \alpha^2(j)
\end{aligned}
$$

since $E\{\varepsilon(n-k-i)\varepsilon(n-k-j)\} = \sigma^2\delta_{ij}$. Furthermore,

$$
\begin{aligned}
N^{-1} \sum_{n=k+1}^{N} y^2(n-k)E(\varepsilon^2(n) \mid \mathscr{F}_{n-1}) & = \sigma^2 N^{-1} \sum_{n=k+1}^{N} y^2(n-k) \text{ a.s.} \\
& \overset{\text{a.s.}}{\to} \sigma^4 \sum_{j=0}^{\infty} \alpha^2(j),
\end{aligned}
$$

using Theorem 1 of [2]. The conditions of our Theorem 1 are clearly satisfied and

$$
(19) \qquad
\begin{aligned}
& \limsup_{N\to\infty} (2N \log\log N)^{-\frac{1}{2}} \sum_{n=k+1}^{N} y(n-k)\varepsilon(n) = \sigma^2 c^{\frac{1}{2}} \text{ a.s.}, \\
& \liminf_{N\to\infty} (2N \log\log N)^{-\frac{1}{2}} \sum_{n=k+1}^{N} y(n-k)\varepsilon(n) = -\sigma^2 c^{\frac{1}{2}} \text{ a.s.}
\end{aligned}
$$

where $c = \sum_{j=0}^{\infty} \alpha^2(j)$.

Now let

$$c^*(j) = N^{-1} \sum_{n=1}^{N-j} y(n)y(n+j),\ j \geqq 0,\ c^*(-j) = c^*(j).$$

We have

$$\sum_{n=k+1}^{N} y(n-k)\varepsilon(n) = \sum_{j=0}^{q} \beta(j) \sum_{n=k+1}^{N} y(n-k)y(n-j)$$
(20)
$$= N c^*(k) + \sum_{j=1}^{q} \beta(j)\{N c^*(k-j) - R_N\},$$

where $R_N = \sum_{n=N+1}^{N+j} y(n-k)y(n-j)$. Also, since the R_N are identically distributed (as R, say) with $ER^2 < \infty$,

$$\sum_N P(|R_N| \geqq \varepsilon(N \log\log N)^{\frac{1}{2}}) \leqq \sum_N P(|R| \geqq \varepsilon N^{\frac{1}{2}}) < \infty$$

for any $\varepsilon > 0$. Thus, from the Borel-Cantelli lemma,

$$R_N/(N \log\log N)^{\frac{1}{2}} \xrightarrow{\text{a.s.}} 0$$

as $N \to \infty$ and hence, using (19) and (20)

$$\limsup_{N\to\infty} N^{\frac{1}{2}}(2\log\log N)^{-\frac{1}{2}} \sum_{j=0}^{q} \beta(j)c^*(k-j) = \sigma^2 c^{\frac{1}{2}} \text{ a.s.},$$
(21)
$$\liminf_{N\to\infty} N^{\frac{1}{2}}(2\log\log N)^{-\frac{1}{2}} \sum_{j=0}^{q} \beta(j)c^*(k-j) = -\sigma^2 c^{\frac{1}{2}} \text{ a.s.}$$

Next, for any fixed j we find that

$$c^*(j) - c(j) = (\bar{x} - \mu)^2 - N^{-1}\left\{\sum_{n=N-j+1}^{N} x(n) + \sum_{n=1}^{j-1} x(n) - j(\bar{x} + \mu)\right\}(\bar{x} - \mu),$$

and with the aid of Theorem 4 it is easily seen that

(22) $$\lim_{N\to\infty} N^{\frac{1}{2}}(\log\log N)^{-\frac{1}{2}}(c^*(j) - c(j)) = 0 \text{ a.s.},$$

so that using (21) and (22) we obtain

$$\limsup_{N\to\infty} N^{\frac{1}{2}}(2\log\log N)^{-\frac{1}{2}} \sum_{j=0}^{q} \beta(j)c(k-j) = \sigma^2 c^{\frac{1}{2}} \text{ a.s.},$$
(23)
$$\liminf_{N\to\infty} N^{\frac{1}{2}}(2\log\log N)^{-\frac{1}{2}} \sum_{j=0}^{q} \beta(j)c(k-j) = -\sigma^2 c^{\frac{1}{2}} \text{ a.s.}$$

The result of the theorem follows upon noting that, from Theorem 1 of [2], $c(0) \xrightarrow{\text{a.s.}} \sigma^2 c$.

Proof of Theorem 3. First we note that $\{\sum_{n=1}^{N}(y(n)\varepsilon(n) - \sigma^2), \mathscr{F}_N, N \geqq 1\}$ is a zero mean martingale since

$$E(y(n)\varepsilon(n) | \mathscr{F}_{n-1}) = E(\varepsilon^2(n) | \mathscr{F}_{n-1}) + \sum_{j=1}^{\infty} \alpha(j)\varepsilon(n-j)E(\varepsilon(n) | \mathscr{F}_{n-1})$$
$$= \sigma^2.$$

Now,

$$E\{(y(n)\varepsilon(n) - \sigma^2)^2 | \mathscr{F}_{n-1}\}$$
$$= E(y^2(n)\varepsilon^2(n) | \mathscr{F}_{n-1}) - \sigma^4$$
$$= \sum_{j=0}^{\infty} \sum_{k=0}^{\infty} \alpha(j)\alpha(k)E\{\varepsilon(n-j)\varepsilon(n-k)\varepsilon^2(n) | \mathscr{F}_{n-1}\} - \sigma^4$$
$$= \sigma^2 \left\{\sum_{j=1}^{\infty} \alpha(j)\varepsilon(n-j)\right\}^2 + 2\sum_{j=1}^{\infty} \alpha(j)\varepsilon(n-j)E\{\varepsilon^3(n)\} - \sigma^4 + \mu_4$$
$$= \sigma^2\{y(n) - \varepsilon(n)\}^2 + 2E\{\varepsilon^3(1)\}\{y(n) - \varepsilon(n)\} - \sigma^4 + \mu_4,$$

$$N^{-1} \sum_{n=1}^{N} E\{(y(n)\varepsilon(n) - \sigma^2)^2 | \mathscr{F}_{n-1}\}$$

$$= \mu_4 - \sigma^4 + \sigma^2 N^{-1} \sum_{n=1}^{N} \{y(n) - \varepsilon(n)\}^2 \tag{24}$$
$$+ 2E\{\varepsilon^3(1)\}N^{-1} \sum_{n=1}^{N} \{y(n) - \varepsilon(n)\}.$$

Also, $\{y(n)\}$ and $\{E(y^2(n) | \mathscr{F}_{n-1})\}$ are stationary sequences and $E\ y^2(n) < \infty$ so an application of the lemma gives

$$N^{-1} \sum_{n=1}^{N} [y^2(n) - E(y^2(n) | \mathscr{F}_{n-1})] \overset{\text{a.s.}}{\to} 0. \tag{25}$$

Furthermore, as already noted under weaker conditions,

$$N^{-1} \sum_{n=1}^{N} y^2(n) \overset{\text{a.s.}}{\to} \sigma^2 c$$

and thus, from (25),

$$N^{-1} \sum_{n=1}^{N} E(y^2(n) | \mathscr{F}_{n-1}) \overset{\text{a.s.}}{\to} \sigma^2 c. \tag{26}$$

But,

$$E(y^2(n) | \mathscr{F}_{n-1}) = \sum_{j=0}^{\infty} \sum_{k=0}^{\infty} \alpha(j)\alpha(k) E\{\varepsilon(n-j)\varepsilon(n-k) | \mathscr{F}_{n-1}\}$$
$$= \{y(n) - \varepsilon(n)\}^2 + \sigma^2,$$

and hence from (26),

$$N^{-1} \sum_{n=1}^{N} \{y(n) - \varepsilon(n)\}^2 \overset{\text{a.s.}}{\to} \sigma^2(c-1). \tag{27}$$

It then follows from (24) and (27) that

$$N^{-1} \sum_{n=1}^{N} E\{(y(n)\varepsilon(n) - \sigma^2)^2 \,|\, \mathscr{F}_{n-1}\} \overset{\text{a.s.}}{\to} \mu_4 + \sigma^4(c-2),$$

since $N^{-1} \sum_{n=1}^{N} y(n) = \bar{x} - \mu \overset{\text{a.s.}}{\to} 0$ (see Theorem 4) and $N^{-1} \sum_{n=1}^{N} \varepsilon(n) \overset{\text{a.s.}}{\to} 0$ by the strong law of large numbers. It is then easily seen that we can apply Theorem 1 and

$$\limsup_{N\to\infty} (2N \log\log N)^{-\frac{1}{2}} \sum_{n=1}^{N} (y(n)\varepsilon(n) - \sigma^2) = [\mu_4 + \sigma^4(c-2)]^{\frac{1}{2}} \text{ a.s.}$$

with a corresponding result for the lim inf. The result of the theorem then follows using the same type of argument as in the proof of Theorem 2.

References

[1] HANNAN, E. J. (1970) *Multiple Time Series.* Wiley, New York.

[2] HANNAN, E. J. AND HEYDE, C. C. (1972) On limit theorems for quadratic functions of discrete time series. *Ann. Math. Statist.* **43,** 2058–2066.

[3] STOUT, W. F. (1970) The Hartman-Wintner law of the iterated logarithm for martingales. *Ann. Math. Statist.* **41,** 2158–2160.

Z. Wahrscheinlichkeitstheorie verw. Geb. 25, 83–95 (1973)

On the Uniform Metric in the Context of Convergence to Normality

C. C. Heyde

1. Introduction and Results

There are two basic parts to this paper. In the first part we suppose that X_i, $i=1, 2, 3, \ldots$ is a sequence of independent and identically distributed random variables. We write $S_n = \sum_{i=1}^{n} X_i$, $n \geqq 1$, and suppose that the variables are centered so that $EX_i = 0$ if $E|X_i| < \infty$. Let $\{B_n\}$ with $B_n \to \infty$, $(B_{n+1}/B_n) \to 1$ as $n \to \infty$ be a sequence of positive constants and write

$$F_n(x) = P(S_n \leqq B_n x)$$

and

$$\Delta_n = \sup_{-\infty < x < \infty} |F_n(x) - \Phi(x)|$$

where

$$\Phi(x) = \frac{1}{\sqrt{2\pi}} \int_{-\infty}^{x} e^{-\frac{1}{2}u^2} du.$$

We shall establish the following theorem.

Theorem 1. *If*

$$\sum_{n=1}^{\infty} n^{-1} \Delta_n = \sum_{n=1}^{\infty} n^{-1} \sup_{-\infty < x < \infty} |F_n(x) - \Phi(x)| < \infty, \tag{1}$$

then $EX_i^2 < \infty$. That is, the X_i belong to the domain of normal attraction of the normal distribution.

Remarks. (1) Note that in formulating the above theorem it has not been assumed that the X_i belong to the domain of attraction of the normal distribution. This assumption has been made in previous work on the problem, for example in Heyde [7] where it is shown that if $\sum_{n=3}^{\infty} n^{-1} (\log\log n)\, \Delta_n < \infty$ then $EX_i^2 < \infty$.

(2) It is known from results of Friedman, Katz and Koopmans [3] and Heyde [5] that if $EX_i^2 = \sigma^2 < \infty$, then the series

$$\sum_{n=1}^{\infty} n^{-1} \sup_{-\infty < x < \infty} |P(S_n \leqq x \sigma \sqrt{n}) - \Phi(x)|$$

may not converge. In fact, it will converge (Heyde [5]) if and only if

$$EX_i^2 \log(1 + |X_i|) < \infty .$$

R. Maller et al. (eds.), *Selected Works of C.C. Heyde*, Selected Works in Probability and Statistics, DOI 10.1007/978-1-4419-5823-5_33, © Springer Science+Business Media, LLC 2010

However, if the normalization by $\sigma\sqrt{n}$ is replaced by that by $B_n=\sigma_n\sqrt{n}$ where

$$\sigma_n^2=\int_{|x|<\sqrt{n}} x^2\,dP(X_i\leqq x)-\Big[\int_{|x|<\sqrt{n}} x\,dP(X_i\leqq x)\Big]^2$$

then

$$\sum_{n=1}^{\infty} n^{-1}\sup_{-\infty<x<\infty}|P(S_n\leqq B_n x)-\Phi(x)|<\infty.$$

Theorem 1 provides a converse to this last result (which is essentially due to Friedman *et al.* [3]; see also Theorem A of [6]).

(3) Theorem 1 is also of interest in connection with the derivation of laws of iterated logarithm type on the basis of an appropriate rate of convergence to normality as measured by the uniform metric. For details see Section 3 of [7] and Section 5 of [6].

In the second part of this paper we shall be concerned with an inequality of Osipov and Petrov. Our setting is the same as before with a sum of independent and identically distributed random variables X_i, $i=1,2,3,\dots$ which are centered so that $EX_i=0$ if $E|X_i|<\infty$. $F(x)$ is the distribution function of the X_i and $\{C_n\}$ denotes a sequence of positive constants. We shall discuss the general inequality of Osipov and Petrov [10] (see also Feller [2]) which we specialize for the present context to give

$$\begin{aligned}\Delta_n(C_n)&=\sup_x|P(S_n\leqq C_n x)-\Phi(x)|\\ &\leqq nP(|X_i|>\tau_n)+\frac{K_0 n\int_{|x|\leqq\tau_n}|x|^3\,dF(x)}{B_n^3}\\ &\quad+\frac{n\left|\int_{|x|\leqq\tau_n}x\,dF(x)\right|}{B_n\sqrt{2\pi}}+\frac{1}{2\sqrt{2\pi e}}\left|1-\frac{B_n^2}{C_n^2}\right|\max\left(1,\frac{B_n^2}{C_n^2}\right)=\delta_n(C_n)\end{aligned}\tag{2}$$

where B_n is given by

$$B_n^2=n\Big\{\int_{|x|\leqq\tau_n}x^2\,dF(x)-\Big[\int_{|x|\leqq\tau_n}x\,dF(x)\Big]^2\Big\}$$

and we shall choose τ_n as $\sqrt{n}$ if $EX_i^2<\infty$ and as B_n otherwise. K_0 is a universal constant. It will be our object to show in a variety of ways that δ_n is a very good bound for Δ_n. In fact, in many cases of importance we shall show that Δ_n and δ_n have equivalent asymptotic behaviour. This exercise is of special interest since no useful lower bound for Δ_n has been found. We shall obtain the following results.

Theorem 2. $\Delta_n(C_n)\to 0$ *if and only if* $\delta_n(C_n)\to 0$.

Theorem 3. *Let* $0<\delta<1$. *Then, the following four conditions are equivalent:*

(i) $\int_{-\infty}^{\infty} x^2\,dF(x)<\infty$ *and* $\int_{|x|>z} x^2\,dF(x)=O(z^{-\delta})$ *as* $z\to\infty$,

(ii) $\inf_{C_n}\Delta_n(C_n)=O(n^{-\delta/2})$ *as* $n\to\infty$,

(iii) $\Delta_n(B_n)=O(n^{-\delta/2})$ *as* $n\to\infty$,

(iv) $\delta_n(B_n)=O(n^{-\delta/2})$ *as* $n\to\infty$.

Theorem 4. *Let* $0 \leqq \delta < 1$. *Then, the following four conditions are equivalent:*

(i) $E|X_i|^{2+\delta} < \infty$,

(ii) $\sum_1^\infty n^{-1+\delta/2} \inf_{C_n} \Delta_n(C_n) < \infty$,

(iii) $\sum_1^\infty n^{-1+\delta/2} \Delta_n(B_n) < \infty$,

(iv) $\sum_1^\infty n^{-1+\delta/2} \delta_n(B_n) < \infty$.

2. Proof of Theorem 1

Since $\sum n^{-1} \Delta_n < \infty$, it follows that there exists a sequence $\{n_j\}$ of integers such that $\Delta_{n_j} \to 0$ as $j \to \infty$ and hence the X_i belong to the domain of partial attraction of the normal distribution. Our first major task is to show that the X_i in fact belong to the domain of attraction of the normal distribution (i.e. that $\Delta_n \to 0$ as $n \to \infty$, by virtue of Polya's theorem). In order to do this we need the following lemma.

Lemma. *Under the assumptions of the theorem, we can choose a subsequence* $\{n_i\}$ *of the integers such that* $(n_{i+1}/n_i) \to 1$ *and* $\Delta_{n_i} \to 0$ *as* $i \to \infty$.

Proof. Let N denote the set of integers n for which $\Delta_n < [\log(n+1)]^{-1}$. N contains an infinite (countable) number of elements for if not there would be a largest, n_L, and writing $u_n = n\log(n+1)$,

$$\sum_{n=n_L}^\infty n^{-1} \Delta_n \geqq \sum_{n=n_L}^\infty u_n^{-1} = \infty$$

contradicting our assumption. Then, writing $\bar{N}$ for the complement of N with respect to the positive integers, we have

$$\infty > \sum_{n=1}^\infty n^{-1} \Delta_n > \sum_{n \in \bar{N}} n^{-1} \Delta_n \geqq \sum_{n \in \bar{N}} u_n^{-1}$$

so that $\sum_{n \in N} u_n^{-1} = \infty$ since

$$\infty = \sum_{n=1}^\infty u_n^{-1} = \sum_{n \in N} u_n^{-1} + \sum_{n \in \bar{N}} u_n^{-1}.$$

Now let $N = \{m_i,\ i = 1, 2, 3, \ldots;\ m_{i+1} > m_i \text{ each } i\}$ and suppose that

$$\liminf_{i \to \infty} u_{m_i}^{-1} u_{m_{i+1}} > 1.$$

Then, there is an $\varepsilon > 0$ and an integer $I = I(\varepsilon)$ such that for all $i \geqq I$,

$$u_{m_{i+1}} > (1+\varepsilon) u_{m_i}$$

and hence

$$\sum_{i=I}^\infty u_{m_i}^{-1} < u_{m_I}^{-1} \sum_{r=0}^\infty (1+\varepsilon)^{-r} < \infty$$

7*

which contracts $\sum_{n\in N} u_n^{-1} = \sum_{i=1}^{\infty} u_{m_i}^{-1} = \infty$. We therefore must have

$$\liminf_{i\to\infty} \frac{u_{m_{i+1}}}{u_{m_i}} = \liminf_{i\to\infty} \frac{m_{i+1}\log(1+m_{i+1})}{m_i\log(1+m_i)} = 1.$$

This implies

$$\liminf_{i\to\infty} m_{i+1}\, m_i^{-1} = 1 \tag{3}$$

for, since $m_{i+1} > m_i$,

$$1 = \liminf_{i\to\infty} \frac{m_{i+1}\log(1+m_{i+1})}{m_i\log(1+m_i)} \geqq \liminf_{i\to\infty} \frac{m_{i+1}}{m_i} \geqq 1.$$

By virtue of (3), the required subsequence $\{n_i\}$ can be obtained by choosing a subsequence from N for which $(n_{i+1}/n_i) \to 1$ holds.

We now resume the proof of the theorem. Let $F(x) = P(X_i \leqq x)$ and suppose that $f(t)$ is the corresponding characteristic function. Firstly we shall prove the theorem for symmetric random variables (in which case $f(t)$ is real valued and symmetric).

Since $f(t)$ is continuous and $f(0)=1$, there is some interval $[-\eta, \eta]$ in which $f(t)$ may be written as $\exp\{-A(t)\}$. Then, choosing the $\{n_i\}$ so that $(n_{i+1}/n_i) \to 1$ and $\Delta_{n_i} \to 0$ as $i \to \infty$ (as can be done, according to the lemma) we have

$$\left[f\left(\frac{t}{B_{n_i}}\right)\right]^{n_i} = \exp\left\{-n_i A\left(\frac{t}{B_{n_i}}\right)\right\} \to e^{-\frac{1}{2}t^2},$$

so that

$$n_i A(t B_{n_i}^{-1}) \to \tfrac{1}{2} t^2 \tag{4}$$

as $i \to \infty$.

Now for fixed u, $0 < u < \eta$, let $n(u) = \min[n_i : B_{n_i}^{-1} \leqq u]$. Then, $B_{n(u)}^{-1} \leqq u < B_{n(u)-1}^{-1}$ and since A is continuous, $A(tu)$ lies between $A(t B_{n(u)}^{-1})$ and $A(t B_{n(u)-1}^{-1})$ for $0 < t \leqq 1$. But, using (4),

$$\frac{A(t B_{n(u)}^{-1})}{A(B_{n(u)-1}^{-1})} \sim t^2 \frac{n(u)-1}{n(u)} \sim t^2$$

as $u \to 0$ and we have

$$\frac{A(tu)}{A(u)} \to t^2$$

as $u \to 0$. Thus, A is regularly varying with exponent 2 at zero ($A(t)$ is of course symmetric in t in view of the assumed symmetry of $f(t)$) and we can write

$$A(t) = \tfrac{1}{2} t^2 H(t) \tag{5}$$

where H is slowly varying at zero. This implies that the X_i belong to the domain of attraction of the normal distribution rather than just its domain of partial attraction (Ibragimov and Linnik [9], Chapter 2; Ibragimov [8], Lemma 2.1).

Now introduce a function $B(t)$ defined for $z > 0$ by

$$B(t) = \begin{cases} (z-t)\, t\, e^{\frac{1}{2}t^2}, & 0 \leqq t \leqq z, \\ 0, & \text{otherwise}, \end{cases}$$

and let $\hat{B}(x)$ denote its Fourier transform,

$$\hat{B}(x)=\int_0^z e^{itx}B(t)\,dt.$$

From two integrations by parts it is readily seen that $|\hat{B}(x)|=O(|x|^{-2})$ as $|x|\to\infty$. Now write $f_n(t)$ for the characteristic function corresponding to $F_n(x)$. Then, noting that $F_n(x)-\Phi(x)$ is integrable and integrating by parts in the equation

$$f_n(t)-e^{-\frac{1}{2}t^2}=\int_{-\infty}^{\infty} e^{itx}\,d[F_n(x)-\Phi(x)]$$

we obtain

$$-\frac{f_n(t)-e^{-\frac{1}{2}t^2}}{it}=\int_{-\infty}^{\infty} e^{itx}[F_n(x)-\Phi(x)]\,dx. \tag{6}$$

Consequently, using Parseval's identity on the two pairs of Fourier transforms, we obtain

$$\begin{aligned} i^{-1}\int_{-\infty}^{\infty} t^{-1}[f_n(t)-e^{-\frac{1}{2}t^2}]\,B(t)\,dt&=i^{-1}\int_0^z t^{-1}[f_n(t)-e^{-\frac{1}{2}t^2}]\,B(t)\,dt\\ &=\int_{-\infty}^{\infty}[F_n(x)-\Phi(x)]\,\hat{B}(x)\,dx. \end{aligned} \tag{7}$$

Thus, from (1) and (7),

$$\begin{aligned} &\sum_1^{\infty} n^{-1}\left|\int_0^z (z-t)(f_n(t)-e^{-\frac{1}{2}t^2})\,e^{\frac{1}{2}t^2}\,dt\right|\\ &=\sum_1^{\infty} n^{-1}\left|\int_{-\infty}^{\infty}[F_n(x)-\Phi(x)]\,\hat{B}(x)\,dx\right|\\ &\leqq\sum_1^{\infty} n^{-1}\,\Delta_n\int_{-\infty}^{\infty}|\hat{B}(x)|\,dx<\infty. \end{aligned} \tag{8}$$

Now, by (5), $f(t)$ is representable in the form

$$f(t)=\exp\{-\tfrac{1}{2}t^2 H(t)\},$$

where $H(t)$ is a slowly varying function as $t\to 0$ and moreover, by Lemma 2.1 of [8],

$$H(t)\sim\int_{|xt|\leqq 1} x^2\,dF(x)$$

as $t\to 0$. Then

$$f_n(t)=\left[f\left(\frac{t}{B_n}\right)\right]^n=\exp\left\{-\frac{nt^2}{2B_n^2}H\left(\frac{t}{B_n}\right)\right\},$$

so that (8) gives

$$\sum_1^{\infty} n^{-1}\left|\int_0^z (z-t)\left\{\exp\left[-\frac{t^2}{2}\left(\frac{n}{B_n^2}H\left(\frac{t}{B_n}\right)-1\right)\right]-1\right\}dt\right|<\infty. \tag{9}$$

However, as $n\to\infty$,

$$\exp\left[-\frac{t^2}{2}\left(\frac{n}{B_n^2}H\left(\frac{t}{B_n}\right)-1\right)\right]-1=\frac{t^2}{2}\left[1-\frac{n}{B_n^2}H\left(\frac{t}{B_n}\right)\right]+R_n(t) \tag{10}$$

where

$$R_n(t)=\sum_{r=2}^{\infty}\left[1-\frac{n}{B_n^2}H\left(\frac{t}{B_n}\right)\right]^r\frac{t^{2r}}{2^r r!}$$

and for n sufficiently large and $0\leqq t\leqq z$,

$$\begin{aligned}|R_n(t)|&\leqq t^4\left[1-\frac{n}{B_n^2}H\left(\frac{t}{B_n}\right)\right]^2\sum_{r=2}^{\infty}\frac{z^{2(r-1)}}{2^r r!}\\&\leqq z^{-2}e^{\frac{1}{2}z^2}t^4\left[1-\frac{n}{B_n^2}H\left(\frac{t}{B_n}\right)\right]^2.\end{aligned} \tag{11}$$

Now, $F_n(x)-\Phi(x)$ obviously belongs to $L_2(-\infty,\infty)$. In fact, $\int_{-\infty}^{\infty}|F_n(x)-\Phi(x)|^\alpha dx$ exists and $\to 0$ as $n\to\infty$ for $\alpha>\frac{1}{2}$ (see Ibragimov and Linnik [9], Theorem 5.2.1, p. 172). It then follows from the Parseval identity that

$$\int_{-\infty}^{\infty}[F_n(x)-\Phi(x)]^2dx=\frac{1}{2\pi}\int_{-\infty}^{\infty}t^{-2}[f_n(t)-e^{-\frac{1}{2}t^2}]^2dt. \tag{12}$$

Furthermore,

$$\sum_1^\infty n^{-1}\int_{-\infty}^{\infty}[F_n(x)-\Phi(x)]^2dx\leqq\sum_1^\infty n^{-1}\sup_x|F_n(x)-\Phi(x)|\int_{-\infty}^{\infty}|F_n(x)-\Phi(x)|\,dx<\infty$$

in view of (1) so that from (12),

$$\sum_1^\infty n^{-1}\int_{-\infty}^{\infty}t^{-2}[f_n(t)-e^{-\frac{1}{2}t^2}]^2dt<\infty$$

and in particular

$$\sum_1^\infty n^{-1}\int_0^z t^{-2}[f_n(t)-e^{-\frac{1}{2}t^2}]^2dt<\infty. \tag{13}$$

But, as $n\to\infty$,

$$\left|\frac{f_n(t)-e^{-\frac{1}{2}t^2}}{t}\right|=\frac{1}{2}e^{-\frac{1}{2}t^2}|t|\left|1-\frac{n}{B_n^2}H\left(\frac{t}{B_n}\right)\right|(1+o(1))$$

so that (13) gives

$$\sum_1^\infty n^{-1}\int_0^z e^{-t^2}t^2\left[1-\frac{n}{B_n^2}H\left(\frac{t}{B_n}\right)\right]^2dt<\infty.$$

We then deduce, using (11), that

$$\sum_1^\infty n^{-1}\int_0^z(z-t)|R_n(t)|\,dt<\infty$$

and, from (9) and (10),

$$\sum_1^\infty n^{-1}\left|\int_0^z (z-t)\,t^2\left[1-\frac{n}{B_n^2}H\left(\frac{t}{B_n}\right)\right]dt\right|<\infty.$$

Now make the transformation $t=zx$; we obtain

$$\sum_1^\infty n^{-1}\left|\int_0^1 (1-x)\,x^2\left[1-\frac{n}{B_n^2}H\left(\frac{zx}{B_n}\right)\right]dx\right|<\infty. \tag{14}$$

Therefore, from (14),

$$\begin{aligned}\sum_1^\infty \frac{1}{B_n^2}&\left|\int_0^1 (1-x)\,x^2\left[H\left(\frac{x}{B_n}\right)-H\left(\frac{2x}{B_n}\right)\right]dx\right|\\ &\leqq\sum_1^\infty n^{-1}\left|\int_0^1 (1-x)\,x^2\left[1-\frac{n}{B_n^2}H\left(\frac{2x}{B_n}\right)\right]dx\right|\\ &\quad+\sum_1^\infty n^{-1}\left|\int_0^1 (1-x)\,x^2\left[1-\frac{n}{B_n^2}H\left(\frac{x}{B_n}\right)\right]dx\right|<\infty,\end{aligned}$$

or equivalently,

$$\sum_1^\infty\left|\int_0^1 (1-x)\,[\log f(2B_n^{-1}x)-4\log f(B_n^{-1}x)]\,dx\right|<\infty. \tag{15}$$

Now,

$$\begin{aligned}&\log f(2B_n^{-1}x)-4\log f(B_n^{-1}x)\\ &\quad=\log[1-\{1-f(2B_n^{-1}x)\}]-4\log[1-\{1-f(B_n^{-1}x)\}]\\ &\quad=-\{1-f(2B_n^{-1}x)\}+4\{1-f(B_n^{-1}x)\}+C_n(x)\end{aligned} \tag{16}$$

where

$$C_n(x)=\sum_{r=2}^\infty r^{-1}[4\{1-f(B_n^{-1}x)\}^r-\{1-f(2B_n^{-1}x)\}^r].$$

Furthermore, as $n\to\infty$ we have for $0<x\leqq 1$,

$$\begin{aligned}&1-f(2B_n^{-1}x)\sim\tfrac{1}{2}(2B_n^{-1}x)^2H(2B_n^{-1}x)\sim 2n^{-1}x^2\\ &1-f(B_n^{-1}x)\sim\tfrac{1}{2}(B_n^{-1}x)^2H(B_n^{-1}x)\sim\tfrac{1}{2}n^{-1}x^2,\end{aligned}$$

while $C_n(0)=0$, so that

$$\sum_1^\infty\int_0^1(1-x)\,|C_n(x)|\,dx<\infty.$$

Consequently, from (15) and (16),

$$\sum_1^\infty\left|\int_0^1(1-x)[4\{1-f(B_n^{-1}x)\}-\{1-f(2B_n^{-1}x)\}]\,dx\right|<\infty.$$

This may be rewritten as (remembering that $f(t)$ is real valued)

$$\sum_1^\infty\left|\int_0^1(1-x)\left[\int_{-\infty}^{\infty}\{4(1-\cos B_n^{-1}xy)-(1-\cos 2B_n^{-1}xy)\}\,dF(y)\right]dx\right|<\infty,$$

which reduces to

$$\sum_1^\infty \int_0^1 (1-x)\left\{\int_{-\infty}^{\infty} (1-\cos B_n^{-1} x y)^2\, dF(y)\right\} dx<\infty$$

and hence

$$\sum_1^\infty \int_0^1 (1-x)\left\{\int_{|y|\leqq \pi B_n} (1-\cos B_n^{-1} x y)^2\, dF(y)\right\} dx<\infty .$$

Now, for $|y|\leqq \pi B_n$, we can find a positive constant C so that

$$1-\cos B_n^{-1} x y \geqq C(B_n^{-1} x y)^2, \quad 0\leqq x\leqq 1,$$

and therefore,

$$\sum_1^\infty B_n^{-4} \int_0^1 x^4(1-x)\left\{\int_{|y|\leqq \pi B_n} y^4\, dF(y)\right\} dx<\infty ,$$

that is,

$$\sum_1^\infty B_n^{-4} \int_{|y|\leqq \pi B_n} y^4\, dF(y)<\infty . \tag{17}$$

$\{B_n\}$ has not been assumed monotone but we can now make this supposition in (17) without loss of generality (since we may discard terms and relabel the remaining ones if necessary). (17) is then Eq. (10) of Heyde [6] and following exactly the argument given in the remainder of Section 2 of [6] we obtain the required result that $EX_i^2<\infty$.

Finally, suppose that the X_i are not symmetric. We consider the sequence Y_i, $i=1, 2, 3, \ldots$ of independent and symmetric random variables, each Y_i having the distribution of the difference between two independent X_i's. Obviously the characteristic function of Y_i is $|f(t)|^2$ and the distribution function of the sum $Z_n=(Y_1+\cdots+Y_n)/B_n\sqrt{2}$ is equal to $F_n(x\sqrt{2})*[1-F_n(-x\sqrt{2}-0)]=G_n(x)$. Hence, if $F_n(x)$ satisfies the condition (1), we have

$$\begin{aligned}
&\sum_1^\infty n^{-1} \sup_x |G_n(x)-\Phi(x)|\\
&\quad=\sum_1^\infty n^{-1} \sup_x |F_n(x\sqrt{2})*(1-F_n(-x\sqrt{2}-0))-\Phi(x\sqrt{2})*(1-\Phi(-x\sqrt{2}))|\\
&\quad\leqq\sum_1^\infty n^{-1} \sup_x |F_n(x\sqrt{2})*(1-F_n(-x\sqrt{2}-0))-\Phi(x\sqrt{2})*(1-F_n(-x\sqrt{2}-0))|\\
&\qquad+\sum_1^\infty n^{-1} \sup_x |\Phi(x\sqrt{2})*(1-F_n(-x\sqrt{2}-0))-\Phi(x\sqrt{2})*(1-\Phi(-x\sqrt{2}))|\\
&\quad<\infty .
\end{aligned}$$

Then, from our results in the symmetric case, we extract the information that $EY_i^2<\infty$ from which it follows that $EX_i^2<\infty$. This completes the proof of the theorem.

3. Proofs of Theorems 2, 3, 4

Proof of Theorem 2. That $\Delta_n(C_n)\to 0$ if $\delta_n(C_n)\to 0$ follows from (2) so we suppose $\Delta_n(C_n)\to 0$ (i.e. the X_i belong to the domain of attraction of the normal distribution) and need to show that $\delta_n(C_n)\to 0$ in order to complete the proof.

Now, as already noted in the proof of Theorem 1, we have that $U(x)=\int_{|u|\leq x} u^2\, dF(u)$ is slowly varying as $x\to\infty$. Clearly $nB_n^{-2}U(B_n)\to 1$ and $B_n\sim C_n$ as $n\to\infty$. Also, from Gnedenko and Kolmogorov [4] p. 172 we have that

$$x^2 P(|X_i|>x)/U(x)\to 0 \tag{18}$$

as $x\to\infty$.

If $EX_i^2<\infty$ we have $\tau_n=\sqrt{n}$ and $nP(|X_1|>\sqrt{n})\to 0$ since the terms of

$$\sum P(|X_i|>\sqrt{n})<\infty$$

are monotone. If $EX_i^2=\infty$ we have $\tau_n=B_n$ and from (18),

$$nP(|X_i|>B_n)\sim B_n^2[U(B_n)]^{-1}P(|X_i|>B_n)\to 0$$

as $n\to\infty$. In either case,

$$nP(|X_i|>\tau_n)\to 0 \quad \text{as } n\to\infty. \tag{19}$$

Next, we have

$$\begin{aligned} nB_n^{-3}\int_{|x|\leq\tau_n}|x|^3\, dF(x) &= nB_n^{-3}\int_0^{\tau_n} x\, dU(x) \\ &= n\tau_n B_n^{-3}U(\tau_n)-nB_n^{-3}\int_0^{\tau_n}U(x)\,dx \end{aligned}$$

and $nB_n^{-3}\int_0^{\tau_n}U(x)\,dx\sim n\tau_n B_n^{-3}U(\tau_n)$ using Theorem 1, p. 273 of Feller [1]. If $EX_i^2<\infty$ we have $\tau_n=\sqrt{n}$, $B_n\sim(nEX_i^2)^{\frac{1}{2}}$ and $U(\tau_n)\to EX_i^2$ as $n\to\infty$ while if $EX_i^2=\infty$ we have $\tau_n=B_n$ and $nB_n^{-2}U(B_n)\to 1$. In either case,

$$nB_n^{-3}\int_{|x|\leq\tau_n}|x|^3\, dF(x)\to 0 \quad \text{as } n\to\infty. \tag{20}$$

Finally, we have

$$\begin{aligned} nB_n^{-1}\Big|\int_{|x|\leq\tau_n}x\,dF(x)\Big| &= nB_n^{-1}\Big|\int_{|x|>\tau_n}x\,dF(x)\Big| \\ &\leq nB_n^{-1}\int_{|x|>\tau_n}|x|\,dF(x) \\ &\leq nB_n^{-1}\tau_n P(|X_i|>\tau_n)+nB_n^{-1}\int_{\tau_n}^{\infty}P(|X_i|>x)\,dx. \end{aligned} \tag{21}$$

The first term on the right hand side of (21) goes to zero as $n\to\infty$ by (19). To deal with the second term we use (18) and, given any $\varepsilon>0$, choose $N=N(\varepsilon)$ so large that

$$P(|X_i|>x)\leq\varepsilon x^{-2}U(x)$$

for $x \geqq N$. Consequently, for n sufficiently large that $\tau_n > N$,

$$nB_n^{-1} \int_{\tau_n}^{\infty} P(|X_i| > x)\, dx \leqq \varepsilon\, n B_n^{-1} \int_{\tau_n}^{\infty} x^{-2}\, U(x)\, dx.$$

Another application of Theorem 1, p. 273 of [1] gives

$$nB_n^{-1} \int_{\tau_n}^{\infty} x^{-2}\, U(x)\, dx \sim n B_n^{-1} \tau_n^{-1}\, U(\tau_n) \to (EX_i^2)^{\frac{1}{2}} \quad \text{if } EX_i^2 < \infty,\ 1$$

otherwise as $n \to \infty$ since either $B_n \sim (n\, EX_i^2)^{\frac{1}{2}}$, $U(\tau_n) \to EX_i^2$, $\tau_n = \sqrt{n}$ in case $EX_i^2 < \infty$ or $\tau_n = B_n$, $n B_n^{-2}\, U(B_n) \to 1$ in case $EX_i^2 = \infty$. Thus, from (21),

$$nB_n^{-1} \Big| \int_{|x| \leqq \tau_n} x\, dF(x) \Big| \to 0 \qquad \text{as } n \to \infty. \tag{22}$$

The required result $\delta_n(C_n) \to 0$ follows from (19), (20) and (22), since $C_n \sim B_n$ as $n \to \infty$.

Proof of Theorem 3. The equivalence of (i) and (ii) follows from Theorem 3.1 of Ibragimov [8]. We thus have (iv) ⇒ (iii) ⇒ (ii) ⇒ (i) and to complete the proof it is just necessary to show that (i) ⇒ (iv).

Now, when (i) holds we have

$$\begin{aligned} \Big| \int_{|x| \leqq \sqrt{n}} x\, dF(x) \Big| &= \Big| \int_{|x| > \sqrt{n}} x\, dF(x) \Big| \\ &\leqq \int_{|x| > \sqrt{n}} |x|\, dF(x) \\ &\leqq n^{-\frac{1}{2}} \int_{|x| > \sqrt{n}} x^2\, dF(x) = O(n^{-(1+\delta)/2}) \end{aligned}$$

and $B_n^2 \sim n\, EX_i^2$ as $n \to \infty$ so that

$$nB_n^{-1} \Big| \int_{|x| \leqq \sqrt{n}} x\, dF(x) \Big| = O(n^{-\delta/2}). \tag{23}$$

Also, putting $R(z) = \int_{|u| > z} u^2\, dF(u)$, we have

$$\begin{aligned} \int_{|x| \leqq \sqrt{n}} |x|^3\, dF(x) &= - \int_0^{\sqrt{n}} u\, dR(u) \\ &= -\sqrt{n}\, R(\sqrt{n}) + \int_0^{\sqrt{n}} R(u)\, du = O(n^{(1-\delta)/2}) \end{aligned}$$

so that

$$B_n^{-3}\, n \int_{|x| \leqq \sqrt{n}} |x|^3\, dF(x) = O(n^{-\delta/2}). \tag{24}$$

Finally,

$$nP(|X_i| > \sqrt{n}) \leqq \int_{|x| > \sqrt{n}} x^2\, dF(x) = O(n^{-\delta/2}), \tag{25}$$

and (iv) follows from (23), (24) and (25).

Proof of Theorem 4. If $EX_i^2<\infty$, the equivalence of (i) and (iii) follows easily from results given in Friedman, Katz and Koopmans [3]. That (iii) implies $EX_i^2<\infty$ follows from Theorem 1. We also have (iv) ⇒ (iii) ⇒ (ii) so it remains to prove that (i) ⇒ (iv) and (ii) ⇒ (i).

First we shall prove that (i) ⇒ (iv). We have

$$\sum_1^\infty n^{\delta/2}\, P(|X_i|>\sqrt{n})<\infty \tag{26}$$

since $E|X_i|^{2+\delta}<\infty$. Also, $B_n\sim(n\,EX_i^2)^{\frac{1}{2}}$ as $n\to\infty$ and using integration by parts,

$$\begin{aligned}
&\sum_{n=1}^\infty n^{(\delta-3)/2}\int_{|x|\le\sqrt{n}}|x|^3\,dF(x)\\
&\le 3\sum_{n=1}^\infty n^{(\delta-3)/2}\sum_{k=1}^{n}\int_{\sqrt{k-1}<x\le\sqrt{k}} x^2\,P(|X_i|>x)\,dx\\
&\le \sum_{n=1}^\infty n^{(\delta-3)/2}\sum_{k=1}^{n} P(|X_i|>\sqrt{k-1})\,[k^{\frac{3}{2}}-(k-1)^{\frac{3}{2}}]\\
&\le C_1\sum_{k=1}^\infty P(|X_i|>\sqrt{k-1})\,k^{\frac{1}{2}}\sum_{n=k}^\infty n^{(\delta-3)/2}\\
&\le C_2\sum_{k=1}^\infty k^{\delta/2}\,P(|X_i|>\sqrt{k-1})<\infty,
\end{aligned} \tag{27}$$

C_1, C_2 denoting suitable positive constants. Furthermore,

$$\begin{aligned}
&\sum_{n=1}^\infty n^{(\delta-1)/2}\Big|\int_{|x|\le\sqrt{n}} x\,dF(x)\Big|\\
&\le \sum_{n=1}^\infty n^{(\delta-1)/2}\int_{|x|>\sqrt{n}}|x|\,dF(x)\\
&= \sum_{n=1}^\infty n^{(\delta-1)/2}\left\{\sqrt{n}\,P(|X_i|>\sqrt{n})+\int_{\sqrt{n}}^\infty P(|X_i|>x)\,dx\right\}\\
&= \sum_{n=1}^\infty n^{\delta/2}\,P(|X_i|>\sqrt{n})+\sum_{n=1}^\infty n^{(\delta-1)/2}\sum_{k=n}^\infty\int_{\sqrt{k}<x\le\sqrt{k+1}} P(|X_i|>x)\,dx\\
&\le \sum_{n=1}^\infty n^{\delta/2}\,P(|X_i|>\sqrt{n})+\sum_{n=1}^\infty n^{(\delta-1)/2}\sum_{k=n}^\infty P(|X_i|>\sqrt{k})(\sqrt{k+1}-\sqrt{k})\\
&\le \sum_{n=1}^\infty n^{\delta/2}\,P(|X_i|>\sqrt{n})+D_1\sum_{k=1}^\infty P(|X_i|>\sqrt{k})\,k^{-\frac{1}{2}}\sum_{n=1}^{k} n^{(\delta-1)/2}\\
&\le D_2\sum_{k=1}^\infty k^{\delta/2}\,P(|X_i|>\sqrt{k})<\infty,
\end{aligned} \tag{28}$$

D_1, D_2 denoting suitable positive constants. (iv) follows from (26), (27), (28).

Finally, we need to prove that (ii) ⇒ (i). Choose $\{C_n^*\}$ so that $\Delta_n(C_n^*) \sim \inf_{C_n} \Delta_n(C_n)$ and fix attention on $\sum n^{-1+\delta/2} \Delta_n(C_n^*) < \infty$. That (i) holds for $\delta = 0$ is given by Theorem 1 and the proof for $\delta > 0$ is obtained by paralleling the proof of Theorem 1. Instead of (8) we now have

$$\sum_1^\infty n^{-1+\delta/2} \left| \int_0^z (z-t)\left(f_n(t) - e^{-\frac{1}{2}t^2}\right) e^{\frac{1}{2}t^2}\, dt \right| < \infty,$$

where

$$f_n(t) = [f(t/C_n^*)]^n$$

and

$$f(t) = \exp\{-\tfrac{1}{2} t^2 H(t)\}$$

with $H(t) \to EX_i^2$ as $t \to 0$. The proof of Theorem 1 can then be followed through in similar fashion up to the stage of obtaining

$$\sum_1^\infty n^{\delta/2} \int_0^1 (1-x) \left\{ \int_{|y| \leq \sigma\sqrt{n}} (1 - \cos(C_n^*)^{-1} x y)^2\, dF(y) \right\} dx < \infty \tag{29}$$

where $\sigma^2 = EX_i^2$. Now, noting that $C_n^* \sim (n\, EX_i^2)^{\frac{1}{2}}$ as $n \to \infty$ we can, for $|y| \leq \pi\, C_n^*$, find a positive constant K such that

$$1 - \cos(C_n^*)^{-1} x y \geq K(n^{-\frac{1}{2}} x)^2, \qquad 0 \leq x \leq 1,$$

and therefore (29) yields

$$\sum_1^\infty n^{(\delta/2)-2} \int_{|y| \leq \sigma\sqrt{n}} y^4\, dF(y) < \infty. \tag{30}$$

Furthermore,

$$\begin{aligned}
\sum_{n=1}^\infty n^{(\delta/2)-2} \int_{|x| \leq \sigma\sqrt{n}} x^4\, dF(x) &\geq \sum_{n=2}^\infty n^{(\delta/2)-2} \sum_{k=1}^{n-1} \int_{\sigma\sqrt{k} < |x| \leq \sigma\sqrt{k+1}} x^4\, dF(x) \\
&\geq \sigma^4 \sum_{k=1}^\infty k^2 P(\sigma\sqrt{k} < |X_i| \leq \sigma\sqrt{k+1}) \sum_{n=k+1}^\infty n^{(\delta/2)-2} \\
&\geq C \sum_{k=1}^\infty k^{1+\delta/2} P(\sigma\sqrt{k} < |X_i| \leq \sigma\sqrt{k+1})
\end{aligned}$$

for some positive constant C, while

$$\begin{aligned}
E|X_i|^{2+\delta} &= \sum_{k=0}^\infty \int_{\sigma\sqrt{k} < |x| \leq \sigma\sqrt{k+1}} |x|^{2+\delta}\, dF(x) \\
&\leq C_1 \sum_{k=1}^\infty k^{1+\delta/2} P(\sigma\sqrt{k} < |X_i| \leq \sigma\sqrt{k+1})
\end{aligned}$$

for some positive constant C_1 and hence $E|X_i|^{2+\delta} < \infty$. This proves the required result (i) for symmetric random variables; the result for the general case is obtained as in the proof of Theorem 1.

References

1. Feller, W.: An introduction to probability theory and its applications, Vol. II. New York: Wiley 1966.
2. Feller, W.: On the Berry-Esseen theorem. Z. Wahrscheinlichkeitstheorie verw. Gebiete **10**, 261–268 (1968).
3. Friedman, N., Katz, M., Koopmans, L.H.: Convergence rates for the central limit theorem. Proc. Nat. Acad. Sci. U.S.A. **56**, 1062–1065 (1966).
4. Gnedenko, B.V. and A.N. Kolmogorov: Limit distributions for sums of independent random variables. Reading, Mass.: Addison-Wesley 1954.
5. Heyde, C.C.: On the influence of moments on the rate of convergence to the normal distribution. Z. Wahrscheinlichkeitstheorie verw. Gebiete **8**, 12–18 (1967).
6. Heyde, C.C.: Some properties of metrics in a study on convergence to normality. Z. Wahrscheinlichkeitstheorie verw. Gebiete **11**, 181–192 (1969).
7. Heyde, C.C.: On the implication of a certain rate of convergence to normality. Z. Wahrscheinlichkeitstheorie verw. Gebiete **16**, 151–156 (1970).
8. Ibragimov, I.A.: On the accuracy of the Gaussian approximation to the distribution function of sums of independent random variables. Theor. Probab. Appl. **11**, 559–576 (1966).
9. Ibragimov, I.A. and Yu.V. Linnik: Independent and stationary dependent random variables (in Russian). Moscow Izd-vo "Nauka" 1965.
10. Osipov, L.V. and V.V. Petrov: On an estimate of the remainder in the central limit theorem. Theor. Probab. Appl. **12**, 281–286 (1967).

C.C. Heyde
Department of Statistics, SGS
Australian National University
Box 4, P.O.
Canberra, A.C.T. 2600
Australia

(Received December 16, 1971)

The Annals of Probability
1973, Vol. 1, No. 3, 428–436

INVARIANCE PRINCIPLES FOR THE LAW OF THE ITERATED LOGARITHM FOR MARTINGALES AND PROCESSES WITH STATIONARY INCREMENTS

BY C. C. HEYDE AND D. J. SCOTT[1]

Australian National University

The main result in this paper is an invariance principle for the law of the iterated logarithm for square integrable martingales subject to fairly mild regularity conditions on the increments. When specialized to the case of identically distributed increments the result contains that of Stout [16] as well as the invariance principle for independent random variables of Strassen [17]. The martingale result is also used to obtain an invariance principle for the iterated logarithm law for a wide class of stationary ergodic sequences and a corollary is given which extends recent results of Oodaira and Yoshihara [10] on ϕ-mixing processes.

1. Introduction. This paper is basically divided into two parts. In the first part a general invariance principle for the law of the iterated logarithm for martingales is obtained. This result (Theorem 1) specializes, in the case of identically distributed increments, to a result (Corollary 2) which contains the invariance principle of Strassen [17] for sums of independent and identically distributed random variables. Corollary 2 also contains Stout's [16] result giving a classical Hartman–Wintner form of iterated logarithm law for stationary ergodic martingales and its minor generalization by Heyde [6]. Our approach to Theorem 1 follows essentially the pattern of that of Theorem 3 of [17] in that it is based on the use of a form of the Skorokhod representation theorem.

In the second part of the paper we make use of Corollary 2 to establish an invariance principle for the law of the iterated logarithm for a certain class of (strictly) stationary processes (Theorem 2). The basic idea is a representation for the increments of the stationary process in terms of the increments of a stationary martingale plus other terms whose sum is negligible under suitable norming. This idea is due to Gordin [4], who has used it to prove a central limit theorem for stationary ergodic sequences and it has been further developed, in that context, by Scott [14]. As a corollary to Theorem 2 we obtain Corollary 3 which extends results of Oodaira and Yoshihara [10], who have given an invariance principle for the law of the iterated logarithm for stationary ϕ-mixing sequences.

Previous work on iterated logarithm results for stationary processes has proceeded via classical methods. Our approach via the Skorokhod representation is

Received March 27, 1972; revised October 6, 1972.

[1] Now at La Trobe University.

AMS 1970 *subject classifications.* Primary: 60B10, 60F15, Secondary: 60G10, 60G45.

Key words and phrases. Invariance principles, iterated logarithm law, martingales, stationary ergodic processes, ϕ-mixing.

R. Maller et al. (eds.), *Selected Works of C.C. Heyde*, Selected Works in Probability and Statistics, DOI 10.1007/978-1-4419-5823-5_34, © Springer Science+Business Media, LLC 2010

more direct, gives ready access to the invariance principle, and avoids the necessity of obtaining detailed estimates of probabilities via, for example, rates of convergence to normality. In principle the result of Theorem 2 could be obtained by classical methods by paralleling the work of Chover [2] but a prerequisite would be estimates of rates of convergence to normality along the lines of those of Heyde [5] for sums of independent random variables.

2. Martingale results. Let $\{S_n, \mathscr{F}_n;\, n \geqq 0\}$ be a martingale on the probability space $(\Omega, \mathscr{A}, p)$ where $\mathscr{F}_0 = \{\phi, \Omega\}$ and $\mathscr{F}_n = \sigma$-field generated by $S_1, S_2, \cdots, S_n$ for $n > 0$. Let $S_0 = X_0 = 0$ almost surely (a.s.) and $S_n = \sum_{k=1}^{n} X_k$ for $n \geqq 1$. Further, let $s_n^2 = ES_n^2 < \infty$.

We consider the metric space (C, ρ) of all real-valued continuous functions on $[0, 1]$ with

$$\rho(x, y) = \sup_{0 \leq t \leq 1} |x(t) - y(t)| \qquad \text{for } x, y \in C\,.$$

Let K be the set of absolutely continuous $x \in C$ such that $x(0) = 0$ and

$$\int_0^1 [\dot{x}(t)]^2\, dt \leqq 1$$

where $\dot{x}$ denotes the derivative of x determined almost everywhere with respect to Lebesgue measure. Define a real function $g(\cdot)$ on $[0, \infty)$ by

$$g(t) = \sup\{n : s_n^2 \leqq t\}\,.$$

Let ξ be a standard Wiener process (Brownian motion) on $[0, \infty)$ and define a sequence of real random functions $\xi_n(\cdot)$ on $[0, 1]$, for $n > g(e)$ by

$$\xi_n(t) = [\phi(s_n^2)]^{-1} \xi(s_n^2 t) \qquad t \in [0, 1]$$

where $\phi(\cdot)$ is a real function on $[e, \infty)$ given by

$$\phi(t) = (2t \log\log t)^{\frac{1}{2}} \qquad t \in [e, \infty)\,.$$

We also define a sequence of real random functions $\eta_n(\cdot)$ on $[0, 1]$, for $n > g(e)$, by

$$\eta_n(t) = [\phi(s_n^2)]^{-1}[S_k + (s_n^2 t - s_k^2)(s_{k+1}^2 - s_k^2)^{-1} X_{k+1}]$$
$$s_k^2 \leqq s_n^2 t \leqq s_{k+1}^2\,, \qquad k = 0, 1, \cdots, n-1\,.$$

THEOREM 1. *If* $s_n^2 \to \infty$ *and*

(1) $$\sum_{n=1}^{\infty} s_n^{-4} E\{X_n^4 I(|X_n| < \delta s_n)\} < \infty \qquad \textit{for some } \delta > 0\,,$$

(2) $$\sum_{n=1}^{\infty} s_n^{-1} E\{|X_n| I(|X_n| \geqq \varepsilon s_n)\} < \infty \qquad \textit{for all } \varepsilon > 0\,,$$

(3) $$s_n^{-2} \sum_{k=1}^{n} X_k^2 \to_{\text{a.s.}} 1 \qquad \textit{as } n \to \infty$$

hold, $I(\cdot)$ *denoting the indicator function, then* $\{\eta_n;\, n > g(e)\}$ *is relatively compact in C and the set of its limit points coincides with K.* ("$\to_{\text{a.s.}}$" means "converges a.s. to.")

COROLLARY 1. *If* $s_n^2 \to \infty$ *as* $n \to \infty$,

(4) $$\sum_{n=1}^{\infty} s_n^{-4} E X_n^4 < \infty$$

and (3) *holds then* $\{\eta_n;\ n > g(e)\}$ *is relatively compact in* C *and the set of its limit points coincides with* K.

COROLLARY 2. *If the* X_n *are identically distributed and satisfy* (3), *then* $\{\eta_n;\ n > e[E(X_1^2)]^{-1}\}$ *is relatively compact in* C *and the set of its limit points coincides with* K.

The condition (3) is, unfortunately, often difficult to check. It holds, for example, if $\{X_n\}$ is a stationary ergodic sequence.

The result of Theorem 1 appears to be new even in the case of independent random variables. Other iterated logarithm type results for martingales have been obtained by Stout [15] and Strassen [18] but they involve random norming and do not seem to be directly comparable with the present results.

3. Martingale proofs. The first step is to define a truncated martingale sequence $\{S_n^*, \mathscr{F}_n^*;\ n \geqq 0\}$. For $\delta > 0$, let $X_0^* = 0$ a.s., $\mathscr{F}_0^* = \{\phi, \Omega\}$ and for $k > 0$,

$$X_k^* = X_k I(|X_k| < \delta s_k) - E\{X_k I(|X_k| < \delta s_k) \,|\, \mathscr{F}_{k-1}^*\}\,,$$

where $\mathscr{F}_k^* = \sigma$-field generated by $X_1^*, \cdots, X_k^*$. Then we let

$$S_n^* = \textstyle\sum_{k=0}^n X_k^* \qquad \text{for } n \geqq 0\,,$$

and $s_n^{*2} = ES_n^{*2} \leqq s_n^2 < \infty$. For $n > g(e)$ we define a sequence of real random functions $\eta_n^*(\cdot)$ on $[0, 1]$ by

$$\eta_n^*(t) = [\phi(s_n^2)]^{-1}[S_k^* + (s_n^2 t - s_k^2)(s_{k+1}^2 - s_k^2)^{-1} X_{k+1}^*]$$
$$s_k^2 \leqq t s_n^2 \leqq s_{k+1}^2,\ k = 0, 1, \cdots, n-1\,.$$

We first observe that

$$\sup_{0 \leq t \leq 1} |\eta_n^*(t) - \eta_n(t)| \rightarrow_{\text{a.s.}} 0 \qquad \text{as } n \rightarrow \infty \tag{5}$$

under the conditions of Theorem 1. This will of course be true if

$$s_n^{-1} \textstyle\sum_{k=1}^n |X_k| I(|X_k| \geqq \delta s_k) \rightarrow_{\text{a.s.}} 0 \qquad \text{as } n \rightarrow \infty \tag{6}$$

and

$$s_n^{-1} \textstyle\sum_{k=1}^n |E\{X_k I(|X_k| < \delta s_k) \,|\, \mathscr{F}_{k-1}^*\}| \rightarrow_{\text{a.s.}} 0 \qquad \text{as } n \rightarrow \infty\,, \tag{7}$$

and these results follow straightforwardly from (2) via the Kronecker lemma and the monotone convergence theorem.

Now we introduce the Skorokhod representation. From Theorem 4.3 of Strassen [18] we may conclude that there exists a sequence $T_0 = 0, T_1, T_2, \cdots$ of nonnegative random variables such that

$$(\bar{S}_1, \bar{S}_2, \cdots, \bar{S}_n) \equiv (\xi(T_1), \xi(T_1 + T_2), \cdots, \xi(\textstyle\sum_{k=1}^n T_k))$$

is distributed as

$$(S_1^*, S_2^*, \cdots, S_n^*) \tag{8}$$

for each $n \geqq 1$. Further, if $\bar{S}_0 = 0$ a.s., $\bar{\mathscr{F}}_0 = \{\phi, \Omega\}$, $\bar{\mathscr{F}}_n$ for $n \geqq 1$ is the σ-field generated by $\bar{S}_1, \cdots, \bar{S}_n$, $\mathscr{G}_0 = \mathscr{F}_0$, $\mathscr{G}_n$ is the σ-field generated by $\xi(t)$ for

$0 \leqq t \leqq \sum_{k=1}^{n} T_k$ for $n \geqq 1$ and $\bar{S}_n = \sum_{k=0}^{n} \bar{X}_k$ for $n \geqq 0$, then

$$E\{T_n \mid \mathscr{G}_{n-1}\} = E\{\bar{X}_n{}^2 \mid \bar{\mathscr{F}}_{n-1}\} \quad \text{a.s.} \qquad \text{for } n \geqq 1 . \tag{9}$$

Also, for $r > 1$,

$$E\{T_n{}^r \mid \mathscr{G}_{n-1}\} \leqq L_r E\{\bar{X}_n{}^{2r} \mid \bar{\mathscr{F}}_{n-1}\} \quad \text{a.s.} \qquad \text{for } n \geqq 1 , \tag{10}$$

where L_r is a constant depending only on r.

Define $\bar{\eta}_n(\cdot)$ on $[0, 1]$, for $n > g(e)$, by

$$\bar{\eta}_n(t) = [\phi(s_n{}^2)]^{-1}[\bar{S}_k + (s_n{}^2 t - s_k{}^2)(s_{k+1}^2 - s_k{}^2)^{-1}\bar{X}_{k+1}]$$
$$s_k{}^2 \leqq s_n{}^2 t \leqq s_{k+1}^2,\ k = 0, 1, \cdots, n-1 ,$$

and $\eta(\cdot)$ on $[0, \infty)$ by

$$\eta(t) = \bar{S}_n + \bar{X}_{n+1}(t - s_n{}^2)(s_{n+1}^2 - s_n{}^2)^{-1} \qquad s_n{}^2 \leqq t \leqq s_{n+1}^2 .$$

Then

$$\eta(t) = \phi(s_n{}^2)\bar{\eta}_n(s_n{}^{-2}t) \qquad \text{for } t \in [0, s_n{}^2]$$

for any $n > g(e)$.

Our proof follows the work of Strassen in [17]. In fact, in view of (5) and (8) above and the results of [17], Theorem 1 will follow provided we can show that $s_{n+1}^2/s_n{}^2 \to 1$ and $\rho(\bar{\eta}_n, \xi_n) \to_{\text{a.s.}} 0$ and this last condition is implied by

$$P(\lim_{t\to\infty} [\phi(t)]^{-1} \sup_{\tau \leqq t} |\eta(\tau) - \xi(\tau)| = 0) = 1 . \tag{11}$$

In order to prove that $s_{n+1}^2/s_n{}^2 \to 1$, we first note that (2) gives, via an application of the Kronecker Lemma,

$$s_n{}^{-1} \sup_{k \leqq n} |X_k| I(|X_k| \geqq \varepsilon s_k) \to_{\text{a.s.}} 0 , \qquad \text{for any } \varepsilon > 0 ,$$

so that upon squaring we have *a fortiori*

$$s_n{}^{-2} \sup_{k \leqq n} X_k{}^2 I(|X_k| \geqq \varepsilon s_n) \to_{\text{a.s.}} 0 .$$

Thus

$$s_n{}^{-2} \sup_{k \leqq n} X_k{}^2 \leqq \varepsilon^2 + s_n{}^{-2} \sup_{k \leqq n} X_k{}^2 I(|X_k| \geqq \varepsilon s_n) ,$$

and since ε may be chosen arbitrarily small,

$$s_n{}^{-2} \sup_{k \leqq n} X_k{}^2 \to_{\text{a.s.}} 0 . \tag{12}$$

From (3) and Theorm 1 of Pratt [12] we then deduce that

$$\lim_{n\to\infty} s_n{}^{-2} \sup_{k \leqq n} EX_k{}^2 \leqq \lim_{n\to\infty} E s_n{}^{-2} \sup_{k \leqq n} X_k{}^2 = 0 , \tag{13}$$

and $s_{n+1}^2/s_n{}^2 \to 1$ follows.

We next proceed to the proof of (11) for which we need to know more about the behavior of the sequence $T_1, T_2, \cdots$.

Lemma 1. *Under the conditions of Theorem* 1,

$$s_n{}^{-2} \sum_{k=1}^{n} E\{T_k \mid \mathscr{G}_{k-1}\} \to_{\text{a.s.}} 1 \qquad \textit{as } n \to \infty .$$

Proof. Using (8) and (9), we need only show

$$s_n{}^{-2} \sum_{k=1}^{n} E\{X_k{}^{*2} \mid \mathscr{F}_{k-1}^*\} \to_{\text{a.s.}} 1 \qquad \text{as } n \to \infty . \tag{14}$$

Now

$$(15)\qquad s_n^{-2}\sum_{k=1}^n E\{X_k^{*2}\mid \mathscr{F}_{k-1}^*\} = s_n^{-2}\sum_{k=1}^n [E\{X_k^2 I(|X_k|<\delta s_k)\mid \mathscr{F}_{k-1}^*\} - (E\{X_k I(|X_k|<\delta s_k)\mid \mathscr{F}_{k-1}^*\})^2]$$

and

$$(16)\qquad s_n^{-2}\sum_{k=1}^n (E\{X_k I(|X_k|<\delta s_k)\mid \mathscr{F}_{k-1}^*\})^2 \leqq s_n^{-1}\delta \sum_{k=1}^n E\{|X_k| I(|X_k|\geqq \delta s_n)\mid \mathscr{F}_{k-1}^*\} \to_{\text{a.s.}} 0 \quad \text{as } n\to\infty .$$

Also, using Proposition IV. 6.1, page 147 of Neveu [8],

$$(17)\qquad s_n^{-2}\sum_{k=1}^n (X_k^2 I(|X_k|<\delta s_k) - E\{X_k^2 I(|X_k|<\delta s_k)\mid \mathscr{F}_{k-1}^*\}) \to_{\text{a.s.}} 0 \quad \text{as } n\to\infty$$

since

$$\sum_{n=1}^\infty s_n^{-4} E[X_n^2 I(|X_n|<\delta s_n) - E\{X_n^2 I(|X_n|<\delta s_n)\mid \mathscr{F}_{n-1}^*\}]^2 \leqq \sum_{n=1}^\infty s_n^{-4} E[X_n^4 I(|X_n|<\delta s_n)] < \infty$$

using (1). Consequently, from (15), (16) and (17), it suffices to show that

$$s_n^{-2}\sum_{k=1}^n X_k^2 I(|X_k|<\delta s_k) \to_{\text{a.s.}} 1 \qquad \text{as } n\to\infty$$

or equivalently, in view of (3),

$$(18)\qquad s_n^{-2}\sum_{k=1}^n X_k^2 I(|X_k|\geqq \delta s_k) \to_{\text{a.s.}} 0 .$$

However,

$$s_n^{-2}\sum_{k=1}^n X_k^2 I(|X_k|\geqq \delta s_k) \leqq (s_n^{-1}\sup_{k\leqq n}|X_k|) s_n^{-1}\sum_{k=1}^n |X_k| I(|X_k|\geqq \delta s_k)$$

and (18) follows from (2) and (12).

LEMMA 2. *Under the conditions of Theorem* 1,

$$s_n^{-2}\sum_{k=1}^n (T_k - E\{T_k\mid \mathscr{G}_{k-1}\}) \to_{\text{a.s.}} 0 \qquad \textit{as } n\to\infty .$$

PROOF. Again using Proposition IV. 6.1, page 147 of [8], the result will follow if

$$\sum_{n=1}^\infty s_n^{-4} E(T_n - E\{T_n\mid \mathscr{G}_{n-1}\})^2 < \infty .$$

Now

$$\begin{aligned}
\sum_{n=1}^\infty s_n^{-4} E(T_n - E\{T_n\mid \mathscr{G}_{n-1}\})^2
&\leqq \sum_{n=1}^\infty s_n^{-4} ET_n^2 \\
&\leqq L_2 \sum_{n=1}^\infty s_n^{-4} EX_n^{*4} \qquad \text{(by (10))} \\
&= L_2 \sum_{n=1}^\infty s_n^{-4} E(X_n I(|X_n|<\delta s_n) - E\{X_n I(|X_n|<\delta s_n)\mid \mathscr{F}_{n-1}^*\})^4 \\
&\leqq L_2 \sum_{n=1}^\infty s_n^{-4}[E(X_n^4 I(|X_n|<\delta s_n)) \\
&\qquad + 15\delta^3 s_n^3 E|E\{X_n I(|X_n|<\delta s_n)\mid \mathscr{F}_{n-1}^*\}|] \\
&\leqq L_2 \sum_{n=1}^\infty s_n^{-4}[E(X_n^4 I(|X_n|<\delta s_n)) + 15\delta^3 s_n^3 E(|X_n| I(|X_n|\geqq \delta s_n))] \\
&< \infty
\end{aligned}$$

by (1) and (2).

We are now in a position to complete the proof of Theorem 1. Since $EX_k^2 < \infty$ for any k,

(19) $$g(t) \to \infty \quad \text{as} \quad t \to \infty .$$

Then also

(20) $$\begin{aligned} |ts_{g(t)}^{-2} - 1| &\leqq |s_{g(t)}^{-2} EX_{g(t)+1}^2| \\ &= |s_{g(t)+1}^{-2} EX_{g(t)+1}^2||s_{g(t)+1}^2 s_{g(t)}^{-2}| \\ &\to 0 \qquad \text{as } t \to \infty \end{aligned}$$

by (13) and (19). Similarly,

(21) $$ts_{g(t)+1}^{-2} \to 1 \qquad \text{as } t \to \infty .$$

From (19), (20), (21) and Lemmas 1 and 2 we obtain

$$t^{-1} \sum_{k=1}^{g(t)} T_k \to_{\text{a.s.}} 1 \qquad \text{as } t \to \infty$$

and

$$t^{-1} \sum_{k=1}^{g(t)+1} T_k \to_{\text{a.s.}} 1 \qquad \text{as } t \to \infty .$$

Finally, we note that

$$|\eta(t) - \xi(t)| \leqq \max \{|\xi(\sum_{k=1}^{g(t)} T_k) - \xi(t)|, |\xi(\sum_{k=1}^{g(t)+1} T_k) - \xi(t)|\} .$$

From this point, it is obvious that Strassen's proof on page 217 of [17] may be followed to obtain (11), which completes the proof of Theorem 1.

Corollaries 1 and 2 follow easily from Theorem 1 using standard arguments of the kind of, for example, Neveu [8], pages 153, 154.

4. Stationary process results. We consider a probability space $(\Omega, \mathscr{A}, P)$ with an ergodic automorphism T. Let $L_2(P)$ be the Hilbert space of random variables with finite second moment, and write $(EX^2)^{\frac{1}{2}}$ as $||X||$. Define U on $L_2(P)$ by $UX(\omega) = X(T\omega)$ for $X \in L_2(P)$, $\omega \in \Omega$. Then U is a unitary operator. If $\mathscr{M}_0$ is a σ-field such that $\mathscr{M}_0 \subset \mathscr{A}$ and $\mathscr{M}_0 \subset T^{-1}(\mathscr{M}_0)$, define $\mathscr{M}_k = T^{-k}(\mathscr{M}_0)$, $\mathscr{M}_{-\infty} = \bigcap_{k=-\infty}^{\infty} \mathscr{M}_k$ and $\mathscr{M}_{+\infty} = \sigma$-field generated by $\bigcup_{k=-\infty}^{\infty} \mathscr{M}_k$.

We consider a particular random variable $X_0 \in L_2(P)$. Defining $X_k = U^k X_0$, we see that $\cdots, X_{-1}, X_0, X_1, \cdots$ is a doubly infinite stationary ergodic sequence. Put $X^{(0)} = 0$ and $X^{(n)} = \sum_{k=0}^{n-1} X_k$ for $n \geqq 1$ and define a sequence of random functions $\theta_n(\cdot)$ on $[0, 1]$ by

$$\theta_n(t) = [\phi(||X^{(n)}||^2)]^{-1}(X^{(k)} + (nt - k)X_k)$$
$$k \leqq nt \leqq k + 1, k = 0, 1, \cdots, n - 1 .$$

Also define

$$g = \sup (n: ||X^{(n)}||^2 \leqq e\} .$$

THEOREM 2. *If*

(22) $$\sum_{m=0}^{\infty} (||E\{X_0 | \mathscr{M}_{-m}\}|| + ||X_0 - E\{X_0 | \mathscr{M}_m\}||) < \infty$$

then

$$\lim_{n\to\infty} ||X^{(n)}||/n^{\frac{1}{2}} = \sigma$$

exists for $0 \leqq \sigma < \infty$. *If* $\sigma > 0$ *then* $g < \infty$, $\{\theta_n; n > g\}$ *is relatively compact and the set of its limit points coincides with* K.

PROOF. The key to the proof is a representation

$$X_0 = Y_0 + UZ_0 - Z_0 \tag{23}$$

where Y_0, Z_0 belong to $L_2(P)$ and, writing $Y_k = U^k Y_0$, $Y^{(0)} = 0$ and $Y^{(n)} = \sum_{k=1}^{n-1} Y_k$ for $k \geqq 1$, $\{Y^{(n)}, \mathscr{M}_{n-1}; n \geqq 0\}$ is a stationary ergodic zero-mean martingale. This representation, which is due to Gordin, holds under the condition (22); see Scott [14] for full details.

Setting $||Y_0|| = \sigma$ we have $||Y^{(n)}|| = \sigma n^{\frac{1}{2}}$. Also, from Minkowski's inequality and (23),

$$|||X^{(n)}|| - \sigma n^{\frac{1}{2}}|/n^{\frac{1}{2}} \leqq 2||Z_0||/n^{\frac{1}{2}} \to 0$$

since $Z_0 \in L_2(P)$ and thus

$$\lim_{n\to\infty} ||X^{(n)}||/n^{\frac{1}{2}} = \sigma\,.$$

We henceforth assume $\sigma > 0$. Then

$$\lim_{n\to\infty} ||X^{(n)}||/\sigma n^{\frac{1}{2}} = 1 \tag{24}$$

and $g < \infty$. Also, from (24) we readily deduce that

$$\phi(||X^{(n)}||^2)/\phi(n\sigma^2) \to 1\,. \tag{25}$$

Now define a sequence of random functions $\zeta_n(\cdot)$ on $[0, 1]$ by

$$\zeta_n(t) = [\phi(n\sigma^2)]^{-1}[Y^{(k)} + (nt - k)Y_k]$$
$$k \leqq nt \leqq k+1,\ k = 0, 1, \cdots, n-1\,.$$

Then by Corollary 2 we know that $\{\zeta_n; n > e/\sigma^2\}$ is relatively compact and the set of its limit points coincides with K. Thus, the proof will be complete if we can show that

$$\sup_{0\leq t\leq 1} |\theta_n(t) - \zeta_n(t)| \to_{\text{a.s.}} 0\,. \tag{26}$$

We first consider

$$\sup_{0\leq t\leq 1} |\zeta_n(t) - \phi(n\sigma^2)[\phi(||X^{(n)}||^2)]^{-1}\zeta_n(t)| \leqq |1 - \phi(n\sigma^2)[\phi(||X^{(n)}||^2)]^{-1}| \sup_{0\leq t\leq 1} |\zeta_n(t)|\,. \tag{27}$$

Now

$$\limsup_{n\to\infty} \sup_{0\leq t\leq 1} |\zeta_n(t)| = \sup_{x\in K} \sup_{0\leq t\leq 1} |x(t)| \quad \text{a.s.}$$
$$\leqq 1 \quad \text{a.s.}$$

From this and (25) and (27) we conclude that

$$\sup_{0\leq t\leq 1} |\zeta_n(t) - \phi(n\sigma^2)[\phi(||X^{(n)}||^2)]^{-1}\zeta_n(t)| \to_{\text{a.s.}} 0\,. \tag{28}$$

Now, writing $Z_k = U^k Z_0$, consider

$$\sup_{0\leq t\leq 1} |\theta_n(t) - \phi(n\sigma^2)[\phi(||X^{(n)}||^2)]^{-1}\zeta_n(t)|$$
$$= \sup_{1\leq k\leq n} [\phi(||X^{(n)}||^2)]^{-1}|Z_k - Z_0| \qquad \text{(by (23))}$$
$$\leqq 2\sup_{0\leq k\leq n} [\phi(||X^{(n)}||^2)]^{-1}|Z_k|$$

which goes almost surely to zero provided

$$\sup_{0\leq k\leq n} [\phi(n\sigma^2)]^{-1}|Z_k| = (n\sigma^2 \log\log n\sigma^2)^{-\frac{1}{2}} \sup_{0\leq k\leq n} |Z_k| \to_{\text{a.s.}} 0$$

(by (25)) and thus if

$$(n \log\log n)^{-\frac{1}{2}}|Z_n| \to_{\text{a.s.}} 0 .$$

But, this last result follows from the Borel–Cantelli Lemma since $Z_0 \in L_2(P)$. Thus

$$\sup_{0\leq t\leq 1} |\theta_n(t) - \phi(n\sigma^2)[\phi(||X^{(n)}||^2)]^{-1}\zeta_n(t)| \to_{\text{a.s.}} 0$$

and taking this with (28), (26) follows and the proof is complete.

As an application of Theorem 2, we shall see how it extends some recent results of Oodaira and Yoshihara [9], [10]. The papers [9], [10] in turn improve results of Iosifescu [7] and Reznik [13] as well as the (strictly) stationary case of results of Philipp [11].

Let $\{X_j, -\infty < j < \infty\}$ be a stationary process defined on a probability space $(\Omega, \mathscr{A}, P)$. Write $\mathscr{M}_k$ for the σ-field generated by $\cdots, X_{k-1}, X_k$; $\mathscr{M}_\infty$ for the σ-field generated by $\cdots, X_{-1}, X_0, X_1, \cdots$ and $\mathscr{M}_k^\infty$ for the σ-field generated by $X_k, X_{k+1}, \cdots$. The sequence $\{X_j\}$ is called *ϕ-mixing* if for each $k(-\infty < k < \infty)$ and each $n(n \geqq 1)$,

$$\sup_{B\in\mathscr{M}_{k+n}^\infty} \operatorname{ess\,sup} |P(B\,|\,\mathscr{M}_k) - P(B)| \leqq \phi_n \downarrow 0 \qquad \text{as } n \to \infty .$$

COROLLARY 3. *If $\{X_n\}$ is a stationary ϕ-mixing process with $EX_0 = 0$, $E|X_0|^{2+\delta} < \infty$ some $\delta \geqq 0$ and $\sum_{n=1}^\infty [\phi_n]^{(1+\delta)/(2+\delta)} < \infty$, then*

$$\lim_{n\to\infty} ||X^{(n)}||^2/n = EX_0^2 + 2\sum_{i=1}^\infty EX_0X_i = \sigma^2 \geqq 0 .$$

If $\sigma > 0$ then $g < \infty$ and $\{\theta_n; n > g\}$ is relatively compact and the set of its limit points coincides with K.

PROOF. First note that $\{X_n\}$ is stationary and *ϕ-mixing* and hence ergodic. Further, there exists an ergodic transformation T on $\mathscr{M}_\infty$ such that $X_k(\omega) = X_0(T^k\omega)$, $\omega \in \Omega$ (e.g. Doob [3], Chapter X) and, since $\mathscr{M}_0$ is the σ-field generated by $\cdots, X_{-1}, X_0$ our present situation fits within the framework of Theorem 2.

Next we need to check (22). This is a simple matter using Lemma 1, page 170 of Billingsley [1]. We have

$$\sum_{m=0}^\infty (||E\{X_0\,|\,\mathscr{M}_{-m}\}|| + ||X_0 - E\{X_0\,|\,\mathscr{M}_m\}||) = \sum_{m=0}^\infty ||E\{X_0\,|\,\mathscr{M}_{-m}\}||$$

and

$$\begin{aligned} E[E\{X_0\,|\,\mathscr{M}_{-m}\}]^2 &= E[X_0E\{X_0\,|\,\mathscr{M}_{-m}\}] \\ &\leqq 2\phi_m^{(1+\delta)/(2+\delta)}[E|E\{X_0\,|\,\mathscr{M}_{-m}\}|^{(2+\delta)/(1+\delta)}]^{(1+\delta)/(2+\delta)}[E|X_0|^{2+\delta}]^{1/(2+\delta)} \\ &\leqq 2\phi_m^{(1+\delta)/(2+\delta)}[E[E\{X_0\,|\,\mathscr{M}_{-m}\}]^2]^{\frac{1}{2}}[E|X_0|^{2+\delta}]^{1/(2+\delta)} \end{aligned}$$

so that

$$||E\{X_0\,|\,\mathscr{M}_{-m}\}|| \leqq 2\phi_m^{(1+\delta)/(2+\delta)}[E|X_0|^{2+\delta}]^{1/(2+\delta)}$$

and (22) holds under the conditions of the corollary.

Finally, from Lemma 3, page 172 of [1],

$$\lim_{n\to\infty} ||X^{(n)}||^2/n = EX_0^2 + 2 \sum_{i=1}^{\infty} E(X_0X_i)$$

and the corollary follows from Theorem 2.

In Theorem 1 of Oodaira and Yoshihara [10], an invariance principle for the law of the iterated logarithm is given under the conditions of Corollary 3 plus the requirements that $\int_{|x|\geqq N} x^2dP(X_0 \leqq x) = O((\log N)^{-5})$ as $N \to \infty$ if $\delta = 0$ or $\phi_n = O(n^{-1-\varepsilon})$ for some $\varepsilon > (1+\delta)^{-1}$ if $\delta > 0$.

REFERENCES

[1] BILLINGSLEY, P. (1968). *Convergence of Probability Measures.* Wiley, New York.

[2] CHOVER, J. (1967). On Strassen's version of the log log law. *Z. Wahrscheinlichkeitstheorie und Verw. Gebiete* **8** 83–90.

[3] DOOB, J. L. (1953). *Stochastic Processes.* Wiley, New York.

[4] GORDIN, M. I. (1969). The central limit theorem for stationary processes. *Soviet Math. Dokl.* **10** 1174–1176.

[5] HEYDE, C. C. (1969). Some properties of metrics in a study on convergence to normality. *Z. Wahrscheinlichkeitstheorie und Verw. Gebiete* **11** 181–192.

[6] HEYDE, C. C. (1971). An iterated logarithm result for martingales and its application in estimation theory for autoregressive processes. To appear in *J. Appl. Probability* **9**.

[7] IOSIFESCU, M. (1968). The law of the iterated logarithm for a class of dependent random variables. *Theor. Probability Appl.* **13** 304–313.

[8] NEVEU, J. (1965). *Mathematical Foundations of the Calculus of Probability.* Holden-Day, San Francisco.

[9] OODAIRA, H. and YOSHIHARA, K. (1971a). The law of the iterated logarithm for stationary processes satisfying mixing conditions. *Kōdai Math. Sem. Rep.* **23** 311–334.

[10] OODAIRA, H. and YOSHIHARA, K. (1971b). Note on the law of the iterated logarithm for stationary processes satisfying mixing conditions. *Kōdai Math. Sem. Rep.* **23** 335–342.

[11] PHILIPP, W. (1969). The law of the iterated logarithm for mixing stochastic processes. *Ann. Math. Statist.* **40** 1985–1991.

[12] PRATT, J. W. (1960). On interchanging limits and integrals. *Ann. Math. Statist.* **31** 74–77.

[13] REZNIK, M. Kh. (1968). The law of the iterated logarithm for some classes of stationary processes. *Theor. Probability Appl.* **13** 606–621.

[14] SCOTT, D. J. (1971). Central limit theorems for martingales and for processes with stationary increments, using a Skorokhod representation approach. To appear in *Adv. Appl. Probability.*

[15] STOUT, W. F. (1970). A martingale analogue of Kolmogorov's law of the iterated logarithm. *Z. Wahrscheinlichkeitstheorie und Verw. Gebiete* **15** 279–290.

[16] STOUT, W. F. (1970). The Hartman-Wintner law of the iterated logarithm for martingales. *Ann. Math. Statist.* **41** 2158–2160.

[17] STRASSEN, V. (1964). An invariance principle for the law of the iterated logarithm. *Z. Wahrscheinlichkeitstheorie und Verw. Gebiete* **3** 211–226.

[18] STRASSEN, V. (1967). Almost sure behavior of sums of independent random variables and martingales. *Proc. Fifth Berkeley Symp. Math. Statist. Prob.* **2** 315–343. Univ. of California Press.

DEPARTMENT OF STATISTICS
AUSTRALIAN NATIONAL UNIVERSITY
BOX 4, P.O. CANBERRA, A.C.T. 2600
AUSTRALIA

The Annals of Probability
1974, Vol. 2, No. 2, 328-332

AN ITERATED LOGARITHM RESULT FOR AUTOCORRELATIONS OF A STATIONARY LINEAR PROCESS

BY C. C. HEYDE

Stanford University and Australian National University

Let $r(j)$ denote the jth autocorrelation based on a sample of N consecutive observations on a stationary linear stochastic process. Under mild regularity conditions on the process, an iterated logarithm result is given for the convergence of $r(j)$ as $N \to \infty$ to the corresponding process autocorrelation $\rho(j)$.

1. Introduction. In this paper we shall be concerned with a stationary linear stochastic process $\{x(n)\}$ which may be represented in the form

$$x(n) - \mu = \sum_{j=-\infty}^{\infty} \alpha(j)\varepsilon(n-j), \qquad \sum_{j=-\infty}^{\infty} \alpha^2(j) < \infty,$$

where the $\varepsilon(n)$ are independent and identically distributed with mean zero and variance σ^2. Most of the standard inferential results for this process (e.g. Hannan [2]) are based on the autocorrelations

$$r(j) = \frac{\sum_{n=1}^{N-j} \{x(n) - \bar{x}\}\{x(n+j) - \bar{x}\}}{\sum_{n=1}^{N} \{x(n) - \bar{x}\}^2}, \qquad j \geqq 0,$$

$x(1), x(2), \cdots, x(N)$ being a sample of N consecutive observations on the process $\{x(n)\}$ and $\bar{x}$ denoting the sample mean. The prime result is that $r(j)$ converges almost surely (a.s.) to $\rho(j) = \gamma(j)/\gamma(0)$ where $\gamma(j)$ is given by

$$\gamma(j) = \sigma^2 \sum_{n=-\infty}^{\infty} \alpha(n)\alpha(n+j).$$

The most common application of this theory would be in the estimation of parameters in an autoregressive or moving average model or a mixture of both models.

It is our object here to give the following iterated logarithm result which provides information on the rate of a.s. convergence of $r(j)$ to $\rho(j)$.

THEOREM. *Suppose that $E\varepsilon^4(n) < \infty$ and $\sum_{n=1}^{\infty} [\sum_{|r| \geqq n} \alpha^2(r)]^{\frac{1}{2}} < \infty$. Then, for any fixed $j \geqq 1$,*

$$r(j) = \rho(j) + A(j)\zeta_j(N)N^{-\frac{1}{2}}(2 \log\log N)^{\frac{1}{2}}$$

where $\zeta_j(N)$ has a.s. *as its set of limit points the interval $[-1, 1]$, in particular* $\limsup_{N\to\infty} \zeta_j(N) = 1$ a.s. *and* $\liminf_{N\to\infty} \zeta_j(N) = -1$ a.s., *while*

$$A(j) = [\{1 + 2\rho^2(j)\} \sum_{s=-\infty}^{\infty} \rho^2(s) + \sum_{s=-\infty}^{\infty} \rho(s+j)\{\rho(s-j) - 4\rho(j)\rho(s)\}]^{\frac{1}{2}}.$$

This result provides a detailed supplement to those obtained in [4] for the

Received December 7, 1972; revised June 21, 1973.

AMS 1970 *subject classifications.* Primary 62M10; Secondary 60F15.

Key words and phrases. Iterated logarithm law, stationary linear processes, estimation of autocorrelations, time series estimation.

R. Maller et al. (eds.), *Selected Works of C.C. Heyde*, Selected Works in Probability and Statistics, DOI 10.1007/978-1-4419-5823-5_35, © Springer Science+Business Media, LLC 2010

linear combination of autocorrelations which arises in estimation of parameters in an autoregressive model. Here we have an iterated logarithm result for each individual component.

2. Proof of theorem. This rests on use of an invariance principle for the law of the iterated logarithm for processes with stationary increments which was established in [5].

Our setting is that of a probability space $(\Omega, \mathscr{B}, P)$ with ergodic measure preserving transformation T such that $\varepsilon(n)\{\omega\} = \varepsilon(T^n\omega)$, $\omega \in \Omega$ (since the $\varepsilon(n)$ are independent and identically distributed and hence stationary and ergodic). Let $\mathscr{F}_n$ denote the σ-field generated by $\varepsilon(m)$, $m \leqq n$. Then, $\mathscr{F}_k = T^{-k}(\mathscr{F}_0)$ and the framework of Theorem 2 of [5] is applicable.

We shall first establish the following lemma which is of independent interest.

LEMMA. *Suppose that* $\sum_{n=1}^{\infty} [\sum_{|r|\geqq n} \alpha^2(r)]^{\frac{1}{2}} < \infty$. *Then,*

$$\bar{x} = \mu + \zeta(n) \sum_{j=-\infty}^{\infty} \gamma(j) N^{-\frac{1}{2}} (2 \log \log N)^{\frac{1}{2}}$$

where $\zeta(N)$ *has* a.s. *as its set of limit points the interval* $[-1, 1]$.

PROOF. Put $y(n) = x(n) - \mu$. We have

$$N^{-1} \operatorname{Var} \bar{x} = \sum_{i=-(N-1)}^{N-1} (1 - |i| N^{-1}) \operatorname{Cov}\{x(0), x(i)\}$$
$$\to \operatorname{Var} x(0) \sum_{i=-\infty}^{\infty} \rho(i) = \sum_{i=-\infty}^{\infty} \gamma(i)$$

as $N \to \infty$. Also,

$$E\{y(0) \mid \mathscr{F}_{-m}\} = \sum_{r=-\infty}^{\infty} \alpha(r) E\{\varepsilon(-r) \mid \mathscr{F}_{-m}\} = \sum_{r=m}^{\infty} \alpha(r)\varepsilon(-r)$$

and

$$\sum_{m=1}^{\infty} [E\{E(y(0) \mid \mathscr{F}_{-m})\}^2]^{\frac{1}{2}} = \sum_{m=1}^{\infty} [\sum_{r=m}^{\infty} \alpha^2(r)]^{\frac{1}{2}} < \infty .$$

Similarly,

$$\sum_{m=1}^{\infty} [E\{y(0) - E(y(0) \mid \mathscr{F}_m)\}^2]^{\frac{1}{2}} = \sum_{m=1}^{\infty} [\sum_{r=-\infty}^{-m-1} \alpha^2(r)]^{\frac{1}{2}} < \infty .$$

The result of the lemma then follows immediately from Theorem 2 of [5].

We now resume the proof of the theorem. Put $X_n = y(n)y(n+j) - \rho(j)y^2(n)$ and note that $\{X_n\}$ is a stationary sequence with zero mean and finite variance under the conditions of the theorem. Further, writing

$$c^*(j) = N^{-1} \sum_{n=1}^{N-j} y(n)y(n+j) ,$$

it is well known that (e.g. Hannan [1] page 40)

$$N \operatorname{Cov}\{c^*(k), c^*(k+j)\} \to \sum_{s=-\infty}^{\infty} \{\gamma(s)\gamma(s+j) + \gamma(s+j+k)\gamma(s-k)\}$$
$$+ k_4 \gamma(k)\gamma(k+j)$$

as $N \to \infty$ where k_4 is the fourth cumulant of $\varepsilon(n)$. Thus,

$$N^{-1} \operatorname{Var} (\sum_{n=1}^{N} X_n)$$
$$= N^{-1} E[N\{c^*(j) - Ec^*(j) - \rho(j)(c^*(0) - Ec^*(0))\}$$
$$+ \sum_{n=N-j+1}^{N} \{y(n)y(n+j) - Ey(n)y(n+j)\}]^2$$
$$\to \lim_{N\to\infty} NE[\{c^*(j) - Ec^*(j)\} - \rho(j)\{c^*(0) - Ec^*(0)\}]^2$$

$$\begin{aligned} &= \lim_{N\to\infty} N \operatorname{Var} c^*(j) + \rho^2(j) \lim_{N\to\infty} N \operatorname{Var} c^*(0) \\ &\qquad - 2\rho(j) \lim_{N\to\infty} N \operatorname{Cov}(c^*(j), c^*(0)) \\ &= \{1 + 2\rho^2(j)\} \textstyle\sum_{s=-\infty}^{\infty} \gamma^2(s) + \sum_{s=-\infty}^{\infty} \gamma(s+j)\{\gamma(s-j) - 4\rho(j)\gamma(s)\} \\ &= \gamma^2(0) A^2(j)\,, \end{aligned}$$

using the fact that $\operatorname{Var}[\sum_{n=N-j+1}^{N} y(n)y(n+j)]$ is independent of N.

In order to apply Theorem 2 of [5] we need to show that

$$\textstyle\sum_{m=1}^{\infty} [E\{E(X_0 \mid \mathscr{F}_{-m})\}^2]^{\frac12} < \infty \quad \text{and} \quad \sum_{m=1}^{\infty} [E\{X_0 - E(X_0 \mid \mathscr{F}_{m})\}^2]^{\frac12} < \infty\,.$$

We have

$$\begin{aligned} E(X_0 \mid \mathscr{F}_{-m}) &= \textstyle\sum_{r=-\infty}^{\infty} \sum_{s=-\infty}^{\infty} \alpha(r)\alpha(s) \\ &\qquad \times \{E(\varepsilon(-r)\varepsilon(j-s) \mid \mathscr{F}_{-m}) - \rho(j) E(\varepsilon(-r)\varepsilon(-s) \mid \mathscr{F}_{-m})\} \\ &= \textstyle\sum_{r=m}^{\infty} \sum_{s\neq j+r,\, s=m+j}^{\infty} \alpha(r)\alpha(s)\varepsilon(-r)\varepsilon(j-s) \\ &\qquad + \textstyle\sum_{r=m}^{\infty} \alpha(r)\alpha(r+j)\{\varepsilon^2(-r) - \sigma^2\} \\ &\qquad - \rho(j) \textstyle\sum_{r=m}^{\infty} \sum_{s\neq r,\, s=m}^{\infty} \alpha(r)\alpha(s)\varepsilon(-r)\varepsilon(-s) \\ &\qquad - \rho(j) \textstyle\sum_{r=m}^{\infty} \alpha^2(r)\{\varepsilon^2(-r) - \sigma^2\}\,, \end{aligned}$$

and after some further algebra

$$\begin{aligned} (1) \qquad E\{E(X_0 \mid \mathscr{F}_{-m})\}^2 &= \{E\varepsilon^4(0) - 3\sigma^4\} \textstyle\sum_{r=m}^{\infty} \alpha^2(r)[\alpha(r+j) - \rho(j)\alpha(r)]^2 \\ &\qquad + \sigma^4 \textstyle\sum_{r=m}^{\infty} \alpha^2(r) \sum_{s=m}^{\infty} [\alpha(s+j) - \rho(j)\alpha(s)]^2 \\ &\qquad + \sigma^4 [\textstyle\sum_{r=m}^{\infty} \alpha(r)\{\alpha(r+j) - \rho(j)\alpha(r)\}]^2\,. \end{aligned}$$

It is then easily checked that $\sum_{m=1}^{\infty} [E\{E(X_0 \mid \mathscr{F}_{-m})\}^2]^{\frac12} < \infty$ under the condition

$$\textstyle\sum_{m=1}^{\infty} \sum_{r=m}^{\infty} \alpha^2(r) = \sum_{r=1}^{\infty} r\alpha^2(r) < \infty\,.$$

Next,

$$E\{X_0 - E(X_0 \mid \mathscr{F}_{m})\}^2 = EX_0^2 - E\{E(X_0 \mid \mathscr{F}_{m})\}^2$$

and a routine calculation gives

$$EX_0^2 = \{E\varepsilon^4(0) - 3\sigma^4\} \textstyle\sum_{r=-\infty}^{\infty} \alpha^2(r)[\alpha(r+j) - \rho(j)\alpha(r)]^2 + \gamma^2(0) - \gamma^2(j)\,,$$

so that using (1) (with $-m$ replaced by m) and after some algebra,

$$\begin{aligned} E\{X_0 - E(X_0 \mid \mathscr{F}_{m})\}^2 &= \{E\varepsilon^4(0) - 3\sigma^4\} \textstyle\sum_{r=-\infty}^{-m-1} \alpha^2(r)[\alpha(r+j) - \rho(j)\alpha(r)]^2 \\ &\qquad - \sigma^4 [\textstyle\sum_{r=-\infty}^{-m-1} \alpha(r)\{\alpha(r+j) - \rho(j)\alpha(r)\}]^2 \\ &\qquad + \sigma^2\gamma(0) \textstyle\sum_{r=-\infty}^{-m-1} [\alpha(r+j) - \rho(j)\alpha(r)]^2 \\ &\qquad + \sigma^2 \textstyle\sum_{r=-\infty}^{-m-1} \alpha^2(r)\{\gamma(0)(1 - \rho^2(j)) \\ &\qquad - \sigma^2 \textstyle\sum_{s=-\infty}^{-m-1} [\alpha(s+j) - \rho(j)\alpha(s)]^2\}\,. \end{aligned}$$

Thus, $\sum_{m=1}^{\infty} [E\{X_0 - E(X_0 \mid \mathscr{F}_{m})\}^2]^{\frac12} < \infty$ if the following four conditions hold:

(i) $\sum_{m=1}^{\infty} [\sum_{r=-\infty}^{-m-1} \alpha^2(r)\{\alpha(r+j) - \rho(j)\alpha(r)\}^2]^{\frac12} < \infty$,

(ii) $\sum_{m=1}^{\infty} |\sum_{r=-\infty}^{-m-1} \alpha(r)\{\alpha(r+j) - \rho(j)\alpha(r)\}| < \infty$,

(iii) $\sum_{m=1}^{\infty} [\sum_{r=-\infty}^{-m-1} \{\alpha(r+j) - \rho(j)\alpha(r)\}^2]^{\frac12} < \infty$,

(iv) $\sum_{m=1}^{\infty} [\sum_{r=-\infty}^{-m-1} \alpha^2(r)]^{\frac12} < \infty$.

Furthermore, we readily find that (i) and (ii) are satisfied for $\sum_{m=1}^{\infty}\sum_{r=-\infty}^{-m-1}\alpha^2(r) < \infty$. On the other hand, (iii) and (iv) are satisfied when $\sum_{m=1}^{\infty}[\sum_{r=-\infty}^{-m-1}\alpha^2(r)]^{\frac{1}{2}} < \infty$.

Consequently, the conditions of Theorem 2 of [5] are satisfied and

$$(2N\log\log N)^{-\frac{1}{2}}\sum_{n=1}^{N}\{y(n)y(n+j)-\rho(j)y^2(n)\}$$

has a.s. as its set of limit points the interval $[-\gamma(0)A(j), \gamma(0)A(j)]$. But, the sequence $\{y(n)y(n+j)\}$ is stationary with finite variance so a simple application of the Borel–Cantelli lemma gives

$$\lim_{N\to\infty}(N\log\log N)^{-\frac{1}{2}}\sum_{n=N-j+1}^{N}y(n)y(n+j) = 0 \quad \text{a.s.}$$

and thus

$$N^{\frac{1}{2}}(2\log\log N)^{-\frac{1}{2}}\{c^*(j)-\rho(j)c^*(0)\}$$

has a.s. as its set of limit points the interval $[-\gamma(0)A(j), \gamma(0)A(j)]$.

Now, if

$$c(j) = N^{-1}\sum_{n=1}^{N-j}\{x(n)-\bar{x}\}\{x(n+j)-\bar{x}\}, \qquad j \geqq 0,$$

we have

$$c^*(j)-c(j) = (\bar{x}-\mu)^2 - N^{-1}\{\sum_{n=N-j+1}^{N}x(n)+\sum_{n=1}^{j-1}x(n)-j(\bar{x}+\mu)\}(\bar{x}-\mu),$$

and with the aid of the lemma it is easily seen that for any fixed $k \geqq 0$,

$$\lim_{N\to\infty}N^{\frac{1}{2}}(\log\log N)^{-\frac{1}{2}}\{c^*(k)-c(k)\} = 0 \quad \text{a.s.}$$

The result of the theorem then follows upon noting that, from the ergodic theorem, $c(0) \to_{\text{a.s.}} \gamma(0)$ as $N \to \infty$.

3. Remarks.

1. Since the result of the theorem does not explicitly involve $E\varepsilon^4(n)$ it seems possible that it will continue to hold in the absence of $E\varepsilon^4(n) < \infty$. This is certainly the case with the corresponding central limit result.

2. It is easy to see that similar methods can be applied to investigate the rate of convergence of $c(j)$ to $\gamma(j)$. In this case Theorem 2 of [5] would be applied to the stationary ergodic sequence $\{Y_n\}$ where $Y_n = y(n)y(n+j)-\gamma(j)$. The result that emerges is, again under $E\varepsilon^4(n) < \infty$ and $\sum_{n=1}^{\infty}[\sum_{|r|\geq n}\alpha^2(r)]^{\frac{1}{2}} < \infty$,

$$c(j) = \gamma(j) + B(j)\eta_j(N)N^{-\frac{1}{2}}(2\log\log N)^{\frac{1}{2}}.$$

Here $\eta_j(N)$ has a.s. as its set of limit points the interval $[-1, 1]$, while $B(j)$ is given by

$$B(j) = [\sum_{s=-\infty}^{\infty}\{\gamma^2(s)+\gamma(s+j)\gamma(s-j)\}+k_4\gamma^2(j)]^{\frac{1}{2}},$$

k_4 being the fourth cumulant of $\varepsilon(n)$. In this case we explicitly need the finite fourth moment condition.

3. If a restriction is made to the case where the process $\{x(n)\}$ is purely nondeterministic ($\alpha(n) = 0$, $n < 0$), the assumption that the $\varepsilon(n)$ are independent and identically distributed can be replaced by a more general martingale condition along the lines of [3] and [4]. However, it is then necessary to make

certain unattractive assumptions on third and fourth moments and also to impose slightly stronger series conditions on the α's than that used here.

4. From an examination of the proof of the theorem it seems likely that, at least in the purely nondeterministic case, the result will continue to hold under only $\sum_{n=1}^{\infty} \sum_{|r| \geq n} \alpha^2(r) = \sum_{r=-\infty}^{\infty} |r|\alpha^2(r) < \infty$. A convergence rate result for $\bar{x} - \mu$ considerably weaker than that provided by the lemma is all that is required. For example, $N^{\frac{1}{2}}(\bar{x} - \mu) \to_{\text{a.s.}} 0$ would suffice.

REFERENCES

[1] HANNAN, E. J. (1960). *Time Series Analysis*. Methuen, London.

[2] HANNAN, E. J. (1970). *Multiple Time Series*. Wiley, New York.

[3] HANNAN, E. J. and HEYDE, C. C. (1972). On limit theorems for quadratic functions of discrete time series. *Ann. Math. Statist.* **43** 2058–2066.

[4] HEYDE, C. C. (1973). An iterated logarithm result for martingales and its application in estimation theory for autoregressive processes. *J. Appl. Probability* **10** 146–157.

[5] HEYDE, C. C. and SCOTT, D. J. (1973). Invariance principles for the law of the iterated logarithm for martingales and processes with stationary increments. *Ann. Probability* **1** 428–436.

THE AUSTRALIAN NATIONAL UNIVERSITY
DEPARTMENT OF STATISTICS, S. G. S.
BOX 4, POST OFFICE
CANBERRA, A. C. T. 2600
AUSTRALIA

Adv. Appl. Prob. **6**, 421–433 (1974)
Printed in Israel

ON ESTIMATING THE VARIANCE OF THE OFFSPRING DISTRIBUTION IN A SIMPLE BRANCHING PROCESS

C. C. HEYDE, *Australian National University*

Abstract

A single realization $\{Z_n, 0 \leqq n \leqq N+1\}$ of a supercritical Galton-Watson process (so called) is considered and it is required to estimate the variance of the offspring distribution. A prospective estimator $N^{-1}\Sigma_{j=1}^{N}(Z_{j+1} - \hat{m}Z_j)^2 Z_j^{-1}$ is proposed, where $\hat{m} = Z_{N+1}/Z_N$, and is shown to be strongly consistent on the non-extinction set. A central limit result and an iterated logarithm result are provided to give information on the rate of convergence of the estimator. It is also shown that the estimation results are robust in the sense that they continue to apply unchanged in the case where immigration occurs. Martingale limit theory is employed at each stage in obtaining the limit results.

BRANCHING PROCESSES; GALTON-WATSON PROCESSES; ESTIMATION; CONSISTENCY; CENTRAL LIMIT THEOREM; ITERATED LOGARITHM LAW; MARTINGALES; ROBUSTNESS

1. Introduction

Let $Z_0 = 1, Z_1, Z_2, \cdots$ denote a supercritical simple branching process of the kind that is commonly but inaccurately termed Galton-Watson process (see [8]). Suppose that $1 < EZ_1 = m$ and $0 < \text{var } Z_1 = \sigma^2 < \infty$. In this case it is well known that there exists a non-degenerate random variable W such that $\lim_{n\to\infty} m^{-n} Z_n = W$ almost surely (a.s.). It is also well known, and easily deducible from the above a.s. convergence result, how to estimate the mean m of the offspring distribution on the basis of a single realization $\{Z_n, 0 \leqq n \leqq N+1\}$. In fact, $Z_{N+1} Z_N^{-1} \overset{\text{a.s.}}{\to} m$ on the non-extinction set $\{W > 0\}$. Alternatively, the maximum likelihood estimator $\Sigma_{j=1}^{N+1} Z_j / \Sigma_{j=0}^{N} Z_j$ may be used and, again on the non-extinction set $\{W > 0\}$, $\Sigma_{j=1}^{N+1} Z_j / \Sigma_{j=0}^{N} Z_j \overset{\text{a.s.}}{\to} m$ (see Heyde [4]). The maximum likelihood estimator, however, does not appear to have any special optimality properties in this context.

In this paper we shall be concerned with the estimation of σ^2, the variance of the offspring distribution, on the basis of the sample $\{Z_n, 0 \leqq n \leqq N+1\}$. This issue does not appear to have been discussed previously in the literature, save that Nagaev [10] observed that in the critical case, $m = 1$, $(Z_{N+1} - Z_N)^2 / Z_N$ could be used to estimate σ^2. We propose the estimator

$$N^{-1} \sum_{j=1}^{N} (Z_{j+1} - \hat{m} Z_j)^2 Z_j^{-1}$$

Received 25 July 1973.

R. Maller et al. (eds.), *Selected Works of C.C. Heyde*, Selected Works in Probability and Statistics,
DOI 10.1007/978-1-4419-5823-5_36, © Springer Science+Business Media, LLC 2010

where $\hat{m} = Z_{N+1}/Z_N$, show that it is strongly consistent for σ^2 on the non-extinction set $\{W > 0\}$ and obtain a central limit result and an iterated logarithm result which give information on its rate of convergence to σ^2. We also show that the estimation procedure is robust in the sense that the results all continue to hold if an immigration component is introduced into the branching process.

2. Results

Theorem 1. Let $\hat{m} = Z_{N+1}Z_N^{-1}$. Then, on the non-extinction set $\{W > 0\}$,

$$N^{-1}\sum_{j=1}^{N}(Z_{j+1} - \hat{m}Z_j)^2 Z_j^{-1} \xrightarrow{\text{a.s.}} \sigma^2 \tag{1}$$

as $N \to \infty$.

Proof. There is no essential loss of generality in giving the proof for the case $P(Z_1 = 0) = 0$ in which case $Z_n \to \infty$ a.s. as $n \to \infty$. This we shall do for the sake of convenience. When $P(Z_1 = 0) > 0$ it is well known that $Z_n \to \infty$ a.s. on the set $\{W > 0\}$.

Let $\mathscr{F}_n$ be the σ-field generated by $Z_1, \cdots, Z_n$ and put

$$U_{n+1} = (Z_{n+1} - mZ_n)^2 Z_n^{-1} - \sigma^2. \tag{2}$$

Then, since we can write

$$Z_{n+1} = Z_1^{(1)} + Z_1^{(2)} + \cdots + Z_1^{(Z_n)} \tag{3}$$

where the $Z_1^{(i)}$ are independent and identically distributed (i.i.d.) each with the distribution of Z_1, we have

$$E(U_{n+1} \mid \mathscr{F}_n) = E\{Z_n^{-1}((Z_1^{(1)} - m) + \cdots + (Z_1^{(Z_n)} - m))^2 \mid \mathscr{F}_n\} - \sigma^2 = 0$$

and $E|U_{n+1}| \leqq 2\sigma^2$ so that the U_n are martingale differences. We prove (1) via a strong law of large numbers for martingales.

First note that, using (2) and (3),

$$U_{n+1} \stackrel{\mathscr{D}}{=} Z_n^{-1} S_{Z_n}^2 - \sigma^2 \tag{4}$$

where $S_n = \sum_{i=1}^n X_i$, $n \geqq 1$, say, the X_i being i.i.d. each with the distribution of $Z_1 - m$ ("$\mathscr{D}$" denotes "in distribution"). Then, using Theorem 1 of Heyde and Leslie [6],

$$\begin{aligned} P(|U_{n+1}| > x) &= P(|Z_n^{-1} S_{Z_n}^2 - \sigma^2| > x) \\ &\leqq \begin{cases} 1, & 0 \leqq x \leqq \sigma^2, \\ P(S_{Z_n}^2 > Z_n x), & x > \sigma^2, \end{cases} \\ &\leqq \begin{cases} 1, & 0 \leqq x \leqq \sigma^2, \\ 2[1 - \Phi(x^{\frac{1}{2}}\sigma^{-1}) + d_n], & x > \sigma^2, \end{cases} \end{aligned} \tag{5}$$

where $\{d_n\}$ is a certain sequence of positive constants with $\Sigma_n d_n < \infty$ and $\Phi(\cdot)$ is the distribution function of the unit normal law, $N(0, 1)$.

Now we put $U'_n = U_n I(|U_n| \leqq n)$, $n \geqq 1$, and then, using (5),

$$\begin{aligned}
&\sum_{n=1}^{\infty} n^{-2} E(U'_n - E(U'_n \mid \mathscr{F}_{n-1}))^2 \\
&\qquad \leqq \sum_{n=1}^{\infty} n^{-2} E(U'_n)^2 \\
&\qquad \leqq 2 \sum_{n=1}^{\infty} n^{-2} \int_{0<x\leqq n} x P(|U_n| > x)\, dx \\
&\qquad \leqq 2 \sum_{n=1}^{[\sigma^2]} n^{-2} \int_0^n x\, dx \\
&\qquad + 2 \sum_{n=[\sigma^2]+1}^{\infty} n^{-2} \left\{ \int_0^{\sigma^2} x\, dx + 2 \int_{\sigma^2}^{n} x(1 - \Phi(x^{\frac{1}{2}} \sigma^{-1}) + d_{n-1})\, dx \right\} \\
&\qquad \leqq [\sigma^2] + 4\sigma^4 \sum_{n=1}^{\infty} n^{-2} + 2 \sum_{n=0}^{\infty} d_n < \infty
\end{aligned}$$

after some algebra. Here $[\sigma^2]$ denotes the integer part of σ^2. Thus, using the standard Kolmogorov convergence criterion for martingales (e.g., Loève [9], p. 387),

$$n^{-1} \sum_{k=1}^{n} \{U'_k - E(U'_k \mid \mathscr{F}_{k-1})\} \xrightarrow{\text{a.s.}} 0 \tag{6}$$

as $n \to \infty$.

Next, again using (5) we have

$$\sum_{k=1}^{\infty} P(U_k \neq U'_k) = \sum_{k=1}^{\infty} P(|U_k| > k) < \infty$$

provided that

$$\sum_{k=1}^{\infty} \{1 - \Phi(k^{\frac{1}{2}} \sigma^{-1})\} < \infty$$

which certainly holds since

$$1 - \Phi(k^{\frac{1}{2}} \sigma^{-1}) \sim (2\pi)^{-\frac{1}{2}} \sigma k^{-\frac{1}{2}} e^{-\frac{1}{2}\sigma^2 k}$$

as $k \to \infty$. Thus, the sequences $\{U_k\}$ and $\{U'_k\}$ are tail equivalent and hence from (6),

$$n^{-1} \sum_{k=1}^{n} \{U_k - E(U'_k \mid \mathscr{F}_{k-1})\} \xrightarrow{\text{a.s.}} 0. \tag{7}$$

Now we shall show that

$$E(U_n' \mid \mathscr{F}_{n-1}) \overset{\text{a.s.}}{\to} 0 \tag{8}$$

as $n \to \infty$. Then of course

$$n^{-1} \sum_{k=1}^{n} E(U_k' \mid \mathscr{F}_{k-1}) \overset{\text{a.s.}}{\to} 0.$$

so that from (7),

$$n^{-1} \sum_{k=1}^{n} U_k \overset{\text{a.s.}}{\to} 0$$

which gives

$$N^{-1} \sum_{k=1}^{N} (Z_{k+1} - m Z_k)^2 Z_k^{-1} \overset{\text{a.s.}}{\to} \sigma^2. \tag{9}$$

To prove (8) we note that, using the martingale property,

$$\begin{aligned} E(U_n' \mid \mathscr{F}_{n-1}) &= -E(U_n I(|U_n| > n) \mid \mathscr{F}_{n-1}) \\ &= \sigma^2 P(|U_n| > n \mid \mathscr{F}_{n-1}) \\ &\quad - Z_{n-1}^{-1} E((Z_n - m Z_{n-1})^2 I(|U_n| > n) \mid \mathscr{F}_{n-1}) \end{aligned}$$

and

$$P(|U_n| > n \mid \mathscr{F}_{n-1}) \overset{\text{a.s.}}{\to} 0$$

since

$$\sum_{n=1}^{\infty} EP(|U_n| > n \mid \mathscr{F}_{n-1}) = \sum_{n=1}^{\infty} P(|U_n| > n) < \infty$$

as shown above while, for $n > \sigma^2$,

$$\begin{aligned} &Z_{n-1}^{-1} E((Z_n - m Z_{n-1})^2 I(|U_n| > n) \mid \mathscr{F}_{n-1}) \\ &\leqq Z_{n-1}^{-1} E((Z_n - m Z_{n-1})^2 I((Z_n - m Z_{n-1})^2 > n Z_{n-1}) \mid \mathscr{F}_{n-1}) \\ &\overset{\text{a.s.}}{\to} 0 \end{aligned}$$

as $n \to \infty$, recalling (4) and since $\{S_k^2/k\}$ is uniformly integrable. The uniform integrability of $\{S_k^2/k\}$ is a simple consequence of Theorem 5.4 of Billingsley [1].

Finally, we have

$$N^{-1}\sum_{k=1}^{N}(Z_{k+1}-\hat{m}Z_k)^2Z_k^{-1}$$

$$(10)\qquad = N^{-1}\sum_{k=1}^{N}(Z_{k+1}-mZ_k)^2Z_k^{-1}+2N^{-1}(m-\hat{m})\sum_{k=1}^{N}(Z_{k+1}-mZ_k)$$

$$+N^{-1}(m-\hat{m})^2\sum_{k=1}^{N}Z_k.$$

The first term on the right-hand side of (10) tends a.s. to σ^2 by (9) so to complete the proof of (1) we need to show that the second and third terms on the right-hand side of (10) tend a.s. to zero.

To deal with the first of these, we note that

$$E(Z_{j+1}\,|\,\mathscr{F}_j) = mZ_j$$

so that $\{\sum_{j=1}^{N}(Z_{j+1}-mZ_j), \mathscr{F}_{N+1}\}$ is a martingale. Also,

$$E((Z_{j+1}-mZ_j)^2\,|\,\mathscr{F}_j) = Z_j\sigma^2$$

so that

$$E(Z_{j+1}-mZ_j)^2 = \sigma^2 m^j$$

and

$$\sum_{j=1}^{\infty} m^{-j}j^{-\frac{3}{2}}E(Z_{j+1}-mZ_j)^2 < \infty$$

which ensures, via the Kolmogorov condition for martingales,

$$(11)\qquad m^{-\frac{1}{2}N}N^{-\frac{3}{4}}\sum_{j=1}^{N}(Z_{j+1}-mZ_j)\xrightarrow{\text{a.s.}}0.$$

Furthermore, Theorem 2 of [6] shows that

$$(12)\qquad \hat{m}-m = Z_{N+1}Z_N^{-1}-m = \sigma\zeta(N)(2Z_N^{-1}\log N)^{\frac{1}{2}}$$

where $\zeta(N)$ has its set of a.s. limit points confined to $[-1,1]$ with $\limsup_{N\to\infty}\zeta(N) = +1$ a.s., $\liminf_{N\to\infty}\zeta(N) = -1$ a.s. Then, using (11) and (12),

$$N^{-1}(\hat{m}-m)\sum_{k=1}^{N}(Z_{k+1}-mZ_k)$$

$$(13)\qquad = \sigma\{\zeta(N)(2\log N)^{\frac{1}{2}}N^{-\frac{1}{4}}\}\{(m^NZ_N^{-1})^{\frac{1}{2}}\}$$

$$\times\left\{m^{-\frac{1}{2}N}N^{-\frac{3}{4}}\sum_{j=1}^{N}(Z_{j+1}-mZ_j)\right\}\xrightarrow{\text{a.s.}}0.$$

For the third term on the right-hand side of (10) we have, using (12) again,

$$N^{-1}(m-\hat{m})^2 \sum_{j=1}^{N} Z_j$$

(14)

$$= 2\sigma^2\{\zeta^2(N)\, N^{-1}\log N\}\, \{m^N Z_N^{-1}\} \left\{m^{-N} \sum_{j=1}^{N} Z_j\right\} \xrightarrow{\text{a.s.}} 0$$

since $m^{-N} \sum_{j=1}^{N} Z_j \xrightarrow{\text{a.s.}} mW/(m-1)$ by Theorem 3 of Heyde [4]. The proof of Theorem 1 is then completed by using (9), (13) and (14) in (10).

Next we proceed to a central limit result which provides a rate of convergence result for the estimator of Theorem 1.

Theorem 2. Suppose that $EZ_1^4 < \infty$. Then, as $N \to \infty$,

$$P\left(N^{\frac{1}{2}}(2\sigma^4)^{-1} \left\{N^{-1} \sum_{j=1}^{N} (Z_{j+1} - \hat{m} Z_j)^2 Z_j^{-1} - \sigma^2 \right\} \leqq x \,\middle|\, Z_N > 0\right) \to \Phi(x).$$

Proof. Again we shall just give the proof for the case $P(Z_1 = 0) = 0$. We first set out to prove

$$N^{\frac{1}{2}}(2\sigma^4)^{-1} \left\{N^{-1} \sum_{j=1}^{N} (Z_{j+1} - m Z_j)^2 Z_j^{-1} - \sigma^2 \right\} \xrightarrow{\mathscr{D}} N(0,1)$$

and then show that we can replace m by its estimator $\hat{m}$. The central limit result is obtained from a central limit theorem for martingales (Theorem 2 of Brown [3]) applied to the martingale $\{\sum_{j=1}^{N} U_j, \mathscr{F}_N\}$.

Using (3) and writing $T_{Z_n} = (Z_1^{(1)} - m) + \cdots + (Z_1^{(Z_n)} - m)$,

$$\begin{aligned} E(U_{n+1}^2 \,|\, \mathscr{F}_n) &= E((Z_n^{-1} T_{Z_n}^2 - \sigma^2)^2 \,|\, \mathscr{F}_n) \\ &= E(Z_n^{-2} T_{Z_n}^4 \,|\, \mathscr{F}_n) - \sigma^4 \end{aligned}$$

and

$$\begin{aligned} &E(Z_n^{-2} T_{Z_n}^4 \,|\, \mathscr{F}_n) \\ &= E\left\{Z_n^{-2}\left(\sum_{i=1}^{Z_n} (Z_1^{(i)} - m)^4 + 6 \sum_{Z_n \geqq i > j \geqq 1} (Z_1^{(i)} - m)^2 (Z_1^{(j)} - m)^2\right) \,\middle|\, \mathscr{F}_n\right\} \\ &= Z_n^{-1} E(Z_1 - m)^4 + 3\sigma^4(1 - Z_n^{-1}) \xrightarrow{\text{a.s.}} 3\sigma^4 \end{aligned}$$

as $n \to \infty$ so that

(15) $$n^{-1} \sum_{j=1}^{n} E(U_j^2 \,|\, \mathscr{F}_{j-1}) \xrightarrow{\text{a.s.}} 2\sigma^4.$$

Also, since $EZ_n^{-1} \leqq \gamma^n$ for some γ, $0 < \gamma < 1$ (see the proof of Theorem 2 of [5]),

(16) $$n^{-1} \sum_{j=1}^{n} E(U_j^2) \to 2\sigma^4$$

Thus, writing

$$s_n^2 = E\left(\sum_{j=1}^{n} U_j\right)^2 = \sum_{j=1}^{n} EU_j^2,$$

we have

(17) $$s_n^{-2} \sum_{j=1}^{n} E(U_j^2 \mid \mathscr{F}_{j-1}) \xrightarrow{\text{a.s.}} 1$$

as $n \to \infty$.

Next, we need to show that a Lindeberg condition is satisfied. In fact, we need

$$s_n^{-2} \sum_{j=1}^{n} E(U_j^2 I(|U_j| \geqq \varepsilon s_n)) \to 0$$

as $n \to \infty$ for all $\varepsilon > 0$ and this is clearly satisfied if

$$s_n^{-2} \sum_{j=1}^{n} E(U_j^2 I(|U_j| \geqq \varepsilon s_j)) \to 0$$

and hence, in view of (16), if

(18) $$E(U_n^2 I(|U_n| \geqq \eta n^{\frac{1}{2}})) \to 0$$

as $n \to \infty$ for all $\eta > 0$.

Now, using (4) and for n sufficiently large,

(19) $$\begin{aligned} &E(U_n^2 I(|U_n| \geqq \eta n^{\frac{1}{2}})) \\ &= E((Z_{n-1}^{-1} S_{Z_{n-1}}^2 - \sigma^2)^2 I(|Z_{n-1}^{-1} S_{Z_{n-1}}^2 - \sigma^2| > \eta n^{\frac{1}{2}})) \\ &\leqq 2E(Z_{n-1}^{-2} S_{Z_{n-1}}^4 I(Z_{n-1}^{-2} S_{Z_{n-1}}^4 > \eta^2 n)) \\ &\quad + 2\sigma^4 P(S_{Z_{n-1}}^2 > \eta\, n^{\frac{1}{2}} Z_{n-1}). \end{aligned}$$

The first term on the right-hand side of (19) tends to zero as $n \to \infty$ in view of the uniform integrability of $\{S_k^4/k^2\}$, which in turn follows, for example, from Theorem 5 of Brown [2] and Theorem 5.4 of Billingsley [1]. That the second term on the right-hand side of (19) tends to zero as $n \to \infty$ follows immediately from Theorem 1 of Heyde and Leslie [6].

The required conditions are thus satisfied ((17) and (18)) for the application of Theorem 2 of [3] which gives

$$s_N^{-2} \sum_{j=1}^{N} U_j \xrightarrow{\mathscr{D}} N(0,1),$$

that is,

(20) $$(2\sigma^4)^{-1} N^{\frac{1}{2}} \left\{N^{-1} \sum_{j=1}^{N} (Z_{j+1} - m Z_j)^2 Z_j^{-1} - \sigma^2\right\} \xrightarrow{\mathscr{D}} N(0,1).$$

Finally, we need to show that we can replace m by $\hat{m}$ in (20). We have

$$(2\sigma^4)^{-1}N^{\frac{1}{2}}\left\{N^{-1}\sum_{j=1}^{N}(Z_{j+1}-\hat{m}Z_j)^2Z_j^{-1}-\sigma^2\right\}$$

$$= (2\sigma^4)^{-1}N^{\frac{1}{2}}\left\{N^{-1}\sum_{j=1}^{N}(Z_{j+1}-mZ_j)^2Z_j^{-1}-\sigma^2\right\}$$

$$+\ 2(2\sigma^4)^{-1}N^{-\frac{1}{2}}(m-\hat{m})\sum_{j=1}^{N}(Z_{j+1}-mZ_j)$$

$$+\ (2\sigma^4)^{-1}N^{-\frac{1}{2}}(m-\hat{m})^2\sum_{j=1}^{N}Z_j$$

and in view of (20) it certainly suffices to show that

$$N^{-\frac{1}{2}}(m-\hat{m})\sum_{j=1}^{N}(Z_{j+1}-mZ_j)\overset{\text{a.s.}}{\to}0 \tag{21}$$

and

$$N^{-\frac{1}{2}}(m-\hat{m})^2\sum_{j=1}^{N}Z_j\overset{\text{a.s.}}{\to}0. \tag{22}$$

The full force of these a.s. convergence results will be used later.

To prove (21) we first note that, in view of (12) and $Z_N/m^N\overset{\text{a.s.}}{\to}W$,

$$m^{-\frac{1}{2}N}(\log N)^{-1}(Z_{N+1}-mZ_N)\overset{\text{a.s.}}{\to}0.$$

Thus, using the Toeplitz Lemma (e.g., Loève [9], p. 238),

$$\left(\sum_{j=1}^{N}m^{\frac{1}{2}j}\log j\right)^{-1}\sum_{j=1}^{N}(Z_{j+1}-mZ_j)\overset{\text{a.s.}}{\to}0$$

and

$$N^{-\frac{1}{2}}(\hat{m}-m)\sum_{j=1}^{N}(Z_{j+1}-mZ_j)$$

$$=\sigma\left\{\zeta(N)N^{-\frac{1}{2}}(2\log N)^{\frac{1}{2}}m^{-\frac{1}{2}N}\left(\sum_{j=1}^{N}m^{\frac{1}{2}j}\log j\right)\right\}$$

$$\times\{(m^NZ_N^{-1})^{\frac{1}{2}}\}\left\{\left(\sum_{j=1}^{N}m^{\frac{1}{2}j}\log j\right)^{-1}\sum_{j=1}^{N}(Z_{j+1}-mZ_j)\right\}$$

$$\overset{\text{a.s.}}{\to}0$$

since

$$m^{-\frac{1}{2}N}N^{-\frac{1}{2}}(\log N)^{\frac{1}{2}}\sum_{j=1}^{N} m^{\frac{1}{2}j}\log j$$

$$\sim\ m^{-\frac{1}{2}N}N^{-\frac{1}{2}}(\log N)^{\frac{1}{2}}\int_1^N \exp\{x\log m^{\frac{1}{2}}\}\log x\,dx$$

$$\sim\ (\log m^{\frac{1}{2}})^{-1}(\log N)^{\frac{3}{2}}N^{-\frac{1}{2}}\to 0.$$

The result (22) is proved by noting that

$$N^{-\frac{1}{2}}(m-\hat{m})^2\sum_{j=1}^{N} Z_j$$

$$= 2\sigma^2\{\zeta^2(N)N^{-\frac{1}{2}}\log N\}\{m^N Z_N^{-1}\}\left\{m^{-N}\sum_{j=1}^{N} Z_j\right\}\stackrel{\text{a.s.}}{\to} 0$$

as with (14). This completes the proof of the theorem.

Another type of rate of convergence result for the estimation procedure of Theorem 1 is provided by the following iterated logarithm result which gives, in a sense, a second term in the asymptotic expansion.

Theorem 3. Suppose that $EZ_1^{4+\delta}<\infty$ for some $\delta>0$. Then, on the non-extinction set $\{W>0\}$,

$$N^{-1}\sum_{j=1}^{N}(Z_{j+1}-\hat{m}Z_j)^2Z_j^{-1} = \sigma^2+\eta(N)(4\sigma^4N^{-1}\log\log N)^{\frac{1}{2}}$$

where $\eta(N)$ has its set of a.s. limit points confined to $[-1,1]$ with $\limsup_{N\to\infty}\eta(N)=+1$ a.s., $\liminf_{N\to\infty}\eta(N)=-1$ a.s.

Proof. Again we shall just give the proof for the case $P(Z_1=0)=0$. The result is obtained with the aid of an iterated logarithm result for martingales (Theorem 1 of Heyde and Scott [7]) applied to the martingale $\{\Sigma_{j=1}^N U_j, \mathscr{F}_N\}$.
Again putting

$$s_n^2 = E\left(\sum_{j=1}^{n} U_j\right)^2 = \sum_{j=1}^{n} EU_j^2,$$

what we need to check in order to apply Theorem 1 of [7] is that

$$\sum_{n=1}^{\infty} s_n^{-4}E(U_n^4 I(|U_n|<\delta s_n))<\infty \text{ for some } \delta>0, \tag{23}$$

$$\sum_{n=1}^{\infty} s_n^{-1}E(|U_n|I(|U_n|\geqq\varepsilon s_n))<\infty \text{ for all } \varepsilon>0, \tag{24}$$

and

$$s_n^{-2}\sum_{k=1}^{n} U_k^2\stackrel{\text{a.s.}}{\to} 1 \qquad \text{as } n\to\infty. \tag{25}$$

That (23) holds is easily seen from (16) and a minor modification of the argument used to show that

$$\sum_{n=1}^{\infty} n^{-2} E(U_n')^2 < \infty$$

in the proof of Theorem 1.

To show that (24) holds we firstly note that, using (4) and the c_r- inequality (e.g., Loève [9], p. 155),

$$\begin{aligned} E\left|U_n\right|^{2+\frac{1}{2}\delta} &= E\left|Z_{n-1}^{-1} S_{Z_{n-1}}^2 - \sigma^2\right|^{2+\frac{1}{2}\delta} \\ &\leqq 2^{1+\frac{1}{2}\delta}\{E(Z_{n-1}^{-\frac{1}{2}}\left|S_{Z_{n-1}}\right|)^{4+\delta} + \sigma^{4+\delta}\} \leqq C < \infty \end{aligned}$$

for some $C > 0$ where we have employed Theorem 5 of [2] to uniformly bound $E(Z_{n-1}^{-\frac{1}{2}}\left|S_{Z_{n-1}}\right|)^{4+\delta}$. Then, using Markov's inequality,

$$P(\left|U_n\right| \geqq x) \leqq x^{-2-\frac{1}{2}\delta}\, E\left|U_n\right|^{2+\frac{1}{2}\delta} \leqq C x^{-2-\frac{1}{2}\delta}. \tag{26}$$

In view of (16), the result (24) holds since, for any $\eta > 0$,

$$\begin{aligned} &\sum_{n=1}^{\infty} n^{-\frac{1}{2}} E(\left|U_n\right| I(\left|U_n\right| \geqq \eta n^{\frac{1}{2}})) \\ &\qquad \leqq \eta \sum_{n=1}^{\infty} P(\left|U_n\right| \geqq \eta n^{\frac{1}{2}}) + \sum_{n=1}^{\infty} n^{-\frac{1}{2}} \int_{\eta\sqrt{n}}^{\infty} P(\left|U_n\right| > x)\, dx \\ &\qquad \leqq \eta C \sum_{n=1}^{\infty} (\eta n^{\frac{1}{2}})^{-2-\frac{1}{2}\delta} + C \sum_{n=1}^{\infty} n^{-\frac{1}{2}} \int_{\eta\sqrt{n}}^{\infty} x^{-2-\frac{1}{2}\delta}\, dx < \infty. \end{aligned}$$

To prove (25) we need to show that

$$n^{-1} \sum_{j=1}^{n} \{U_j^2 - E(U_j^2 \mid \mathscr{F}_{j-1})\} \xrightarrow{\text{a.s.}} 0 \tag{27}$$

and then the required result follows using (16) and (17). The result (27) is obtained by paralleling much of the argument used to derive the strong law for martingales in the proof of Theorem 1. In fact, now defining $U_n' = U_n^2 I(U_n^2 \leqq n)$, $n \geqq 1$, we can mimic the previous argument up to the point of obtaining

$$n^{-1} \sum_{k=1}^{n} \{U_k^2 - E(U_k' \mid \mathscr{F}_{k-1})\} \xrightarrow{\text{a.s.}} 0. \tag{28}$$

The only extra piece of information we need in order to achieve this is that $\sum_{n=1}^{\infty} n^2 d_n < \infty$ and this holds since d_n decreases geometrically under the present moment conditions ([5], Theorem 2). Next we have, using (26),

$$\sum_{n=1}^{\infty} n^{-1}E(E(U_n^2 I(U_n^2 > n) \mid \mathscr{F}_{n-1}))$$

(29)
$$= \sum_{n=1}^{\infty} n^{-1}E(U_n^2 I(U_n^2 > n))$$
$$\leqq \sum_{n=1}^{\infty} P(U_n^2 > n) + \sum_{n=1}^{\infty} n^{-1}\int_n^{\infty} P(U_n^2 > x)\,dx$$
$$\leqq C \sum_{n=1}^{\infty} n^{-1-\frac{1}{2}\delta} + C\sum_{n=1}^{\infty} n^{-1}\int_n^{\infty} x^{-1-\frac{1}{2}\delta}dx < \infty.$$

Thus, using the Kronecker lemma together with (29),

$$n^{-1}\sum_{k=1}^{n} E(U_k^2 I(U_k^2 > k) \mid \mathscr{F}_{k-1}) \overset{\text{a.s.}}{\to} 0$$

and hence (27) follows via (28).

Theorem 1 of [7] then gives

$$\limsup_{N\to\infty} \frac{\sum_{j=1}^{N}(Z_{j+1} - mZ_j)^2 Z_j^{-1} - N\sigma^2}{(2s_N^2 \log\log s_N^2)^{\frac{1}{2}}} = 1 \text{ a.s.},$$

$$\liminf_{N\to\infty} \frac{\sum_{j=1}^{N}(Z_{j+1} - mZ_j)^2 Z_j^{-1} - N\sigma^2}{(2s_N^2 \log\log s_N^2)^{\frac{1}{2}}} = -1 \text{ a.s.},$$

and hence, using (16), (21) and (22),

$$\limsup_{N\to\infty} \frac{\sum_{j=1}^{N}(Z_{j+1} - \hat{m}Z_j)^2 Z_j^{-1} - N\sigma^2}{(4\sigma^4 N \log\log N)^{\frac{1}{2}}} = 1 \text{ a.s.},$$

$$\liminf_{N\to\infty} \frac{\sum_{j=1}^{N}(Z_{j+1} - \hat{m}Z_j)^2 Z_j^{-1} - N\sigma^2}{(4\sigma^4 N \log\log N)^{\frac{1}{2}}} = -1 \text{ a.s.}$$

The required result then follows.

3. The process with immigration

The properties of the estimator

$$\hat{\sigma}^2(\hat{m}) = N^{-1}\sum_{j=1}^{N}(Z_{j+1} - \hat{m}Z_j)^2 Z_j^{-1},$$

where $\hat{m} = Z_{N+1}Z_N^{-1}$, are robust to the extent that $\hat{\sigma}^2(\hat{m})$ continues to estimate the variance of the offspring distribution and the results of Theorems 2 and 3 continue to hold even if the branching process is one with immigration. Only minor modifications of the proofs of Theorems 1, 2 and 3 are necessary to establish these results.

Again we define

$$U_{n+1} = (Z_{n+1} - m Z_n)^2 Z_n^{-1} - \sigma^2.$$

In the case with immigration we can write

$$Z_{n+1} = Z_1^{(1)} + \cdots + Z_1^{(Z_n)} + I_{n+1}$$

where there $Z_1^{(i)}$ are i.i.d. each with the offspring distribution and are independent of Z_n while the I's are i.i.d. each with the immigration distribution, I_{n+1} being independent of Z_n. The sequence $\{U_n, \mathscr{F}_n\}$ is no longer a sequence of martingale differences since

$$\begin{aligned} E(U_{n+1} \mid \mathscr{F}_n) &= E\{Z_n^{-1}((Z_1^{(1)} - m) + \cdots + (Z_1^{(Z_n)} - m) + I_{n+1})^2 \mid \mathscr{F}_n\} - \sigma^2 \\ &= Z_n^{-1} E I^2 \end{aligned}$$

but we can of course form a sequence of martingale differences by taking

$$\{U_n - E(U_n \mid \mathscr{F}_{n-1}), \mathscr{F}_n\}.$$

The required results are obtained by working with the martingale

$$\left\{\sum_{j=1}^{n} [U_j - E(U_j \mid \mathscr{F}_{j-1})], \mathscr{F}_n\right\}$$

and then showing that the conditional expectations can be discarded. For example, to obtain the strong consistency result $\hat{\sigma}^2(\hat{m}) \overset{\text{a.s}}{\to} \sigma^2$ it is a matter of showing

$$N^{-1} \sum_{j=1}^{N} \{U_j - E(U_j \mid \mathscr{F}_{j-1})\} \overset{\text{a.s.}}{\to} 0 \tag{30}$$

while

$$N^{-1} \sum_{j=1}^{N} E(U_j \mid \mathscr{F}_{j-1}) \overset{\text{a.s.}}{\to} 0 \tag{31}$$

and

$$\hat{\sigma}^2(m) - \hat{\sigma}^2(\hat{m}) \overset{\text{a.s.}}{\to} 0. \tag{32}$$

The proof of (30) parallels the proof of the corresponding result in the proof of Theorem 1. Theorem 3 of [6] is used instead of Theorem 1 of [6]. The proof of (31) follows simply from the Kronecker lemma since $\sum_{j=1}^{\infty} j^{-1} Z_j^{-1} < \infty$ a.s. as a consequence of $\sum_{j=1}^{\infty} j^{-1} E Z_j^{-1} < \infty$. The proof of (32) again follows that of the corresponding result in the proof of Theorem 1. Theorem 4 of [6] is used to represent the difference $\hat{m} - m$ in this case. Similar considerations also apply for the central limit and iterated logarithm results but the details do not bear repetition and are omitted.

It is worth remarking that the estimator $\hat{m}$ also possesses robustness properties of the type outlined above. The paper [6] provides the results which lead to this conclusion.

References

[1] Billingsley, P. (1968) *Convergence of Probability Measures.* Wiley, New York.

[2] Brown, B. M. (1970) Characteristic functions, moments and the central limit theorem. *Ann. Math. Statist.* **41,** 658–664.

[3] Brown, B. M. (1971) Martingale central limit theorems. *Ann. Math. Statist.* **42,** 59–66.

[4] Heyde, C. C. (1970) Extension of a result of Seneta for the super-critical Galton-Watson process. *Ann. Math. Statist.* **41,** 739–742.

[5] Heyde, C. C. and Brown, B. M. (1971) An invariance principle and some convergence rate results for branching processes. *Z. Wahrscheinlichkeitsth.* **20,** 271–278.

[6] Heyde, C. C. and Leslie, J. R. (1971) Improved classical limit analogues for Galton-Watson processes with or without immigration. *Bull. Austral. Math. Soc.* **5,** 145–155.

[7] Heyde, C. C. and Scott, D. J. (1973) Invariance principles for the law of the iterated logarithm for martingales and processes with stationary increments. *Ann. Prob.* **1,** 428–436.

[8] Heyde, C. C. and Seneta, E. (1972) The simple branching process, a turning point test and a fundamental inequality: A historical note on I. J. Bienaymé. *Biometrika,* **59,** 680–683.

[9] Loève, M. (1963) *Probability Theory.* (3rd ed.) Van Nostrand, Princeton, N. J.

[10] Nagaev, A. V. (1967) On estimating the number of direct descendants of a particle in a branching process. *Theor. Probability Appl.* **12,** 314–320.

The Annals of Probability
1975, Vol. 3, No. 5, 903-907

A NONUNIFORM BOUND ON CONVERGENCE TO NORMALITY

BY C. C. HEYDE
Australian National University

Various asymptotically correct bounds on the uniform metric for distance between distribution functions in the central limit theorem for sums of independent and identically distributed random variables have previously been given. It is shown in the present paper that corresponding nonuniform bounds can be given for the difference between distribution functions. These results have much wider applicability, such as for obtaining probabilities of moderate deviation or for dealing with L_p metrics, $1 \leqq p \leqq \infty$.

1. Introduction and results. Let X_i, $i = 1, 2, \cdots$ be independent and identically distributed random variables with $EX_1 = 0$, $EX_1^2 = 1$, and distribution function F. Write $S_n = \sum_{i=1}^n X_i$, $n \geqq 1$, $G(x) = P(|X_1| \leqq x)$, $F_n(x) = P(S_n \leqq xn^{\frac{1}{2}})$, and denote by Φ the distribution function of the unit normal law. We shall establish the following theorem.

THEOREM 1. *Suppose $E|X_1|^{2+\alpha} < \infty$ for some $0 \leqq \alpha \leqq 1$. Then for all x,*

$$|F_n(x) - \Phi(x)| < (1 + |x|^{2+\alpha})^{-1} c_n(\alpha), \tag{1}$$

where

$$c_n(\alpha) = C\{n^{-\frac{1}{2}} \int_0^{n^{\frac{1}{2}}} y^3 \, dG(y) + n^{-\alpha/2} \int_{n^{\frac{1}{2}}}^{\infty} y^{2+\alpha} \, dG(y)\}, \tag{2}$$

C being a universal constant. Furthermore, writing $\Delta_n = \sup_x |F_n(x) - \Phi(x)|$, the following results hold.

(i) *The conditions $\sum n^{-1}\Delta_n < \infty$ and $\sum n^{-1}c_n(0) < \infty$ are equivalent and hold if and only if $EX_1^2 \log(1 + |X_1|) < \infty$.*

(ii) *For $0 < \delta < 1$, the conditions $\sum n^{-1+\delta/2}\Delta_n < \infty$ and $\sum n^{-1+\delta/2}c_n(0) < \infty$ are equivalent and hold if and only if $E|X_1|^{2+\delta} < \infty$.*

(iii) *For $0 < \delta < 1$, the conditions $\Delta_n = O(n^{-\delta/2})$ and $c_n(0) = O(n^{-\delta/2})$ as $n \to \infty$ are equivalent and hold if and only if $\int_z^\infty x^2 \, dG(x) = O(z^{-\delta})$ as $z \to \infty$.*

This theorem extends results of Heyde [2] and of Ibragimov [5] which deal with the uniform metric in (i), (ii) and (iii). The result is useful for many purposes, such as for obtaining probabilities of moderate deviation (e.g. to generalize Theorem 4 of Davis [1]), for obtaining results on L_p metrics, $1 \leqq p < \infty$, for departure from normality or for estimating differences of the kind $Eb(S_n) - Eb(n^{\frac{1}{2}}Y)$ for large n and suitable functions b, Y having a unit normal distribution. A sample application, establishing a similarity of behaviour of the L_p metrics,

Received November 4, 1974; revised March 16, 1975.
AMS 1970 *subject classifications.* Primary 60F05, 60G50; Secondary 60F10.
Key words and phrases. Central limit theorem, convergence rates, independent random variables, nonuniform convergence, probabilities of moderate deviations, L_p metrics.

$1 \leqq p \leqq \infty$, is given in the following corollary which extends Theorem 2 of Heyde [3].

COROLLARY 2. *Put*

$$||F_n - \Phi||_p = (\int_{-\infty}^{\infty} |F_n(x) - \Phi(x)|^p \, dx)^{1/p}, \qquad 1 \leqq p < \infty,$$
$$||F_n - \Phi||_\infty = \sup_x |F_n(x) - \Phi(x)|.$$

Then, for $0 \leqq \delta < 1$ *and* $1 \leqq p \leqq \infty$, *the following two conditions are equivalent*:

(a) $EX_1^2 \log(1 + |X_1|) < \infty$ *if* $\delta = 0$, $E|X_1|^{2+\delta} < \infty$ *if* $0 < \delta < 1$,

(b) $\sum n^{-1+\delta/2} ||F_n - \Phi||_p < \infty$.

It should be remarked that the complementary result to this corollary which deals with the case of $||F_n - \Phi||_p = O(n^{-\delta/2})$, $0 < \delta < 1$, has been obtained, by quite different methods, in Theorem 4.3 of Ibragimov [5].

2. Proofs. To prove Theorem 1 we start by taking $x \geqq 2$ for definiteness. The case $x \leqq -2$ can then be dealt with by replacing the X_i's by $-X_i$'s and the case $|x| < 2$ will follow from a bound on the uniform metric of Osipov and Petrov [7].

Put

$$\begin{aligned} X_k^{(n)}(x) &= X_k \qquad \text{if} \quad |X_k| \leqq n^{\frac{1}{2}} x, \\ &= 0 \qquad \text{otherwise,} \end{aligned}$$

and write $S_n^{(n)}(x) = \sum_{k=1}^n X_k^{(n)}(x)$. We have

$$\begin{aligned} F_n(x) &= P(S_n \leqq n^{\frac{1}{2}}x, S_n^{(n)}(x) \leqq n^{\frac{1}{2}}x) + P(S_n \leqq n^{\frac{1}{2}}x, S_n^{(n)}(x) > n^{\frac{1}{2}}x) \\ &\leqq P(S_n^{(n)}(x) \leqq n^{\frac{1}{2}}x) + nP(|X_1| > n^{\frac{1}{2}}x) \end{aligned}$$

and similarly

$$1 - F_n(x) \leqq P(S_n^{(n)}(x) > n^{\frac{1}{2}}x) + nP(|X_1| > n^{\frac{1}{2}}x)$$

so that

$$F_n(x) \geqq P(S_n^{(n)}(x) \leqq n^{\frac{1}{2}}x) - nP(|X_1| > n^{\frac{1}{2}}x)$$

and hence

$$|F_n(x) - \Phi(x)| \leqq |P(S_n^{(n)}(x) \leqq n^{\frac{1}{2}}x) - \Phi(x)| + nP(|X_1| > n^{\frac{1}{2}}x). \tag{3}$$

Now put $\mu_n(x) = EX_1^{(n)}(x)$, $\sigma_n^2(x) = \operatorname{Var} X_1^{(n)}(x)$. We have, writing

$$y_n = [\sigma_n(x)]^{-1} n^{-\frac{1}{2}} [n^{\frac{1}{2}}x - n\mu_n(x)],$$

$$\begin{aligned} &|P(S_n^{(n)}(x) \leqq n^{\frac{1}{2}}x) - \Phi(x)| \\ &\quad \leqq |P([\sigma_n(x)]^{-1}(S_n^{(n)}(x) - n\mu_n(x)) \leqq n^{\frac{1}{2}}y_n) - \Phi(y_n)| \\ &\qquad + |\Phi(y_n) - \Phi([\sigma_n(x)]^{-1}x)| + |\Phi([\sigma_n(x)]^{-1}x) - \Phi(x)|. \end{aligned} \tag{4}$$

But, using Theorem 2 of Nagaev [6],

$$\begin{aligned} &|P([\sigma_n(x)]^{-1}(S_n^{(n)}(x) - n\mu_n(x)) \leqq n^{\frac{1}{2}}y_n) - \Phi(y_n)| \\ &\quad \leqq CE|X_1^{(n)}(x) - \mu_n(x)|^3 n^{-\frac{1}{2}} [\sigma_n(x)]^{-3} (1 + |y_n|^3)^{-1} \\ &\quad \leqq 8CE|X_1^{(n)}(x)|^3 n^{-\frac{1}{2}} |x - n^{\frac{1}{2}}\mu_n(x)|^{-3}, \end{aligned} \tag{5}$$

C denoting a universal constant.

Now, for $x \geqq 2$ we have

$$|\mu_n(x)| = |\int_{|y|>n^{\frac{1}{2}}x} y\, dF(x)| \leqq \int_{n^{\frac{1}{2}}x}^{\infty} y\, dG(x) \leqq n^{-\frac{1}{2}}x^{-1} \int_{n^{\frac{1}{2}}x}^{\infty} y^2\, dG(y) \leqq n^{-\frac{1}{2}}x^{-1},$$

so that

$$n^{\frac{1}{2}}|\mu_n(x)| \leqq x^{-1} \leqq \tfrac{1}{2}x, \qquad |x - n^{\frac{1}{2}}\mu_n(x)|^3 \geqq \tfrac{1}{8}x^3, \tag{6}$$

and hence, from (5),

$$\begin{aligned} |P([\sigma_n(x)]^{-1}(S_n^{(n)}(x) - n\mu_n(x)) \leqq n^{\frac{1}{2}}y_n) - \Phi(y_n)| \\ &\leqq 64CE|X_1^{(n)}(x)|^3 n^{-\frac{1}{2}}x^{-3} \\ &= 64Cn^{-\frac{1}{2}}x^{-3}\{\int_0^{n^{\frac{1}{2}}} y^3\, dG(y) + \int_{n^{\frac{1}{2}}}^{n^{\frac{1}{2}}x} y^3\, dG(y)\} \\ &\leqq 64Cn^{-\frac{1}{2}}x^{-3}\{\int_0^{n^{\frac{1}{2}}} y^3\, dG(y) + (n^{\frac{1}{2}}x)^{1-\alpha} \int_{n^{\frac{1}{2}}}^{\infty} y^{2+\alpha}\, dG(y)\} \\ &\leqq 64Cx^{-(2+\alpha)}\{n^{-\frac{1}{2}} \int_0^{n^{\frac{1}{2}}} y^3\, dG(y) + n^{-\alpha/2} \int_{n^{\frac{1}{2}}}^{\infty} y^{2+\alpha}\, dG(y)\}. \end{aligned} \tag{7}$$

Further, for $x \geqq 1$,

$$\begin{aligned} nP(|X_1| > n^{\frac{1}{2}}x) &\leqq x^{-(2+\alpha)}n^{-\alpha/2} \sup_{u \geqq n^{\frac{1}{2}}} u^{2+\alpha}P(|X_1| > u) \\ &\leqq x^{-(2+\alpha)}n^{-\alpha/2} \sup_{u \geqq n^{\frac{1}{2}}} \int_u^{\infty} y^{2+\alpha}\, dG(y) \\ &\leqq x^{-(2+\alpha)}n^{-\alpha/2} \int_{n^{\frac{1}{2}}}^{\infty} y^{2+\alpha}\, dG(y). \end{aligned} \tag{8}$$

It now remains to bound the terms $|\Phi(y_n) - \Phi([\sigma_n(x)]^{-1}x)|$ and $|\Phi([\sigma_n(x)]^{-1}x) - \Phi(x)|$ from (4). Since $e^{\frac{1}{2}u^2} > \frac{1}{2}u^2$ for $u > 0$ we have

$$\begin{aligned} |\Phi(y_n) - \Phi([\sigma_n(x)]^{-1}x)| &\leqq (2\pi)^{-\frac{1}{2}} \int_{(x-n^{\frac{1}{2}}|\mu_n(x)|)[\sigma_n(x)]^{-1}}^{x[\sigma_n(x)]^{-1}} e^{-\frac{1}{2}u^2}\, du \\ &\leqq 2(2\pi)^{-\frac{1}{2}} \int_{(x-n^{\frac{1}{2}}|\mu_n(x)|)[\sigma_n(x)]^{-1}}^{x[\sigma_n(x)]^{-1}} u^{-2}\, du \\ &\leqq 4(2\pi)^{-\frac{1}{2}}n^{\frac{1}{2}}x^{-2}|\mu_n(x)| \end{aligned} \tag{9}$$

using (6). Also,

$$n^{\frac{1}{2}}|\mu_n(x)| \leqq n^{\frac{1}{2}} \int_{n^{\frac{1}{2}}x}^{\infty} y\, dG(y) \leqq x^{-1} \int_{n^{\frac{1}{2}}x}^{\infty} y^2\, dG(y) \leqq x^{-1} \int_{2n^{\frac{1}{2}}}^{\infty} y^2\, dG(y) \tag{10}$$

and hence, from (9) and (10),

$$\begin{aligned} |\Phi(y_n) - \Phi([\sigma_n(x)]^{-1}x)| &\leqq 4(2\pi)^{-\frac{1}{2}}x^{-3} \int_{2n^{\frac{1}{2}}}^{\infty} y^2\, dG(y) \\ &\leqq 4(2\pi)^{-\frac{1}{2}}x^{-(2+\alpha)} \int_{2n^{\frac{1}{2}}}^{\infty} y^2\, dG(y) \\ &\leqq 4(2\pi)^{-\frac{1}{2}}x^{-(2+\alpha)}n^{-\alpha/2} \int_{n^{\frac{1}{2}}}^{\infty} y^{2+\alpha}\, dG(y). \end{aligned} \tag{11}$$

Next, since $e^{\frac{1}{2}u^2} > \frac{1}{8}u^4 \geqq \frac{1}{4}u^3$ for $u \geqq 2$, we have for $x \geqq 2$,

$$\begin{aligned} |\Phi([\sigma_n(x)]^{-1}x) - \Phi(x)| &\leqq 4(2\pi)^{-\frac{1}{2}} \int_x^{x[\sigma_n(x)]^{-1}} u^{-3}\, du \\ &= 2(2\pi)^{-\frac{1}{2}}x^{-2}[1 - \sigma_n^2(x)] \\ &\leqq 4(2\pi)^{-\frac{1}{2}}x^{-2} \int_{n^{\frac{1}{2}}x}^{\infty} y^2\, dG(y) \\ &\leqq 4(2\pi)^{-\frac{1}{2}}x^{-(2+\alpha)}n^{-\alpha/2} \int_{n^{\frac{1}{2}}x}^{\infty} y^{2+\alpha}\, dG(y) \\ &\leqq 4(2\pi)^{-\frac{1}{2}}x^{-(2+\alpha)}n^{-\alpha/2} \int_{2n^{\frac{1}{2}}}^{\infty} y^{2+\alpha}\, dG(y). \end{aligned} \tag{12}$$

Hence, using (3), (4), (7), (8), (11) and (12), we have for $|x| \geqq 2$ that there exists a universal constant C such that, a fortiori,

$$|F_n(x) - \Phi(x)| \leqq C(1 + |x|^{2+\alpha})^{-1}\{n^{-\frac{1}{2}} \int_0^{n^{\frac{1}{2}}} y^3\, dG(y) + n^{-\alpha/2} \int_{n^{\frac{1}{2}}}^{\infty} y^{2+\alpha}\, dG(y)\}. \tag{13}$$

Furthermore, the inequality of Osipov and Petrov [7] can be specialized to give (see (2) of Heyde [4] wherein we take $\tau_n = C_n = n^{\frac12}$; K_0 is a universal constant)

$$\begin{aligned} |F_n(x) - \Phi(x)| \leq nP(|X_1| > n^{\frac12}) &+ K_0 n^{-\frac12}[\sigma_n(1)]^{-3} \textstyle\int_0^{n^{\frac12}} y^3\,dG(y) \\ &+ n^{\frac12}(2\pi)^{-\frac12}[\sigma_n(1)]^{-1}|\textstyle\int_{|y|\leq n^{\frac12}} y\,dF(y)| \\ &+ (2\pi e)^{-\frac12} \textstyle\int_{n^{\frac12}}^{\infty} y^2\,dG(y)\,, \end{aligned} \tag{14}$$

and noting (8), (10) and $\sigma_n(1) \to 1$ as $n \to \infty$, we observe that the right hand side of (14) is upper bounded by the right hand side of (13) for suitable universal C and $|x| < 2$. This ensures that (1) also holds for $|x| < 2$ and hence completes the proof of (1). In order to complete the remainder of the proof we first obtain a simple bound for $c_n(0)$.

Write $L(y) = \int_y^\infty x^2\,dG(x)$. Then

$$\begin{aligned} n^{-\frac12} \textstyle\int_0^{n^{\frac12}} y^3\,dG(y) &= -n^{-\frac12} \textstyle\int_0^{n^{\frac12}} y\,dL(y) \\ &\leq n^{-\frac12} \textstyle\int_0^{n^{\frac12}} L(y)\,dy\,, \end{aligned}$$

and, since $L(y)\downarrow$ as y increases,

$$c_n(0) \leq 2Cn^{-\frac12} \textstyle\int_0^{n^{\frac12}} L(y)\,dy\,. \tag{15}$$

Now suppose $0 \leq \delta < 1$ and $EX_1^2 \log(1 + |X_1|) < \infty$ if $\delta = 0$, $E|X_1|^{2+\delta} < \infty$ if $0 < \delta < 1$. Here and below we use C to denote a positive universal constant which may differ from one expression to the next. Using (15), we have

$$\begin{aligned} \textstyle\sum_{n=1}^\infty n^{-1+\delta/2}c_n(0) &\leq C \textstyle\sum_{n=1}^\infty n^{-(3-\delta)/2} \sum_{k=1}^n \int_{(k-1)^{\frac12}}^{k^{\frac12}} L(y)\,dy \\ &\leq C \textstyle\sum_{n=1}^\infty n^{-(3-\delta)/2} \sum_{k=1}^n L((k-1)^{\frac12})(k^{\frac12} - (k-1)^{\frac12}) \\ &\leq C \textstyle\sum_{k=1}^\infty k^{-\frac12} L((k-1)^{\frac12}) \sum_{n=k}^\infty n^{-(3-\delta)/2} \\ &\leq C \textstyle\sum_{k=1}^\infty k^{-1+\delta/2} L((k-1)^{\frac12}) \\ &= C \textstyle\sum_{k=1}^\infty k^{-1+\delta/2} \sum_{n=k}^\infty E\{X_1^2 I((n-1)^{\frac12} < |X_1| \leq n^{\frac12})\} \\ &\leq C \textstyle\sum_{n=1}^\infty n^{\delta/2} E\{X_1^2 I((n-1)^{\frac12} < |X_1| \leq n^{\frac12})\} \\ &\leq CE|X_1|^{2+\delta} < \infty\,, \qquad \text{if} \quad 0 < \delta < 1\,, \quad \text{or}, \\ &\leq C \textstyle\sum_{n=1}^\infty \log(n+1) E\{X_1^2 I((n-1)^{\frac12} < |X_1| \leq n^{\frac12})\} \\ &\leq CEX_1^2 \log(1 + |X_1|) < \infty\,, \qquad \text{if} \quad \delta = 0\,. \end{aligned}$$

Conversely, if $0 \leq \delta < 1$ and $\sum n^{-1+\delta/2}c_n(0) < \infty$, then (2) gives $\sum n^{-1+\delta/2}L(n^{\frac12}) < \infty$ and

$$\begin{aligned} \textstyle\sum_{n=1}^\infty n^{-1+\delta/2}L(n^{\frac12}) &= \textstyle\sum_{n=1}^\infty n^{-1+\delta/2} \sum_{k=n}^\infty E\{X_1^2 I(k^{\frac12} < |X_1| \leq (k+1)^{\frac12})\} \\ &= \textstyle\sum_{k=1}^\infty E\{X_1^2 I(k^{\frac12} < |X_1| \leq (k+1)^{\frac12})\} \sum_{n=1}^k n^{-1+\delta/2} \\ &\geq C \textstyle\sum_{k=1}^\infty k^{\delta/2} E\{X_1^2 I(k^{\frac12} < |X_1| \leq (k+1)^{\frac12})\} \\ &\geq CE|X_1|^{2+\delta}\,, \qquad \text{if} \quad 0 < \delta < 1\,, \quad \text{or} \\ &\geq C \textstyle\sum_{k=1}^\infty \log k E\{X_1^2 I(k^{\frac12} < |X_1| \leq (k+1)^{\frac12})\} \\ &\geq CEX_1^2 \log(1 + |X_1|)\,, \qquad \text{if} \quad \delta = 0\,. \end{aligned}$$

The equivalence of $\sum n^{-1+\delta/2}\Delta_n < \infty$ and $EX_1^2 \log(1 + |X_1|) < \infty$ if $\delta = 0$,

$E|X_1|^{2+\delta} < \infty$ if $0 < \delta < 1$, follows from Heyde [2] and hence the proof of (i) and (ii) is complete.

Finally, to prove (iii) we first suppose that $L(y) = O(y^{-\delta})$ as $y \to \infty$. Then, using (15), $c_n(0) = O(n^{-\delta/2})$. On the other hand, if $c_n(0) = O(n^{-\delta/2})$ then certainly, from (2),

$$L(n^{\frac{1}{2}}) = \int_{n^{\frac{1}{2}}}^{\infty} y^2\,dG(y) = O(n^{-\delta/2})$$

which yields $L(y) = O(y^{-\delta})$ as $y \to \infty$ as required. The equivalence of $\Delta_n = O(n^{-\delta/2})$ and $L(z) = O(z^{-\delta})$ as $z \to \infty$ was obtained by Ibragimov [5]. This completes the proof of Theorem 1.

To establish Corollary 2 we first observe that the equivalence of (a) and (b) for $p = \infty$ follows from Theorem 1 as does the result that (a) implies (b) for $1 \leq p < \infty$ since (a) gives $\sum n^{-1+\delta/2}c_n(0) < \infty$ and $||F_n - \Phi||_p = O(c_n(0))$. To see that (b) implies (a) note that the proof of the necessity part of the theorem of [2] is still applicable with minor modification (involving applications of Hölder's and Minkowski's inequalities if $p > 1$).

Acknowledgment. I am indebted to J. R. Leslie for a helpful comment.

REFERENCES

[1] Davis, J. A. (1968). Convergence rates for probabilities of moderate deviations. *Ann. Math. Statist.* **39** 2016–2028.

[2] Heyde, C. C. (1967). On the influence of moments on the rate of convergence to the normal distribution. *Z. Wahrscheinlichkeitstheorie und Verw. Gebiete* **8** 12–18.

[3] Heyde, C. C. (1969). Some properties of metrics in a study on convergence to normality. *Z. Wahrscheinlichkietstheorie und Verw. Gebiete* **11** 181–192.

[4] Heyde, C. C. (1973). On the uniform metric in the context of convergence to normality. *Z. Wahrscheinlichkeitstheorie und Verw. Gebiete* **25** 83–95.

[5] Ibragimov, I. A. (1966). On the accuracy of the Gaussian approximation to the distribution function of sums of independent random variables. *Theor. Probability Appl.* **11** 559–576.

[6] Nagaev, S. V. (1965). Some limit theorems for large deviations. *Theor. Probability Appl.* **10** 214–235.

[7] Osipov, L. V. and Petrov, V. V. (1967). On an estimate of the remainder in the central limit theorem. *Theor. Probability Appl.* **12** 281–286.

Department of Statistics
Australian National University
P.O. Box 4 Canberra
A.C.T. 2600
Australia

Biometrika (1975), **62**, 1, *p*. 49
Printed in Great Britain

Remarks on efficiency in estimation for branching processes

By C. C. HEYDE
Department of Statistics, S.G.S., Australian National University, Canberra

SUMMARY

This paper is concerned with efficiency in the estimation of the mean θ of the offspring distribution of a supercritical Galton–Watson branching process on the basis of a sample of n consecutive generation sizes. First, a direct comparison is made between the maximum likelihood estimator and the simple ratio of generation sizes estimator. Next, a new definition of asymptotic efficiency of an estimator is given, generalizing that of Rao (1973). It is shown that, for offspring distributions belonging to the class of power series distributions, the maximum likelihood estimator is efficient in this new sense. The paper concludes with some remarks on the implications of this theory in the estimation of the growth rate in a pure birth process.

Some key words: Asymptotic efficiency; Birth processes; Branching process estimation; Galton–Watson processes; Maximum likelihood estimation.

1. Introduction

Let $\{Z_0 = 1, Z_1, ..., Z_n\}$ be a sample of consecutive generation sizes from a so-called supercritical Galton–Watson process with $1 < \theta = E(Z_1) < \infty$ and $\sigma^2 = \text{var}(Z_1) < \infty$. We shall discuss the estimation of θ on the basis of this sample. Our starting point will be a comparison of two estimators that have previously been proposed for θ, namely the maximum likelihood estimator $\hat{\theta}_n = (Y_n - 1)Y_{n-1}^{-1}$, where $Y_n = Z_0 + ... + Z_n$ (Harris, 1948; Heyde, 1970) and the simple ratio estimator $\tilde{\theta}_n = Z_n Z_{n-1}^{-1}$ (Nagaev, 1967; Heyde, 1974). We shall then discuss the efficiency of estimation of θ for a particular class of offspring distributions. We do this by proposing a generalization of Rao's concept of efficiency (Rao, 1973, p. 348). This theory is of special interest since a random norming turns out to be appropriate and this highlights the inadequacy of direct, constant norming, generalizations of classical theory to the Markov case. The paper concludes with some remarks on the estimation of the growth rate in a pure birth process.

2. Comparison of maximum likelihood and ratio estimators

For the purposes of this section we shall suppose for convenience that $\text{pr}(Z_1 = 0) = 0$, so that $Z_n \to \infty$ almost surely as $n \to \infty$. If $\text{pr}(Z_1 = 0) > 0$, it is well known that $Z_n \to \infty$ almost surely on the nonextinction set and the results which we shall obtain hold conditionally on nonextinction.

The estimate $\hat{\theta}$ has been studied by Jagers (1973) and in an unpublished thesis of J. P. Dion. It is known that

$$Y_{n-1}^{\frac{1}{2}}(\hat{\theta}_n - \theta) \to N(0, \sigma^2) \tag{1}$$

R. Maller et al. (eds.), *Selected Works of C.C. Heyde*, Selected Works in Probability and Statistics,
DOI 10.1007/978-1-4419-5823-5_38,

in distribution, while for $\tilde{\theta}$ we have, from a result due to Bühler (1969),

$$Z_{n-1}^{\frac{1}{2}}(\tilde{\theta}_n - \theta) \to N(0, \sigma^2) \tag{2}$$

in distribution. Further, it is well known that $\theta^{-n} Z_n \to W$ almost surely with W non-degenerate and almost surely positive (Harris, 1963, p. 13) and, from Theorem 3 of Heyde (1970), $\theta^{-n} Y_n \to \theta W/(\theta - 1)$, so that $Z_n Y_n^{-1} \to (\theta - 1)\theta^{-1}$ and (2) can be rewritten as

$$(\theta - 1)^{\frac{1}{2}} \theta^{-\frac{1}{2}} Y_{n-1}^{\frac{1}{2}}(\tilde{\theta}_n - \theta) \to N(0, \sigma^2) \tag{3}$$

in distribution. We can thus conclude that in an obvious, albeit not entirely standard, sense the asymptotic efficiency of $\tilde{\theta}_n$ relative to that of $\hat{\theta}_n$ is $1 - \theta^{-1}$.

The random normings in (1) and (3) can be replaced by constant normings without difficulty. Note that

$$Y_{n-1}(\hat{\theta}_n - \theta) = \sum_{i=1}^{n} (Z_i - \theta Z_{i-1}) = \sum_{i=1}^{Y_{n-1}} \eta_i,$$

where the η_i's are independent and identically distributed, each with the distribution of $Z_1 - \theta$. Also,

$$Z_{n-1}(\tilde{\theta}_n - \theta) = Z_n - \theta Z_{n-1} = \sum_{i=1}^{n-1} \xi_{in},$$

where the ξ_{in}'s are independent and identically distributed, each with the distribution of $Z_1 - \theta$ and are, furthermore, independent of Z_{n-1}. A minor adjustment to the proof of Theorem 17·2 of Billingsley (1968) yields the limits in distribution

$$\sigma^{-1}(\theta-1)^{-\frac{1}{2}}\theta^{\frac{1}{2}n}(\hat{\theta}_n - \theta) \to W^{-\frac{1}{2}}N(0,1), \quad \sigma^{-1}\theta^{-\frac{1}{2}}\theta^{\frac{1}{2}n}(\tilde{\theta}_n - \theta) \to W^{-\frac{1}{2}}N(0,1),$$

where in each case the $W^{-\frac{1}{2}}$ is independent of the $N(0, 1)$; see also Jagers (1973).

The estimators $\hat{\theta}_n$ and $\tilde{\theta}_n$ of course do not exhaust the possibilities for simple estimators of θ. Another example is $\theta_n^* = Z_n^{1/n}$. From the almost sure limit $\theta^{-n} Z_n \to W$ we readily deduce that θ_n^* is strongly consistent for θ. Furthermore, from the iterated logarithm analogue given in Theorem 2 of Heyde & Leslie (1971) it is a straightforward matter to show that almost surely $n(\theta_n^* - \theta) \to \theta \log W$ as $n \to \infty$. Although the asymptotic properties of θ_n^* are not directly comparable to those of $\hat{\theta}_n$ and $\tilde{\theta}_n$, it is clearly substantially inferior, providing basically a linear rate of convergence in contrast to a geometric one.

3. The general asymptotic efficiency question

It is, of course, useful if a more objective analysis than that of § 2 can be given for the performance of estimators. To that end we propose a definition of asymptotic efficiency which generalizes that of Rao (1973, pp. 348, 349).

Suppose that we have a sample $X_1, \ldots, X_n$ of consecutive observations from some stochastic process whose distribution depends on a single parameter θ, $\theta \in \Theta \subset R$. Let $L_n(\theta)$ be the likelihood function associated with $X_1, \ldots, X_n$ and suppose that $L_n(\theta)$ is differentiable with respect to θ and that $E\{(d \log L_n/d\theta)^2\} < \infty$ for each n. We write $\mathscr{F}_k$ for the σ-field generated by $X_1, \ldots, X_k$ and suppose that, taking $L_0 = 1$, we have almost surely for $n \geqslant 1$

$$E\left(\frac{d \log L_n}{d\theta} \middle| \mathscr{F}_{n-1}\right) = \frac{d \log L_{n-1}}{d\theta}. \tag{4}$$

A sufficient condition for (4) is that, if $P_n(X_1, \ldots, X_n) = L_n(\theta)$ is the joint probability function of $X_1, \ldots, X_n$, then

$$\sum_{X_n} P_n(X_1, \ldots, X_n)$$

can be differentiated under the summation sign; there is a corresponding definition and requirement in the continuous case. The condition (4) is precisely that

$$\left\{d\log L_n/d\theta = \sum_{i=1}^{n} u_i, \mathscr{F}_n, n \geqslant 1\right\}$$

is a zero mean martingale.

Next, we set

$$I_n(\theta) = \sum_{k=1}^{n} E\left\{\left(\frac{d\log L_k}{d\theta} - \frac{d\log L_{k-1}}{d\theta}\right)^2 \middle| \mathscr{F}_{k-1}\right\}. \tag{5}$$

This is a form of conditional information which reduces to the standard Fisher information in the case where the X_i's are independent. We can think of

$$E\left[\left(\frac{d\log L_k}{d\theta} - \frac{d\log L_{k-1}}{d\theta}\right)^2 \middle| \mathscr{F}_{k-1}\right] = E\left[\left(\frac{d\log L_k}{d\theta}\right)^2 \middle| \mathscr{F}_{k-1}\right] - \left(\frac{d\log L_{k-1}}{d\theta}\right)^2 = E\left(u_k^2|\mathscr{F}_{k-1}\right)$$

as the information contained in $X_1, \ldots, X_k$ which is not contained in $X_1, \ldots, X_{k-1}$, for given $X_1, \ldots, X_{k-1}$.

Definition. *We shall say that a consistent estimator T_n of θ is asymptotically efficient if*

$$I_n^{\frac{1}{2}}(\theta)\left\{T_n - \theta - \beta(\theta)\, I_n^{-1}(\theta)\frac{d\log L_n(\theta)}{d\theta}\right\} \to 0 \tag{6}$$

in probability as $n \to \infty$ for some β which does not involve the observations.

In the standard case of independent and identically distributed observations, this definition reduces to that of Rao (1973). Furthermore, the motivation behind the general definition is the same as for the independence case. Under the conditions we have imposed

$$\left\{d\log L_n(\theta)/d\theta = \sum_{i=1}^{n} u_i\right\}$$

is a martingale and a Central Limit result for martingales gives that

$$I_n^{-\frac{1}{2}}(\theta)\frac{d\log L_n(\theta)}{d\theta} = \left(\sum_{i=1}^{n} u_i\right)\Big/\left\{\sum_{i=1}^{n} E(u_i^2|\mathscr{F}_{i-1})\right\}^{\frac{1}{2}} \tag{7}$$

converges in distribution to $N(0, 1)$ under certain regularity conditions. This result, together with (6), will ensure that

$$I_n^{\frac{1}{2}}(\theta)\,(T_n - \theta) \to N\{0, \beta^2(\theta)\}$$

in distribution. Interestingly, the random norming $I_n^{\frac{1}{2}}(\theta)$ in (6) and (7) gives rise to much more general Central Limit results than are provided by the constant norming

$$[E\{I_n(\theta)\}]^{\frac{1}{2}} = \left\{E\left(\sum_{i=1}^{n} u_i\right)^2\right\}^{\frac{1}{2}} = \left\{\sum_{i=1}^{n} E(u_i^2)\right\}^{\frac{1}{2}}.$$

If the condition (6) is strengthened to convergence in the mean of order two, then equivalent forms are

$$\operatorname{corr}\left[I_n^{\frac{1}{2}}(\theta)\,(T_n - \theta), \{I_n(\theta)\}^{-\frac{1}{2}}\frac{d\log L_n(\theta)}{d\theta}\right] \to 1, \tag{8}$$

or alternatively,

$$E\left\{(T_n - \theta)\frac{d\log L_n(\theta)}{d\theta}\right\}\Big/\left(E\{I_n(\theta)\,(T_n-\theta)^2\}\,E\left[(I_n(\theta)^{-1}\left\{\frac{d\log L_n(\theta)}{d\theta}\right\}^2\right]\right)^{\frac{1}{2}} \to 1. \tag{9}$$

A detailed discussion of the general properties of asymptotically efficient estimators will be given elsewhere.

To check asymptotic efficiency via the definition (6), the vital piece of information is a suitably tractable expression for $d\log L_n(\theta)/d\theta$. For the branching process application we shall consider the class of offspring distributions depending on a single parameter which we can write in the form

$$\operatorname{pr}(Z_1 = k) = p_k(\theta) \quad (k = 0, 1, \ldots),$$

where $\theta = E(Z_1)$. We shall further suppose that the following three conditions are satisfied:

(i) the set S of k's for which $p_k(\theta)$ is positive is independent of θ, $1 < \theta < \infty$;

(ii) for all k, $(d/d\theta)\, p_k(\theta)$ exists for all $1 < \theta < \infty$;

(iii) $\sum_{k_1 \in S} \cdots \sum_{k_n \in S} \prod_{i=1}^{n} \operatorname{pr}(Z_1 = k_i | Z_0 = k_{i-1}) \quad (k_0 = 1)$ can be differentiated under the summation signs.

In this situation, since

$$\operatorname{pr}(Z_1 = k_1, \ldots, Z_n = k_n) = \prod_{i=1}^{n} \operatorname{pr}(Z_1 = k_i | Z_0 = k_{i-1}),$$

we have

$$\frac{d\log L_n(\theta)}{d\theta} = \sum_{i=1}^{n} \frac{d}{d\theta} \log Q(Z_{i-1}, Z_i), \tag{10}$$

where $Q(k_{i-1}, k_i) = \operatorname{pr}(Z_1 = k_i | Z_0 = k_{i-1})$.

Now,

$$Q(k_{i-1}, k_i) = \textstyle\sum' p_{j_1} p_{j_2} \cdots p_{j_{k_{i-1}}},$$

where the sum is over $j_1, \ldots, j_{k_{i-1}} \in S$, $j_1 + \ldots + j_{k_{i-1}} = k_i$, so that

$$\frac{dQ(k_{i-1}, k_i)}{d\theta} = \sum{}' \left(\sum_{r=1}^{k_{i-1}} \frac{d\log p_{j_r}}{d\theta} \right) p_{j_1} \cdots p_{j_{k_{i-1}}}. \tag{11}$$

To proceed further we consider the class of Galton–Watson processes whose offspring distributions are power series distributions. That is, the class where

$$p_j = a_j \lambda^j \{f(\lambda)\}^{-1} \quad (j = 0, 1, \ldots; \lambda > 0),$$

where $a_j \geqslant 0$ and $f(\lambda) = \Sigma a_j \lambda^j$. For this class we readily find that

$$\theta = \lambda f'(\lambda)\{f(\lambda)\}^{-1}, \quad \sigma^2 = \{(d/d\theta)\log\lambda\}^{-1}$$

and

$$(d/d\theta)\, p_j = \sigma^{-2}(j - \theta)\, p_j. \tag{12}$$

Power series distributions are in fact characterized by the property (12). Details and references concerning the class of power series distributions may be obtained from Ord (1972, Chapter 6) and Johnson & Kotz (1969, Chapter 2). Examples are the geometric, binomial, Poisson, negative binomial and logarithmic distributions. An exception, notable from the point of view of the theory of the Galton–Watson process, is the distribution with linear functional probability generating function; for fixed b, $f(s) = 1 - b(1-p)^{-1} + bs(1-ps)^{-1}$. Here $\theta = b(1-p)^{-2}$.

Using (12) in (11) we find that

$$(d/d\theta)\log Q(Z_{i-1}, Z_i) = \sigma^{-2}(Z_i - \theta Z_{i-1}), \tag{13}$$

so that

$$\frac{d\log L_n(\theta)}{d\theta} = \sigma^{-2}\sum_{i=1}^{n}(Z_i - \theta Z_{i-1})$$
$$= \sigma^{-2}(Y_n - 1 - \theta Y_{n-1}) = \sum_{i=1}^{n} u_i, \tag{14}$$

say. Notice that, for each i,

$$E(u_i|\mathscr{F}_{i-1}) = E(Z_i|Z_{i-1}) - \theta Z_{i-1} = 0,$$

which is the martingale property and

$$E(u_i^2|\mathscr{F}_{i-1}) = E\{(Z_i - \theta Z_{i-1})^2\,|Z_{i-1}\} = \sigma^2 Z_{i-1},$$

so that

$$I_n(\theta) = \sigma^{-2}Y_{n-1} \tag{15}$$

and hence

$$\frac{d\log L_n(\theta)}{d\theta} = I_n(\theta)\,(\hat\theta_n - \theta). \tag{16}$$

It is then clear that the maximum likelihood estimator $\hat\theta_n$ is asymptotically efficient in the case of power series distributions, (6) being satisfied with $\beta(\theta) = 1$.

Of course if there was another asymptotically efficient estimator T_n of θ for which (6) was satisfied with a $\beta(\theta) < 1$, then this estimator would be preferred to $\hat\theta_n$ since, in view of (1), we would have $I_n^{\frac{1}{2}}(\theta)\,(T_n - \theta) \to N\{0, \beta^2(\theta)\}$, while $I_n^{\frac{1}{2}}(\theta)\,(\hat\theta_n - \theta) \to N(0, 1)$, both in distribution.

It certainly does not seem possible to establish the nonexistence of such a T_n which is better than $\hat\theta_n$ unless $\{I_n(\theta)\,(\hat\theta_n - \theta)^2\}$ is uniformly integrable or equivalently, as $n \to \infty$,

$$E\{I_n(\theta)\,(\hat\theta_n - \theta)^2\} \to 1$$

using Theorem 5·4 of Billingsley (1968), since $I_n(\theta)\,(\hat\theta_n - \theta)^2 \to \chi_1^2$ in distribution as a consequence of (1). In fact, we need even more conditions. If, indeed, $\{I_n(\theta)\,(\hat\theta_n - \theta)^2\}$ is uniformly integrable and $\{T_n\}$ satisfies (6) strengthened to convergence in the mean of order two, then

$$E\{I_n(\theta)\,(T_n - \theta)^2\} - \beta^2(\theta) \to 0 \tag{17}$$

and

$$E\{I_n(\theta)\,(T_n - \theta)^2\} - 2\beta(\theta)\,E\{(T_n - \theta)\,I_n(\theta)\,(\hat\theta_n - \theta)\} + \beta^2(\theta) \to 0. \tag{18}$$

Furthermore, if T_n is unbiased for θ and the right-hand side of

$$\theta = \sum_{k_1 \in S} \cdots \sum_{k_n \in S} T_n(k_1, \ldots, k_n) \prod_{i=1}^{n} Q(k_{i-1}, k_i)$$

can be differentiated under the summations, then from (11), (12) and (15),

$$1 = E\{T_n I_n(\theta)\,(\hat\theta_n - \theta)\} = E\{(T_n - \theta)\,I_n(\theta)\,(\hat\theta_n - \theta)\}. \tag{19}$$

We thus obtain from (17) and (18) that $\beta(\theta) = 1$ and hence $I_n^{\frac{1}{2}}(\theta)\,(T_n - \hat\theta_n) \to 0$ in the mean of order two.

Notice that the alternative conditions (8) and (9) provide a method of assessing an estimator by computing an appropriate correlation or analogue. For example, when

$$T_n = \tilde{\theta}_n = Z_n Z_{n-1}^{-1}$$

we have, using (19),

$$E\{I_n^{\frac{1}{2}}(\theta)(\tilde{\theta}_n-\theta) I_n^{\frac{1}{2}}(\theta)(\hat{\theta}_n-\theta)\}/[E\{I_n(\theta)(\tilde{\theta}_n-\theta)^2\} E\{I_n(\theta)(\hat{\theta}_n-\theta)^2\}]^{\frac{1}{2}}$$
$$= [E\{I_n(\theta)(\tilde{\theta}_n-\theta)^2\} E\{I_n(\theta)(\hat{\theta}_n-\theta)^2\}]^{-\frac{1}{2}},$$

and, using Theorem 5·3 of Billingsley (1968),

$$\liminf E\{I_n(\theta)(\tilde{\theta}_n-\theta)^2\} \geqslant \theta(\theta-1)^{-1}$$

since $I_n^{\frac{1}{2}}(\theta)(\tilde{\theta}_n-\theta) \to N\{0, \theta(\theta-1)^{-1}\}$ in distribution, while

$$\liminf E\{I_n(\theta)(\hat{\theta}_n-\theta)^2\} \geqslant 1,$$

since $I_n^{\frac{1}{2}}(\theta)(\hat{\theta}_n-\theta) \to N(0,1)$ in distribution, so that

$$\limsup E\{I_n^{\frac{1}{2}}(\tilde{\theta}_n-\theta) I_n^{\frac{1}{2}}(\theta)(\hat{\theta}_n-\theta)\}/[E\{I_n(\theta)(\tilde{\theta}_n-\theta)^2\} E\{I_n(\theta)(\hat{\theta}_n-\theta)^2\}]^{\frac{1}{2}} \leqslant (\theta-1)^{\frac{1}{2}}\theta^{-\frac{1}{2}}.$$

4. Estimation of the growth rate in a pure birth process

Let X_t be the population size at time t in a pure birth process with rate parameter λ. That is, X_t is a Markov process for which

$$\mathrm{pr}(X_{t+h} = j \mid X_t = i) = \begin{cases} i\lambda h + o(h) & (j = i+1), \\ 1-\lambda h + o(h) & (j = i), \\ o(h) & \text{otherwise}, \end{cases}$$

$i = 1, 2, \ldots$; $\lambda > 0$. Suppose that $\mathrm{pr}(X_0 = q) = 1$, where q is a fixed positive integer. Various sampling schemes have been considered for the estimation of λ in this model and the reader is referred to the recent work of Keiding (1974) for references and details. We shall here discuss the case where the process is sampled at equidistant time points $0, \tau, \ldots, k\tau$. Then, as is well known (Harris, 1963, pp. 103–4), the observations form a Galton–Watson process $Z_n = X_{n\tau}$ with geometric offspring distribution

$$\mathrm{pr}(Z_1 = i \mid Z_0 = 1) = e^{-\lambda\tau}(1-e^{-\lambda\tau})^{i-1} \quad (i = 1, 2, \ldots)$$

and $\mathrm{pr}(Z_1 = 0 \mid Z_0 = 1) = 0$. This is a power series distribution for which we have

$$E(Z_1 \mid Z_0 = 1) = e^{\lambda\tau}, \quad \mathrm{var}(Z_1 \mid Z_0 = 1) = e^{\lambda\tau}(e^{\lambda\tau}-1).$$

Furthermore, it is clear from results quoted above that the maximum likelihood estimator of λ is

$$\hat{\lambda}_{k\tau} = \tau^{-1}\log\{(X_\tau + \ldots + X_{k\tau})/(X_0 + \ldots + X_{(k-1)\tau})\}.$$

Also, from Keiding (1974), subject to a minor correction made in (21) below, almost surely

$$\hat{\lambda}_{k\tau} \to \lambda, \tag{20}$$

$$\tau(X_0 + \ldots + X_{(k-1)\tau})^{\frac{1}{2}}(\hat{\lambda}_{k\tau} - \lambda) \to N(0, 1-e^{-\lambda\tau}) \tag{21}$$

in distribution.

Writing $L_n(\lambda)$ for the likelihood function, we find in this case that

$$\frac{d\log L_k(\lambda)}{d\lambda} = \frac{\tau e^{-\lambda\tau}}{1-e^{-\lambda\tau}}(Y_{k\tau}-q-e^{\lambda\tau}Y_{(k-1)\tau})$$

$$= \frac{\tau e^{-\lambda\tau}}{1-e^{-\lambda\tau}}Y_{(k-1)\tau}(e^{\tau\hat{\lambda}_{k\tau}}-e^{\lambda\tau}),$$

while $I_k(\lambda) = \tau^2(1-e^{-\lambda\tau})^{-1}\,Y_{(k-1)\tau}$. Then,

$$I_k^{\frac{1}{2}}(\lambda)\left\{\hat{\lambda}_{k\tau}-\lambda-I_k^{-1}(\lambda)\frac{d\log L_k(\lambda)}{d\lambda}\right\} = \tau(1-e^{-\lambda\tau})^{-\frac{1}{2}}\,Y_{(k-1)\tau}^{\frac{1}{2}}\{\hat{\lambda}_{k\tau}-\lambda-\tau^{-1}(e^{\tau(\hat{\lambda}_{k\tau}-\lambda)}-1)\}$$

$$\to 0$$

in probability, using (20) and (21). This gives the asymptotic efficiency of $\hat{\lambda}_{k\tau}$, for fixed τ. Nevertheless one can, not unexpectedly, obtain a more efficient estimate if the process is observed continuously. Let

$$S_t = \int_0^t X_u\,du.$$

If the process starts with q individuals at time zero and is observed continuously up to time t, the maximum likelihood estimator of λ is $(X_t-q)\,S_t^{-1}$ and for $t = k\tau$, the efficiency of this estimator with respect to $\hat{\lambda}_{k\tau}$ as $k\to\infty$ is $(1-e^{-\lambda\tau})^2(\lambda\tau)^{-2}e^{\lambda\tau}$ from Keiding (1974) with modification as in (21) above.

References

Billingsley, P. (1968). *Convergence of Probability Measures.* New York: Wiley.

Bühler, W. (1969). Ein zentraler Grenzwertsatz für verzweigungsprozesse. *Z. Wahr. verw. Geb.* **11**, 139–41.

Harris, T. E. (1948). Branching Processes. *Ann. Math. Statist.* **19**, 474–94.

Harris, T. E. (1963). *The Theory of Branching Processes.* Berlin: Springer-Verlag.

Heyde, C. C. (1970). Extension of a result of Seneta for the supercritical Galton–Watson process. *Ann. Math. Statist.* **41**, 739–42.

Heyde, C. C. (1974). On estimating the variance of the offspring distribution in a simple branching process. *Adv. Appl. Prob.* **6**, 421–33.

Heyde, C. C. & Leslie, J. R. (1971). Improved classical limit analogues for Galton–Watson processes with or without immigration. *Bull. Aust. Math. Soc.* **5**, 145–55.

Jagers, P. (1973). A limit theorem for sums of random numbers of i.i.d. random variables. In *Mathematics. Essays in Honour of Harald Bergström*, pp. 33–9. Gotëborg: Chalmers Institute of Technology.

Johnson, N. L. & Kotz, S. (1969). *Discrete Distributions.* Boston: Houghton Mifflin.

Keiding, N. (1974). Estimation in the birth process. *Biometrika* **61**, 71–80.

Nagaev, A. V. (1967). On estimating the number of direct descendents of a particle in a branching process. *Theory Prob. Appl.* **12**, 314–20.

Ord, J. K. (1972). *Families of Frequency Distributions.* London: Griffin.

Rao, C. R. (1973). *Linear Statistical Inference and its Applications*, 2nd edition. New York: Wiley.

[*Received February* 1974. *Revised October* 1974]

Journal of Mathematical Biology 1, 317—320 (1975)

The Genetic Balance between Random Sampling and Random Population Size

C. C. Heyde and E. Seneta, Canberra

Received October 28, 1974

Summary

This note pertains to a generalized model for random fluctuation of allele frequency, where the population size is permitted to fluctuate randomly from generation to generation. Martingale methods are applied to discuss in two propositions, respectively, necessary and sufficient conditions for $P(Y(1-Y)>0)>0$, where Y is the (almost sure) limiting frequency of one allele. Under an additional restriction a necessary and sufficient condition results. Some simple remarks are made on further description of the distribution of Y.

1. Introduction

We concern ourselves with the standard problem of a single diploid diallelic autosomal locus (genotypes AA, Aa, aa) and non-overlapping generations, in the absence of external pressures, in a random-mating population whose size remains finite, and is denoted by $N_n (\geq 1)$ at the n-th generation, $n \geq 0$. In an extension of a classical fixed size model of S. Wright, one of the authors (Seneta 1974) assumed that the population size may vary, even randomly.

Specifically, the model states:

(a) the process $\{X_n, N_n\}$, $n \geq 0$, is bivariate Markov, X_n denoting the number of a alleles in the n-th generation; and

(b) the distribution of X_{n+1}, given N_{n+1}, N_n, X_n is binomial, and given by

$$\binom{2N_{n+1}}{j} Y_n^j (1-Y_n)^{2N_{n+1}-j}, \quad 0 \leq j \leq 2N_{n+1}, \tag{1}$$

where $Y_n = X_n/(2N_n)$ is the a-allele proportion of alleles in the n-th generation.

Write $\mathscr{F}_n$ for the σ-field generated by $\{(X_0, N_0), (X_1, N_1), \ldots, (X_n, N_n)\}$, $n \geq 0$. It was observed in the cited paper that $\{Y_n, \mathscr{F}_n, n \geq 0\}$ is a martingale, and so $Y_n \to Y$ a. s. as $n \to \infty$ for some random variable Y with $0 \leq Y \leq 1$, and $EY = EY_n$, $n \geq 0$.

One of the main issues of interest is, then: when is the limit distribution trivial i.e. $P(\{Y=0\} \cup \{Y=1\}) = 1$, or equivalently $P(Y(1-Y)=0)=1$, so fixation is certain; and when not, so that a situation of balanced polymorphism occurs with positive probability.

In the case where $\{N_n\}$ is purely deterministically varying, it was noted (Seneta, 1974) that a necessary and sufficient condition for non-triviality is $\sum_{n=0}^{\infty} 1/N_{n+1} < \infty$;

R. Maller et al. (eds.), *Selected Works of C.C. Heyde*, Selected Works in Probability and Statistics,
DOI 10.1007/978-1-4419-5823-5_39, © Springer Science+Business Media, LLC 2010

providing as always the process does not begin in a state of fixation. The more general situation where the sequence $\{N_n\}$ is permitted to be random, in relation to the same question, is the subject of the present note; which is also, again, an illustration of the applicability of martingale methods to the present model. One direct deduction from the result on the deterministic case to a stochastic one in relation to $\{N_n\}$ may be worth mentioning at this point; if $\sum 1/N_{n+1}<\infty$ a.s. on a set A of histories of positive probability, and for almost every given sequence $\{N_n\}$ in the set A, the sequence $\{X_n\}$ is still Markovian, then the limit distribution is not trivial. (This may most easily be seen by postulating the contrary i.e. triviality). Results related to this one are given in Theorem 2 below, which gives other sufficient conditions for non-triviality.

The starting point of our deductions is the relation

$$E\left(Y_{n+1}(1-Y_{n+1}) \mid \mathscr{F}_n\right)=\left(1-E\{(2N_{n+1})^{-1} \mid \mathscr{F}_n\}\right) Y_n(1-Y_n). \tag{2}$$

This follows from the above basic assumptions (a) and (b), since these imply via (1) that

$$E(Y_{n+1} \mid N_{n+1}, \mathscr{F}_n)=Y_n, \operatorname{Var}(Y_{n+1} \mid N_{n+1}, \mathscr{F}_n)=(2N_{n+1})^{-1} Y_n(1-Y_n)$$

whence

$$E\left(Y_{n+1}(1-Y_{n+1}) \mid N_{n+1}, \mathscr{F}_n\right)=\left(1-(2N_{n+1})^{-1}\right) Y_n(1-Y_n).$$

From (2) we note immediately that $E\left(Y_n(1-Y_n)\right) \downarrow$ as n increases (and, indeed, that $\{Y_n(1-Y_n), \mathscr{F}_n, n\geq 0\}$ is a supermartingale).

2. Main Results

Theorem 1: *If* $\sum_{n=0}^{\infty} E(N_{n+1}^{-1} \mid \mathscr{F}_n)=\infty$ *a.s. on some set A of histories of the process* $\left(\text{or equivalently } \sum_{n=0}^{\infty} N_{n+1}^{-1}=\infty \text{ a.s. on } A\right)$ *then fixation occurs on A (i.e. $Y(1-Y)=0$ a.s. on A).*

Proof: Define the martingale $\{U_n, \mathscr{F}_n, n\geq 1\}$ by

$$U_n=\sum_{k=1}^{n}\left[Y_k(1-Y_k)-E\left(Y_k(1-Y_k) \mid \mathscr{F}_{k-1}\right)\right], n\geq 1.$$

Using (2), we have

$$U_n=\sum_{k=1}^{n} Y_k(1-Y_k)-\sum_{k=0}^{n-1}\left[1-E\left((2N_{k+1})^{-1} \mid \mathscr{F}_k\right)\right] Y_k(1-Y_k)$$

$$=Y_n(1-Y_n)-Y_0(1-Y_0)+\sum_{k=0}^{n-1} E\left((2N_{k+1})^{-1} \mid \mathscr{F}_k\right) Y_k(1-Y_k).$$

$\{U_n, \mathscr{F}_n, n\geq 1\}$ is a zero-mean martingale with bounded increments and so, using Proposition IV-6-3 (of Neveu, 1965) $\liminf_{n\to\infty} U_n=-\infty$ a.s., $\limsup_{n\to\infty} U_n=+\infty$ a.s. on the set where U_n does not converge a.s. It follows that U_n must converge a.s. as $n\to\infty$ since $U_n\geq -Y_0(1-Y_0)$ for all n. Since $Y_n\to Y$ a.s., we have $Y_n(1-Y_n)\to Y(1-Y)$ a.s. and it follows that

$$\sum_{n=0}^{\infty} E(N_{n+1}^{-1} \mid \mathscr{F}_n) Y_n(1-Y_n)<\infty \text{ a. s.} \tag{3}$$

It is clear from (3) that if $\sum_{n=0}^{\infty} E(N_{n+1}^{-1} \mid \mathscr{F}_n)=\infty$ a. s. on A then $Y_n(1-Y_n)\to 0$ on A, i. e. fixation occurs on A. That $\sum_{n=0}^{\infty} N_{n+1}^{-1}=\infty$ a.s. on A if and only if $\sum_{n=0}^{\infty} E(N_{n+1}^{-1} \mid \mathscr{F}_n)=\infty$ a. s. on A follows from another application of Proposition IV-6-3 of Neveu. This completes the proof.

It is worth noting that in complementary form the result of Theorem 1 says that we must have $\sum_{n=0}^{\infty} E(N_{n+1}^{-1} \mid \mathscr{F}_n)<\infty$ a. s. on the set $\{Y(1-Y)>0\}$ (and equivalently $\sum_{n=0}^{\infty} N_{n+1}^{-1}<\infty$ a. s. on $\{Y(1-Y)>0\}$).

It has not been possible to obtain a converse to Theorem 1. The following partial result does however, cover many cases of interest.

Theorem 2: *If $E(N_{n+1}^{-1} \mid \mathscr{F}_n)\le\alpha_n$ a. s. for all n where $\{\alpha_n\}$ is a sequence of positive constants with $\sum \alpha_n<\infty$, or if $N_n\ge\beta_n$ a. s. for all n where $\{\beta_n\}$ is a sequence of positive constants with $\sum \beta_n^{-1}<\infty$, then $P(Y(1-Y)>0)>0$.*

Proof: We use the martingale $\{U_n, \mathscr{F}_n, n\ge 1\}$ introduced in the proof of Theorem 1. First note that, under the conditions of the theorem,

$$E\sum_{k=0}^{\infty} E((2N_{k+1})^{-1} \mid \mathscr{F}_k) Y_k(1-Y_k)<\infty.$$

Then, for any fixed $n\ge 0$ and $N>n$ we have from (2) that

$$\begin{aligned} E\sum_{k=n}^{N} Y_k(1-Y_k) E((2N_{k+1})^{-1} \mid \mathscr{F}_k) &= E[U_{N+1}-U_n-Y_{N+1}(1-Y_{N+1})+Y_n(1-Y_n)] \\ &= -EY_{N+1}(1-Y_{N+1})+EY_n(1-Y_n) \\ &\to -EY(1-Y)+EY_n(1-Y_n) \end{aligned}$$

as $N\to\infty$ by dominated convergence since $Y_{N+1}(1-Y_{N+1})\to Y(1-Y)$ a.s. as $N\to\infty$. The monotone convergence theorem then gives

$$E\sum_{k=n}^{\infty} Y_k(1-Y_k) E((2N_{k+1})^{-1} \mid \mathscr{F}_k)=-EY(1-Y)+EY_n(1-Y_n) \tag{4}$$

and hence

$$E\sum_{k=n}^{\infty} Y_k(1-Y_k) E((2N_{k+1})^{-1} \mid \mathscr{F}_k)\le EY_n(1-Y_n)$$

with equality if and only if $EY(1-Y)=0$ (i. e. $P(Y(1-Y)>0)=0$).

Now suppose that $E((2N_{k+1})^{-1} \mid \mathscr{F}_k)\le\alpha_k$ a. s. for all k where $\{\alpha_k\}$ is a sequence of constants with $\sum \alpha_k<\infty$. We can choose n so large that $\sum_{k=n}^{\infty}\alpha_k<1$ and then

$$E \sum_{k=n}^{\infty} Y_k(1-Y_k) E\left((2N_{k+1})^{-1} \mid \mathscr{F}_k\right)$$

$$\leq \sum_{k=n}^{\infty} E\left(Y_k(1-Y_k)\right) \alpha_k$$

$$\leq E\left(Y_n(1-Y_n)\right) \sum_{k=n}^{\infty} \alpha_k < E Y_n (1-Y_n)$$

since we have noted $E(Y_k(1-Y_k)) \downarrow$ as k increases. This assures, via (4), that $E(Y(1-Y))>0$, i.e. $P(Y(1-Y)>0)>0$. Similar reasoning applies if $N_n > \beta_n$ a. s. for all n where $\{\beta_n\}$ is a sequence of positive constants with $\sum \beta_n^{-1} < \infty$. This completes the proof.

Two corollaries of Theorems 1 and 2 may be worth mentioning. First, if $E(N_{n+1}^{-1} \mid \mathscr{F}_n) = E(N_{n+1}^{-1})$, then $P(Y(1-Y)>0)>0$ if and only if $\sum E(N_{n+1}^{-1}) < \infty$, a direct extension of the purely deterministic situation of $\{N_n\}$, which can also be deduced directly from (2). Secondly, if N_{n+1} is totally independent of $\mathscr{F}_n$, $n \geq 0$, i. e. of the past history of the bivariate process, which covers the situation of so-called "randomly fluctuating environments", then clearly this same necessary and sufficient condition applies.

3. The Limit Distribution

By noting that for $n \geq 0$, and y fixed at 0 or 1,

$$\{Y_n = y\} \Rightarrow \{Y_{n+1} = y\} \Rightarrow \{Y = y\}$$

we obtain

$$P(Y_n = y) \uparrow \alpha(y) \leq P(Y = y), \tag{5}$$

so for any given n, $P(Y_n = y)$ may serve as a lower bound for $P(Y=y)$. Consequently, also $1-\alpha(0)-\alpha(1) \geq P(0<Y<1)$. That strict inequality will sometimes apply in (5) may be expected from other contexts. One also has available the inequalities valid for any random variable Y on $[0,1]$, whose mean is $c: P(Y=1) \leq c$, $P(|Y-c| \geq t) \leq t^{-2} \max\{c^2, (1-c)^2\}$, these being basically Markov's and Chebyshev's inequalities, respectively. In our case, $c = EY = EY_0$ is generally known.

References

Neveu, J.: Mathematical Foundations of the Calculus of Probability. San Francisco: Holden-Day 1965.

Seneta, E.: A note on the balance between random sampling and population size. (On the 30th anniversary of G. Malécot's paper.) Genetics 77, 607—610 (1974).

Dr. C. C. Heyde
Dr. E. Seneta
Department of Statistics, S.G.S.
Australian National University
P.O. Box 4
Canberra, A.C.T. 2600
Australia

Printed in Austria
Druck: Paul Gerin, A-1021 Wien

BULL. AUSTRAL. MATH. SOC. 60G45
VOL. 14 (1976), 435-447. (60F15, 60B10)

On a unified approach to the law of the iterated logarithm for martingales

P.G. Hall and C.C. Heyde

There are two distinct approaches in the literature to framing a version of the law of the iterated logarithm for martingales. One involves norming by constants, using the martingale variance and the other involves norming by random variables, using the sums of conditional variances of the increments, given their past. In this paper a portmanteau approach is provided, still based on the Skorokhod representation of the martingale, but involving normalization by more general random variables. This extends the functional forms of all the previously existing results.

1. Introduction and results

The literature contains various laws of the iterated logarithm for martingales which are not directly comparable. There are versions based on norming by constants (for example, Heyde and Scott [1]) and versions based on norming by random variables (for example, Strassen [7], Stout [5]). The former use variances and the latter, conditional variances. Here we employ a Skorokhod representation approach, based on the law of the iterated logarithm for brownian motion, to relate the earlier results.

Let ϕ be the real-valued function on (e, ∞) defined by

$$\phi(t) = (2t \log \log t)^{\frac{1}{2}} .$$

If $W(t)$ $(t \geq 0)$ is standard brownian motion then (Lévy [3])

Received 1 March 1976.

R. Maller et al. (eds.), *Selected Works of C.C. Heyde*, Selected Works in Probability and Statistics, DOI 10.1007/978-1-4419-5823-5_40, © Springer Science+Business Media, LLC 2010

$$(1)\qquad \limsup_{t\to\infty} W(t)/\phi(t) = +1 \quad \text{almost surely,}$$
$$\liminf_{t\to\infty} W(t)/\phi(t) = -1 \quad \text{almost surely.}$$

If $\{t_n,\ n \geq 1\}$ is a sequence of constants increasing to ∞ then (1) implies that

$$(2)\qquad \limsup_{n\to\infty} W(t_n)/\phi(t_n) \leq +1 \quad \text{almost surely,}$$
$$\liminf_{n\to\infty} W(t_n)/\phi(t_n) \geq -1 \quad \text{almost surely.}$$

If the points of the sequence $\{t_n\}$ are close enough together then equality holds throughout (2) with probability 1. In fact, if $\{T_n,\ n \geq 1\}$ is any sequence of random variables satisfying

$$(3)\qquad T_n > e\ ,\quad T_n \uparrow \infty \text{ almost surely and } T_{n+1}^{-1}T_n \xrightarrow{\text{a.s.}} 1\ ,$$

then (1) implies that

$$(4)\qquad \limsup_{n\to\infty} W(T_n)/\phi(T_n) = 1 \quad \text{almost surely}$$
$$\liminf_{n\to\infty} W(T_n)/\phi(T_n) = -1 \quad \text{almost surely,}$$

and so the variables $S_n = W(T_n)$ satisfy the law of the iterated logarithm with a random norming.

Now let $\{S_n,\ F_n,\ n \geq 1\}$ be a zero-mean, square-integrable martingale on the probability space $(\Omega,\ F,\ P)$, where $F_0 = \{\phi,\ \Omega\}$ and F_n is the σ-field generated by S_j , $0 \leq j \leq n$. Let $S_0 = X_0 = 0$ and $S_n = \sum_1^n X_k$. We shall write

$$V_n^2 = \sum_1^n E\left(X_k^2 \mid F_{k-1}\right)\ ,\quad U_n^2 = \sum_1^n X_k^2\ ,\quad s_n^2 = ES_n^2 = EV_n^2\ ,$$

and $\{W_n,\ n \geq 1\}$ will denote a non-decreasing sequence of positive random variables such that $W_1^2 > e$. By extending the original probability space if necessary, we may suppose that there exists a brownian motion W and a non-decreasing sequence of non-negative variables $\{T_n, n \geq 1\}$ defined on our probability space such that $S_n = W(T_n)$ almost surely for all n (see the statement of the Skorokhod representation theorem in Strassen [7], Theorem 4.3). If the T_n satisfy condition (3) then (4) immediately gives us a law of the iterated logarithm for $\{S_n, n \geq 0\}$.

Unfortunately the T_n are quite difficult to compute from the martingale $\{S_n, F_n\}$, and the law is not very meaningful unless we can replace T_n by a variable W_n^2 which can be expressed as a function of the differences $X_1, X_2, \ldots$; W_n^2 should satisfy

(5) $$T_n^{-1} W_n^2 \xrightarrow{\text{a.s.}} 1 .$$

Strassen [6] formulated a functional law of the iterated logarithm which extends the law in (4). Consider the metric space (C, ρ) of all real-valued continuous functions on $[0, 1]$ with

$$\rho(x, y) = \sup_{0 \leq t \leq 1} |x(t) - y(t)| \quad \text{for } x, y \in C .$$

Let K be the set of absolutely continuous $x \in C$ such that $x(0) = 0$ and

$$\int_0^1 \dot{x}(t)^2 dt \leq 1$$

where $\dot{x}$ denotes the derivative of x determined almost everywhere with respect to Lebesgue measure. For $u \in [0, 1]$ define

$$l = l(n, u) = \max\left\{j \leq n \mid W_j^2 \leq u W_n^2\right\} \quad (= 0 \text{ if the set is empty})$$

and

$$\mu_n(u) = \left[\phi\left(W_n^2\right)\right]^{-1} \left[S_l + \left(u W_n^2 - W_l^2\right) \left(W_{l+1}^2 - W_l^2\right)^{-1} X_{l+1}\right] ,$$

where $S_n = W(T_n)$ and $X_n = S_n - S_{n-1}$ $(S_0 = 0)$.

THEOREM A. *If* (3) *and* (5) *hold then* $\{\mu_n, n \geq 1\}$ *is relatively compact in* $C[0, 1]$ *and the set of its almost sure limit points coincides with* K .

(Here we do not assume that $\{S_n, F_n, n \geq 0\}$ is a martingale, although we do in all the work which follows.)

If we are to use Theorem A to establish a law of the iterated logarithm for martingales, we must find conditions under which T_n can be approximated by functions W_n^2 of the martingale differences. It is convenient to work with martingales whose differences are truncated, and hence our technique is to approximate to $\{S_n\}$ by a truncated martingale $\{S_n^*\} = \{W(T_n^*)\}$, and then find functions W_n^2 which are close to T_n^* . Our main result is given in the following theorem.

THEOREM 1. *Let* $\{Z_n, n \geq 1\}$ *be a sequence of non-negative random variables and suppose that* Z_n *and* W_n *are* F_{n-1}*-measurable. If*

$$\lim_{n\to\infty} \left[\phi\left(W_n^2\right)\right]^{-1} \sum_1^n \{X_j I(|X_j| > Z_j) - E(X_j I(|X_j| > Z_j) \mid F_{j-1})\} = 0 \tag{6}$$

almost surely,

$$\lim_{n\to\infty} W_n^{-2} \sum_1^n \left[E\left(X_j^2 I(|X_j| \leq Z_j) \mid F_{j-1}\right) - \{E(X_j I(|X_j| \leq Z_j) \mid F_{j-1})\}^2\right] = 1 \tag{7}$$

almost surely,

$$\sum_1^\infty W_j^{-4} E\left(X_j^4 I(|X_j| \leq Z_j) \mid F_{j-1}\right) < \infty \quad \textit{almost surely,} \tag{8}$$

and

$$\lim_{n\to\infty} W_{n+1}^{-1} W_n = 1 \quad \textit{almost surely and} \quad W_n \to \infty \quad \textit{almost surely,} \tag{9}$$

then the conclusions of Theorem A *hold.*

Two obvious candidates for W_n for use in this theorem are $V_n \vee 2$ and $U_{n-1} \vee 2$. These lead to equivalent results if $\lim_{n\to\infty} V_n^{-2} U_n^2 = 1$ almost surely, which would commonly be the case; but interestingly enough this would not universally be true. When V_n is used in the norming the condition (7) simplifies usefully, giving the following result.

COROLLARY 1. *Let* $\{Z_n, n \geq 1\}$ *be as in Theorem* 1. *The conclusions of Theorem* A *hold for* $W_n = V_n \vee 2$ *if*

$$(10)\quad \lim_{n\to\infty} \left[\phi\left(V_n^2\right)\right]^{-1} \sum_1^n \{X_j I(|X_j| > Z_j) - -E(X_j I(|X_j| > Z_j) \mid F_{j-1})\} = 0 \quad \textit{almost surely},$$

$$(11)\quad \lim_{n\to\infty} V_n^{-2} \sum_1^n E\left(X_j^2 I(|X_j| > Z_j) \mid F_{j-1}\right) = 0 \quad \textit{almost surely},$$

$$(12)\quad \sum_1^\infty V_j^{-4} E\left(X_j^4 I(|X_j| \leq Z_j) \mid F_{j-1}\right) < \infty \quad \textit{almost surely},$$

and

$$(13)\quad \lim_{n\to\infty} V_{n+1}^{-1} V_n = 1 \quad \textit{almost surely}, \quad V_n \to \infty \quad \textit{almost surely}.$$

The next corollary extends the main result (Theorem 1) of Heyde and Scott [1] which deals with the case $\eta^2 = 1$ almost surely. It highlights the restrictiveness of a constant norming form.

COROLLARY 2. *If for some almost surely finite and non-zero random variable* η^2 *we have*

$$(14)\quad s_n^{-2} U_n^2 \xrightarrow{\text{a.s.}} \eta^2 \quad \textit{and} \quad s_n \to \infty ,$$

$$(15)\quad \sum_1^\infty s_j^{-1} E[|X_j| I(|X_j| > \varepsilon s_j)] < \infty , \quad \textit{for all} \quad \varepsilon > 0 ,$$

and

$$(16)\quad \sum_1^\infty s_j^{-4} E\left[X_j^4 I(|X_j| \leq \delta s_j)\right] < \infty , \quad \textit{for some} \quad \delta > 0 ,$$

then the conclusions of Theorem A *hold for* $W_n = U_{n-1} \vee 2$.

The final corollary presents a portmanteau form for the sufficient conditions.

COROLLARY 3. *Let* f *and* g *be positive non-decreasing functions on* $(0, \infty)$ *such that* $t^{-1}f(t) \downarrow 0$, $t^{-1}g(t) \downarrow 0$ *as* t *increases and*

$$\int_1^\infty t^{-2}f(t)dt < \infty . \tag{17}$$

If $V_n \to \infty$ *almost surely as* $n \to \infty$ *and*

$$\sum_1^\infty \left[g\left(V_j^2\right)\right]^{-1} E\left(X_j^2 I\left(X_j^2 > f\left(V_j^2\right)\right) \mid F_{j-1}\right) < \infty \text{ almost surely,} \tag{18}$$

then the conclusions of Theorem A *hold for* $W_n = V_n \vee 2$.

This result can be compared with Corollary 4.5 of Strassen [7] which deals with the case $f(t) = g(t) = t(\log t)^{-5}$, $t > e$, and with Theorem 5.1 of Jain, Jogdeo and Stout [2] which deals with the case $f(t) = t(\log t)^{-1}(\log\log t)^{-6}$, $g(t) = t(\log\log t)^{-2}$, $t > e$. These results, however, are concerned with the integral test for upper and lower functions which, even in the case of independent variables, requires the imposition of slightly more stringent conditions than for the classical law of the iterated logarithm itself. Theorem 3 of Stout [5] essentially deals with the case $f(t) = g(t) = o\left(t(\log\log t)^{-1}\right)$, $t > e$, but obtains the classical and not the functional law of the iterated logarithm.

2. Proofs

Proof of Theorem A. Define $\mu(u)$ on $[0, \infty)$ by

$$\mu(u) = S_p + \left(u - W_p^2\right)\left(W_{p+1}^2 - W_p^2\right)^{-1} X_{p+1} ,$$

where

$$p = p(u) = \max\left\{j \mid W_j^2 \le u\right\} .$$

Then

$$\mu_n(u) = \left[\phi\left(W_n^2\right)\right]^{-1}\mu\left(W_n^2 u\right) ,$$

and so in view of Corollary 1 of Strassen [6], it suffices to prove that

(19) $$\lim_{t\to\infty} [\phi(t)]^{-1} \sup_{u\le t} |\mu(u)-W(u)| = 0 \text{ almost surely.}$$

Now, (3) and (5) imply that $W_{n+1}^{-2}W_n^2 \xrightarrow{\text{a.s.}} 1$ and hence

$$1 \ge u^{-1}W_{p(u)}^2 \ge W_{p(u)+1}^{-2}W_{p(u)}^2 \xrightarrow{\text{a.s.}} 1 \text{ as } u\to\infty ,$$

so that

$$u^{-1}W_{p(u)}^2 \xrightarrow{\text{a.s.}} 1 \text{ as } u\to\infty .$$

Similarly,

$$u^{-1}W_{p(u)+1}^2 \xrightarrow{\text{a.s.}} 1 \text{ as } u\to\infty .$$

Combining these and (5),

$$u^{-1}T_{p(u)} \xrightarrow{\text{a.s.}} 1 \text{ and } u^{-1}T_{p(u)+1} \xrightarrow{\text{a.s.}} 1 \text{ as } u\to\infty .$$

Since

$$|\mu(u)-W(u)| \le \max\{|W(T_{p(u)})-W(u)|, |W(T_{p(u)+1})-W(u)|\} ,$$

then result (19) follows as in Strassen's proof on page 217 of [6].

Proof of Theorem 1. The proof is based on that of Strassen [6] and follows that of Heyde and Scott [1].

Define

(20) $$\tilde{X}_j = X_j I\left(c_j < |X_j| \le Z_j\right) + \tfrac{1}{2}X_j I\left(|X_j| \le c_j\right) + \tfrac{1}{2}\,\mathrm{sgn}\left(X_j\right)c_j\left(1+Z_j|X_j|^{-1}\right)I\left(|X_j| > Z_j\right)$$

where $\{c_j, j\ge 1\}$ is a monotone sequence of positive constants with $c_j \to 0$ as $j\to\infty$ so fast that

$$\sum_1^\infty c_j < \infty , \quad \sum_1^\infty c_j Z_j W_j^{-2} < \infty \text{ almost surely.}$$

(If $Z_j(\omega) < c_j$, let $I(c_j < |X_j| \le Z_j)(\omega) = 0$.) Set also

$$X_j^* = \tilde{X}_j - E(\tilde{X}_j \mid F_{j-1}) .$$

It is easily checked that $\tilde{X}_1, \ldots, \tilde{X}_n$ and hence $X_1^*, \ldots, X_n^*$ also generate the σ-field F_n.

Write $S_n^* = \sum_1^n X_j^*$, $V_n^{*2} = \sum_1^n E\left(X_j^{*2} \mid F_{j-1}\right)$, and if $u \in [0, 1]$, let

$$\mu_n^*(u) = \left[\phi\left(W_n^2\right)\right]^{-1}\left[S_l^* + \left(uW_n^2 - W_l^2\right)\left(W_{l+1}^2 - W_l^2\right)^{-1} X_{l+1}^*\right] .$$

We use (20) to obtain

$$|X_j - X_j^* - \{X_j I(|X_j| > Z_j) - E(X_j I(|X_j| > Z_j) \mid F_{j-1})\}| \le 2c_j ,$$

and then

$$\sup_{0 \le u \le 1} |\mu_n(u) - \mu_n^*(u)| \le \left[\phi\left(W_n^2\right)\right]^{-1} \sup_{1 \le r \le n} \left|\sum_{j=1}^r (X_j - X_j^*)\right|$$

$$\le \left[\phi\left(W_n^2\right)\right]^{-1} \sup_{1 \le r \le n} \left|\sum_1^r \{X_j I(|X_j| > Z_j) - E(X_j I(|X_j| > Z_j) \mid F_{j-1})\}\right| +$$

$$+ \left[\phi\left(W_n^2\right)\right]^{-1} \sum_1^n 2c_j$$

$$\xrightarrow{\text{a.s.}} 0$$

as $n \to \infty$, in view of (6).

Next we introduce the Skorokhod representation (see Strassen [7], Theorem 4.3). By extending the original probability space if necessary, we may suppose that there exists a brownian motion W and a sequence $\{T_n^*, n \ge 1\}$ of non-negative random variables defined on our probability space such that $S_n^* = W(T_n^*)$ almost surely for all n. Let $t_n = T_n^* - T_{n-1}^*$, $n \ge 1$ $(T_0 = 0)$. If G_n is the σ-field generated by $X_1, \ldots, X_n$ and $W(u)$ for $u < T_n^*$, then t_n is G_n-measurable,

$$E(t_n \mid G_{n-1}) = E\left(X_n^{*2} \mid G_{n-1}\right) = E\left(X_n^{*2} \mid F_{n-1}\right) \text{ almost surely,}$$

and for some constant L,

$$E\left(t_n^2 \mid G_{n-1}\right) \le LE\left(X_n^{*4} \mid G_{n-1}\right) = LE\left(X_n^{*4} \mid F_{n-1}\right) \quad \text{almost surely.}$$

In view of (9), (21) and Theorem A, it suffices to prove that

$$W_n^{-2} T_n^* \xrightarrow{\text{a.s.}} 1 \ . \tag{22}$$

To this end we first show that

$$T_n^* - V_n^{*2} = o\left(W_n^2\right) \quad \text{almost surely as} \quad n \to \infty \ . \tag{23}$$

Since

$$\begin{aligned} E\left(X_j^{*4} \mid F_{j-1}\right) &= E\left(\tilde{X}_j^4 \mid F_{j-1}\right) - 4E\left(\tilde{X}_j^3 \mid F_{j-1}\right)E\left(\tilde{X}_j \mid F_{j-1}\right) \\ &\quad + 6E\left(\tilde{X}_j^2 \mid F_{j-1}\right)\left[E\left(\tilde{X}_j \mid F_{j-1}\right)\right]^2 - 3\left[E\left(\tilde{X}_j \mid F_{j-1}\right)\right]^4 \\ &\le 11E\left(\tilde{X}_j^4 \mid F_{j-1}\right) \\ &\le 11E\left(X_j^4 I\left(|X_j| \le Z_j\right) \mid F_{j-1}\right) + 11c_j^4 \ , \end{aligned}$$

we have from (8) that

$$\sum_1^\infty W_j^{-4} E\left(X_j^{*4} \mid F_{j-1}\right) < \infty \quad \text{almost surely}$$

and hence, using Proposition IV.6.2, page 148 of Neveu [4],

$$\sum_1^n \left[t_j - E\left(t_j \mid G_{j-1}\right)\right] = o\left(W_n^2\right) \quad \text{almost surely,}$$

which is equivalent to (23).

Next, we have

$$\left|E\left(\tilde{X}_j^2 \mid F_{j-1}\right) - E\left(X_j^2 I\left(|X_j| \le Z_j\right) \mid F_{j-1}\right)\right| \le c_j^2 \ , \tag{24}$$

$$E\left(X_j^{*2} \mid F_{j-1}\right) = E\left(\tilde{X}_j^{*2} \mid F_{j-1}\right) - \left(E\left(\tilde{X}_j \mid F_{j-1}\right)\right)^2 \ , \tag{25}$$

and

$$\left|E\left(\tilde{X}_j \mid F_{j-1}\right) - E\left(X_j I\left(|X_j| \le Z_j\right) \mid F_{j-1}\right)\right| \le c_j \ . \tag{26}$$

Then, using (26),

$$(27)\quad \sum_1^n \left[\{E(\tilde{X}_j \mid F_{j-1})\}^2 - \{E(X_j I(|X_j| \le Z_j) \mid F_{j-1})\}^2\right] =$$

$$= o\left(W_n^2\right) \text{ almost surely}$$

since

$$\sum_1^n c_j |E(X_j I(|X_j| \le Z_j) \mid F_{j-1})| = o\left(W_n^2\right) \text{ almost surely,}$$

by virtue of $\sum_1^\infty c_j Z_j W_j^{-2} < \infty$ almost surely and Kronecker's Lemma. Hence, from (7) in conjunction with (24), (25), and (27), we have

$$V_n^{*2} - W_n^2 = o\left(W_n^2\right) \text{ almost surely.}$$

Combined with (23), this establishes (22) and completes the proof of Theorem 1.

Proof of Corollary 1. Take $W_n = V_n \vee 2$ for all n. Conditions (6), (8), and (9) translate immediately into (10), (12), and (13). To check condition (7) it suffices to show that

$$(28)\quad \lim_{n\to\infty} V_n^{-2} \sum_1^n \{E(X_j I(|X_j| \le Z_j) \mid F_{j-1})\}^2 = 0 \text{ almost surely.}$$

That (28) is indeed satisfied follows from

$$E(X_j I(|X_j| \le Z_j) \mid F_{j-1}) = -E(X_j I(|X_j| > Z_j) \mid F_{j-1}) \text{ almost surely,}$$

$$\{E(X_j I(|X_j| > Z_j) \mid F_{j-1})\}^2 \le E\left(X_j^2 I(|X_j| > Z_j) \mid F_{j-1}\right) \text{ almost surely,}$$

and (11).

Proof of Corollary 2. We use Theorem 1 and take $Z_j = \delta s_j$, $W_j = U_{j-1} \vee 2$, $j > 1$.

The condition (15), together with Kronecker's Lemma, gives

$$s_n^{-1} \sum_1^n |X_j| I(|X_j| > \varepsilon s_j) \xrightarrow{\text{a.s.}} 0 ,$$

(29)

$$s_n^{-1} \sum_1^n E(|X_j| I(|X_j| > \varepsilon s_j) \mid F_{j-1}) \xrightarrow{\text{a.s.}} 0 ,$$

as $n \to \infty$ for any $\varepsilon > 0$, so that in particular, for any $\varepsilon > 0$,

$$s_n^{-1} \sup_{j \le n} |X_j| \le \varepsilon + s_n^{-1} \sum_1^n |X_j| I(|X_j| > \varepsilon s_j) \xrightarrow{\text{a.s.}} \varepsilon .$$

Hence

$$s_n^{-2} \sup_{j \le n} X_j^2 \xrightarrow{\text{a.s.}} 0 , \tag{30}$$

and in view of (14),

$$U_n^{-2} \sup_{j \le n} X_j^2 \xrightarrow{\text{a.s.}} 0 ,$$

which implies (9). We can now observe from (29) that (6) holds, while (16) ensures (8).

Finally, (16) yields via an application of Proposition IV.6.1 of [4],

$$\lim_{n\to\infty} s_n^{-2} \sum_1^n \left[X_j^2 I(|X_j| \le \delta s_j) - E\left(X_j^2 I(|X_j| \le \delta s_j) \mid F_{j-1}\right)\right] = 0 \text{ almost surely,} \tag{31}$$

while

$$s_n^{-2} \sum_1^n X_j^2 I(|X_j| > \delta s_j) \le \left(s_n^{-2} \sup_{j \le n} |X_j|\right) s_n^{-1} \sum_1^n |X_j| I(|X_j| > \delta s_j) \xrightarrow{\text{a.s.}} 0 \tag{32}$$

using (29) and (30). We deduce from (31) and (32) that

$$\lim_{n\to\infty} s_n^{-2} \sum_1^n E\left(X_j^2 I(|X_j| \le \delta s_j) \mid F_{j-1}\right) = \eta^2 \text{ almost surely,}$$

and condition (7) follows in view of (29). This completes the proof.

Proof of Corollary 3. We take $Z_j^2 = f\left(V_j^2\right)$ and check the conditions of Corollary 1.

A sufficient condition for (10) is, using Proposition IV.6.2 of [4],

$$\sum_1^\infty \left[\phi\left(V_j^2\right)\right]^{-2} E\left(X_j^2 I\left(X_j^2 > f\left(V_j^2\right)\right) \mid F_{j-1}\right) < \infty \text{ almost surely,}$$

which holds in view of (18) since $t^{-1}g(t) \to 0$ as $t \to \infty$. Furthermore, that (11) holds follows immediately from (18) and an application of Kronecker's Lemma.

To check (12), we have

$$\begin{aligned}\sum_2^\infty V_j^{-4} E\left(X_j^4 I\left(X_j^2 \le f\left(V_j^2\right)\right) \mid F_{j-1}\right) &\le \sum_2^\infty V_j^{-4} f\left(V_j^2\right)\left(V_j^2 - V_{j-1}^2\right)\\ &\le \sum_2^\infty \int_{V_{j-1}^2}^{V_j^2} t^{-2} f(t)\,dt\\ &= \int_{V_1^2}^\infty t^{-2} f(t)\,dt\\ &< \infty \text{ almost surely,}\end{aligned}$$

using the fact that $t^{-1}f(t) \downarrow 0$ and (17).

Finally, $V_n \to \infty$ almost surely,

$$1 - V_n^{-2} E\left(X_n^2 \mid F_{n-1}\right) = V_n^{-2} V_{n-1}^2 \le 1 ,$$

while

$$\begin{aligned}0 \le V_n^{-2} E\left(X_n^2 \mid F_{n-1}\right) &=\\ &= V_n^{-2} E\left(X_n^2 I\left(X_n^2 \le f\left(V_n^2\right)\right) \mid F_{n-1}\right) + V_n^{-2} E\left(X_n^2 I\left(X_n^2 > f\left(V_n^2\right)\right) \mid F_{n-1}\right)\\ &\le V_n^{-2} f\left(V_n^2\right) + V_n^{-2} E\left(X_n^2 I\left(X_n^2 > f\left(V_n^2\right)\right) \mid F_{n-1}\right)\\ &\xrightarrow{\text{a.s.}} 0 ,\end{aligned}$$

since $t^{-1}f(t) \downarrow 0$ and (11) holds. This verifies condition (13) and completes the proof.

References

[1] C.C. Heyde and D.J. Scott, "Invariance principles for the law of the iterated logarithm for martingales and processes with stationary increments", *Ann. Probability* 1 (1973), 428-436.

[2] Naresh C. Jain, Kumar Jogdeo and William F. Stout, "Upper and lower functions for martingales and mixing processes", *Ann. Probability* 3 (1975), 119-145.

[3] Paul Lévy, *Processus stochastiques et mouvement brownien* (Suivi d'une note de M. Loève. Deuxième édition revue et augmentée. Gauthier-Villars, Paris, 1965).

[4] Jacques Neveu, *Mathematical foundations of the calculus of probability* (translated by Amiel Feinstein. Holden-Day, San Francisco, California; London; Amsterdam; 1965).

[5] William F. Stout, "A martingale analogue of Kolmogorov's law of the iterated logarithm", *Z. Wahrscheinlichkeitstheorie und verw. Gebiete* 15 (1970), 279-290.

[6] V. Strassen, "An invariance principle for the law of the iterated logarithm", *Z. Wahrscheinlichkeitstheorie und verw. Gebiete* 3 (1964), 211-226.

[7] Volker Strassen, "Almost sure behavior of sums of independent random variables and martingales", *Proc. Fifth Berkeley Sympos. Math. Statistics Probability*, Volume II, Part 1, 315-343 (University of California Press, Berkeley and Los Angeles, 1967).

Mathematical Institute,
University of Oxford,
St Giles,
Oxford,
England;

Division of Mathematics and Statistics,
CSIRO,
Canberra,
ACT.

Reprinted from THEORETICAL POPULATION BIOLOGY
All Rights Reserved by Academic Press, New York and London

Vol. 11, No. 2, April 1977
Printed in Belgium

Commonwealth of Australia
Commonwealth Scientific and Industrial Organisation

The Effect of Selection on Genetic Balance when the Population Size is Varying

C. C. HEYDE

CSIRO Division of Mathematics and Statistics, Canberra, Australia

Received August 3, 1976

It is known that a reformulation for variable population size of the classical Sewall Wright model for balance between two genotypes can lead, under some circumstances, to a situation of balanced polymorphism when there is no selection present. In this note it is shown that the presence of selection prohibits the possibility of balance and assures ultimate homozygosity with probability one.

1. INTRODUCTION

We consider the classical situation of a finite population of haploid individuals in which the generations are nonoverlapping. Suppose that there are M individuals in the population which are either of genotype a or A. Suppose also that a has a small selective advantage σ over A. Then, in the classical Sewall Wright formulation, each generation is produced by binomial sampling with probabilities proportional to the numbers of a and A individuals in the previous generation multiplied by their respective selective values. Indeed, if there are i individuals of type a in a particular generation, the probability that there will be j in the next is

$$p_{ij} = \binom{M}{j} [(1+\sigma)i/(M+\sigma i)]^j \, [(M-i)/(M+\sigma i)]^{M-j}, \qquad 0 \leqslant j \leqslant M. \quad (1)$$

It is well known that if X_n denotes the number of a alleles in the nth generation, then $\{X_n\}$ is a finite homogeneous Markov chain with transition probabilities given by (1). Further, the states 0 and M are absorbing while the others are transient, so that the probability of ultimate fixation of one allele or the other is one. This holds for $\sigma \geqslant 0$.

One of the major criticisms levelled at the above model concerns the assumption of a constant population size M, which must, in a sense, restrict modes of long-term behaviour. We can, however, following Seneta (1974) and Heyde and Seneta (1975), who have dealt with the case $\sigma = 0$, relax this assumption

ISSN 0040-5809

R. Maller et al. (eds.), *Selected Works of C.C. Heyde*, Selected Works in Probability and Statistics,
DOI 10.1007/978-1-4419-5823-5_41, © Springer Science+Business Media, LLC 2010

in a general way while retaining the essential features of the Wright model. This then allows consideration of, for example, a fluctuating environment.

Let M_n be a random variable denoting the population size at time n (with $M_n > 0$ for all n) and assume further that (a) the process $\{X_n, M_n, n \geqslant 0\}$ is bivariate Markov and (b) the distribution of X_{n+1} given M_{n+1}, M_n, X_n is binomial with probability given by

$$\binom{M_{n+1}}{j} [(1+\sigma) X_n/(M_n + \sigma X_n)]^j \, [(M_n - X_n)/(M_n + \sigma X_n)]^{M_{n+1}-j}, \qquad 0 \leqslant j \leqslant M_{n+1}.$$

The question now is to consider the behaviour of X_n in this general case. Indeed, if $\sigma = 0$ (no selection) it is known that ultimate homozygosity is not in general certain, and a discussion of circumstances under which balanced polymorphism obtains is given in Seneta (1974) and Heyde and Seneta (1975). Roughly speaking, balanced polymorphism can occur if the population size increases sufficiently rapidly. It is the purpose of this note to show that *if selection is present* ($\sigma > 0$), *there must be ultimate homozygosity with probability one.* That is, a state of balanced polymorphism is no longer possible, regardless of the behaviour of the population size.

An intuitive explanation for this phenomenon is not difficult to provide. It is elementary and well known that fixation obtains in a deterministic infinite population haploid selection model. Specifically, for genotype $A[a]$ having fitness $1 + \sigma[1]$, $A[a]$ will fix if $\sigma > 0$ $[\sigma < 0]$. On the other hand, for a fixed population size stochastic model (with no mutation) the probability of ultimate fixation is one as in any Markov chain with a bounded number of states. Thus, for large or small populations, the deterministic and stochastic influences both drive towards an outcome of fixation when $\sigma \neq 0$. This situation may be contrasted with the case $\sigma = 0$, where the deterministic forces lead to neutral outcomes (all points are invariants).

2. Mathematical Details

We confine our attention to the case $\sigma > 0$, the case $\sigma = 0$ having been discussed previously, as is mentioned above.

Write F_n for the σ field generated by $\{X_0, M_0, \ldots, M_n\}$, $n \geqslant 0$, i.e., the history of the process up to time n. Put $Y_n = X_n/M_n$. We can easily see that

$$E(Y_{n+1} \mid F_n) = \frac{(1+\sigma) Y_n}{1 + \sigma Y_n} = Y_n + \frac{\sigma Y_n(1 - Y_n)}{1 + \sigma Y_n} \geqslant Y_n,$$

so that $\{Y_n, F_n, n \geqslant 1\}$ is a submartingale and, since $0 \leqslant Y_n \leqslant 1$, the martingale convergence theorem assures the existence of a random variable Y such that $Y_n \to Y$ a.s. as $n \to \infty$.

In order to identify Y, we take expectations in the above equation to obtain

$$EY_{n+1} = EY_n + \sigma E[Y_n(1 - Y_n)/(1 + \sigma Y_n)].$$

Then, letting $n \to \infty$ and using dominated convergence we obtain

$$\sigma E[Y(1 - Y)/(1 + \sigma Y)] = 0,$$

and since $0 \leqslant Y \leqslant 1$ and $\sigma > 0$ it follows that $Y(1 - Y) = 0$ a.s. The introduction of selection thus assures ultimate homozygosity with probability one.

In the classical case, where M_n is constant for all n, there have been a number of attempts, dating back to Fisher (1930), to obtain accurate bounds on $P(Y = 0) = 1 - P(Y = 1)$ (see Moran, 1960 and Arnold, 1968). The problem is much more difficult in the case of stochastically varying M_n and we merely note that

$$EY_n \uparrow EY = P(Y = 1)$$

as $n \to \infty$.

References

Arnold, B. C. 1968. A modification of a result due to Moran, *J. Appl. Prob.* **5**, 220–223.

Fisher, R. A. 1930. "The Genetical Theory of Natural Selection," Clarendon, Oxford.

Heyde, C. C., and Seneta, E. 1975. The genetic balance between random sampling and random population size, *J. Math. Biol.* **1**, 317–320.

Moran, P. A. P. 1960. The survival of a mutant gene under selection II, *J. Austral. Math. Soc.* **1**, 485–491.

Seneta, E. 1974. A note on the balance between random sampling and population size, *Genetics* **77**, 607–610.

Printed by the St Catherine Press Ltd., Tempelhof 37, Bruges, Belgium

J. Appl. Prob. **14**, 758–775 (1977)
Printed in Israel

ON CENTRAL LIMIT AND ITERATED LOGARITHM SUPPLEMENTS TO THE MARTINGALE CONVERGENCE THEOREM

C. C. HEYDE, *C.S.I.R.O. Division of Mathematics and Statistics, Canberra*

Abstract

Let $\{S_n, n \geqq 1\}$ be a zero, mean square integrable martingale for which $\lim_{n\to\infty} ES_n^2 < \infty$ so that $S_n \to S_\infty$ a.s., say, by the martingale convergence theorem. The paper is principally concerned with obtaining central limit and iterated logarithm results for $B_n(S_n - S_\infty)$ where the multipliers $B_n \uparrow \infty$ a.s. An example on the Pólya urn scheme is given to illustrate the results.

RATES OF CONVERGENCE: MARTINGALE CONVERGENCE THEOREM; CENTRAL LIMIT THEOREM; ITERATED LOGARITHM LAW; PÓLYA URN SCHEME

1. Introduction

A serious criticism that can be levelled at probability theory, and with some justification, is that many of the limit results obtained have their relevance obscured by lack of information on the sample sizes necessary for their use to be appropriate. Furthermore, the problem of rates of convergence in general is exacerbated by the hierarchy of limit theory that now exists.

The most basic results in probability theory seem to be those of the form $Z_n \xrightarrow{\text{a.s.}} Z$ as $n \to \infty$ and the single most powerful tool for establishing results of this kind would appear to be the martingale convergence theorem. However, rate results for the martingale convergence theorem are notably absent from the literature. It is the object of this paper to contribute to the rectification of this deficiency.

The analogy which we shall principally pursue is with the classical strong law of large numbers (SLLN) and the associated central limit theorem (CLT) and law of the iterated logarithm (LIL) which are rate of convergence results about the SLLN. Indeed, if the X_i, $i = 1, 2, 3, \cdots$ are independent and identically distributed with $E|X_1| < \infty$, $EX_1 = \mu$ and $S_n = \sum_{i=1}^n X_i$, then the SLLN gives $n^{-1}S_n \xrightarrow{\text{a.s.}} \mu$ as $n \to \infty$. If, additionally, $\operatorname{Var} X_1 = \sigma^2 < \infty$, then the CLT gives

$$n^{1/2}\sigma^{-1}(n^{-1}S_n - \mu) \xrightarrow{d} N(0,1)$$

and the LIL gives

$$n^{-1}S_n - \mu = \zeta(n)(2n^{-1}\log\log n)^{1/2}$$

Received in revised form 1 March 1977.

R. Maller et al. (eds.), *Selected Works of C.C. Heyde*, Selected Works in Probability and Statistics, DOI 10.1007/978-1-4419-5823-5_42, © Springer Science+Business Media, LLC 2010

where $\zeta(n)$ has its set of a.s. limit points confined to $[-1,1]$ with $\limsup_{n\to\infty}\zeta(n)=+1$ a.s., $\liminf_{n\to\infty}\zeta(n)=-1$ a.s. It seems important to write these limit results in a form which highlights their role as rate results about the SLLN, rather than in the more familiar way which suppresses this relationship.

In this paper we shall be concerned with $\{\sum_{i=1}^{n} X_i, F_n, n\geqq 1\}$, a zero mean, square integrable martingale. Square integrability is needed for the most widely applicable rate results so it is assumed at the outset. Now suppose $\lim_{n\to\infty} E(\sum_{i=1}^{n} X_i)^2=\sum_{i=1}^{\infty} EX_i^2<\infty$. Then, the martingale convergence theorem gives $\sum_{i=1}^{n} X_i \xrightarrow{\text{a.s.}} \sum_{i=1}^{\infty} X_i$, and in asking for rate results on this convergence we look for multipliers $\{B_n, n\geqq 1\}$, possibly random but increasing to infinity (a.s.), such that $B_n \sum_{i=n}^{\infty} X_i$ has behaviour of the type of the CLT or LIL. This will be the subject of Section 2. Of course these results involve a tail sum of martingale differences and the ordinary CLT and LIL for martingales do not bear directly on this context. Section 3 provides an example, involving the Pólya urn scheme.

2. Central limit and iterated logarithm results

For cases where the X_i are not independent, the literature on this problem amounts to a variety of special results for particular contexts. For example, if $\{Z_0=1, Z_1, Z_2,\cdots\}$ denotes a supercritical Bienaymé–Galton–Watson process whose offspring distribution has mean $m(>1)$ and finite variance σ^2, then an application of the martingale convergence theorem gives $m^{-n}Z_n \xrightarrow{\text{a.s.}} W$, where W turns out to be a non-degenerate random variable. It is further known that, conditional on non-extinction,

$$m^nZ_n^{-1/2}(m^{-n}Z_n-W)\xrightarrow{\text{d}} N(0,\sigma^2(m^2-m)^{-1}).$$

This is a result of the type under discussion. It rests, however, on results for sums of independent and identically distributed random variables since Z_n-m^nW can be represented as a sum of Z_n independent random variables, each with the distribution of $1-W$ (which has mean zero and variance $\sigma^2(m^2-m)^{-1}$). Full details can be found in Chapter I of Athreya and Ney (1972).

For the case of independent X_i there are some nice duality results in the literature. Chow and Teicher (1973) have obtained the following LIL's.

Proposition A. Let $\{X_n, n\geqq 1\}$ be independent and identically distributed random variables with $EX_n=0$ and $EX_n^2=1$ and $\{a_n, n\geqq 1\}$ be real constants.

(a) If $\sum_1^n a_j^2\to\infty$ as $n\to\infty$ and $(a_n^2/\sum_1^n a_j^2)\leqq C/n$, $n\geqq 1$, for some C in $(0,\infty)$, then

$$\limsup_{n\to\infty}\left(2\sum_{j=1}^{n} a_j^2 \log\log \sum_{j=1}^{n} a_j^2\right)^{-1/2}\sum_{j=1}^{n} a_jX_j=1 \text{ a.s.}$$

(b) If $\sum_1^\infty a_j^2<\infty$ and $(a_n^2/\sum_n^\infty a_j^2)\leqq C/n$, $n\geqq 1$, for some C in $(0,\infty)$, then

$$\limsup_{n\to\infty}\left(2\sum_{j=n}^{\infty}a_j^2\log\log\left(\sum_{j=n}^{\infty}a_j^2\right)^{-1}\right)^{-1/2}\sum_{j=n}^{\infty}a_jX_j=1 \text{ a.s.}$$

The proofs given by Chow and Teicher for these results are entirely classical, using truncations and exponential inequalities.

Much more important from the point of view of general theory is the paper of Barbour (1974). Barbour works with independent random variables but sketches a method which is widely applicable, well beyond this case. The method has its origins in work of Whitt (1972) in a rather different context. Barbour's basic result is the following duality form for the Lindeberg–Feller version of the CLT.

Proposition B. Let $\{X_n, n \geqq 1\}$ be independent random variables with zero means and finite variances.

(a) Suppose that $s_n^2 = \Sigma_{k=1}^n EX_k^2 \to \infty$ as $n \to \infty$. If

(i) $$s_n^{-2}EX_n^2 \to 0 \quad \text{as} \quad n\to\infty,$$

and

(ii) $$s_n^{-2}\sum_{k=1}^{n}E(X_k^2 I(|X_k| > \varepsilon s_n)) \to 0, \ \forall \varepsilon > 0,$$

then

$$s_n^{-1}\sum_{k=1}^{n}X_k \xrightarrow{d} N(0,1).$$

(b) Suppose that $s_n^2 = \Sigma_{k=n}^\infty EX_k^2 \to 0$ as $n \to \infty$. If

(i) $$s_n^{-2}EX_n^2 \to 0 \quad \text{as} \quad n\to\infty,$$

and

(ii) $$s_n^{-2}\sum_{k=n}^{\infty}E(X_k^2 I(|X_k| > \varepsilon s_n)) \to 0, \qquad \forall \varepsilon > 0,.$$

then

$$s_n^{-1}\sum_{k=n}^{\infty}X_k \xrightarrow{d} N(0,1).$$

Barbour also purports to sketch a proof of a duality result for the Kolmogorov form of the LIL but this is unfounded as the necessary prerequisite of an ordinary functional LIL under the Kolmogorov condition has yet to be established. Nevertheless, useful basic methodology is provided and we shall take this up for the martingale case.

First a remark about notation. Let J be an interval of the form $[0, L]$ for some $L < \infty$ or $[0, \infty)$. Let $D(J)$ be the space of right-continuous real-valued functions on J with left limits. We shall understand that $D(J)$ is endowed with the Skorokhod J_1 topology in the absence of another prescription.

The basic setting for Barbour's methodology is T, the subspace of $D[0,\infty)$ consisting of those functions x which satisfy

(A) $$\limsup_{t\to\infty} t^{-1}|x(t)| = 0,$$

(B) $$\int_1^\infty u^{-2}|x(u)|\,du < \infty,$$

(C) $$\int_0^1 u^{-1}|x(u)|\,du < \infty.$$

Let m be the metric on T such that $m(x,y)$ is the infimum of those $\varepsilon > 0$ for which there exists some continuous strictly increasing function $\lambda : [0,\infty) \to [0,\infty)$ with $\lambda(0) = 0$, such that

(A′) $$\sup_{t\geqq 0} |x(t) - y(\lambda(t))|/(t \vee 1) < \varepsilon,$$

(B′) $$\int_1^\infty u^{-2}|x(u) - y(\lambda(u))|\,du < \varepsilon,$$

(C′) $$\int_0^1 u^{-1}|x(u) - y(\lambda(u))|\,du < \varepsilon,$$

(D) $$\sup_{t\neq s}\left|\log\frac{\lambda(t)-\lambda(s)}{t-s}\right| < \varepsilon.$$

Let T_1 and m_1 be defined similarly but omitting the restrictions C, C'. We shall use a star to denote the corresponding subspaces of $D^*[0,\infty)$, the space of left-continuous functions with right limits on $[0,\infty)$.

We consider the mapping $g : T \to T_1^*$ or: $T^* \to T_1$ defined by

$$g(x)(0) = 0; \quad g(x)(s) = -sx(s^{-1}) + \int_{s^{-1}}^\infty u^{-2}x(u)\,du, \qquad 0 < s < \infty.$$

That is, $g(x)(s)$ is the stochastic integral $\int_{s^{-1}}^\infty u^{-1}dx(u)$ defined by its integration by parts formula. It is easily checked that the mapping is continuous (e.g. as with Lemmas 2.1 and 3.7 of Whitt (1972)) and that for $x \in T \cup T^*$ with $x(0) = 0$, $g(g(x)) = x$. The mapping provides a device for reading results for tail sums off from those of ordinary sums (and vice versa) and we shall explore its ramifications for the case of martingale differences.

Recall that $\{\sum_{i=1}^n X_i, F_n, n \geqq 1\}$ is a zero mean, square integrable martingale with $\sum_{i=1}^\infty EX_i^2 < \infty$. We set $S_n = \sum_{k=n}^\infty X_k$ and $s_n^2 = ES_n^2 = \sum_{k=n}^\infty EX_k^2$. We shall initially be dealing with the case where $s_n^{-2}\sum_{k=n}^\infty X_k^2 \xrightarrow{\text{a.s.}} \eta^2$, η^2 being some positive and a.s. finite random variable.

Let $\{Z_n, n \geq 1\}$ be a monotone sequence of positive random variables defined by $Z_n^2 = \max_{k \leq n} s_k^{-2}(E(\eta^2 | F_{k-1}))^{-1}$. Then Z_n is F_{n-1}-measurable and $Z_n^2 \sim s_n^{-2}\eta^{-2}$ a.s. $n \to \infty$ since $E(\eta^2 | F_{n-1}) \xrightarrow{\text{a.s.}} \eta^2$ as $n \to \infty$. Put

(1) $$Y_n(t) = Z_n f(n) S_{k(n,t)+1}, \qquad t \geq 0,$$

where $f(n)$ is a normalization function and

(2) $$k(n, t) = \max [j \mid tZ_j^2 \leq Z_n^2].$$

Then, formally,

(3) $$g(Y_n)(s) = Z_n^{-1} f(n) \sum_{j=1}^{k(n,s^{-1})} X_j Z_j^2, \qquad s > 0,$$

and we shall apply standard martingale results to $g(Y_n)$.

At this stage we need to be specific about the conditions imposed. We shall establish the following theorem which, for the purposes of the exposition, is described in terms of a duality result.

Theorem 1. Let $\{\Sigma_1^n X_i, F_n, n \geq 1\}$ be a zero mean square integrable martingale.

(a) Write $W_n^2 = \Sigma_1^n X_i^2$ and $s_n^2 = EW_n^2$. Suppose that $s_n \to \infty$ and

(i) $$s_n^{-2} \sum_1^n X_i^2 = s_n^{-2} W_n^2 \xrightarrow{\text{P}} \eta^2$$

where η^2 is some a.s. finite and non-zero random variable. If, in addition to (i) we have

(ii) $$s_n^{-2} E\left(\max_{1 \leq k \leq n} X_k^2 \right) \to 0 \quad \text{as} \quad n \to \infty,$$

then

(4) $$W_n^{-1} \sum_1^n X_i \xrightarrow{\text{d}} N(0,1) \quad \text{and} \quad s_n^{-1} \sum_1^n X_i \xrightarrow{\text{d}} \eta' N(0,1)$$

where η' is independent of $N(0,1)$ and distributed as η.

On the other hand, if the convergence in (a)(i) is strengthened to a.s. convergence and in addition we have

(iii) $$\sum_1^\infty s_j^{-1} E\,[|X_j| I(|X_j| > \varepsilon s_j)] < \infty, \qquad \forall \varepsilon > 0,$$

and

(iv) $$\sum_1^\infty s_j^{-4} E\,[X_j^4 I(|X_j| \leq \delta s_j)] < \infty \quad \text{for some} \quad \delta > 0,$$

then, writing $\phi(t) = (2 \log \log (t \vee 3))^{1/2}$,

$$(W_n\phi(W_n))^{-1}\sum_{j=1}^{n} X_j = \theta_1(n)$$

has for its set of a.s. limit points the closed interval $[-1, 1]$ with

$$\limsup_{n\to\infty} \theta_1(n) = +1 \text{ a.s.,} \qquad \liminf_{n\to\infty} \theta_1(n) = -1 \text{ a.s.}$$

(b) Write $W_n^2 = \Sigma_n^\infty X_i^2$ and $s_n^2 = EW_n^2$. Suppose that $s_n \to 0$ and

(i)
$$s_n^{-2}\sum_{n}^{\infty} X_j^2 = s_n^{-2} W_n^2 \xrightarrow{\text{P}} \eta^2$$

where η^2 is some a.s. finite and non-zero random variable. If, in addition to (b)(i) we have

(ii)
$$s_n^{-2} E\left(\max_{k\geq n} X_k^2\right) \to 0 \quad \text{as} \quad n\to\infty,$$

then

(5)
$$W_n^{-1}\sum_{n}^{\infty} X_i \xrightarrow{\text{d}} N(0,1) \quad \text{and} \quad s_n^{-1}\sum_{n}^{\infty} X_i \xrightarrow{\text{d}} \eta' N(0,1)$$

where η' is independent of $N(0,1)$ and distributed as η.

On the other hand, if the convergence in (b)(i) is strengthened to a.s. convergence and in addition we have

(iii)
$$\sum_{1}^{\infty} s_j^{-1} E[|X_j| I(|X_j| > \varepsilon s_j)] < \infty, \qquad \forall \varepsilon > 0,$$

and

(iv)
$$\sum_{1}^{\infty} s_j^{-4} E[X_j^4 I(|X_j| \leqq \delta s_j)] < \infty \quad \text{for some} \quad \delta > 0,$$

then

$$(W_n\phi(W_n^{-1}))^{-1}\sum_{j=n}^{\infty} X_j = \theta_2(n)$$

has for its set of a.s. limit points the closed interval $[-1,1]$ with

$$\limsup_{n\to\infty} \theta_2(n) = +1 \text{ a.s.,} \qquad \liminf_{n\to\infty} \theta_2(n) = -1 \text{ a.s.}$$

The conditions given in the statement of this theorem are frequently not the most convenient for applications and some useful variants are given in the following corollaries.

Corollary 1. (a) Sufficient conditions for (a)(i) and (a)(ii) in Theorem 1 are (a)(i′) $$s_n^{-2}\sum_1^n E(X_i^2|F_{i-1})\xrightarrow{P}\eta^2$$

and

(a)(ii′) $$s_n^{-2}\sum_1^n E(X_i^2 I(|X_i|>\varepsilon s_n))\to 0,\qquad \forall\varepsilon>0.$$

(b) Sufficient conditions for (b)(i) and (b)(ii) in Theorem 1 are

(b)(i′) $$s_n^{-2}\sum_n^\infty E(X_i^2|F_{i-1})\xrightarrow{P}\eta^2$$

and

(b)(ii′) $$s_n^{-2}\sum_n^\infty E(X_i^2 I(|X_i|>\varepsilon s_n))\to 0,\qquad \forall\varepsilon>0.$$

Corollary 2. (a)[(b)]. Put $V_n^2=\Sigma_1^n E(X_i^2|F_{i-1})[=\Sigma_n^\infty E(X_i^2|F_{i-1})]$. A sufficient condition for the replacement of W_n by V_n in Part (a)[(b)] of Theorem 1 is that $\Sigma_1^\infty s_n^{-2}(X_n^2-E(X_n^2|F_{n-1}))$ converges a.s.

As a preliminary to the proofs of these results we shall establish the following lemma.

Lemma 1. Let $\{a_k, k\geqq 1\}$ and $\{b_k, k\geqq 1\}$ be sequences of real numbers with $a_k\geqq 0$, $\forall k$.

(i) If $\Sigma_1^\infty a_k<\infty$ and $(a_k/\Sigma_{j=k}^\infty a_j)\to l$, some $0\leqq l<1$, as $k\to\infty$, then

$$\left(\sum_{j=n}^\infty a_j\right)\sum_{k=1}^n\left[a_k\left(\sum_{j=k}^\infty a_j\right)^{-2}\right]\to 1\quad\text{as}\quad n\to\infty.$$

(ii) If $a_n\uparrow\infty$ and the series $\Sigma_1^\infty a_kb_k$ converges, then $\Sigma_{k=n}^\infty b_k=o(a_n^{-1})$ as $n\to\infty$.

Proof. (i) Write $c_n=(\Sigma_{j=n}^\infty a_j)^{-1}$, so that what we have to show is

(6) $$c_n^{-1}\sum_{k=1}^n c_k^2(c_k^{-1}-c_{k+1}^{-1})\to 1.$$

However, (6) can be rewritten as

$$c_n^{-1}\sum_{k=1}^n c_kc_{k+1}^{-1}(c_{k+1}-c_k)\to 1=c_n^{-1}\sum_{k=1}^{n-1}(c_{k+1}-c_k)+c_n^{-1}c_1$$

or equivalently

(7) $$(1-c_nc_{n+1}^{-1})+c_n^{-1}\sum_{k=1}^{n-1}(c_kc_{k+1}^{-1}-1)(c_{k+1}-c_k)\to 0.$$

The result (7) follows via the Toeplitz lemma since $1-c_kc_{k+1}^{-1}\to l$ as $k\to\infty$.

(ii) Let $K_n=\sup_{N\geqq M\geqq n}|\Sigma_{k=n}^M a_kb_k|$, so that $K_n\to 0$ as $n\to\infty$.

From Abel's lemma,

$$\left|\sum_{k=n}^{N} b_k\right| \leqq K_n / a_n$$

and the required result follows upon letting $N \to \infty$.

Proof of Theorem 1. The results of Part (a) of the theorem are known and are just stated for the purpose of exhibiting the duality. The central limit result follows, for example, from Corollaries 2.1.1 and 2.2.1 of Hall (1977). The iterated logarithm result is essentially that of Hall and Heyde (1976). We shall thus confine our attention to the proof of Part (b) of the theorem.

We begin by using invariance principle versions of Part (a) and the methodology proposed by Barbour. For this purpose we assume a.s. convergence in (b)(i). This restriction will later be removed for the central limit case via other methods.

The first thing we need to check is that $g(Y_n)$ as given by (3) belongs to T under the conditions of (b). As a prelude to this we observe that $s_{n+1}^2 s_n^{-2} \to 1$ as $n \to \infty$. This is easily deduced from (b)(ii) for the central limit case. For the iterated logarithm case, on the other hand, we have for any $\varepsilon > 0$,

$$s_n^{-1}|X_n| \leqq \varepsilon + s_n^{-1}|X_n| I(|X_n| > \varepsilon s_n) \xrightarrow{\text{a.s.}} \varepsilon$$

using (b)(iii) so that $s_n^{-2} X_n^2 \xrightarrow{\text{a.s.}} 0$ and the desired result follows readily using (b)(i).

Next, note that

$$\frac{X_n^2}{W_n^2} = \frac{W_n^2 - W_{n+1}^2}{W_n^2} = 1 - \frac{s_{n+1}^{-2}\, W_{n+1}^2}{s_n^{-2}\, W_n^2}\, \frac{s_{n+1}^2}{s_n^2} \xrightarrow{\text{a.s.}} 0$$

and hence, using (b)(i), Lemma 1(i) and $Z_n^2 \sim W_n^{-2}$ a.s.,

$$Z_n^{-2} \sum_{1}^{n} X_k^2 Z_k^4 \xrightarrow{\text{a.s.}} 1. \tag{8}$$

With the aid of (8) and recalling that $Z_{k(n,s^{-1})}^2 \sim s Z_n^2$ as $s \to \infty$ we can deduce that

$$s^{-3/4} g(Y_n)(s) \xrightarrow{\text{a.s.}} 0 \quad \text{as} \quad s \to \infty. \tag{9}$$

Indeed, in view of (8) the result (9) holds if

$$\sum_{j=1}^{k(n,s^{-1})} X_j Z_j^2 \Big/ \left(\sum_{j=1}^{k(n,s^{-1})} X_j^2 Z_j^4 \right)^{3/4} \xrightarrow{\text{a.s.}} 0 \quad \text{as} \quad s \to \infty. \tag{10}$$

However, (10) follows, using the Kronecker lemma, if

$$\sum_{1}^{\infty} \left[X_n Z_n^2 \Big/ \left(\sum_{1}^{n} X_j^2 Z_j^4 \right)^{3/4} \right] \quad \text{converges a.s.}$$

and hence if, using (8) again,

(11) $$\sum_1^\infty X_n s_n^{-1/2} \quad \text{converges a.s.}$$

Then, the martingale convergence theorem gives (11) provided that

$$\sum_1^\infty s_n^{-1} E X_n^2 < \infty$$

as is easily checked since $s_n \geqq s_{n+1}$ and

$$\sum_1^\infty s_n^{-1}(s_n^2 - s_{n+1}^2) = \sum_1^\infty s_n^{-1}(s_n + s_{n+1})(s_n - s_{n+1}) \leqq 2\sum_1^\infty (s_n - s_{n+1})$$
$$= 2s_1 < \infty$$

and the proof of (9) is complete.

The prerequisite conditions for $g(Y_n)$ to belong to T are now easily verified. The relation (9) ensures that (A) and (B) hold, while (C) is satisfied as the defining integral reduces to a finite sum.

We now take $f(n) = 1$ and focus attention on the central limit part of (b). We begin by applying a minor modification of Corollary 2.1.1 of Hall (1977) to the martingale array $\{s_n \sum_{j=1}^k X_j Z_j^2, F_k, 1 \leqq k \leqq n, n \geqq 1\}$. For $z \geqq 0$ let

$$\xi_n(z) = Z_n^{-1} \sum_{j=1}^{k(n,z^{-1})} X_j Z_j^2.$$

This formulation is not quite that of Hall but the modifications to the summation can be simply effected as in the proof of his Theorem 3.1. We take, in the notation of Hall's result, $F_{nj} = F_j$. In view of (8), we need only check that $s_n^2 \max_{1 \leqq j \leqq n} X_j^2 Z_j^4 \xrightarrow{P} 0$ and hence that $s_n^2 \max_{1 \leqq j \leqq n} X_j^2 s_j^{-4} \xrightarrow{P} 0$. Furthermore, since $s_n \downarrow 0$ and $s_{n+1} s_n^{-1} \to 1$ we can, given $\varepsilon > 0$, choose a sequence $\{j(n)\}$ such that $s_n^2 / s_{j(n)}^2 \to \varepsilon' \leqq \varepsilon$ as $n \to \infty$. Then,

$$0 \leqq \limsup_{n\to\infty} s_n^2 E\left(\max_{1 \leqq j \leqq n} X_j^2 s_j^{-4}\right)$$
$$\leqq \limsup_{n\to\infty} s_n^2 E\left(\max_{1 \leqq j < j(n)} X_j^2 s_j^{-4}\right) + \limsup_{n\to\infty} s_n^2 E\left(\max_{j(n) \leqq j \leqq n} X_j^2 s_j^{-4}\right)$$
$$\leqq \limsup_{n\to\infty} s_n^2 \sum_{j=1}^{j(n)-1} (s_j^2 - s_{j+1}^2) s_j^{-4} + \limsup_{n\to\infty} s_n^{-2} E\left(\max_{j \geqq j(n)} X_j^2\right) \leqq \varepsilon'$$

in view of (b)(ii) and

$$\sum_{j=1}^{j(n)-1} (s_j^2 - s_{j+1}^2) s_j^{-4} \leqq \sum_{j=1}^{j(n)-1} (s_j^2 - s_{j+1}^2) s_j^{-2} s_{j+1}^{-2} = \sum_{j=1}^{j(n)-1} (s_{j+1}^{-2} - s_j^{-2}) = s_{j(n)}^{-2}.$$

The corollary in question then gives

$$(\xi_n, s_n^2 Z_n^2) \xrightarrow{d} (W, (\eta')^{-2})$$

where W is standard Brownian motion and η' has the same distribution as η and is independent of W. The basic setting is $D[0,1]$ and, defining $\eta_n(z) = s_n^{-1} Z_n^{-1} \xi_n(z)$, we have

(12) $$\xi_n \xrightarrow{d} W \quad \text{and} \quad \eta_n \xrightarrow{d} \eta' W$$

in this setting. However, the extension of the results (12) to $D[0, T], 0 < T < \infty$, and indeed to $D[0, \infty)$ is entirely routine (e.g. see McLeish (1974), pp. 625–626).

Of course $\xi_n = g(Y_n)$ and η_n is also an element of T so it remains to check that the results (12) continue to apply in (T, m). This however follows from the argument of Müller ((1968), p. 176, 177; see also Whitt (1972)) since the Kolmogorov inequality continues to hold for martingales. Then, using the continuous mapping theorem (e.g. Billingsley (1968), Theorem 5.1),

$$g(\xi_n) = g(g(Y_n)) = Y_n \xrightarrow{d} g(W) \stackrel{d}{=} W,$$

while

$$g(\eta_n) = s_n^{-1} Z_n^{-1} g(g(Y_n)) = s_n^{-1} Z_n^{-1} Y_n \xrightarrow{d} g(\eta' W) \stackrel{d}{=} \eta' W$$

and the results (5) follow immediately.

It should be noted that the functional limit results just obtained offer scope for a variety of other important results via the continuous mapping theorem. From the convergence $Y_n \xrightarrow{d} W$ in T_1^* we can extract $Y_n \xrightarrow{d} W$ in $(D^*[0,1], d_0)$ and hence in $(D^*[0,1], d)$, and d_0 being the metrics discussed by Billingsley ((1968), pp. 111–113) which generate the Skorokhod topology. Then, for example, the mapping $h: D^*[0,1] \to R$ defined by $h(x)(t) = \sup_{0 \le t \le 1} |x(t)|$ is continuous and the continuous mapping theorem gives $h(Y_n) \xrightarrow{d} h(W)$, or equivalently,

$$W_n^{-1} \sup_{k \geq n} \left| \sum_{j=k}^{\infty} X_j \right| \xrightarrow{d} \sup_{0 \le t \le 1} \left| W(t) \right|.$$

Next we take $f(n) = (2 \log \log (s_n^{-1} \vee 3))^{-1/2}$ in (1) and focus attention on the iterated logarithm part of (b). It is our object to show that $\{g(Y_n), n \geqq 1\}$ has compact closure a.s. with set of limit points K, K being the set of absolutely continuous real functions on $[0, \infty)$ for which $x(0) = 0$ and $\int_0^\infty [\dot{x}(t)]^2 \, dt \leqq 1$, $\dot{x}$ denoting the derivative of x determined almost everywhere with respect to Lebesgue measure. The iterated logarithm part of (b) is then easily extracted from the result that $\{Y_n = g(g(Y_n)),\ n \geqq 1\}$ has compact closure a.s. with set of limit points $g(K) = K$, while $Z_n W_n \phi(W_n^{-1})[\phi(Z_n^{-1})]^{-1} \xrightarrow{\text{a.s.}} 1$. That K belongs to T and T^* follows readily from

$$|x(t)| = \left| \int_0^t \dot{x}(u)\, du \right| \leqq \left\{ \int_0^t du \int_0^t [\dot{x}(u)]^2\, du \right\}^{1/2} \leqq t^{1/2}.$$

We shall apply Theorem 1 of Hall and Heyde (1976) which deals with the setting $C[0, 1]$ but whose scope can be extended to cover the problem in hand. The martingale to which the theorem is applied is $\{\sum_{j=1}^n X_j Z_j^2, F_n, n \geqq 1\}$ and for the Z_j and W_j of the theorem we use $\delta s_j Z_j^2$ and Z_j respectively. The checking of the conditions is routine and will not be exhibited. It follows quite closely upon the proof of Corollary 2 of Hall and Heyde (1976) and makes repeated use of (8) and the F_{j-1}-measurability of Z_j. Furthermore, the extension of this setting from $C[0, 1]$ to $C[0, \infty)$ is a standard exercise. For example, the argument provided in the proof of Theorem 1.5.15 of Vervaat (1972) can be employed. The topology for $C[0, \infty)$ is that of uniform convergence on compacta. Then, writing for $0 \leqq u < \infty$, $k = k(n, u^{-1})$ as given by (2) and

$$\mu_n(u) = [Z_n \phi(Z_n)]^{-1} \left[\sum_{j=1}^k X_j Z_j^2 + (u Z_n^2 - Z_k^2)(Z_{k+1}^2 - Z_k^2)^{-1} X_{k+1} Z_{k+1}^2 \right],$$

we have that $\{\mu_n, n \geqq 1\}$ is relatively compact in $C[0, \infty)$ and the set of its a.s. limit points coincides with K.

Next, we move to $D[0, \infty)$ and note that the Skorokhod J_1 topology relativized to $C[0, \infty)$ coincides there with the topology of uniform convergence on compacta. We can use the metric

$$\psi(x, y) = \int_0^\infty \min\{1, \rho_{[0,t]}(x|_{[0,t]}, Y|_{[0,t]})\} e^{-t}\, dt$$

for $x, y \in D[0, \infty)$ where $\rho_{[0,t]}$ is a metric on $D[0, t]$ which generates the J_1 topology on $D[0, t]$ and $z|_{[0,t]}$ denotes the restriction of z to $[0, t]$ (Vervaat (1972), p. 23).

In $D[0, \infty)$ we set

$$\nu_n(u) = [Z_n \phi(Z_n)]^{-1} \sum_{j=1}^k X_j Z_j^2, \qquad 0 \leqq u < \infty.$$

Then, $\psi(\nu_n, \mu_n) \xrightarrow{\text{a.s.}} 0$ as $n \to \infty$ provided that

$$\int_0^\infty \min\left\{1, \sup_{0 \leqq u \leqq t} \left| \nu_n(u) - \mu_n(u) \right| \right\} e^{-t}\, dt \xrightarrow{\text{a.s.}} 0 \quad \text{as} \quad n \to \infty, \tag{13}$$

by choosing a version of $\rho_{[0,t]}$ which is dominated by the uniform metric. Further, (13) holds if

$$\int_0^\infty \min\left\{1, [Z_n\phi(Z_n)]^{-1} \sup_{0\le u\le t} \left|X_{k+1}\right| Z_{k+1}^2\right\} e^{-t}\,dt \xrightarrow{\text{a.s.}} 0 \quad \text{as} \quad n\to\infty$$

which is easily verified since, using (8),

$$[Z_n\phi(Z_n)]^{-2} \sup_{0\le u\le t} X_{k+1}^2 Z_{k+1}^4 = O\left(Z_n^{-2} \sum_{j=1}^{k(n,t^{-1})+1} X_j^2 Z_j^4\right) = o(t)\text{ a.s.}$$

as $t\to\infty$. It then follows that $\{\nu_n, n\geqq 1\}$ is relatively compact in $(D[0,\infty), \psi)$ and the set of its a.s. limit points coincides with K.

Finally, we move to (T, m). The Borel σ-field associated with (T, m) coincides with the σ-field on T generated by the coordinate projections. This can be verified using obvious modifications of the arguments on pp. 121–122 of Billingsley (1968) (see also Lemma 3.5 of Whitt (1972)). Further, similar arguments show that the Borel σ-field associated with $D[0,\infty)$ for the Skorokhod topology is that generated by the coordinate projections. Thus, relative compactness of $\{\nu_n, n\geqq 1\}$ in $\{D[0,\infty), \psi\}$ implies the same property in (T, m). Since $K\subset T$ and $f(n)\sim[\phi(Z_n)]^{-1}$ a.s. as $n\to\infty$, the proof of the iterated logarithm part of (b) is complete.

In the previous work we have dealt with the case of a.s. convergence in (b)(i); we now weaken this to convergence in probability for the purposes of proving the relevant central limit results.

We choose a monotone sequence $\{k_n, n\geqq 1\}$ of positive integers increasing so fast that $s_n^{-2}s_{n+k_n+1}^2\to 0$ as $n\to\infty$. Further, we introduce the array of random variables $\{X_{nj}, j\geqq 1, n\geqq 1\}$ defined by $X_{nj} = s_n^{-1}X_{n+j}$. Then, from (b)(i),

$$\sum_{j=1}^{k_n} X_{nj}^2 = s_n^{-2}\sum_{j=n}^{n+k_n} X_j^2 \xrightarrow{\text{p}} \eta^2 \tag{14}$$

since $s_n^{-2}\sum_{j=n+k_n+1}^\infty X_j^2 \xrightarrow{L_1} 0$ in view of the choice of $\{k_n\}$. Also, using (b)(ii),

$$E\left(\max_{j\le k_n} X_{nj}^2\right) = s_n^{-2}E\left(\max_{n\le j\le n+k_n} X_j^2\right) \leqq s_n^{-2}E\left(\max_{j\ge n} X_j^2\right)\to 0. \tag{15}$$

The σ-fields $F_{nj}=F_{n+j}$ satisfy the nesting condition $F_{nj}\subseteq F_{n+1,j}$ so the conditions of Corollaries 2.1.1 and 2.2.1 of Hall (1977) are satisfied and

$$\sum_{j=1}^{k_n} X_{nj} = s_n^{-1}\sum_{j=n}^{n+k_n} X_j \xrightarrow{\text{d}} \eta' N(0,1)$$

$$\left\{\sum_{j=1}^{k_n} X_{nj}^2\right\}^{-1/2} \sum_{j=1}^{k_n} X_{nj} = \left\{\sum_{j=n}^{n+k_n} X_j^2\right\}^{-1/2} \sum_{j=n}^{n+k_n} X_j \xrightarrow{\text{d}} N(0,1).$$

The desired results (4) are then immediate since $s_n^{-2}\sum_{j=n+k_n+1}^\infty X_j^2\xrightarrow{L_1}0$. This completes the proof of the theorem.

It is interesting to note that a minor reformulation of results of Hall (1977) provides a convenient general result from which the central limit results of both Parts (a) and (b) of Theorem 1 can be derived. This is given in the following theorem.

Theorem 2. Let $\{X_{nj}, F_{nj}, j \geqq 1\}$ denote a martingale difference sequence for each $n \geqq 1$ which is subject to the condition

$$\sum_{j=1}^{\infty} E(X_{nj}^2) < \infty, \qquad \forall n. \tag{16}$$

If $F_{nj} \subseteq F_{n+1,j}$ for all $j \geqq 1$ and in addition

$$\sum_1^{\infty} X_{nj}^2 \xrightarrow{\mathrm{P}} \eta^2 \ (>0 \text{ a.s.}) \tag{17}$$

and

$$E\left(\max_j X_{nj}^2\right) \to 0 \tag{18}$$

as $n \to \infty$, then

$$\sum_1^{\infty} X_{nj} \xrightarrow{\mathrm{d}} \eta' N(0,1) \quad \text{and} \quad \left(\sum_1^{\infty} X_{nj}^2\right)^{-1/2} \sum_1^{\infty} X_{nj} \xrightarrow{\mathrm{d}} N(0,1), \tag{19}$$

η' being independent of $N(0,1)$ and distributed as η.

These results follow from Corollaries 2.1.1 and 2.2.1 of Hall (1977) in similar fashion to those just discussed in the last part of the proof of Theorem 1. The virtue of this formulation was pointed out to the author by P. G. Hall. The central limit results of Part (a) of Theorem 1 can then be extracted by taking $X_{nj} = (\Sigma_{k=1}^{n} EX_k^2)^{-1/2} X_j$, $1 \leqq j \leqq n$; $X_{nj} = 0$, $j > n$, and $F_{nj} = F_j$. Those of Part (b) follow by taking $X_{nj} = (\Sigma_{k=n}^{\infty} EX_k^2)^{-1/2} X_{n+j}$, $F_{nj} = F_{n+j}$, $j \geqq 1$, $n \geqq 1$, as in the above proof.

Proof of Corollary 1. Here we shall confine attention to the proof of Part (b). That for Part (a) is entirely analogous.

Take arbitrary $\delta > 0$. Then,

$$s_n^{-2} E\left(\max_{k \geqq n} X_k^2\right) \leqq s_n^{-2} E\left(\max_{k \geqq n} X_k^2 I(|X_k| \leqq \delta s_n)\right) + s_n^{-2} E\left(\max_{k \geqq n} X_k^2 I(|X_k| > \delta s_n)\right)$$

$$\leqq \delta^2 + s_n^{-2} \sum_{k=n}^{\infty} E(X_k^2 I(|X_k| > \delta s_n)) \to \delta^2 \quad \text{as} \quad n \to \infty$$

which gives (b)(ii). To obtain (b)(i) we first note that (b)(ii′) ensures

$$s_n^{-2} \sum_{k=n}^{\infty} [X_k^2 I(|X_k| > \varepsilon s_n) - E(X_k^2 I(|X_k| > \varepsilon s_n) | F_{k-1})] \xrightarrow{L_1} 0, \tag{20}$$

while

(21) $$s_n^{-2} \sum_{k=n}^{\infty} [X_k^2 I(|X_k| \leqq \varepsilon s_n) - E(X_k^2 I(|X_k| \leqq \varepsilon s_n) | F_{k-1})] \xrightarrow{L_2} 0$$

provided that

$$s_n^{-4} \sum_{k=n}^{\infty} E[X_k^4 I(|X_k| \leqq \varepsilon s_n)] \to 0.$$

That this last result holds, however, is easily verified since for arbitrary δ, $\varepsilon > \delta > 0$,

$$\begin{aligned} s_n^{-4} \sum_{k=n}^{\infty} E[X_k^4 I(|X_k| \leqq \varepsilon s_n)] &= s_n^{-4} \sum_{k=n}^{\infty} E[X_k^4 I(|X_k| \leqq \delta s_n)] \\ &\quad + s_n^{-4} \sum_{k=n}^{\infty} E[X_k^4 I(\delta s_n < |X_k| \leqq \varepsilon s_n)] \\ &\leqq \delta^2 + \varepsilon^2 s_n^{-2} \sum_{k=n}^{\infty} E[X_k^2 I(|X_k| > \delta s_n)] \to \delta^2 \end{aligned}$$

as $n \to \infty$. The result (b)(i) then follows from (20) and (21) together with (b)(i′).

It can be further shown, and is worth noting, that the pair of conditions (a)(i′), (a)(ii′) [(b)(i′), (b)(ii′)] is equivalent to the pair (a)(i), (a)(ii) [(b)(i), (b)(ii)] if $E\eta^2 = 1$. We omit the details.

Proof of Corollary 2. Part (a) is an immediate consequence of the Kronecker lemma and Part (b) of Lemma 1 (ii).

Next we show that Propositions A and B follow as a consequence of Theorem 1.

Proof of Proposition A. We use the martingale $\{\Sigma_{j=1}^n a_j X_j, n \geqq 1\}$.

First take the assumptions of Part (a) of Proposition A. To check the a.s. convergence form of (a)(i) of Theorem 1 we observe that it suffices to show $(\Sigma_1^n a_j^2)^{-1} \Sigma_1^n a_j^2 (X_j^2 - 1) \xrightarrow{\text{a.s.}} 0$, and this follows readily from Theorem 2 of Heyde (1968) since $(a_n^2 / \Sigma_1^n a_j^2) \leqq C/n$, $n \geqq 1$.

To check (a)(iii), we note that

$$\begin{aligned} &\sum_1^{\infty} \left(\sum_1^n a_j^2 \right)^{-1/2} \cdot E\left\{ |a_j X_j| I\left(|X_j| > \varepsilon \left(\sum_1^n a_j^2 \right)^{1/2} \right) \right\} \\ &\leqq C^{1/2} \sum_1^{\infty} n^{-1/2} E\{|X_1| I(|X_1| > \varepsilon C^{-1/2} n^{1/2})\} \\ &= C^{1/2} E\left\{ |X_1| \sum_1^{\infty} n^{-1/2} I(|X_1| > \varepsilon C^{-1/2} n^{1/2}) \right\} \\ &\leqq C^{1/2} E\left\{ |X_1| \sum_1^{[C\varepsilon^{-2} X_1^2]} n^{-1/2} \right\} \end{aligned}$$

where $[x]$ denotes the integer part of x

$$\leqq C^{1/2} E\left\{|X_1| \int_0^{C\varepsilon^{-2}X_1^2} x^{-1/2}\, dx\right\}.$$
$$\leqq 2C\varepsilon^{-1} EX_1^2 < \infty.$$

Finally, to check (a)(iv) we write $u_n = \Sigma_1^n a_k^2/a_n^2$ so that the required result holds if $\Sigma_1^\infty u_n^{-2} E\{X_1^4 I(n^{1/2} < |X_1| \leqq \delta u_n^{1/2})\} < \infty$ and $\Sigma_1^\infty u_n^{-2} E\{X_1^4 I(|X_1| \leqq n^{1/2})\} < \infty$. However, the former holds since

$$\sum_1^\infty u_n^{-2} E\{X_1^4 I(n^{1/2} < |X_1| \leqq \delta u_n^{1/2})\} \leqq \delta^4 \sum_1^\infty P(|X_1| > n^{1/2})$$
$$= \delta^4 E \sum_1^\infty I(|X_1| > n^{1/2}) \leqq \delta^4 EX_1^2 < \infty,$$

and the latter since

$$\sum_1^\infty u_n^{-2} E\{X_1^4 I(|X_1| \leqq n^{1/2})\}$$
$$\leqq C^2 \sum_1^\infty n^{-2} E\{X_1^4 I(|X_1| \leqq n^{1/2})\}$$
$$\leqq C^2 \sum_{n=1}^\infty n^{-2} \sum_{k=1}^n kE\{X_1^2 I((k-1)^{1/2} < |X_1| \leqq k^{1/2})\}$$
$$= C^2 \sum_{k=1}^\infty kE\{X_1^2 I((k-1)^{1/2} < |X_1| \leqq k^{1/2})\} \sum_{n=k}^\infty n^{-2}$$
$$\leqq C^2 EX_1^2 < \infty.$$

This completes the proof of Part (a) of Proposition A, the random norming being replaced by constants in view of (a)(i).

To prove Part (b) of Proposition A we begin by checking the a.s. convergence form of (b)(i) or equivalently, $\Sigma_n^\infty a_j^2(X_j^2 - 1)/\Sigma_n^\infty a_j^2 \xrightarrow{\text{a.s.}} 0$ as $n \to \infty$. Write $u_n = \Sigma_n^\infty a_j^2/a_n^2$ and set

$$W_j = X_j^2 - 1 \quad \text{if} \quad |X_j^2 - 1| \leqq u_j$$
$$= 0 \qquad \text{otherwise.}$$

Then, $P(W_n \neq X_n^2 - 1 \text{ i.o.}) = 0$ since

$$\sum_1^\infty P(W_n \neq X_n^2 - 1) = \sum_1^\infty P(|X_1^2 - 1| > u_n) \leqq \sum_1^\infty P(C|X_1^2 - 1| > n)$$
$$\leqq E \sum_1^\infty I(C|X_1^2 - 1| > n)$$
$$\leqq CE|X_1^2 - 1| < \infty$$

and hence $\Sigma_n^\infty a_j^2(X_j^2-1)/\Sigma_n^\infty a_j^2 \xrightarrow{\text{a.s.}} 0$ if $\Sigma_n^\infty a_j^2 W_j/\Sigma_n^\infty a_j^2 \xrightarrow{\text{a.s.}} 0$. Also, $EW_n \to 0$ as $n\to\infty$ and hence $\Sigma_n^\infty a_j^2 EW_j/\Sigma_n^\infty a_j^2 \to 0$ so it remains to prove $\Sigma_n^\infty a_j^2(W_j-EW_j)/\Sigma_n^\infty a_j^2 \xrightarrow{\text{a.s.}} 0$ which holds, by virtue of Part (ii) of Lemma 1, if $\Sigma_1^\infty u_n^{-1}(W_n-EW_n)$ converges a.s. However, the Kolmogorov convergence criterion gives the a.s. convergence of $\Sigma_1^\infty u_n^{-1}(W_n-EW_n)$ provided that $\Sigma_1^\infty u_n^{-2} EW_n^2 < \infty$ and this last result holds since

$$\sum_1^\infty u_n^{-2} EW_n^2 = \sum_1^\infty u_n^{-2} E\{(X_1^2-1)^2 I(|X_1^2-1| \leqq u_n)\} < \infty$$

using the same argument as for the checking of (a)(iv) above. The remainder of the proof of Part (b) involving checking of the series Conditions (b)(iii) and (b)(iv) of Theorem 1 completely parallels the corresponding work for (a)(iii) and (a)(iv) above and is omitted.

Proof of Proposition B. The results are immediate from Corollary 1 and Theorem 1. In this case we also have $\eta = 1$ a.s.

3. An example

In order to give some idea of the scope of the results of the paper and the relevance of the conditions therein we shall give an example dealing with the well-known Pólya urn scheme.

Suppose we have an urn initially containing b black and r red balls. At each draw we remove a ball at random and then replace it, together with c balls of the colour drawn. Let b_n, r_n denote, respectively, the numbers of black and red balls in the urn after the nth drawing and write $Y_n = b_n/(b_n+r_n)$, $n \geqq 0$, the proportion of black balls. Note also that $b_n + r_n = b + r + nc$.

Now it is well known that $\{Y_n, n \geqq 0\}$ is a martingale with respect to the past history σ-fields, that the martingale convergence theorem is applicable since $0 < Y_n < 1$, and indeed that the limit random variable Y, say, has a beta distribution with parameters b/c and r/c (e.g. Feller (1966), pp. 211, 226). Here we shall be concerned with the rate of convergence of Y_n to Y.

Write $K_n = c/(b+r+(n+1)c)$ and let F_n denote the σ-field generated by Y_j, $0 \leqq j \leqq n$. It is easy to see that, conditional on Y_n, the distribution of the increment $X_{n+1} = Y_{n+1} - Y_n$ is given by

$$\begin{aligned} X_{n+1} &= (1-Y_n)K_n \quad \text{with probability} \quad Y_n \\ &= -Y_nK_n \quad \text{with probability} \quad 1-Y_n. \end{aligned}$$

Then,

$$E(X_{n+1}^2 | F_n) = Y_n(1-Y_n)K_n^2 \sim Y(1-Y)K_n^2 \quad \text{a.s.} \quad \text{as} \quad n\to\infty, \tag{22}$$

while

(23) $EX_{n+1}^2 = K_n^2 EY_n(1-Y_n) \sim K_n^2 EY(1-Y) = K_n^2 br(b+r)^{-1}(b+r+c)^{-1}$

as $n \to \infty$ using dominated convergence since $Y_n(1-Y_n) \leqq \frac{1}{4}$. Furthermore,

$$E(X_{n+1}^4 | F_n) = Y_n(1-Y_n)K_n^4\{(1-Y_n)^3 + Y_n^3\} \leqq \tfrac{1}{2}K_n^4. \tag{24}$$

Next, using (23),

$$s_n^2 = \sum_{j=n}^{\infty} EX_j^2 = \sum_{j=n-1}^{\infty} K_j^2 EY_j(1-Y_j) \sim n^{-1} EY(1-Y) \quad \text{as} \quad n \to \infty. \tag{25}$$

Also, $\sum_1^\infty s_n^{-2}\{X_n^2 - E(X_n^2|F_{n-1})\}$ converges a.s. if, using the Kolmogorov convergence criterion,

$$\sum_1^\infty s_n^{-4} EX_n^4 < \infty \tag{26}$$

(Proposition IV–6–2 of Neveu (1970)), and (26) holds by virtue of (24) and (25). Thus, the provisions of Corollary 2 [(b)] hold and we shall state the results in terms of normalization using

$$V_n^2 = \sum_{j=n}^{\infty} E(X_j^2 | F_{j-1}) \sim n^{-1} Y(1-Y) \text{ a.s.} \tag{27}$$

Then, using (25) and (27), $s_n^{-2}V_n^2 \xrightarrow{\text{a.s.}} Y(1-Y)/EY(1-Y)$, while for $k \geqq n$,

$$EX_k^2 I(|X_k| > \varepsilon s_n) = \int_{\{|X_k|>\varepsilon s_n\}} X_k^2\, dP < \varepsilon^{-2}s_n^{-2} \int_{\{|X_k|>\varepsilon s_n\}} X_k^4\, dP$$
$$< \tfrac{1}{2}\varepsilon^{-2}s_n^{-2}K_k^4$$

using (24), so that

$$s_n^{-2} \sum_{k=n}^{\infty} E(X_k^2 I(|X_k| > \varepsilon s_n)) < \tfrac{1}{2}\varepsilon^{-2}s_n^{-4} \sum_{k=n}^{\infty} K_k^4 = O(n^{-1}) \quad \text{as} \quad n \to \infty$$

and the results of Corollary 1(b) are applicable.

To utilize the iterated logarithm results in Theorem 1 it remains to check Conditions (b)(iii) and (b)(iv). That (b)(iii) holds follows from (24) and (25) since

$$E|X_n| I(|X_n| > \varepsilon s_n) = \int_{\{|X_n|>\varepsilon s_n\}} |X_n|\, dP < \varepsilon^{-3}s_n^{-3} EX_n^4 = O(n^{-5/2})$$

while (b)(iv) also follows from (24) and (25). Theorem 1 thus yields, for example

$$n^{1/2}(Y_n - Y) \xrightarrow{\text{d}} [Y'(1-Y')]^{1/2} N(0,1)$$

where $Y'(1-Y')$ is distributed as $Y(1-Y)$ and is independent of the $N(0,1)$ random variable, while

$$\limsup_{n\to\infty} n^{1/2}(2\log\log n)^{-1/2}(Y_n - Y) = Y^{1/2}(1-Y)^{1/2} \text{ a.s.}$$

$$\liminf_{n\to\infty} n^{1/2}(2\log\log n)^{-1/2}(Y_n - Y) = -Y^{1/2}(1-Y)^{1/2} \text{ a.s.}$$

Acknowledgements

Thanks are due to J. A. Hartigan for stimulating this research and to B. M. Brown and P. G. Hall for helpful comments.

References

Athreya, K. B. and Ney, P. E. (1972) *Branching Processes.* Springer-Verlag, Berlin.

Barbour, A. D. (1974) Tail sums of convergent series of independent random variables. *Proc. Camb. Phil. Soc.* **75**, 361–364.

Billingsley, P. (1968) *Convergence of Probability Measures.* Wiley, New York.

Chow, Y. S. and Teicher, H. (1973) Iterated logarithm laws for weighted averages. *Z. Wahrscheinlichkeitsth.* **26**, 87–94.

Feller, W. (1966) *An Introduction to Probability Theory and its Applications*, Vol. II. Wiley, New York.

Hall, P. G. (1977) Martingale invariance principles. *Ann. Prob.* To appear.

Hall, P. G. and Heyde, C. C. (1976) On a unified approach to the law of the iterated logarithm for martingales. *Bull. Austral. Math. Soc.* **14**, 435–447.

Heyde, C. C. (1968) On almost sure convergence for sums of independent random variables. *Sankhyā* A**30**, 353–358.

McLeish, D. L. (1974) Dependent central limit theorems and invariance principles. *Ann. Prob.* **2**, 620–628.

Müller, D. W. (1968) Verteilungs-Invarianzprinzipien für das starke Gesetz der grossen Zahl. *Z. Wahrscheinlichkeitsth.* **10**, 173–192.

Neveu, J. (1970) *Bases Mathématiques du Calcul des Probabilités*, 2nd edn. Masson, Paris.

Vervaat, W. (1972) *Success Epochs in Bernoulli Trials.* Math. Centre Tracts 42, Mathematisch Centrum, Amsterdam.

Whitt, W. (1972) Stochastic Abelian and Tauberian theorems. *Z. Wahrscheinlichkeitsth.* **22**, 251–267.

The Annals of Probability
1978, Vol. 6, No. 5, 870–875

A LOG LOG IMPROVEMENT TO THE RIEMANN HYPOTHESIS FOR THE HAWKINS RANDOM SIEVE

BY C. C. HEYDE

CSIRO Division of Mathematics and Statistics, Canberra

This paper is concerned with the Hawkins random sieve which is a probabilistic analogue of the sieve of Eratosthenes. Analogues of the prime number theorem, Mertens' theorem and the Riemann hypothesis have previously been established for the Hawkins sieve. In the present paper we give a more delicate analysis using iterated logarithm results for both martingales and tail sums of martingale differences to deduce a considerably improved log log replacement for the Riemann hypothesis result.

1. Introduction. The random sieve introduced by Hawkins [2] is analogous to the sieve of Eratosthenes and produces a random sequence with asymptotic properties similar in many ways to the primes. It is constructed as follows: Let $A_1 = \{2, 3, 4, \cdots\}$.

STAGE 1. Put $X_1 = \min A_1$. From the set $A_1 \backslash \{X_1\}$ each number in turn is (independently of the others) deleted with probability X_1^{-1} or not deleted with probability $1 - X_1^{-1}$. The set of elements of $A_1 \backslash \{X_1\}$ which remain is denoted by A_2.

STAGE n. Put $X_n = \min A_n$. From the set $A_n \backslash \{X_n\}$ each number in turn is (independently of the others) deleted with probability X_n^{-1} or not deleted with probability $1 - X_n^{-1}$. The set of elements of $A_n \backslash \{X_n\}$ which remain is denoted by A_{n+1}.

Define

$$Y_n = \prod_{1 \le k \le n} (1 - X_k^{-1})^{-1} .$$

Wunderlich [8], [9] has obtained the results

$$\lim_{n\to\infty} (n \log n)^{-1} X_n = 1 \quad \text{a.s.} \qquad \text{and} \qquad \lim_{n\to\infty} (\log n)^{-1} Y_n = 1 \quad \text{a.s.},$$

which are analogues of the prime number theorem and Mertens' theorem respectively, while Neudecker and Williams [6] have obtained an analogue of the Riemann hypothesis. Write

$$\operatorname{li}(x) = \lim_{\delta \downarrow 0} \left(\int_0^{1-\delta} + \int_{1+\delta}^{x}\right) \frac{dz}{\log z} \sim \frac{x}{\log x} \qquad \text{as} \quad x \to \infty ,$$

and recall that the "real" Riemann hypothesis about the zeros of the Riemann

Received April 14, 1977.
AMS 1970 *subject classifications.* Primary 60F15, 10H30; Secondary 60G45, 60J05.
Key words and phrases. Random sieve, prime numbers, Riemann hypothesis, martingale iterated logarithm laws.

R. Maller et al. (eds.), *Selected Works of C.C. Heyde*, Selected Works in Probability and Statistics,
DOI 10.1007/978-1-4419-5823-5_43, © Springer Science+Business Media, LLC 2010

zeta function is equivalent to

$$\mathrm{li}\,(p_n) = n + O(n^{\frac{1}{2}+\varepsilon})$$

for any $\varepsilon > 0$ where p_n denotes the nth prime. Neudecker and Williams showed that

$$L\,\mathrm{li}\,(X_n L^{-1}) = n + O(n^{\frac{1}{2}+\varepsilon}) \quad \text{a.s.}\,,$$

for any $\varepsilon > 0$ where L denotes the (random) limit

$$L = \lim_{n\to\infty} X_n \exp(-Y_n)\,,$$

which they show to exist and be nonzero a.s.

It is our object in this paper to sharpen this result to a log log form as given in the following theorem.

THEOREM. *For the Hawkins random sieve,*

$$\limsup_{n\to\infty} (2n \log\log n)^{-\frac{1}{2}} |L\,\mathrm{li}\,(X_n L^{-1}) - n| \leqq 3 \quad \text{a.s.}$$

The proof of this theorem rests heavily on the use of some recent results ([4], [5]) on the law of the iterated logarithm for martingales and for tail sums of martingale differences. The result, of course, suggests a candidate for the bound in the "real" Riemann hypothesis. It should be remarked that a similar type of log log error term also emerges for a structurally simpler diffusion analogue of the Hawkins sieve (Williams [7]).

2. Proof of the theorem. The proof requires a sharpening of the analysis of [6] whose notation we retain. The starting point is the fact that the process $\{(X_n, Y_n), n \geqq 1\}$ is Markovian with

$$P(X_{n+1} - X_n = j \mid F_n) = Y_n^{-1}(1 - Y_n^{-1})^{j-1}\,, \qquad j \geqq 1\,,$$

F_n being the σ-field generated by (X_j, Y_j), $j \leqq n$.

Set

$$Z_n = X_n - 1\,, \qquad U_{n+1} = (Z_{n+1} - Z_n)Y_n^{-1}\,, \qquad n \geqq 1\,.$$

In [6], Proposition 2, it is shown that if $H_n = \log Z_n - Y_n$, $n \geqq 1$, then

$$H_n = \log L + O(n^{-\frac{1}{2}+\varepsilon}) \quad \text{a.s.}$$

for any $\varepsilon > 0$, and we shall strengthen this to show that

$$(1) \qquad H_n = \log L + \delta(n)(2n^{-1}\log\log n)^{\frac{1}{2}} \quad \text{a.s.}\,,$$

where $\limsup_{n\to\infty} \delta(n) = +1$ a.s., $\liminf_{n\to\infty} \delta(n) = -1$ a.s.

Writing $\alpha_n = Y_n Z_n^{-1}$, $\beta_n = Y_n Z_{n+1}^{-1}$, we have from [6] that $\alpha_n = O(n^{-1})$ a.s., $\beta_n = O(n^{-1})$ a.s. as $n \to \infty$ and then

$$\begin{aligned} H_{n+1} - H_n &= \log(1 + \alpha_n U_{n+1}) - \beta_n \\ &= \alpha_n U_{n+1} - \beta_n + R_n \end{aligned}$$

where

$$|R_n| \leqq \alpha_n^2 U_{n+1}^2 \quad \text{a.s.}$$

Further,

$$|\alpha_n - \beta_n| = |\alpha_n \beta_n U_{n+1}| \leqq \alpha_n^2 |U_{n+1}|$$

so that

$$H_{n+1} - H_n = \alpha_n(U_{n+1} - 1) + S_n ,$$

where

(2) $$|S_n| \leqq \alpha_n^2(|U_{n+1}| + U_{n+1}^2) = O(n^{-2} \log n) \quad \text{a.s.} ,$$

since, from [6], $|U_{n+1}| = O(\log n)$ a.s. Then

$$\begin{aligned} H_{n+1} &= \sum_{k=1}^{n} (H_{k+1} - H_k) - 2 \\ &= \sum_{k=1}^{n} \alpha_k(U_{k+1} - 1) + \sum_{k=1}^{n} S_k - 2 \end{aligned}$$

and, since $H_n \to_{\text{a.s.}} \log L$ (finite a.s.) as $n \to \infty$ and $\sum_1^\infty |S_k| < \infty$ a.s. in view of (2),

(3) $$H_n - \log L = \sum_{k=n}^{\infty} \alpha_k(U_{k+1} - 1) + \sum_{k=n}^{\infty} S_k .$$

Now, from (2),

(4) $$\begin{aligned} |(2n^{-1} \log \log n)^{-\frac{1}{2}} \textstyle\sum_{k=n}^{\infty} S_k| &= O(n^{\frac{1}{2}}(\log \log n)^{-\frac{1}{2}} \textstyle\sum_{k=n}^{\infty} k^{-2} \log k) \\ &= o(1) \quad \text{a.s.} \end{aligned}$$

Furthermore,

(5) $$\sum_{k=n}^{\infty} \alpha_k(U_{k+1} - 1) = \sum_{k=n}^{\infty} k^{-1}(U_{k+1} - 1) + \sum_{k=n}^{\infty} (\alpha_k - k^{-1})(U_{k+1} - 1)$$

and it is easily seen that

$$E(U_{n+1} - 1 \mid F_n) = 0 \quad \text{a.s.} ,$$

so that the $\{U_{n+1} - 1, n \geqq 1\}$ are martingale differences. To prove that

(6) $$\lim_{n\to\infty} n^{\frac{1}{2}}(\log \log n)^{-\frac{1}{2}} \sum_{k=n}^{\infty} (\alpha_k - k^{-1})(U_{k+1} - 1) = 0 \quad \text{a.s.} ,$$

it suffices, upon noting that the α_n are F_n-measurable so that the $\{(\alpha_k - k^{-1})(U_{k+1} - 1), k \geqq 1\}$ are still martingale differences, to show that (e.g. Lemma 1 of Heyde [4])

$$\sum_{k=1}^{\infty} k^{\frac{1}{2}}(\alpha_k - k^{-1})(U_{k+1} - 1) \quad \text{converges a.s.}$$

This holds if (e.g. Doob [1], page 320)

$$\sum_{k=1}^{\infty} k(\alpha_k - k^{-1})^2 E[(U_{k+1} - 1)^2 \mid F_k] < \infty \quad \text{a.s.} ,$$

and hence if

(7) $$\sum_{k=1}^{\infty} k(\alpha_k - k^{-1})^2 < \infty \quad \text{a.s.} ,$$

since $E[(U_{k+1} - 1)^2 \mid F_k] = 1 - Y_k^{-1}$ a.s. But,

$$\alpha_k - k^{-1} = Z_k^{-1}(Y_k - k^{-1}X_k + k^{-1}) = o((\log \log k)(k \log k)^{-1}) \quad \text{a.s.} ,$$

using the theorem of Heyde [3] and hence (7) and consequently (6) holds.

We now apply the iterated logarithm result of Theorem 1(b) of Heyde [4] to the tail sum $\sum_{k=n}^{\infty} k^{-1}(U_{k+1}-1)$ of martingale differences. We have

$$s_n^2 = \sum_{k=n}^{\infty} k^{-2}E[E\{(U_{k+1}-1)^2 \mid F_k\}] = \sum_{k=n}^{\infty} k^{-2}E(1-Y_k^{-1}) \sim n^{-1} \quad \text{a.s.}$$

as $n \to \infty$ and

$$(8) \qquad \sum_{n=1}^{\infty} n^{-1}[(U_{n+1}-1)^2 - E\{(U_{n+1}-1)^2 \mid F_n\}] \quad \text{converges a.s.},$$

again using [1], page 320, since $E(U_{n+1}-1)^4 \leqq 25$ for all $n \geqq 1$. This last result follows from

$$\begin{aligned} E(U_{n+1}-1)^4 &= E[E\{(U_{n+1}-1)^4 \mid F_n\}] \\ &= E\left[E\left\{\left(\frac{X_{n+1}-X_n}{Y_n}-1\right)^4 \middle| F_n\right\}\right] \\ &= E[\textstyle\sum_{j=1}^{\infty}(jY_n^{-1}-1)^4 Y_n^{-1}(1-Y_n^{-1})^{j-1}] \\ &\leqq E[\textstyle\sum_{j=1}^{Y_n} Y_n^{-1}(1-Y_n^{-1})^{j-1} + \sum_{j=Y_n+1}^{\infty}(jY_n^{-1})^4 Y_n^{-1}(1-Y_n^{-1})^{j-1}] \\ &\leqq 1 + EY_n^{-5}\textstyle\sum_{j=1}^{\infty} j^4(1-Y_n^{-1})^{j-1} \\ &\leqq 1 + 24EY_n^{-5}[1-(1-Y_n^{-1})]^{-5} \leqq 25 . \end{aligned}$$

The conditions required for the use of Theorem 1(b), Corollary 1 and Corollary 2 of [4] are then simply verified. In particular,

$$s_n^{-2}\sum_{k=n}^{\infty} k^{-2}E\{(U_{k+1}-1)^2 \mid F_k\} \to_{\text{a.s.}} 1 ,$$

$E[|U_j-1|I(j^{-1}|U_j-1|>\varepsilon s_j)] \leqq 25(\varepsilon j s_j)^{-3}$, and

$$E(U_j-1)^2 I(j^{-1}|U_j-1|>\varepsilon s_n) \leqq 25(\varepsilon j s_n)^{-2}$$

so that

$$\sum_{j=1}^{\infty} s_j^{-1}E[j^{-1}|U_j-1|I(j^{-1}|U_j-1|>\varepsilon s_j)] < \infty , \qquad \text{and}$$
$$s_n^{-2}\sum_{j=n}^{\infty} E[j^{-2}(U_j-1)^2 I(j^{-1}|U_j-1|>\varepsilon s_n)] \to 0$$

as $n \to \infty$. We then obtain that the lim sup as $n \to \infty$ of $(2n^{-1}\log\log n)^{-\frac{1}{2}}\sum_{k=n}^{\infty} k^{-1}(U_{k+1}-1)$ is $+1$ a.s. while the lim inf is -1 a.s. The result (1) then follows via (3), (4), (5) and (6).

We now exponentiate (1) and rewrite it in the form

$$(9) \qquad Z_n = Le^{Y_n}\{1+\eta(n)(2n^{-1}\log\log n)^{\frac{1}{2}}\}$$

where $\limsup_{n\to\infty}\eta(n) = +1$ a.s., $\liminf_{n\to\infty}\eta(n) = -1$ a.s.

Next we have

$$Z_{n+1}Y_{n+1}^{-1} - Z_nY_n^{-1} = U_{n+1} - Y_{n+1}^{-1}, \qquad n \geqq 1 ,$$

so that summation gives

$$(10) \qquad Z_nY_n^{-1} = \sum_{k=2}^{n}(U_k-1) + n - \sum_{k=1}^{n} Y_k^{-1} .$$

The martingale law of the iterated logarithm (e.g. Heyde and Scott [5], Corollary 1) applies to $\sum_{k=2}^{n}(U_k-1)$ to give

$$\limsup_{n\to\infty}(2n\log\log n)^{-\frac{1}{2}}\sum_{k=2}^{n}(U_k-1) = +1 \quad \text{a.s.},$$
$$\liminf_{n\to\infty}(2n\log\log n)^{-\frac{1}{2}}\sum_{k=2}^{n}(U_k-1) = -1 \quad \text{a.s.},$$

since

$$\sum_{k=1}^{n} E(U_k - 1)^2 = \sum_{k=1}^{n} (1 - EY_k^{-1}) \sim n \qquad \text{as} \quad n \to \infty$$

and $E(U_k - 1)^4 \leqq 25$ while (8) and $E[(U_{k+1} - 1)^2 | F_k] = 1 - Y_k^{-1}$ a.s. ensure that $n^{-1} \sum_{k=1}^{n} (U_k - 1)^2 \to_{\text{a.s.}} 1$ as $n \to \infty$. Then, from (10),

$$\text{(11)} \qquad \begin{aligned} &\limsup_{n\to\infty} (2n \log\log n)^{-\frac{1}{2}}(Z_n Y_n^{-1} - n + \textstyle\sum_{k=1}^{n} Y_k^{-1}) = +1 \quad \text{a.s.}, \\ &\liminf_{n\to\infty} (2n \log\log n)^{-\frac{1}{2}}(Z_n Y_n^{-1} - n + \textstyle\sum_{k=1}^{n} Y_k^{-1}) = -1 \quad \text{a.s.}, \end{aligned}$$

and using (9) and (11) we obtain

$$\begin{aligned} &\limsup_{n\to\infty} (2n \log\log n)^{-\frac{1}{2}} |Le^{Y_n} Y_n^{-1} - n + \textstyle\sum_{k=1}^{n} Y_k^{-1}| \\ &\quad \leqq \limsup_{n\to\infty} (2n \log\log n)^{-\frac{1}{2}} |Le^{Y_n} Y_n^{-1} - Z_n Y_n^{-1}| \\ &\qquad + \limsup_{n\to\infty} (2n \log\log n)^{-\frac{1}{2}} |Z_n Y_n^{-1} - n + \textstyle\sum_{k=1}^{n} Y_k^{-1}| \\ &\quad \leqq 2 \quad \text{a.s.} \end{aligned}$$

The remainder of the proof parallels that of [6]. We extend the random function Y from $\{1, 2, 3, \cdots\}$ to $(1, \infty)$ by linear interpolation,

$$Y_t = Y_n + (t - n)(Y_{n+1} - Y_n), \qquad n \leqq t \leqq n + 1.$$

Then, it is easily checked that

$$LY_t^{-1} e^{Y_t} = \int_1^t (1 - Y_s^{-1})\, ds + f(t)$$

where $f(t) = (2t \log\log (t \vee 3))^{\frac{1}{2}} \alpha(t)$ with $\limsup_{t\to\infty} |\alpha(t)| \leqq 2$ a.s. Furthermore,

$$\begin{aligned} [L \operatorname{li} (\exp Y_s)]_1^t &= \int_1^t (1 - Y_s^{-1})\, d(LY_s^{-1} e^{Y_s}) \\ &= t - 1 + \int_1^t f'(s)(1 - Y_s^{-1})^{-1}\, ds \\ &= t - 1 + f(t)(1 - Y_t^{-1})^{-1} - f(1)(1 - Y_1^{-1})^{-1} \\ &\quad - \int_1^t f(s) Y_s'(Y_s - 1)^{-2}\, ds, \end{aligned}$$

while

$$|f(s) Y_s'(Y_s - 1)^{-2}| = O((\log\log s)^{\frac{1}{2}} s^{-\frac{1}{2}} (\log s)^{-2}) \qquad \text{as} \quad s \to \infty$$

so that

$$(t^{\frac{1}{2}} \log\log t)^{-1} \int_1^t |f(s) Y_s'(Y_s - 1)^{-2}|\, ds = o(1) \qquad \text{as} \quad t \to \infty$$

and hence

$$\text{(12)} \qquad \limsup_{t\to\infty} (2t \log\log t)^{-\frac{1}{2}} |L \operatorname{li} (e^{Y_t}) - t| \leqq 2 \quad \text{a.s.}$$

The result of the theorem then follows from (9) and (12) since (9) gives

$$\limsup_{n\to\infty} (2n \log\log n)^{-\frac{1}{2}} |L \operatorname{li} (Z_n L^{-1}) - L \operatorname{li} (e^{Y_n})| \leqq 1 \quad \text{a.s.}$$

REFERENCES

[1] Doob, J. L. (1953). *Stochastic Processes.* Wiley, New York.

[2] Hawkins, D. (1958). The random sieve. *Math. Mag.* **31** 1-3.

[3] Heyde, C. C. (1976). On asymptotic behaviour for the Hawkins random sieve. *Proc. Amer. Math. Soc.* **56** 277-280.

[4] HEYDE, C. C. (1977). On central limit and iterated logarithm supplements to the martingale convergence theorem. *J. Appl. Probability* **14** 758–775.

[5] HEYDE, C. C. and SCOTT, D. J. (1973). Invariance principles for the law of the iterated logarithm for martingales and processes with stationary increments. *Ann. Probability* **1** 428–436.

[6] NEUDECKER, W. and WILLIAMS, D. (1974). The "Riemann hypothesis" for the Hawkins random sieve. *Composito Math.* **29** 197–200.

[7] WILLIAMS, D. (1974). A diffusion process motivated by the sieve of Eratosthenes. *Bull. London Math. Soc.* **6** 155–164.

[8] WUNDERLICH, M. C. (1974). A probabilistic setting for prime number theory. *Acta Arith.* **26** 59–81.

[9] WUNDERLICH, M. C. (1976). The prime number theorem for random sequences. *J. Number Theory* **8** 369–371.

DIVISION OF MATHEMATICS AND STATISTICS
CSIRO
P.O. BOX 1965, CANBERRA CITY
A.C.T. 2601
AUSTRALIA

Stochastic Processes and their Applications 8 (1978) 1–9.

ON AN OPTIMAL ASYMPTOTIC PROPERTY OF THE MAXIMUM LIKELIHOOD ESTIMATOR OF A PARAMETER FROM A STOCHASTIC PROCESS

C.C. HEYDE

CSIRO Division of Mathematics and Statistics, P.O. Box 1965, Canberra City, A.C.T. 2601, Australia

Received 13 October 1977
Revised 24 April 1978

This paper is concerned with the estimation of a parameter of a stochastic process on the basis of a single realization. It is shown, under suitable regularity conditions, that the maximum likelihood estimator is the best consistent asymptotically normal estimator in the sense of having minimum asymptotic variance. It also produces the best limiting probability of concentration in symmetric intervals. An application is given for the problem of estimating the mean of the offspring distribution in a Galton–Watson branching process.

Estimation for stochastic processes
best uniformly asymptotically normal estimator
martingale limit theorems
maximum likelihood
limiting probability of concentration

Introduction

In the case of estimating a parameter on the basis of observations which are independent and identically distributed or come from a stationary ergodic Markov chain, it is known that the maximum likelihood estimator (MLE) enjoys certain optimality properties. One of the most significant of these is a result of Schmetterer [13] (extending one of Rao [10]) that, subject to suitable regularity conditions, the MLE is the best consistent continuously asymptotically normal estimator in the sense of having minimum asymptotic variance. In this paper we shall obtain an analogous result for the much more complex case where the sample is from a general stochastic process. The result is an augmented and corrected version of that presented in Heyde [6]. We also show that the MLE produces the best asymptotic probability of concentration in symmetric intervals. The results follow from ideas of Weiss and Wolfowitz [15].

We consider a sample $X_1, X_2, \ldots, X_n$ of consecutive observations from some stochastic process whose distribution depends on a single parameter, θ, $\theta \in \Theta$, Θ being an open interval. Let $L_n(\theta)$ be the likelihood function associated with

R. Maller et al. (eds.), *Selected Works of C.C. Heyde*, Selected Works in Probability and Statistics, DOI 10.1007/978-1-4419-5823-5_44, © Springer Science+Business Media, LLC 2010

$X_1, \ldots, X_n$ and suppose that $L_n(\theta)$ is differentiable with respect to θ and $E_\theta(\mathrm{d}\log L_n(\theta)/\mathrm{d}\theta)^2 < \infty$ for each n. Suppose in addition that if $P_n(X_1, \ldots, X_n)$ $(=L_n(\theta))$ is the joint probability (density) function of $X_1, \ldots, X_n$, then $\sum_{x_n} P_n(x_1, \ldots, x_n)(\int P_n(x_1, \ldots, x_n)\,\mathrm{d}x_n)$ can be differentiated twice with respect to θ under the summation (integration) sign. We shall write F_k for the σ-field generated by $X_1, \ldots, X_k$, $k \geq 1$, take F_0 as the trivial σ-field and set $L_0 = 1$. Then, writing

$$\frac{\mathrm{d}\log L_n(\theta)}{\mathrm{d}\theta} = \sum_{i=1}^{n} u_i(\theta),$$

we have $E_\theta(u_i(\theta) \mid F_{i-1}) = 0$ a.s., so that $\{\mathrm{d}\log L_n(\theta)/\mathrm{d}\theta, F_n, n \geq 1\}$ is a (square integrable) martingale. In addition, we set

$$I_n(\theta) = \sum_{i=1}^{n} E_\theta(u_i^2(\theta) \mid F_{i-1})$$

and note that, under the conditions imposed above, $v_i(\theta) = \mathrm{d}u_i(\theta)/\mathrm{d}\theta$ satisfies

$$E_\theta(u_i^2(\theta) \mid F_{i-1}) = -E_\theta(v_i(\theta) \mid F_{i-1}) \text{ a.s.}$$

Also, writing

$$J_n(\theta) = \sum_{i=1}^{n} v_i(\theta)$$

we note that $\{J_n(\theta) + I_n(\theta), F_n, n \geq 1\}$ is a martingale.

The quantity $I_n(\theta)$ is a form of conditional information which reduces to the standard Fisher information in the case where the X_i are independent random variables. The rôle of $I_n(\theta)$ in the stochastic process estimation context is a vital one and aspects of this are discussed in Heyde [5], [6], Heyde and Feigin [7]. The importance of $I_n(\theta)$ is easily seen from the following expansion. We suppose that θ is the true parameter value. Then, we can use Taylor's expansion to write for $\theta' \in \Theta$,

$$\begin{aligned}\frac{\mathrm{d}}{\mathrm{d}\theta'}\log L_n(\theta') &= \sum_{i=1}^{n} u_i(\theta') \\ &= \sum_{i=1}^{n} u_i(\theta) + (\theta' - \theta)\sum_{i=1}^{n} v_i(\theta_n^*) \\ &= \sum_{i=1}^{n} u_i(\theta) - (\theta' - \theta) I_n(\theta) + (\theta' - \theta)(J_n(\theta_n^*) + I_n(\theta)) \qquad (1)\end{aligned}$$

where $\theta_n^* = \theta + \gamma(\theta' - \theta)$ with $\gamma = \gamma(n, \theta)$ satisfying $|\gamma| < 1$. Further, since

$$[I_n(\theta)]^{-1} \sum_{i=1}^{n} u_i(\theta) \xrightarrow{\text{a.s.}} 0$$

provided $I_n(\theta) \xrightarrow{\text{a.s.}} \infty$ as $n \to \infty$ (a martingale strong law, e.g. see Heyde and Feigin [7]), we see that the likelihood equation has a root $\hat{\theta}_n$ which is strongly consistent for θ if $I_n(\theta) \xrightarrow{\text{a.s.}} \infty$ and

$$\limsup_{n\to\infty} [I_n(\theta)]^{-1}|I_n(\theta)+J_n(\theta_n^*)|<1 \text{ a.s.}$$

for $|\theta_n^* - \theta| < \delta$ sufficiently small. Suppose that these conditions hold.

In order to utilize most effectively the MLE from large samples it is necessary for $\sum_{i=1}^n u_i(\theta)$ to converge in distribution, when appropriately normalized, to some proper limit law. Furthermore, it can be arranged in many cases that this limit law is normal, which is most convenient for confidence interval purposes. The most effective general norming seems to be provided by $I_n^{1/2}(\theta)$ and, indeed, $[I_n(\theta)]^{-1/2}\sum_{i=1}^n u_i(\theta)$ converges in distribution to normality under quite wide ranging circumstances (e.g. Hall [2], Scott [14]).

Now suppose that

$$[I_n(\theta)]^{-1/2} \sum_{i=1}^n u_i(\theta) \xrightarrow{d} N(0,1).$$

If, in addition to the above conditions, $J_n(\theta_n^{**}) = -I_n(\theta)(1+o(1))$ in probability as $n\to\infty$ for $\theta_n^{**} = \theta + s(\hat\theta_n - \theta)$ with any s satisfying $|s| \le 1$, we have from (1) that

$$I_n^{1/2}(\theta)(\hat\theta_n - \theta) \xrightarrow{d} N(0,1).$$

Then, if $T_n = T_n(X_1, \ldots, X_n)$ is any consistent estimator of θ for which $I_n^{1/2}(\theta)(T_n - \theta) \xrightarrow{d} N(0, \beta^2(\theta))$ where $\beta(\theta)$ is bounded and continuous in θ, we shall show, under suitable additional regularity conditions, that $\beta^2(\theta) \ge 1$. That is, the MLE is optimal within this class since $\beta(\theta) \equiv 1$ when T_n is the MLE.

Of course it is desirable if comparisons can be made between the MLE and those other consistent estimators for which $I_n^{1/2}(\theta)(T_n - \theta)$ converges in distribution to a proper law (not necessarily normal) or perhaps does not even converge. Such comparisons have been made in the case where the $\{X_i\}$ form a stationary ergodic Markov chain by using the concept of asymptotic efficiency in the Wolfowitz sense (see Roussas [12, Chapter 5]). Minor modifications of the standard theory as presented in [12] leads to comparisons of the kind

$$\begin{aligned}&\lim_{n\to\infty} P_\theta(-c < (E_\theta I_n(\theta)^{1/2}(\hat\theta_n - \theta) < b) \ge \\ &\quad \ge \limsup_{n\to\infty} P_\theta(-c + w(\theta) < (E_\theta I_n(\theta))^{1/2}(T_n - \theta) < W(\theta) + b)\end{aligned}$$

for arbitrary positive b and c and certain $W(\theta) \ge w(\theta)$. This inequality holds under conditions which are quite similar to those of this paper but (at present) require that $I_n(\theta)(E_\theta I_n(\theta))^{-1} \xrightarrow{p} 1$ as $n\to\infty$. It is, however, the cases where this last condition is not satisfied that pose the real interest and challenge in the treatment of stochastic process estimation. Our arguments in this paper are principally concerned with the commonly occurring circumstances under which $I_n(\theta)(E_\theta I_n(\theta))^{-1} \xrightarrow{p} \eta(\theta)$ as $n \to \infty$, η being a random variable in general.

These difficulties can be avoided if we restrict consideration to symmetric concentration intervals. Then, using results of Weiss and Wolfowitz [15], we shall show that

$$\lim_{n\to\infty} P_\theta(-c < (E_\theta I_n(\theta))^{1/2}(\hat{\theta}_n - \theta) < c) \geq$$
$$\geq \limsup P_\theta(-c < (E_\theta I_n(\theta))^{1/2}(T_n - \theta) < c) \quad \text{for all } c > 0.$$

Principal result

We suppose that $(X_1, \ldots, X_n)$ possesses a density $P_n(X_1, \ldots, X_n)(=L_n(\theta))$, which is continuous in θ, with respect to a σ-finite measure μ_n and that the following two assumptions are satisfied.

Assumption 1. (i) $I_n(\theta) \xrightarrow{\text{a.s.}} \infty$, $I_n(\theta)/E_\theta I_n(\theta) \xrightarrow{p} \eta(\theta)(>0$ a.s.) for some r.v. η, and $J_n(\theta)/I_n(\theta) \xrightarrow{p} -1$ as $n \to \infty$, the convergences in probability being uniform in compacts of θ.

(ii) $[I_n(\theta)]^{-1/2} \, \mathrm{d} \log L_n(\theta)/\mathrm{d}\theta \xrightarrow{d} N(0, 1)$ (mixing), continuously in θ.

Assumption 2. For $\delta > 0$, supose $|\theta_n - \theta_0| \leq \delta/(E_{\theta_0} I_n(\theta_0))^{1/2}$. Then,

(i) $E_{\theta_n} I_n(\theta_n) = E_{\theta_0} I_n(\theta_0)(1 + \mathrm{o}(1))$ as $n \to \infty$, and

(ii) $I_n(\theta_n) = I_n(\theta_0)(1 + \mathrm{o}(1))$ a.s. as $n \to \infty$,

(iii) $J_n(\theta_n) = J_n(\theta_0) + \mathrm{o}(I_n(\theta_0))$ a.s. as $n \to \infty$.

These assumptions are not severe and conditions may be imposed directly on the stochastic process to ensure them. The mixing convergence in Assumption 1 calls for some comment. Here the mixing convergence means

$$\lim_{n\to\infty} E_\theta\left(\exp\left\{it(I_n(\theta))^{-1/2}\frac{\mathrm{d}\log L_n(\theta)}{\mathrm{d}\theta}\right\} \mid B\right) = \exp -\tfrac{1}{2}t^2$$

for all real t and all measurable B with $P(B) > 0$. The two parts of Assumption 1 ensure that for $g(x, y)$ any continuous function of two variables,

$$g\left\{(I_n(\theta))^{-1/2}\frac{\mathrm{d}\log L_n(\theta)}{\mathrm{d}\theta}, \frac{I_n(\theta)}{E_\theta I_n(\theta)}\right\} \xrightarrow{d} g(N(0, 1), \eta(\theta))$$

where N and η are independent (e.g. Eagleson [1]). Further, mixing versions of the standard martingale central limit results hold without additional conditions (McLeish [8, p. 628], Scott [14]).

We are now in a position to establish the following result.

Theorem. *Fix $C>0$, let $\theta_n=\theta+2C(E_\theta I_n(\theta))^{-1/2}$ and write $P_\theta^{(n)}$ for the probability measure corresponding to the likelihood $L_n(\theta)$. Suppose that T_n is any estimator of θ which satisfies the condition that for any $\theta\in\Theta$,*

$$\lim_{n\to\infty}\{P_\theta^{(n)}((E_\theta I_n(\theta))^{1/2}(T_n-\theta)>-C)-P_{\theta_n}^{(n)}((E_\theta I_\theta(\theta))^{1/2}(T_n-\theta_n)>-C)\}=0.$$

If there is a maximum likelihood estimator $\hat\theta_n$ which is consistent for any $\theta\in\Theta$, then under Assumptions 1 and 2,

$$\lim_{n\to\infty}P_\theta^{(n)}(-C<(E_\theta I_n(\theta))^{1/2}(\hat\theta_n-\theta)<C)$$
$$\geq\limsup_{n\to\infty}P_\theta^{(n)}(-C<(E_\theta I_n(\theta))^{1/2}(T_n-\theta)<C).$$

Furthermore, if $I_n^{1/2}(\theta)(T_n-\theta)\xrightarrow{d}N(0,\gamma^2(\theta))$ (mixing), continuously in θ, where $\gamma(\theta)$ is bounded, then $\gamma^2(\theta)\geq 1$.

The result of this theorem resolves the question of the best limiting probability of concentration in symmetric intervals but not in the general case. For the general case, I.V. Basawa and D.J. Scott in a preprint entitled "Efficient estimation for stochastic processes" have obtained the following inequality:

$$\lim_{n\to\infty}P_\theta^{(n)}(-\delta_1<(E_\theta I_n(\theta))^{1/2}(T_n-\theta)<\delta_2)\leq M_\theta(\delta_2)-M_\theta(-\delta_1),$$

when the limit exists, where

$$M_\theta(\delta_2)=P(\delta_2N(0,1)\eta^{1/2}(\theta)-\tfrac12\delta_2^2\eta(\theta)\leq c_2)$$

and c_2 is the (unique) median of $\delta_2N(0,1)\eta^{1/2}(\theta)+\frac12\delta_2^2\eta(\theta)$, η and N being independent, while

$$M_\theta(-\delta_1)=P(-\delta_1N(0,1)\eta^{1/2}(\theta)-\tfrac12\delta_1^2\eta(\theta)\leq c_1)$$

where c_1 is the (unique) median of $-\delta_1N(0,1)\eta^{1/2}(\theta)+\frac12\delta_1^2\eta(\theta)$. They also show that the upper bound in the inequality is not attained for the MLE $\hat\theta_n$ unless $\eta(\theta)=1$ a.s. The methods used are similar to those employed in Roussas [12], Chapter 5.

Proof of Theorem. The results follow from a straightforward application of Theorem 3.1 of Weiss and Wolfowitz [15]. We just need to verify Conditions A and B involved therein.

From (1) we have

$$0=\sum_{i=1}^{n}u_i(\theta)+(\hat\theta_n-\theta)J_n(\theta_n^*),\tag{2}$$

where $\theta_n^* = \theta + \gamma(n, \theta)(\hat{\theta}_n - \theta)$ with $|\gamma(n, \theta)| < 1$, so that Assumptions 1 and 2 give

$$\lim_{n\to\infty} P_\theta^{(n)}((E_\theta I_n(\theta))^{1/2}(\hat{\theta}_n - \theta) < y) = \lim_{n\to\infty} P_{\theta_n}^{(n)}((E_\theta I_n(\theta))^{1/2}(\hat{\theta}_n - \theta) < y) \quad (3)$$

$$= \mathbf{P}(\eta^{1/2}(\theta) N(0, 1) < y)$$

for any y, $-\infty < y < \infty$, where η and N are independent. This verifies Condition A.

To check Condition B we define

$$\Lambda_n = \log\,[L_n(\theta_n)/L_n(\theta)] \text{ on } \{L_n(\theta_n)/L_n(\theta) > 0\}$$

and suppose that Λ_n is arbitrarily (but measurably) defined on $\{L_n(\theta_n)L_n(\theta) = 0\}$. Then we may use Taylor's expansion to write

$$\Lambda_n = (\theta_n - \theta) \sum_{i=1}^{n} u_i(\theta) + \tfrac{1}{2}(\theta_n - \theta)^2 J_n(\theta_n^{**})$$

for $\theta_n^{**} \in [\theta, \theta_n]$, and using (2),

$$\Lambda_n = -2C(E_\theta I_n(\theta))^{-1/2}(\hat{\theta}_n - \theta) J_n(\theta_n^*) + 2C^2(E_\theta I_n(\theta))^{-1} J_n(\theta_n^{**}).$$

Thus, employing Assumptions 1 and 2 again,

$$\{\Lambda_n < 0\} = \{\hat{\theta}_n < \theta + C(E_\theta I_n(\theta))^{-1/2}(1 + o_p(1))\}$$

(the $o_p(1)$ denoting a term which tends in probability to zero as $n \to \infty$) while

$$\lim_{n\to\infty} P_\theta^{(n)}(\Lambda_n = 0) = \lim_{n\to\infty} P_{\theta_n}^{(n)}(\Lambda_n = 0) = 0,$$

so that Condition B is satisfied. The first part of the result of the theorem then follows from Theorem 3.1 of [15].

To establish the second part of the theorem we note that the mixing form of convergence results prescribed for T_n, together with Assumption 1, ensures that

$$\lim_{n\to\infty} P_\theta^{(n)}(-C < (E_\theta I_n(\theta))^{1/2}(T_n - \theta) < C) =$$

$$= \mathbf{P}(-C < \gamma(\theta)\eta^{1/2}(\theta) N(0, 1) < C),$$

where η and N are independent and the required result follows in view of (3).

Some remarks on applications

Various structural requirements need to be imposed on the process $\{X_i\}$ in order to allow the various assumptions to be checked. The imposition of stationarity, for example, leads to useful simplification in the requirements. However, we shall here restrict attention to the case where the stochastic process $\{X_i\}$ is a time-homogeneous Markov process whose distribution belongs to a *conditional exponential family*.

The concept of a conditional exponential family, which is discussed in some detail in Heyde and Feigin [7], is a generalization of that of the exponential family for the case of independent X_i. For $\{X_i\}$ belonging to a conditional exponential family,

$$\frac{\mathrm{d}\log L_n(\theta)}{\mathrm{d}\theta}=I_n(\theta)(\hat{\theta}_n-\theta) \tag{4}$$

and $I_n(\theta)=\phi(\theta)\sum_{i=1}^n H(X_{i-1})$ where ϕ does not involve the X_i and H does not involve θ. Furthermore, (Heyde and Feigin [7]), the result (4) holds if and only if the conditional probability (density) function of X_k given X_{k-1}, $f(X_k \mid X_{k-1}, \theta)$ satisfies

$$\frac{\mathrm{d}}{\mathrm{d}\theta}\log f(x \mid y, \theta)=\phi(\theta)H(y)[m(x, y)-\theta]$$

where

$$E_\theta[m(X_i, X_{i-1}) \mid F_{i-1}]=\theta \text{ a.s.}$$

For the case of the conditional exponential family we easily see that Assumption 2 holds if $\phi(\theta)$ is continuous and differentiable.

As a particular example we shall consider the estimation of the mean θ of the offspring distribution of a supercritical Galton–Watson branching process. Here we have $1<\theta=E_\theta(X_1 \mid X_0=1)$ and we shall suppose that $\sigma^2=\operatorname{var}_\theta(X_1 \mid X_0=1)<\infty$. In this case it is known (Heyde and Feigin [7]) that the conditional exponential family consists of the family of power series distributions. These are the distributions for which

$$P_j=\mathbf{P}(X_1=j \mid X_0=1)=a_j\lambda^j[f(\lambda)]^{-1}, \quad j=0, 1, 2, \ldots, \lambda>0,$$

where $a_j \geqslant 0$ and $f(\lambda)=\sum_{j=0}^\infty a_j\lambda^j$. Then,

$$\theta=\lambda f'(\lambda)[f(\lambda)]^{-1}, \qquad \sigma^2=[(\mathrm{d}/\mathrm{d}\theta)\log\lambda]^{-1}$$

and

$$\frac{\mathrm{d}}{\mathrm{d}\theta}\log f(x \mid y,\theta)=\sigma^{-2}(x-y\theta),$$

so that

$$\frac{\mathrm{d}\log L_n(\theta)}{\mathrm{d}\theta}=\sigma^{-2}Y_{n-1}(\hat{\theta}_n-\theta)$$

where $Y_k=\sum_{j=0}^k X_j$ and $\hat{\theta}_n=(Y_n-Y_0)Y_{n-1}^{-1}$. For this family, $\theta=\theta(\lambda)$ is known to be a non-negative monotone increasing function of λ (Patil [9]) so that the parametrization can be equally well expressed in terms of θ.

Suppose $X_0=1$ and $\mathbf{P}(X_1=0)=0$ for definiteness and convenience. The latter gives $X_n \xrightarrow{\text{a.s.}} \infty$ as $n\to\infty$. If $\mathbf{P}(X_1=0)>0$, it is well known that $X_n \xrightarrow{\text{a.s.}} \infty$ on the non-extinction set and the results which we shall describe hold conditionally on non-extinction.

It follows from Theorem 3 of Heyde [4] that $Y_n(E_\theta Y_n)^{-1} \xrightarrow{\text{a.s.}} \eta(\theta)$ with η non-degenerate and a.s. positive. Further,

$$I_n(\theta) = \sigma^{-2}(\theta) Y_{n-1},$$

$$J_n(\theta) = (\mathrm{d}\sigma^2(\theta)/\mathrm{d}\theta)(Y_n - 1 - \theta Y_{n-1}) - \sigma^{-2}(\theta) Y_{n-1},$$

and Assumption 1(i) is satisfied since convergence in probability holds uniformly for $\theta \geq 1 + \varepsilon$, any $\varepsilon > 0$, via calculation of $E_\theta(Y_n(E_\theta Y_n)^{-1} - \eta)^2$ (e.g. Harris [3], Lemma 7.1). That

$$(E_\theta I_n(\theta))^{-1/2} \frac{\mathrm{d} \log L_n(\theta)}{\mathrm{d}\theta} \xrightarrow{d} N(0, 1) \quad \text{(mixing)}$$

follows from Corollary 2 of Scott [14] and uniformity of convergence for $\beta \geq 1 + \varepsilon$, any $\varepsilon > 0$, can be gleaned from a detailed examination of the proofs of Theorem 1 and Corollary 2 of the same paper. Assumption 1(ii) is therefore satisfied. Also, $\sigma^2(\theta)$ is continuous in θ and differentiable so that Assumption 2 holds, as noted previously.

The application of our theorem in this context substantially clarifies Heyde's [5] investigation of the class of estimators which are called asymptotically efficient for θ.

Acknowledgment

Helpful comments from P.D. Feigin, P. Jagers and a referee on an earlier version of this paper are gratefully acknowledged.

References

[1] G.K. Eagleson, Some simple conditions for limit theorems to be mixing, Teoriya Veroyat. i ee Primenen 21 (1976) 653–660.
[2] P.G. Hall, Martingale invariance principles, Ann. Probab. 5 (1977) 875–887.
[3] T.E. Harris, Branching processes, Ann. Math. Statist. 19 (1948) 474–494.
[4] C.C. Heyde, Extension of a result of Seneta for the supercritical Galton–Watson process, Ann. Math. Statist. 41 (1970) 739–742.
[5] C.C. Heyde, Remarks on efficiency in estimation for branching processes, Biometrika 62 (1975) 49–55.
[6] C.C. Heyde, An optimal property of maximum likelihood with application to branching process estimation, Bull. Int. Statist. Inst. 47 (1977) (to appear).
[7] C.C. Heyde and P.D. Feigin, On efficiency and exponential families in stochastic process estimation, in: G.P. Patil, S. Kotz and J.K. Ord, eds., Statistical Distributions in Scientific Work (Reidel, Dordrecht, Vol. 1, 1975) 227–240.

(8) D.L. McLeish, Dependent central limit theorems and invariance principles, Ann. Probab. 2 (1974) 620–628.
[9] G.P. Patil, Certain properties of the generalized power series distributions, Ann. Inst. Statist. Math. 14 (1962) 179–182.
[10] C.R. Rao, Criteria of estimation in large samples, Sankhyā, Ser. A, 25 (1963) 189–206.
[11] C.R. Rao, Linear statistical Inference and its Applications (Wiley, New York, 1965).
[12] G.G. Roussas, Contiguity of Probability Measures (Cambridge University Press, Cambridge, 1972).
[13] L. Schmetterer, On the asymptotic efficiency of estimates, in: F.N. David, ed., Research Papers in Statistics. Festschrift for J. Neyman (Wiley, New York, 1966).
[14] D.J. Scott, A central limit theorem for martingales and an application to branching processes, Stoch. Processes Appl. 6 (1977) 241–252.
[15] L. Weiss and J. Wolfowitz, Generalized maximum likelihood estimators, Theory Prob. Appl. 11 (1966) 58–81.

On Asymptotic Posterior Normality for Stochastic Processes

BY

C. C. HEYDE and I. M. JOHNSTONE

Reprinted from

THE JOURNAL OF THE ROYAL STATISTICAL SOCIETY

SERIES B (METHODOLOGICAL)

Volume 41, No. 2, 1979

(pp. 184–189)

PRINTED FOR PRIVATE CIRCULATION

1979

J. R. Statist. Soc. B (1979),
41, *No.* 2, *pp.* 184–189

On Asymptotic Posterior Normality for Stochastic Processes

By C. C. Heyde and I. M. Johnstone

CSIRO, Canberra *Australian National University*

[Received August 1978. Revised October 1978]

Summary

Asymptotic normality of the posterior distribution of a parameter in a stochastic process is shown to hold under conditions which do little more than ensure consistency of a maximum likelihood estimator. Much more stringent conditions are required to ensure asymptotic normality of the MLE. This contrast, which has implications of considerable significance, does not emerge in the classical context of independent and identically distributed observations.

Keywords: PARAMETER ESTIMATION; STOCHASTIC PROCESSES; BAYESIAN METHODS; MAXIMUM LIKELIHOOD ESTIMATOR; ASYMPTOTIC NORMALITY

1. Introduction

In the classical context of independent and identically distributed observations it is well known that, under mild conditions, the maximum likelihood estimator (MLE) is consistent and, with appropriate normalization, asymptotically normally distributed. This "regular" behaviour is much less common when one moves to the more general context of estimation for stochastic processes. Indeed, quite stringent conditions are required to ensure that the MLE, when appropriately normalized, has an asymptotically normal distribution; e.g. Basawa *et al.* (1976), Heyde (1978).

Interestingly, a Bayesian approach to the problem has much greater universality. The conditions for asymptotic normality of the posterior distribution, appropriately scaled, are very similar to those which ensure consistency of the MLE and are much weaker than those required for asymptotic normality of the MLE. This contrast does not appear in the i.i.d. case, where the conditions for asymptotic normality of the MLE simplify to (essentially) a finite variance requirement and hence it appears to have escaped the attention which it deserves. Our methodology is similar to that of Walker (1969) who has discussed the i.i.d. case but the conditions have been simplified.

A Bayesian approach to some questions of "hypothesis testing" in a large sample situation also has an important advantage. It can provide, via the posterior distribution, a single categorical answer in each application while Neyman–Pearson based methodology can, in many cases, lead to complex questions of optimality of procedure in which categorical answers do not seem possible; e.g. Basawa and Scott (1976, 1977), Feigin (1978), Sweeting (1978).

2. The Setting and Results

Let $(X_1, X_2, \ldots, X_n)$ be a sample of consecutive observations from a stochastic process whose distribution depends on a single unknown parameter $\theta \in \Theta$, Θ denoting an interval. We suppose that $(X_1, \ldots, X_n)$ possesses a density $P_n(X_1, \ldots, X_n \mid \theta)$ with respect to a σ-finite measure μ_n. Let $\hat{\theta}_n = \hat{\theta}_n(X_1, \ldots, X_n)$ be an MLE—a value of θ maximizing the likelihood function $P_n(X_1, \ldots, X_n \mid \theta)$ (uniqueness of $\hat{\theta}_n$ is not essential).

We put $L_n(\theta) = \log P_n(X_1, \ldots, X_n \mid \theta)$ and use P_θ and E_θ to denote probabilities and expectations computed under the assumption that θ is the true parameter value.

Let θ have an absolutely continuous prior distribution over Θ, with density $\pi(\theta)$ with respect to Lebesgue measure. Then θ has an absolutely continuous posterior distribution with density

$$\pi_n(\theta \mid X_1, \ldots, X_n) = \pi(\theta)\, P_n(X_1, \ldots, X_n \mid \theta)/P_n(X_1, \ldots, X_n),$$

where $P_n(X_1, \ldots, X_n) = \int_\Theta \pi(\theta)\, P_n(X_1, \ldots, X_n \mid \theta)\, d\theta$.

Suppose that θ_0 is an interior point of Θ. We impose the following regularity conditions throughout:

(A1) The prior density $\pi(\theta)$ is continuous and positive at θ_0.

(A2) $L_n(\theta)$ is a.s. twice continuously differentiable with respect to θ in some neighbourhood of $\theta_0 \subset \Theta$, N_0 say.

(A3) $[-L''_n(\theta_0)] \to \infty$ a.s. (P_{θ_0}) as $n \to \infty$.

(A4) For any $\delta > 0$ for which $N_0(\delta) = (\theta_0 - \delta, \theta_0 + \delta) \subset \Theta$ there exists $k(\delta) > 0$ and $h(\delta) \downarrow 0$ as $\delta \downarrow 0$ satisfying

$$\liminf_{n\to\infty} P_{\theta_0}\{\sup_{\theta \notin \Theta \setminus N_0(\delta)} [-L''_n(\theta_0)]^{-1}[L_n(\theta) - L_n(\theta_0)] < -k(\delta)\} \geq 1 - h(\delta).$$

(A5) Given $\varepsilon > 0$, there exists $\delta = \delta(\varepsilon)$ satisfying

$$\liminf_{n\to\infty} P_{\theta_0}\{\sup_{\theta \in N_0(\delta)} |[L''_n(\hat\theta_n)]^{-1}(L''_n(\theta) - L''_n(\hat\theta_n))| < \varepsilon\} \geq 1 - \varepsilon.$$

Conditions (A1) and (A2) require no comment while Condition (A3) essentially ensures that the process does not degenerate. Condition (A4) is little more than a simple sufficient condition for the consistency of $\hat\theta_n$ for θ_0 while Condition (A5) is a continuity condition for L''_n. The Conditions A2–A5 are, in fact, implied by familiar classical conditions of Wald (1948) for consistency of the MLE, provided that it is unique. This can be straightforwardly checked via the conditions *A* in Basawa *et al.* (1976).

We are now in a position to establish the following theorem, the proof of which is deferred until the following section.

Theorem. Suppose that conditions *A* hold. Let σ_n be the positive square root of $\{-L''_n(\hat\theta_n)\}^{-1}$ whenever this exists (which will be true with probability tending to unity as $n \to \infty$ by (A3) and (A5)). Then, if $-\infty \leq b < a \leq \infty$, the posterior probability that $\hat\theta_n + b\sigma_n < \theta < \hat\theta_n + a\sigma_n$, namely

$$\int_{\hat\theta_n + b\sigma_n}^{\hat\theta_n + a\sigma_n} \pi_n(\theta \mid X_1, \ldots, X_n)\, d\theta,$$

tends in P_{θ_0} probability to

$$(2\pi)^{-\frac{1}{2}} \int_a^b e^{-\frac{1}{2}u^2}\, du$$

as $n \to \infty$.

This theorem provides the Bayesian analogue of confidence intervals. These are, of course, typically approached via asymptotic normality results for $\hat\theta_n$ in the Neyman–Pearson theory. Consequently, the assumptions underlying the result of the theorem should be contrasted with those necessary to ensure the asymptotic normality of $\hat\theta_n$, when appropriately normed, in a non-Bayesian situation.

The usual approach to asymptotic normality for $\hat\theta_n$ is via martingale theory and complicated regularity conditions are intrinsic (e.g. Heyde and Feigin (1975), Basawa *et al.* (1976), Heyde (1978). Indeed, under mild differentiability and integrability conditions,

$$\left\{L'_n(\theta) = \sum_{i=1}^{n} u_i(\theta), F_n, n \geq 1\right\},$$

the *u*'s being defined by differencing, is a martingale, F_k denoting the σ-field generated by $X_1, \ldots, X_k$. Asymptotic normality of $\hat\theta_n - \theta$, when suitably normalized, holds under minor additional conditions if and only if a corresponding asymptotic result holds for $L'_n(\theta)$. This, however, involves additional stringent assumptions on the increments $u_i(\theta)$ of the $L'_n(\theta)$ process. Typical are conditions such as

$$s_n^{-2} \sum_1^n E_\theta(u_i^2(\theta) \mid F_{i-1}) \to \eta^2(\theta) \quad \text{in probability,}$$

as $n \to \infty$ for some r.v. $\eta^2 > 0$ a.s. and

$$s_n^{-2} \sum_1^n E_\theta(u_i^2(\theta)\, I(|u_i(\theta)| > \varepsilon s_n)) \to 0$$

as $n \to \infty$ for any $\varepsilon > 0$. Here $s_n^2 = \Sigma_1^n E_\theta u_i^2(\theta)$. Requirements of this kind are indeed stringent by comparison with their rather robust Bayesian counterparts. There are not even any formal requirements concerning existence of moments in the Bayesian formulation.

The theorem is given for the single parameter case to facilitate comparison with corresponding results based on the MLE. These are not yet available in such general form in the multiparameter case. There is, however, no special difficulty in extending the theorem to the multiparameter case and details are provided in Johnstone (1978).

The scope of the theorem may also be extended in another direction, namely to increase somewhat the family of distributions to which the results are applicable. Thus, using a formulation along the lines of Dawid (1970) it is possible to deal with some distributions whose range depends on the parameter. There is, however, a significant cost in terms of formal complications of the regularity conditions.

3. PROOF OF THE THEOREM

Write

$$P_n(X_1, \ldots, X_n \mid \theta) = P_n(X_1, \ldots, X_n \mid \hat{\theta}_n) \exp\{L_n(\theta) - L(\hat{\theta}_n)\}. \tag{1}$$

Whenever $\theta, \hat{\theta}_n \in N_0$ we have by Taylor expansion that

$$L_n(\theta) = L_n(\hat{\theta}_n) + \tfrac{1}{2}(\theta - \hat{\theta}_n)^2 L''_n(\theta_n^*) \tag{2}$$

for $\theta_n^* = \theta + \gamma(\hat{\theta}_n - \theta)$ with $0 < \gamma < 1$. Thus, letting

$$R_n = R_n(\theta, X_1, \ldots, X_n) = \sigma_n^2\{L''_n(\theta_n^*) - L''_n(\hat{\theta}_n)\},$$

we have

$$-(1 - R_n)/\sigma_n^2 = L''_n(\theta_n^*)$$

so that, for $\theta, \hat{\theta}_n \in N_0$,

$$P_n(X_1, \ldots, X_n \mid \theta) = P_n(X_1, \ldots, X_n \mid \hat{\theta}_n) \exp[-\{(\theta - \hat{\theta}_n)^2/2\sigma_n^2\}(1 - R_n)]. \tag{3}$$

Given $\varepsilon > 0$ we choose $\delta = \delta(\varepsilon)$ so small that $N_0(\delta) \subset N_0$ and

$$(1 - \varepsilon) < \inf_{\theta \in N_0(\delta)} \pi(\theta)/\pi(\theta_0) \leqslant \sup_{\theta \in N_0(\delta)} \pi(\theta)/\pi(\theta_0) < 1 + \varepsilon, \tag{4}$$

this last condition being justified by (A1). Now we split the integral defining $P_n(X_1, \ldots, X_n)$ into $I_1^{(n)}(\delta)$ and $I_2^{(n)}(\delta)$, taken over $\Theta \setminus N_0(\delta)$ and $N_0(\delta)$ respectively. Then, using (1),

$$I_1^{(n)}(\delta) = P_n(X_1, \ldots, X_n \mid \hat{\theta}_n) \int_{\Theta \setminus N_0(\delta)} \pi(\theta) \exp\{L_n(\theta) - L_n(\hat{\theta}_n)\}\, d\theta,$$

so that, conditional on $L_n''(\hat{\theta}_n)<0$ and

$$A_n = A_n(\varepsilon) = [\sup_{\Theta\setminus N_0(\delta)} \{-L_n''(\theta_0)\}^{-1}\{L_n(\theta)-L_n(\theta_0)\} < -k(\delta)],$$

$$[P_n(X_1,\ldots,X_n\,|\,\hat{\theta}_n)\,\sigma_n]^{-1} I_1^{(n)}(\delta)$$

$$= \exp\{L_n(\theta_0)-L_n(\hat{\theta}_n)\}\{-L_n''(\hat{\theta}_n)\}^{\frac{1}{2}} \times \int_{\Theta\setminus N_0(\delta)} \pi(\theta)\exp\{L_n(\theta)-L_n(\theta_0)\}\,d\theta$$

$$\leqslant \{-L_n''(\hat{\theta}_n)\}^{\frac{1}{2}} \exp[-k(\delta)\{-L_n''(\theta_0)\}]$$

by the definition of $\hat{\theta}_n$. Now

$$\{-L_n''(\theta_0)\}^{\frac{1}{2}} \exp[-k(\delta)\{-L_n''(\theta_0)\}] \to 0, \quad \text{a.s.}$$

as $n\to\infty$ in view of (A3) and hence, by virtue of (A3), (A5) and $P_{\theta_0}(L_n''(\hat{\theta}_n)<0)\to 1$ as $n\to\infty$, it follows that conditional on A_n and

$$B_n = B_n(\varepsilon) = \{\sup_{\theta\in N_0(\delta)} |[L_n''(\hat{\theta}_n)]^{-1}(L_n''(\theta)-L_n''(\hat{\theta}_n))| < \varepsilon\},$$

we have

$$\{P_n(X_1,\ldots,X_n\,|\,\hat{\theta}_n)\,\sigma_n\}^{-1} I_1^{(n)}(\delta) \to 0, \tag{5}$$

in probability.

To deal with $I_2^{(n)}(\delta)$ we first suppose that $\hat{\theta}_n \in N_0(\delta) \subset N_0$ and use (3) to write

$$\{P_n(X_1,\ldots,X_n\,|\,\hat{\theta}_n)\,\pi(\theta_0)\}^{-1} I_1^{(n)}(\delta) = \int_{N_0(\delta)} \{\pi(\theta)/\pi(\theta_0)\}\exp[-\{(\theta-\hat{\theta}_n)^2/2\sigma_n^2\}(1-R_n)]\,d\theta.$$

Then, putting

$$I_3^{(n)}(\delta) = \int_{N_0(\delta)} \exp[-\{(\theta-\hat{\theta}_n)^2/2\sigma_n^2\}(1-R_n)]\,d\theta,$$

we have from (4) that

$$(1-\varepsilon)\,I_3^{(n)}(\delta) < \{P_n(X_1,\ldots,X_n\,|\,\hat{\theta}_n)\,\pi(\theta_0)\}^{-1} I_2^{(n)}(\delta) < (1+\varepsilon)\,I_3^{(n)}(\delta). \tag{6}$$

If $\sup_{N_0(\delta)}|R_n| < \varepsilon < 1$,

$$\int_{N_0(\delta)} \exp\{-(\theta-\hat{\theta}_n)^2(1+\varepsilon)/2\sigma_n^2\}\,d\theta < I_3^{(n)}(\delta) < \int_{N_0(\delta)} \exp\{-(\theta-\hat{\theta}_n)^2(1-\varepsilon)/2\sigma_n^2\}\,d\theta$$

whilst for $\eta = +\varepsilon$ or $-\varepsilon$,

$$\int_{N_0(\delta)} \exp\{-(\theta-\hat{\theta}_n)^2(1+\eta)/2\sigma_n^2\}\,d\theta$$

$$= (2\pi)^{\frac{1}{2}}\,\sigma_n(1+\eta)^{-\frac{1}{2}}[\Phi\{\sigma_n^{-1}(\theta_0+\delta-\hat{\theta}_n)(1+\eta)^{\frac{1}{2}}\} - \Phi\{\sigma_n^{-1}(\theta_0-\delta-\hat{\theta}_n)(1+\eta)^{\frac{1}{2}}\}], \tag{7}$$

where

$$\Phi(y) = (2\pi)^{-\frac{1}{2}} \int_{-\infty}^{y} e^{-\frac{1}{2}u^2}\,du.$$

Conditions (A3) and (A5) imply that $\sigma_n^{-1}\to\infty$, in probability, and, since $\hat{\theta}_n$ is consistent, the factor in curly brackets in (6) converges to unity in probability. Thus, using (A5) to secure the appropriate bound on R_n, it follows that

$$P_{\theta_0}((2\pi)^{\frac{1}{2}}(1+\varepsilon)^{-\frac{1}{2}} < \sigma_n^{-1} I_3^{(n)}(\delta) < (2\pi)^{\frac{1}{2}}(1-\varepsilon)^{-\frac{1}{2}}\,|\,A_n \cap B_n) \to 1 \tag{8}$$

as $n\to\infty$. The conditions (5), (6) and (8) then give

$$P_{\theta_0}(C_n \mid A_n \cap B_n) \to 1$$

as $n\to\infty$, where

$$C_n = [(1+\varepsilon)^{\frac{1}{2}}(1-\varepsilon) < \{P_n(X_1,\ldots,X_n \mid \hat{\theta}_n)\,\pi(\theta_0)\,(2\pi)^{\frac{1}{2}}\,\sigma_n\}^{-1}\,P_n(X_1,\ldots,X_n) < (1+\varepsilon)\,(1-\varepsilon)^{-\frac{1}{2}}].$$

But, by virtue of (A4) and (A5), and using a bar to denote complementation,

$$\begin{aligned}\liminf_{n\to\infty} P_{\theta_0}(A_n \cap B_n) &= 1 - \limsup_{n\to\infty} P_{\theta_0}(\bar{A}_n \cup \bar{B}_n)\\ &\geqslant 1 - \limsup_{n\to\infty} P_{\theta_0}(\bar{A}_n) - \limsup_{n\to\infty} P_{\theta_0}(\bar{B}_n) \geqslant 1 - h(\delta(\varepsilon)) - \varepsilon,\end{aligned}$$

and hence,

$$\begin{aligned}\liminf_{n\to\infty} P_{\theta_0}(C_n) &\geqslant \liminf_{n\to\infty} P_{\theta_0}(C_n \cap A_n \cap B_n)\\ &= \liminf_{n\to\infty} P_{\theta_0}(C_n \mid A_n \cap B_n)\,P_{\theta_0}(A_n \cap B_n)\\ &\geqslant 1 - h(\delta(\varepsilon)) - \varepsilon. \qquad (9)\end{aligned}$$

The results (9) holds for all ε (and hence $h(\delta)$) small enough and thus,

$$\{P_n(X_1,\ldots,X_n \mid \hat{\theta}_n)\,\sigma_n\}^{-1}\,P_n(X_1,\ldots,X_n) \to (2\pi)^{\frac{1}{2}}\,\pi(\theta_0), \qquad (10)$$

in probability.

To complete the proof, we consider

$$I_2' = \int_{\hat{\theta}_n + b\sigma_n}^{\hat{\theta}_n + a\sigma_n} \pi(\theta)\,P_n(X_1,\ldots,X_n \mid \theta)\,d\theta,$$

which is $I_2^{(n)}(\delta)$ with $N_0(\delta)$ replaced by $(\hat{\theta}_n + b\sigma_n, \hat{\theta}_n + a\sigma_n)$. Similarly, let I_3' be $I_3^{(n)}(\delta)$ with the limits of integration changed to $(\hat{\theta}_n + b\sigma_n, \hat{\theta}_n + a\sigma_n)$.

We use (A1) and the argument up to (6) so that if $\varepsilon > 0$ is given, there exists δ such that if $(\hat{\theta}_n + b\sigma_n, \hat{\theta}_n + a\sigma_n) \subset N_0(\delta)$,

$$(1-\varepsilon)\,I_3' < \{P_n(X_1,\ldots,X_n \mid \hat{\theta}_n)\,\pi(\theta_0)\}^{-1}\,I_2' < (1+\varepsilon)\,I_3',$$

while the right-hand side corresponding to (7) becomes

$$(2\pi)^{\frac{1}{2}}\,\sigma_n(1+\eta)^{-\frac{1}{2}}[\Phi\{a(1+\eta)^{\frac{1}{2}}\} - \Phi\{b(1+\eta)^{\frac{1}{2}}\}].$$

Clearly

$$\lim_{n\to\infty} P_{\theta_0}\{(\hat{\theta}_n + b\sigma_n, \hat{\theta}_n + a\sigma_n) \subset N_0(\delta)\} = 1$$

and our previous argument can be paralleled to obtain

$$\{P_n(X_1,\ldots,X_n \mid \hat{\theta}_n)\,\sigma_n\}^{-1}\,I_2' \to (2\pi)^{\frac{1}{2}}\,\pi(\theta_0)\,\{\Phi(a) - \Phi(b)\}, \qquad (11)$$

in probability. Since

$$\int_{\hat{\theta}_n + b\sigma_n}^{\hat{\theta}_n + a\sigma_n} \pi_n(\theta \mid X_1,\ldots,X_n)\,d\theta = I_2'/P_n(X_1,\ldots,X_n),$$

the result of the theorem follows from (10) and (11).

REFERENCES

BASAWA, I. V., FEIGIN, P. D. and HEYDE, C. C. (1976). Asymptotic properties of maximum likelihood estimators for stochastic processes. *Sankhyā* A, **38**, 259–270.

BASAWA, I. V. and SCOTT, D. J. (1976). Efficient tests for branching processes. *Biometrika*, **63**, 531–536.

—— (1977). Efficient tests for stochastic processes. *Sankhyā* A, **39**, 21–31.

DAWID, A. P. (1970). On the limiting normality of posterior distributions. *Proc. Camb. Phil. Soc.*, **67**, 625–633.

FEIGIN, P. D. (1978). The efficiency criteria problem for stochastic processes. *Stoch. Proc. Appl.*, **6**, 115–127

HEYDE, C. C. (1978). On an optimal property of the maximum likelihood estimator of a parameter from a stochastic process. *Stoch. Proc. Appl.* **8**, 1–9.

HEYDE, C. C. and FEIGIN, P. D. (1975). On efficiency and exponential families in stochastic process estimation. In *Statistical Distributions in Scientific Work* (G. P. Patil, S. Kotz and J. K. Ord, eds), Vol. 1, pp. 227–240. Dordrecht: Reidel.

JOHNSTONE, I. M. (1978). M.Sc. Thesis, Australian National University.

SWEETING, T. J. (1978). On efficient tests for branching processes. *Biometrika*, **65**, 123–127.

WALD, A. (1948). Asymptotic properties of the maximum likelihood estimate of an unknown parameter of a discrete stochastic process. *Ann. Math. Statist.*, **19**, 40–46.

WALKER, A. M. (1969). On the asymptotic behaviour of posterior distributions. *J. R. Statist. Soc.* B, **31**, 80–88.

J. Math. Biology 12, 91–99 (1981)

Journal of
Mathematical
Biology

On the Survival of a Gene Represented in a Founder Population

C. C. Heyde

CSIRO Division of Mathematics and Statistics, Box 1965, P.O. Canberra City, A.C.T. 2601, Australia

Abstract. Gene survival in a population which increases without density dependence is considered using a generalization of the Moran model for haploid individuals. It is shown that situations where ultimate homozygosity is certain and where there is a non-zero probability of balanced polymorphism are both possible. Necessary and sufficient conditions in terms of the mean of the population growth distribution are given which determine which of these situations holds.

Key words: Gene survival – Founder population – Moran model – Balanced polymorphism – Ultimate homozygosity – Martingale methods

1. Introduction and Results

This paper deals with a founder population which is introduced into a new niche and increases without density dependence. We consider two possible alleles a and A at a single locus and shall address the question of the probability that the population ultimately becomes homozygous.

Suppose that we have a haploid population in which individuals die one by one at random and are replaced by new individuals. At time n the population consists of X_n a-alleles and Y_n A-alleles. The next death occurs at time $n + 1$ and the dying individual is replaced by a random number $U_n = k$, say, of new individuals which are of type a or A with probabilities $X_n/(X_n + Y_n)$ and $Y_n/(X_n + Y_n)$ respectively. The number of new individuals has a distribution $\{p_k, k \geqslant 1\}$ independent of the replacement type. The probabilities of an a or A individual dying are also $X_n/(X_n + Y_n)$ and $Y_n/(X_n + Y_n)$, respectively, and hence the transition probabilities for the bivariate Markov chain of the process are

$$P(X_{n+1} = X_n - 1, Y_{n+1} = Y_n + k | X_n, Y_n) = X_n Y_n (X_n + Y_n)^{-2} p_k, \qquad k \geqslant 1,$$

$$P(X_{n+1} = X_n + k, Y_{n+1} = Y_n - 1 | X_n, Y_n) = X_n Y_n (X_n + Y_n)^{-2} p_k, \qquad k \geqslant 1,$$

$$P(X_{n+1} = X_n + k - 1, Y_{n+1} = Y_n | X_n, Y_n) = X_n^2 (X_n + Y_n)^{-2} p_k, \qquad k \geqslant 1,$$

$$P(X_{n+1} = X_n, Y_{n+1} = Y_n + k - 1 | X_n, Y_n) = Y_n^2 (X_n + Y_n)^{-2} p_k, \qquad k \geqslant 1.$$

This overlapping generation model is an extension of the Moran model [e.g. Moran

0303–6812/81/0012/0091/$01.80

(1962), Chapter IV; Ewens (1979), Chapter 3] which corresponds to the case $p_1 = 1, p_k = 0, k > 1$.

We shall write $X_n + Y_n = N_n$, the number of individuals in the population at time n and let $\mathscr{F}_n$ denote the σ-field generated by $X_k, Y_k, 0 \leqslant k \leqslant n$. The initial values X_0 and N_0 will be taken as fixed constants. Note that $N_n = S_n + N_0$ where $S_n = (U_1 - 1) + \cdots + (U_n - 1)$, the numbers of new individuals U_i being independent and identically distributed random variables with $P(U_1 = k) = p_k, k \geqslant 1$.

The evolution of the population will be studied by considering the process $\{Z_n = X_n N_n^{-1}, n \geqslant 0\}$ which keeps track of the proportion of a-individuals in the population. We have

$$E(Z_{n+1}|\mathscr{F}_n) = \sum_{k=1}^{\infty} \Big[(X_n - 1)Z_n(1 - Z_n) + (X_n + k)Z_n(1 - Z_n)$$

$$+ (X_n + k - 1)Z_n^2 + X_n(1 - Z_n)^2\Big](N_n + k - 1)^{-1}p_k$$

$$= Z_n \sum_{k=1}^{\infty} [(2N_nZ_n + k - 1)(1 - Z_n) + (N_nZ_n + k - 1)Z_n$$

$$+ N_n(1 - Z_n)^2](N_n + k - 1)^{-1}p_k = Z_n$$

and hence $\{Z_n, \mathscr{F}_n, n \geqslant 0\}$ is a martingale. Since $0 \leqslant Z_n \leqslant 1$ we have from the martingale convergence theorem that $Z_n \to Z$ a.s. as $n \to \infty$ for some random variable Z with $0 \leqslant Z \leqslant 1$. That is, the proportion of a-alleles in the population approaches some limit with probability one.

Our principal question in this paper concerns the eventual genetic composition of the population. It turns out that certain ultimate homozygosity (i.e. $P(Z = 0) + P(Z = 1) = P(Z(1 - Z) = 0) = 1$) or a prospect of balanced polymorphism (i.e. $P(Z(1 - Z) > 0) > 0$) are both possible and the following theorem gives a complete categorization of the circumstances under which each of these can be obtained.

Theorem. *If* $EU_1 = \sum_{k=1}^{\infty} kp_k = \infty$ *or if* $p_1 = 1$ *(and hence* $U_j = 1$ *and* $N_j = N_0$ *for all* j*) then* $Z(1 - Z) = 0$ a.s. *so that there is ultimate homozygosity with probability one. On the other hand, if* $1 < EU_1 < \infty$, *then* $P(Z(1 - Z) > 0) > 0$, *so that there is a non-zero probability of balanced polymorphism in the population.*

It is interesting to contrast these results with the corresponding ones for a generalization of the Sewall-Wright model (Heyde and Seneta, 1975). Balanced polymorphism is a possibility in both contexts but under circumstances of much less extreme population growth in the present one. Here we have given a complete categorization of circumstances under which balanced polymorphism can be obtained in contrast to partial results in Heyde and Seneta (1975). However, population growth occurs by additions of independent and identically distributed random variables in the present model in contrast to a quite general process for the other case.

The results of this paper are also relevant in the context of the founder principle of Mayr [Mayr (1942), p. 237] and to the problem of the survival of a new mutant. It should be noted that, no matter what distribution is chosen for U_1, there is always a non-zero probability of ultimate homozygosity.

2. Proof of Theorem

We have for $n \geqslant 0$, and after some algebra,

$$E(Z_{n+1}(1 - Z_{n+1})|\mathscr{F}_n) = \frac{X_n Y_n}{N_n^2} \sum_{k=1}^{\infty} [(X_n - 1)(Y_n + k) + (X_n + k)(Y_n - 1)$$

$$+ X_n(X_n + k - 1) + Y_n(Y_n + k - 1)] \frac{p_k}{(N_n + k - 1)^2}$$

$$= Z_n(1 - Z_n) \sum_{k=1}^{\infty} \frac{N_n^2 + 2kN_n - 2N_n - 2k}{(N_n + k - 1)^2} p_k$$

$$= Z_n(1 - Z_n)\left\{1 - \sum_{k=1}^{\infty} \frac{1 + k^2}{(N_n + k - 1)^2} p_k\right\}.$$

Thus,

$$\left\{W_n = \frac{Z_n(1 - Z_n)}{\prod_{j=0}^{n-1}\left(1 - \sum_{k=1}^{\infty} \frac{1 + k^2}{(N_j + k - 1)^2} p_k\right)}, \mathscr{F}_n, n \geqslant 1\right\}$$

is a non-negative martingale with mean $Z_0(1 - Z_0) > 0$. By the martingale convergence theorem, W_n converges a.s. to some random variable W with $EW \leqslant Z_0(1 - Z_0)$. Of course $Z_n(1 - Z_n) \to Z(1 - Z)$ a.s. since $Z_n \to Z$ a.s. and hence $Z(1 - Z) = 0$ at least on the set

$$\left\{\prod_{j=0}^{\infty}\left(1 - \sum_{k=1}^{\infty} \frac{1 + k^2}{(N_j + k - 1)^2} p_k\right) = 0\right\},$$

i.e. on the set

$$\left\{\sum_{j=0}^{\infty} \sum_{k=1}^{\infty} \frac{1 + k^2}{(N_j + k - 1)^2} p_k = \infty\right\}.$$

On the other hand, returning to the martingale $\{W_n, \mathscr{F}_n, n \geqslant 1\}$, we note that if

$$V = \prod_{j=0}^{\infty}\left(1 - \sum_{k=1}^{\infty} \frac{1 + k^2}{(N_j + k - 1)^2} p_k\right),$$

then

$$W = Z(1 - Z)V^{-1} \leqslant \tfrac{1}{4}V^{-1}$$

and if $EV^{-1} < \infty$ the dominated convergence theorem enables us to conclude that

$$EW = E(Z(1 - Z)V^{-1}) = Z_0(1 - Z_0) > 0$$

which ensures that $P(Z(1 - Z) > 0) > 0$. The proof will now be completed with the aid of three lemmas.

Lemma 1.

$$P\left(\sum_{j=0}^{\infty} \sum_{k=1}^{\infty} \frac{1 + k^2}{(N_j + k - 1)^2} p_k < \infty\right) = 0 \text{ or } 1$$

according as

$$\sum_{j=0}^{\infty}\sum_{k=1}^{\infty}(1+k^2)p_kE(N_j+k-1)^{-2}=\infty \text{ or } <\infty.$$

Proof. We use similar ideas to the proof of Lemma 1 and Theorem 1 of Schuh (1980). Write

$$A_j=\sum_{k=1}^{\infty}(1+k^2)(N_j+k-1)^{-2}p_k.$$

We shall first show that $\{\sum_{j=0}^{\infty}A_j<\infty\}\in\mathscr{G}$, the tail σ-field of $U_1-1, U_2-1,\ldots,$ and hence $P(\sum_{j=0}^{\infty}A_j<\infty)=0$ or 1.

Let $\mathscr{G}_n$ be the σ-field generated by U_k-1, $k\geqslant n$. Then,

$$\left[U_1\left(N_0+\sum_{i=2}^{n}(U_i-1)+k-1\right)\right]^{-2}\leqslant\left[N_0+\sum_{i=1}^{n}(U_i-1)+k-1\right]^{-2}$$
$$\leqslant\left[N_0+\sum_{i=2}^{n}(U_i-1)+k-1\right]^{-2}$$

for all $n\geqslant 2$ and hence

$$\left\{\sum_{j=0}^{\infty}A_j<\infty\right\}=\left\{\sum_{j=2}^{\infty}\sum_{k=1}^{\infty}(1+k^2)\left(N_0+\sum_{i=2}^{n}(U_i-1)+k+1\right)^{-2}p_k<\infty\right\}\in\mathscr{G}_2.$$

Applying this reasoning iteratively we obtain

$$\left\{\sum_{j=0}^{\infty}A_j<\infty\right\}\in\mathscr{G}_r\qquad\text{for all } r$$

and hence it belongs to $\mathscr{G}=\bigcap_{r=1}^{\infty}\mathscr{G}_r$.

Now it follows from the Borel-Cantelli lemma that

$$P\left(\sum_{j=0}^{\infty}A_j<\infty\right)=0\qquad\text{implies}\qquad\sum_{j=0}^{\infty}EA_j=\infty$$

and consequently it suffices to show that

$$P\left(\sum_{j=0}^{\infty}A_j<\infty\right)=1\qquad\text{implies}\qquad\sum_{j=0}^{\infty}EA_j<\infty\tag{1}$$

in order to complete the proof.

To establish (1) we first choose $l>0$ so large that $P(\sum_{j=0}^{\infty}A_j\leqslant l)>0$ and define

$$=\min\left[n\,\middle|\,\sum_{j=0}^{n}A_j>l\right]\qquad\text{if}\qquad\sum_{j=0}^{\infty}A_j>l,\quad=\infty\text{ otherwise,}$$

so that $P(n^*=\infty)>0$. We have

$$\int_{\{n^*=n\}}(N_n+k-1)^{-2}\,dP\leqslant(N_0+k-1)^{-2}P(n^*=n)\tag{2}$$

and since $\sum_{j=0}^{n-1} A_j \leqslant l$ on $\{n^* = n\}$,

$$\int_{\{n^*=n\}} \left(\sum_{j=0}^{n-1} A_j\right) dP \leqslant lP(n^* = n). \tag{3}$$

Further, since $\{n^* = n\}$ belongs to the σ-field generated by $U_i - 1$, $i \leqslant n$, we have for $t \geqslant n$,

$$\begin{aligned}\int_{\{n^*=n\}} &(N_0 + S_{t+r} + k - 1)^{-2}\, dP \\ &= \int_{\{n^*=n\}} (N_0 + k - 1 + (U_1 - 1) + \cdots + (U_{t+r} - 1))^{-2}\, dP \\ &\leqslant \int_{\{n^*=n\}} (N_0 + k - 1 + (U_{t+1} - 1) + \cdots + (U_{t+r} - 1))^{-2}\, dP \\ &\leqslant EA_r P(n^* = n)\end{aligned} \tag{4}$$

for all $r = 1, 2, \ldots$.

The results (2), (3) and (4) imply that

$$\int_{\{n^*=n\}} \left(\sum_{j=0}^{N} A_j\right) dP \leqslant \left[\sum_{k=1}^{\infty} \frac{1 + k^2}{(N_0 + k - 1)^2} p_k + l + E\sum_{j=0}^{N} A_j\right] P(n^* = n)$$

for all $0 \leqslant n \leqslant N$ and hence

$$\int_{\{n^* \leqslant N\}} \left(\sum_{j=0}^{N} A_j\right) dP \leqslant \left[\sum_{k=1}^{\infty} \frac{1 + k^2}{(N_0 + k - 1)^2} p_k + l + E\sum_{j=0}^{N} A_j\right] P(n^* \leqslant N). \tag{5}$$

However, as $\sum_{j=0}^{N} A_j \leqslant l$ on $\{n^* > N\}$, we have

$$\begin{aligned}\int_{\{n^* > N\}} \left(\sum_{j=0}^{N} A_j\right) dP &\leqslant lP(n^* > N) \\ &\leqslant \left[l + \sum_{k=1}^{\infty} \frac{1 + k^2}{(N_0 + k - 1)^2} p_k\right] P(n^* > N),\end{aligned} \tag{6}$$

and from (5) and (6),

$$E\sum_{j=0}^{N} A_j \leqslant l + \sum_{k=1}^{\infty} \frac{1 + k^2}{(N_0 + k - 1)^2} p_k + \left(E\sum_{j=0}^{N} A_j\right) P(n^* \leqslant N),$$

so that

$$P(n^* > N) E\sum_{j=0}^{N} A_j \leqslant l + \sum_{k=1}^{\infty} \frac{1 + k^2}{(N_0 + k - 1)^2} p_k < \infty. \tag{7}$$

Finally, we let $N \to \infty$ in (7) and since $P(n^* > N) \to P(n^* = \infty) > 0$ we obtain

$$E\sum_{j=0}^{\infty} A_j = \lim_{N\to\infty} E\sum_{j=0}^{N} A_j < \infty$$

as required. This completes the proof of the lemma.

Lemma 2. $\sum_{j=0}^{\infty}\sum_{k=1}^{\infty}(1+k^2)p_k E(N_j+k-1)^{-2}$ *diverges if* $U_1 = 1$ a.s. *or* $EU_1 = \infty$ *and converges if* $1 < EU_1 < \infty$.

Proof. The result for $U_1 = 1$ a.s. is obvious and we shall subsequently suppose that $P(U_1 > 1) > 0$. The result for $1 < EU_1 < \infty$ follows readily via a minor modification of the proof of Lemma 3 of Athreya and Karlin (1967) but this approach does not enable us to deal with the case $EU_1 = \infty$. We shall adopt a unified approach which enables us to treat both cases.

If X is a non-negative integer valued random variable with probability generating function (p.g.f.) f and $r > 0$, we have for $0 \leqslant s \leqslant 1$,

$$E(s^{r+X}(r+X)^{-1}) = E\int_0^s t^{r+X-1}\,dt$$

$$= \int_0^s Et^{r+X-1}\,dt = \int_0^s t^{r-1}f(t)\,dt$$

using Fubini's theorem. Further,

$$\int_0^1 E(s^{r+X-1}(r+X)^{-1})\,ds = E(r+X)^{-2}$$

and hence

$$E(r+X)^{-2} = \int_0^1 s^{-1}\left(\int_0^s t^{r-1}f(t)\,dt\right)ds$$

$$= \int_0^1\left(\int_t^1 s^{-1}\,ds\right)t^{r-1}f(t)\,dt$$

$$= \int_0^1 (\log t^{-1})t^{r-1}f(t)\,dt. \tag{8}$$

Now $N_j = N_0 + (U_1 - 1) + \cdots + (U_j - 1)$, so that if f is the p.g.f. of $U_1 - 1$ we have from (8) that

$$E(N_j+k-1)^{-2} = \int_0^1 (\log t^{-1})t^{N_0+k-1}f^j(t)\,dt$$

and

$$\sum_{j=0}^{\infty}\sum_{k=1}^{\infty}(1+k^2)p_k E(N_j+k-1)^{-2}$$

$$= \int_0^1 (\log t^{-1})t^{N_0}\left(\sum_{j=0}^{\infty} f^j(t)\right)\left(\sum_{k=1}^{\infty}(1+k^2)p_k t^{k-1}\right)dt$$

$$= \int_0^1 (\log t^{-1})t^{N_0}(1-f(t))^{-1}(2f(t) + t^2f''(t) + 3tf'(t))\,dt$$

since $f(t) = \sum_{k=1}^{\infty} p_k t^{k-1}$.

We observe that

$$\lim_{t \downarrow 0} (\log t^{-1}) t^{N_0} (1 - f(t))^{-1} (2f(t) + t^2 f''(t) + 3t f'(t)) = 0,$$

and as $t \uparrow 1$,

$$t^{N_0} \log t^{-1} \sim 1 - t$$

while

$$2f(t) + t^2 f''(t) + 3t f'(t) \sim C f''(t)$$

for some constant C ($C = 1$ if $EU_1^2 = \infty$ for then $f''(t) \uparrow \infty$ and $f'(t) = o(f''(t))$ as $t \uparrow 1$ and C is given by $Cf''(1) = 2 + f''(1) + 3f'(1)$ if $EU_1^2 < \infty$). Thus,

$$\sum_{j=0}^{\infty} \sum_{k=1}^{\infty} (1 + k^2) p_k E(N_j + k - 1)^{-2} < \infty \text{ or } = \infty$$

according as

$$\int_{\varepsilon}^{1} (1 - t)(1 - f(t))^{-1} f''(t)\, dt < \infty \text{ or } = \infty \tag{9}$$

for some ε, $0 < \varepsilon < 1$.

Now write

$$Q(t) = (1 - f(t))(1 - t)^{-1}.$$

Integration by parts in (9) gives

$$\begin{aligned} &\int_{\varepsilon}^{1} (1 - t)(1 - f(t))^{-1} f''(t)\, dt \\ &= \int_{\varepsilon}^{1} [Q(t)]^{-1} f''(t)\, dt \\ &= 1 - f'(\varepsilon)[Q(\varepsilon)]^{-1} + \int_{\varepsilon}^{1} f'(t) Q^{-2}(t) Q'(t)\, dt \end{aligned} \tag{10}$$

with $f'(t)[Q(t)]^{-1} \to 1$ as $t \uparrow 1$ holding since

$$Q(t) = (1 - f(t))(1 - t)^{-1} = f'(\alpha(t))$$

for some $t < \alpha(t) < 1$ by the mean value theorem and in view of the monotonicity of both Q and f'. Then, using $f'(t) \sim Q(t)$ as $t \uparrow 1$ we have

$$\int_{\varepsilon}^{1} f'(t) Q^{-2}(t) Q'(t)\, dt < \infty \text{ or } = \infty$$

according as

$$\int_{\varepsilon}^{1} Q'(t)[Q(t)]^{-1}\, dt < \infty \text{ or } = \infty.$$

Finally,

$$\int_{\varepsilon}^{1} Q'(t)[Q(t)]^{-1}\,dt = [\log Q(t)]_{\varepsilon}^{1}$$

and the result of the lemma follows via (9) and (10) since $Q(1) = f'(1)$ is finite if and only if $EU_1 < \infty$.

Lemma 3. *$EV^{-1} < \infty$ if $1 < EU_1 < \infty$.*

Proof. We now write

$$A(N_j) = \sum_{k=1}^{\infty} (1 + k^2)(N_j + k - 1)^{-2} p_k$$

so that

$$V = \prod_{j=0}^{\infty} (1 - A(N_j))$$

and

$$V^{-1} \leqslant \exp\left[\sum_{j=0}^{\infty} A(N_j)(1 - A(N_j))^{-1}\right]$$

since $1 - x \geqslant \exp(-x/(1-x))$ for $0 < x < 1$. Furthermore, $A(N_j) \leqslant A(N_0)$ and hence

$$V^{-1} \leqslant \exp\left[(1 - A(N_0))^{-1} \sum_{j=0}^{\infty} A(N_j)\right]. \tag{11}$$

Now let n_i, $i \geqslant 0$, denote the number of zeros in the sequence $\{U_j - 1, j \geqslant 1\}$ between the ith and $(i+1)$st non-zero elements. Then, the n_i are independent and identically distributed random variables with

$$P(n_0 = j) = p_1^j(1 - p_1), \qquad j = 0, 1, 2, \ldots$$

and

$$\sum_{j=0}^{\infty} A(N_j) \leqslant \sum_{j=0}^{\infty} (n_j + 1) A(N_0 + j),$$

so that from (11),

$$EV^{-1} \leqslant \prod_{j=0}^{\infty} E[\exp\{(n_0 + 1)(1 - A(N_0))^{-1} A(N_0 + j)\}]. \tag{12}$$

But,

$$\begin{aligned} Ee^{\lambda(n_0+1)} &= e^{\lambda} \sum_{j=0}^{\infty} e^{\lambda j} p_1^j (1 - p_1) \\ &= (1 - p_1)e^{\lambda}(1 - p_1 e^{\lambda})^{-1} \\ &= e^{\lambda}(1 - p_1(1 - e^{\lambda})(1 - p_1 e^{\lambda})^{-1}), \end{aligned} \tag{13}$$

and hence, using (13) in (12),

$$EV^{-1} \leqslant \exp\left[\sum_{j=0}^{\infty} A(N_0+j)\right] \prod_{j=0}^{\infty} \left(1 - \frac{p_1(1-\exp[A(N_0+j)])}{1-p_1\exp[A(N_0+j)]}\right) < \infty$$

since

$$\begin{aligned}\sum_{j=0}^{\infty} A(N_0+j) &= \sum_{j=0}^{\infty}\sum_{k=1}^{\infty} (1+k^2)(N_0+j+k-1)^{-2}p_k \\ &= \sum_{k=1}^{\infty}\left(\sum_{j=0}^{\infty}(N_0+j+k-1)^{-2}\right)(1+k^2)p_k \\ &\leqslant \sum_{k=1}^{\infty}(1+k^2)(N_0+k-2)^{-1}p_k < \infty\end{aligned}$$

for $1 < EU_1 < \infty$. This establishes Lemma 3.

To complete the proof of the theorem it only remains to note that $Z(1-Z)=0$ a.s. if

$$\sum_{j=0}^{\infty}\sum_{k=1}^{\infty}(1+k^2)(N_n+k-1)^{-2}p_k = \infty \quad \text{a.s.}$$

which holds if $U_1 = 1$ a.s. or $EU_1 = \infty$ using Lemmas 1 and 2. On the other hand, $P(Z(1-Z)>0)>0$ if $EV^{-1}<\infty$ and, according to Lemma 3, this holds if $1 < EU_1 < \infty$.

Acknowledgement. Some helpful comments from P. A. P. Moran are gratefully acknowledged.

References

Athreya, K. B., Karlin, S.: Limit theorems for split times of branching processes. J. Math. Mech. **17**, 257–277 (1967)

Ewens, W. J.: Mathematical population genetics. Berlin: Springer 1979

Heyde, C. C., Seneta, E.: The genetic balance between random sampling and random population size. J. Math. Biol. **1**, 317–320 (1975)

Mayr, E.: Systematics and the origin of species. New York: Columbia Univ. Press 1942; New York: Reprinted by Dover Publications 1964

Moran, P. A. P.: The statistical processes of evolutionary theory. Oxford: Clarendon Press 1962

Schuh, H.-J.: Sums of i.i.d. random variables and an application to the explanation to the explosion criterion for Markov branching processes. Department of Statistics, University of Melbourne Research Report Series, No. 2, 1980

Received June 12, 1980/Revised November 11, 1980

J. Math. Biology (1983) 18: 163–168

Journal of
Mathematical
Biology

An alternative approach to asymptotic results on genetic composition when the population size is varying

C. C. Heyde

CSIRO Division of Mathematics and Statistics, Box 1965, P.O. Canberra City, A.C.T. 2601, Australia

Abstract. This paper is concerned with methodology for studying the long-term genetic composition of a population of haploid individuals in the case where the population size is varying. A general approach requiring a minimum of assumptions is described based on constructing martingales out of expressions for the means of the numbers of allelic types, conditional on the past. Earlier investigations were based on studying the asymptotic behaviour of the proportion of alleles of a certain type in the population. Several applications demonstrate that the approach suggested in the paper provides results which usefully complement ones obtained previously.

Key words: Allelic composition — Ultimate homozygosity — Polymorphism — Growth of conditional means — Martingale methods

1. Introduction

We consider a population whose individual members may have two possible alleles a or A at a single locus. Suppose that the stochastic processes $\{X_n, n \geq 0\}$ and $\{Y_n, n \geq 0\}$ track the numbers of a and A alleles respectively in the population and write $\{N_n = X_n + Y_n, n \geq 0\}$ for the corresponding population size process. In the case of constant N_n, classical models due to Wright–Fisher and to Moran (see e.g. Ewens (1979, Chapter 3)) have been formulated to study the changing allelic composition of the population and various generalizations dealing with varying N_n have been introduced. The asymptotic behaviour of the proportion of a alleles $Z_n = X_n / N_n$ has been the major focus and in particular whether $\{Z_n\}$ converges a.s. and, if it does, when $P(Z_n \to 0 \text{ or } 1) = 1$. See for example Seneta (1974), Heyde and Seneta (1975), Heyde (1977, 1981, 1982, 1983), Daley, Hall and Heyde (1982).

The approach via the study of the process $\{Z_n\}$ has provided comprehensive results but there have been several important deficiencies. Detailed structural assumptions have been necessary for progress and the underlying similarity of the results coming from different models lacks adequate explanation. In the (most common) case where $P(Z_n \to 0 \text{ or } 1) = 1$, only crude information is available on the individual asymptotic behaviour of $\{X_n\}$ and $\{Y_n\}$. There has not been a simple and intuitive method for discriminating between the cases where $P(Z_n \to 0$ or $1) = 1$ and $P(Z_n \to Z) = 1$ with $P(0 < Z < 1) > 0$. Finally, in cases where extinction of the population is possible, the use of $Z_n = X_n / N_n$ raises awkward conditioning problems. These deficiencies can be largely avoided if the processes $\{X_n\}$ and $\{Y_n\}$ are studied directly and separately. In fact, it is the behaviour of the conditional means, given the past, of X_n and Y_n that is crucial. Considerable information can be obtained without the need for any concrete distributional assumptions.

2. Methods

Quite a wide variety of population models give rise to evolutionary equations for conditional means of the form

$$\begin{aligned} E(X_{n+1}|\mathscr{F}_n) &= X_n(1+f_n) \quad \text{a.s.}, \\ E(Y_{n+1}|\mathscr{F}_n) &= Y_n(1+g_n) \quad \text{a.s.}, \end{aligned} \tag{1}$$

where $\mathscr{F}_n$ is the (past history) σ-field generated by $\{X_k, Y_k, 0 \leqslant k \leqslant n\}$, f_n and g_n are $\mathscr{F}_n$-measurable and $1+f_n \geqslant 0$ a.s., $1+g_n \geqslant 0$ a.s. In particular, all the above cited papers deal with models whose behaviour follows the prescription (1). The models of Heyde (1981, 1982, 1983) and Daley et al. (1982) based on extensions of the Moran model all give the form (1) with

$$\begin{aligned} f_n &= [(1+\sigma)EU/(1+\sigma Z_n) - 1]/N_n, \\ g_n &= [EV/(1+\sigma Z_n) - 1]/N_n, \end{aligned} \tag{2}$$

while the Wright–Fisher model extension treated in Heyde (1977) gives (1) with

$$\begin{aligned} f_n &= [(1+\sigma)E(N_{n+1}|\mathscr{F}_n)/(1+\sigma Z_n)N_n] - 1 \\ g_n &= [E(N_{n+1}|\mathscr{F}_n)/(1+\sigma Z_n)N_n] - 1. \end{aligned} \tag{3}$$

(In each case σ represents a selective advantage. For (3) the $\sigma = 0$ case had previously been discussed in Heyde and Seneta (1975) and the work originated with Seneta 1974).)

To analyse the system (1) we note that

$$\left\{ X_n \Big/ \prod_{j=0}^{n-1} (1+f_j), \mathscr{F}_n, n \geqslant 1 \right\}, \qquad \left\{ Y_n \Big/ \prod_{j=0}^{n-1} (1+g_j), \mathscr{F}_n, n \geqslant 1 \right\}$$

are non-negative martingales with means EX_0 and EY_0 respectively. Consequently, the martingale convergence theorem gives

$$X_n \Big/ \prod_{j=0}^{n-1} (1+f_j) \xrightarrow{\text{a.s.}} \alpha, \qquad Y_n \Big/ \prod_{j=0}^{n-1} (1+g_j) \xrightarrow{\text{a.s.}} \beta \tag{4}$$

for some random variables $\alpha \geqslant 0$ a.s. and $\beta \geqslant 0$ a.s. with $E\alpha \leqslant EX_0$, $E\beta \leqslant EY_0$. Of course α and/or β might be a.s. zero but they can often be identified in particular cases as we shall observe below.

The comparison of growth rates of X_n and Y_n is always crucial and we shall distinguish the cases (i) $f_n = g_n$ for all n (balanced) and (ii) $f_n \neq g_n$ for some n (unbalanced).

In the balanced case we note that

$$N_n \Big/ \prod_{j=0}^{n-1} (1+f_j) \xrightarrow{\text{a.s.}} \alpha + \beta$$

and on the set $\{\alpha + \beta > 0\}$,

$$Z_n = X_n / N_n \xrightarrow{\text{a.s.}} \alpha/(\alpha+\beta).$$

In the unbalanced case we note that if $\prod_{j=0}^{\infty} [(1+f_j)/(1+g_j)] = 0$ or ∞ a.s. the growth rates of the numbers of the two alleles are intrinsically different and we expect $Z_n \to 0$ or 1 a.s.

Equations (1) with f_n and g_n given by (2) are of the balanced type if and only if $(1+\sigma)EU = EV$ while for (3) the corresponding condition is $\sigma = 0$. In the case of (2) it is clear that $Z_n \to 0$ or 1 can be expected if $(1+\sigma)EU \neq EV$.

The basic approach via (1) and (4) thus requires a minimum of structural assumptions, much less than for the models in the above cited papers on allelic composition of populations. In the next section we shall show how it can usefully complement results already obtained in the literature using the specific models and the more technically complex and rather less transparent approach via the study of $\{Z_n, n \geq 0\}$.

3. Applications

For our first application we shall take the setting of Daley, Hall and Heyde (1982) (henceforth known as DHH) which is a little more general than the original formulation of Heyde (1981, 1982).

Now it is established in DHH that, provided $\min(EU, EV) > 1$,

$$Z_n = X_n / N_n \xrightarrow{\text{a.s.}} Z \tag{5}$$

and

$$N_n \sim nC(Z) \tag{6}$$

as $n \to \infty$ where

$$C(Z) = [(1+\sigma)ZEU + (1-Z)EV - (1+\sigma Z)]/(1+\sigma Z).$$

We note that if $\{h_j, j \geq 0\}$ is a sequence of positive constants with $h_j \sim j^{-1}\theta$ as $j \to \infty$ then

$$\prod_{j=0}^{n-1} (1+h_j) \sim n^{\theta} \tag{7}$$

as $n \to \infty$, since

$$\log \prod_{j=0}^{n-1} (1+h_j) = \sum_{j=0}^{n-1} \log(1+h_j) \sim \sum_{j=0}^{n-1} h_j \sim \theta \sum_{j=1}^{n-1} j^{-1} \sim \theta \log n,$$

the second last step being a consequence of

$$\lim_{n\to\infty}(\log n)^{-1}\sum_{j=1}^{n-1} j^{-1}(jh_j-\theta)=0.$$

Then, from (1), (2), (5), (6) and (7), we have

$$\begin{aligned}\prod_{j=0}^{n-1}(1+f_j)&\sim n^{a(Z)}\\ \prod_{j=0}^{n-1}(1+g_j)&\sim n^{b(Z)}\end{aligned}\tag{8}$$

as $n\to\infty$ where

$$a(Z)=\left[\frac{(1+\sigma)EU}{1+\sigma Z}-1\right]\Big/C(Z)$$

and

$$b(Z)=\left[\frac{EV}{1+\sigma Z}-1\right]\Big/C(Z).$$

It is clear from (8) that the intrinsic rates of increase of X_n and Y_n are different unless $(1+\sigma)EU=EV$.

From (4) and (8),

$$n^{-a(Z)}X_n\xrightarrow{\text{a.s.}}\alpha(Z),\qquad n^{-b(Z)}Y_n\xrightarrow{\text{a.s.}}\beta(Z).\tag{9}$$

However, we already know from (5) and (6) that

$$\begin{aligned}n^{-1}X_n&\xrightarrow{\text{a.s.}}ZC(Z)\\ n^{-1}Y_n&\xrightarrow{\text{a.s.}}(1-Z)C(Z).\end{aligned}\tag{10}$$

Note that

$$a(Z)\leqslant 1\Leftrightarrow(1+\sigma)EU\leqslant EV \text{ or } Z=1$$

$$b(Z)\leqslant 1\Leftrightarrow(1+\sigma)EU\geqslant EV \text{ or } Z=0.$$

Thus, if $(1+\sigma)EU=EV$ we see that (9) and (10) are identical. On the other hand, if $(1+\sigma)EU\neq EV$, then it follows from Theorem 1 of DHH that $Z=0$ or 1 with probability one and we obtain from (9) and (10) that if $EV>(1+\sigma)EU$

$$n^{-((1+\sigma)EU-1)/(EV-1)}X_n\to\alpha(0),\qquad n^{-1}Y_n\to EV-1\quad\text{a.s. on }\{Z=0\}$$

$$n^{-1}X_n\to EU-1,\quad n^{-1}Y_n\to 0\qquad\text{a.s. on }\{Z=1\},$$

while if $EV<(1+\sigma)EU$,

$$\left.\begin{aligned}n^{-1}X_n&\to 0\\ n^{-1}Y_n&\to EV-1\end{aligned}\right.\quad\text{a.s. on }\{Z=0\},$$

$$\left.\begin{aligned}n^{-1}X_n&\to(1+\sigma)EU-1\\ n^{-(EV-(1+\sigma))/((1+\sigma)EU-(1+\sigma))}Y_n&\to\beta(1)\end{aligned}\right.\quad\text{a.s. on }\{Z=1\}.$$

Of course $\alpha(0)$ and $\beta(1)$ could be zero but, nevertheless, differences in growth rates of X_n and Y_n are highlighted.

A particular advantage of the approach via (1) is that it allows us to straightforwardly cope with the possibility of death without replacement in the model. We shall, however, omit the details since the basic results are unchanged. The approach used previously, based on the study of $X_n/(X_n+Y_n)$, raises awkward conditioning problems if extinction is possible.

As a second application we take $N_n = c_n$, $\{c_n\}$ being an increasing sequence of positive constants, in the model treated by Heyde (1977) which gives rise to (3). Then,

$$1+f_n = \frac{(1+\sigma)c_{n+1}}{c_n+\sigma X_n} \sim \frac{(1+\sigma)c_{n+1}}{c_n(1+\sigma Z)} \quad \text{a.s.},$$

$$1+g_n = \frac{c_{n+1}}{c_n+\sigma X_n} \sim \frac{c_{n+1}}{c_n(1+\sigma Z)} \quad \text{a.s.},$$

it being known that $c_n^{-1}X_n \xrightarrow{\text{a.s.}} Z$ as $n\to\infty$. Thus, from (4),

$$\begin{aligned} [(1+\sigma Z)/(1+\sigma)]^n c_0 c_n^{-1} X_n &\xrightarrow{\text{a.s.}} \alpha(Z) \\ (1+\sigma Z)^n c_0 c_n^{-1} Y_n &\xrightarrow{\text{a.s.}} \beta(Z) \end{aligned} \tag{11}$$

and the intrinsic rates of increase of X_n and Y_n are the same if and only if $\sigma=0$.

First suppose $\sigma=0$; we can identify $\alpha(Z)=c_0Z$ and $\beta(Z)=c_0(1-Z)$ and it is known (Seneta (1974)) that $P(0<Z<1)>0$ if $\sum c_n^{-1}<\infty$ while $P(\{Z=0\}\cup\{Z=1\})=1$ if $\sum c_n^{-1}=\infty$.

Now suppose that $\sigma>0$. In this case we have $P(\{Z=0\}\cup\{Z=1\})=1$ (Heyde (1977)). On $\{Z=0\}$ we have $c_n^{-1}X_n\to 0$, $c_n^{-1}Y_n\to 1$, so that identification with (11) gives $\alpha(0)=0$, $\beta(0)=c_0$. Also, on $\{Z=1\}$ we have $c_n^{-1}X_n\to 1$, $c_n^{-1}Y_n\to 0$ so that, from (11), $\alpha(1)=c_0$. It is not clear whether $\beta(1)=0$ in general but if $(1+\sigma)^n c_n^{-1}\to\infty$ then $\beta(1)=0$. This last result is implied by $Y_N=0$ for some sufficiently large (random) N which holds since, if not, then $Y_n\geq 1$ for all n and hence $(1+\sigma)^n c_n^{-1}Y_n\to\infty$ contradicting (11). Nevertheless, the difference in growth rates of X_n and Y_n is highlighted.

References

Daley, D. J., Hall, P., Heyde, C. C.: Further results on the survival of a gene represented in a founder population. J. Math. Biology **14**, 355–363 (1982)

Ewens, W. J.: Mathematical population genetics, Berlin–Heidelberg–New York: Springer 1979

Heyde, C. C.: The effect of selection on genetic balance when the population size is varying. Theoret. Population Biology **11**, 249–251 (1977)

Heyde, C. C.: On the survival of a gene represented in a founder population. J. Math. Biology **12**, 91–99 (1981)

Heyde, C. C.: The effect of differential reproductive rates on the survival of a gene represented in a founder population. In: (Gani, J., Hannan, E. J., eds.) Essays in statistical science. J. Applied Prob. **19A**, 19–25 (1982)

Heyde, C. C.: On limit theorems for gene survival. Colloquia Mathematica János Bolyai (1983) in press

Heyde, C. C., Seneta, E.: The genetic balance between random sampling and random population size. J. Math. Biology **1**, 317–320 (1975)

Seneta, E.: A note on the balance between random sampling and population size. Genetics **77**, 607–610 (1974)

Received February 14/Revised April 19, 1983

Z. Wahrscheinlichkeitstheorie verw. Gebiete
68, 97–106 (1984)

Zeitschrift für
Wahrscheinlichkeitstheorie
und verwandte Gebiete

On the Asymptotic Equivalence of L_p Metrics for Convergence to Normality

C.C. Heyde [1] and T. Nakata [2]

[1] Department of Statistics, University of Melbourne, Parkville, Vic. 3052, Australia
[2] Chukyo University, 101-2 Yagoto Honmachi, Showa-Ku, Nagoya City, Japan

Summary. This paper is concerned with the rate of convergence to zero of the L_p metrics Δ_{np}, $1 \leqq p \leqq \infty$, constructed out of differences between distribution functions, for departure from normality for normed sums of independent and identically distributed random variables with zero mean and unit variance. It is shown that the Δ_{np} are, under broad conditions, asymptotically equivalent in the strong sense that, for $1 \leqq p$, $p' \leqq \infty$, $\Delta_{np'}/\Delta_{np}$ is universally bounded away from zero and infinity as $n \to \infty$.

1. Introduction and Results

Let X_i, $i=1,2,\ldots$ be independent and identically distributed random variables with zero mean and variance unity and write $S_n=\sum_{i=1}^{n} X_i$, $n \geqq 1$, $F_n(x) = P(S_n \leqq x n^{\frac{1}{2}})$ and $\Phi(x)=(2\pi)^{-\frac{1}{2}} \int_{-\infty}^{x} e^{-\frac{1}{2}u^2} du$, $-\infty < x < \infty$. It is well known that the L_p metrics Δ_{np} given by

$$\Delta_{np}=\left(\int_{-\infty}^{\infty} |F_n(x)-\Phi(x)|^p dx\right)^{1/p}, \qquad 1 \leqq p < \infty.$$

$$\Delta_{n\infty}=\sup_x |F_n(x)-\Phi(x)|, \qquad p=\infty,$$

which measure departure from normality, all converge to zero as $n \to \infty$ (e.g. Ibragimov and Linnik [10], Theorem 5.2.1, p. 141) and it is of interest to compare their rates of convergence. Little attempt has been made to obtain a direct comparison although the literature contains considerable circumstantial evidence in favour of the conjecture that they are all asymptotically equivalent in the strong sense that for any $p' \neq p$, $1 \leqq p$, $p' \leqq \infty$, there are *universal* constants C and D such that

$$0 < C < \liminf_{n \to \infty} \Delta_{np'}/\Delta_{np} \leqq \limsup_{n \to \infty} \Delta_{np'}/\Delta_{np} < D < \infty.$$

(The requirement that the constants be univeral is a special feature of our discussion.) In particular, results date back to Esseen [3] from which it can be deduced that asymptotic equivalence holds if $E|X_1|^3<\infty$ and $EX_1^3\neq 0$ or X_1 has a lattice distribution. More recent circumstantial evidence is of the kind: for $0\leqq\delta<1$, $\sum n^{-1+\delta/2}\Delta_{np}<\infty$ if and only if $EX_1^2\log(1+|X_1|)<\infty$ $(\delta=0)$, $E|X_1|^{2+\delta}<\infty$ $(0<\delta<1)$ (Heyde [7]) and, for $0<\delta<1$, $\Delta_{np}=O(n^{-\delta/2})$ if and only if $EX_1^2<\infty$ and $\int_{|x|>z} x^2 dP(X_1\leqq x)=O(z^{-\delta})$ as $z\to\infty$ (Ibragimov [8]) and further refinements appear in, for example, Rozovskii [12]. In this paper we will show that the metrics are asymptotically equivalent under a very broad range of conditions.

The question of asymptotic equivalence is treated in the following theorem. Here $a_n \asymp b_b$ for sequences of positive numbers $\{a_n\}$, $\{b_n\}$ means that

$$0<\liminf_{n\to\infty} a_n/b_n \leqq \limsup_{n\to\infty} a_n/b_n<\infty.$$

Theorem 1. *All the metrics Δ_{np}, $1\leqq p\leqq\infty$, have the same asymptotic behaviour if any of the following four conditions are satisfied:*

(i) $x^3P(|X_1|>x)\to\infty$ *as* $x\to\infty$,

(ii) $E|X_1|^3<\infty$ *and* $EX_1^3\neq 0$,

(iii) X_1 *has a lattice distribution and* $E|X_1|^3<\infty$,

(iv) $\limsup_{|t|\to\infty}|Ee^{itX_1}|<1$ *and* $E|X_1|^r<\infty$, $EX_1^r\neq\int_{-\infty}^{\infty} x^r d\Phi(x)$

for some integer $r\geqq 3$.

In the case of (i), $\Delta_{np}\asymp\delta_n$ *as* $n\to\infty$ *where*

$$\delta_n=EX_1^2I(|X_1|>n^{\frac{1}{2}})+n^{-1}EX_1^4I(|X_1|\leqq n^{\frac{1}{2}})+n^{-\frac{1}{2}}|EX_1^3I(|X_1|\leqq n^{\frac{1}{2}})|. \tag{1}$$

For (ii) *and* (iii), $\Delta_{np}\asymp n^{-\frac{1}{2}}$, *and for* (iv), $\Delta_{np}\asymp n^{-(r-2)/2}$ *as* $n\to\infty$.

On the way to Theorem 1 we shall, however, obtain the following result of independent interest. Note particularly that there are universal constants in the upper and lower bounds.

Theorem 2. *Let* Y_i, $i=1,2,\ldots,n$ *be independent random variables with* $EY_i=0$, $EY_i^2=\sigma_i^2<\infty$ *and* $\sum_{i=1}^n\sigma_i^2=1$. *Write*

$$\Delta_{np}=\left(\int_{-\infty}^{\infty}\left|P\left(\sum_{i=1}^n Y_i\leqq x\right)-\Phi(x)\right|^p dx\right)^{1/p}, \qquad 1\leqq p<\infty,$$

$$\Delta_{n\infty}=\sup_x\left|P\left(\sum_{i=1}^n Y_i\leqq x\right)-\Phi(x)\right|, \qquad p=\infty.$$

Then, there exists a universal constant $C>0$ *such that*

$$\delta_n\leqq C\left(\Delta_{np}+\sum_{i=1}^n\sigma_i^4\right) \tag{2}$$

where

$$\delta_n = \sum_{i=1}^{n} EY_i^2 I(|Y_i|>1) + \sum_{i=1}^{n} EY_i^4 I(|Y_i|\leqq 1) + \left|\sum_{i=1}^{n} EY_i^3 I(|Y_i|\leqq 1)\right|.$$

On the other hand, if ε_n *is a positive constant then there exists a universal constant* $D>0$ *such that whenever*

$$\sum_{i=1}^{n} EY_i^2 I(|Y_i|>\varepsilon_n) \leqq \tfrac{1}{8},$$

$$\Delta_{np} \leqq D\left(\delta_n + \varepsilon_n + \sum_{i=1}^{n} \sigma_i^4\right). \tag{3}$$

Let $\mathscr{S}$ denote the real line or a subset thereof and write

$$\Delta_{np}(\mathscr{S}) = \left(\int_{\mathscr{S}} |F_n(x) - \Phi(x)|^p dx\right)^{1/p}, \qquad 1 \leqq p < \infty,$$

$$\Delta_{n\infty}(\mathscr{S}) = \sup_{x\in\mathscr{S}} |F_n(x) - \Phi(x)|.$$

Asymptotically equivalent behaviour of the L_p metrics appears to be closely linked to the phenomenon of intervals of the form $(-A, A)$, $A>0$, being convergence determining sets. The subset $\mathscr{S}$ of the real line may be called *convergence determining* for the L_p metrics if $(1\leqq)\ \Delta_{np}/\Delta_{np}(\mathscr{S})$ is universally bounded away from infinity as $n\to\infty$, $1\leqq p\leqq\infty$. A number of results concerning these sets are given in Hall [5], in which a rather different definition is adopted. Here we shall give a modification of one of the results of [5] to indicate the possibilities.

Theorem 3. *Suppose* $x^3 P(|X_1|>x)\to\infty$ *as* $x\to\infty$. *Then, any set* $\mathscr{S}=(-A, A)$, $0<A<\frac{1}{2}$, *is convergence determining for the* L_p *metrics,* $1\leqq p\leqq\infty$. *We have* $\Delta_{np}((-A, A))\uparrow$ *as* $p\uparrow$ *and* $\Delta_{np}((-A, A))\asymp\delta_n$ *as* $n\to\infty$ *where* δ_n *is given by* (1).

2. Proofs

As a preliminary to the proofs of the theorems we shall establish the following Lemma which is also of some independent interest.

Lemma. *Let* $\mathscr{S}$ *denote the real line or a subset thereof and for a function* $u\in L_p(\mathscr{S})$, $p\geqq 1$, *we write*

$$N_p(u;\mathscr{S}) = \left(\int_{\mathscr{S}} |u(x)|^p dx\right)^{1/p}, \qquad 1\leqq p<\infty,$$

$$N_\infty(u;\mathscr{S}) = \sup_{x\in\mathscr{S}} |u(x)|.$$

Then, $p\log N_p(u;\mathscr{S})$ *is a convex function of* p.

This result holds in particular for $u(x)=|F_n(x)-\Phi(x)|$ *and if* $\mathscr{S}=(-A, A)$, $A<\frac{1}{2}$, *then* $\Delta_{np}(A)=\Delta_{np}((-A, A))\uparrow$ *as* $p\uparrow$.

Proof. Let $1 \leqq p' < p$. Then, Schwarz' inequality gives

$$\int_{\mathscr{S}} |u(x)|^p dx \leqq \left(\int_{\mathscr{S}} |u(x)|^{p+p'} dx \int_{\mathscr{S}} |u(x)|^{p-p'} dx\right)^{\frac{1}{2}}$$

and, upon taking logarithms of both sides,

$$\log N_p^p(u; \mathscr{S}) \leqq \tfrac{1}{2} \log N_{p+p'}^{p+p'}(u; \mathscr{S}) + \tfrac{1}{2} \log N_{p-p'}^{p-p'}(u; \mathscr{S})$$

which gives the convexity of $p \log N_p(u; \mathscr{S})$.

For n sufficiently large, $\log \Delta_{np}(\mathscr{S}) < 0$, while $p \log \Delta_{np}(\mathscr{S}) \to -\infty$ as $p \to \infty$. In the case $\mathscr{S} = (-A, A)$, $0 < A < \frac{1}{2}$, $\Delta_{np}(A) = \Delta_{np}((-A, A))$ can be defined for all $p > 0$ and

$$p \log \Delta_{np}(A) = \log \int_{-A}^{A} |F_n(x) - \Phi(x)|^p dx \to \log 2A < 0$$

as $p \to 0$. Then convexity ensures that the slope $\log \Delta_{np}(A)$ of the line joining the origin to $(p, p \log \Delta_{np}(A))$ is nondecreasing in p. This gives $\Delta_{np}(A) \uparrow$ as $p \uparrow$ and the proof of the Lemma is complete.

Proof of Theorem 2. The result (2) can be obtained by modifying the proof of the theorem of Hall and Barbour [6] which deals with the case $p = \infty$. The modifications required to deal with $1 \leqq p < \infty$ are first to use

$$\begin{aligned} &\left| \int_{-\infty}^{\infty} g(w)\{P(W \leqq w) - \Phi(w)\} dw \right| \\ &\qquad \leqq \left(\int_{-\infty}^{\infty} |g(w)|^q dw \right)^{1/q} \Delta_{np}, \qquad q = (1 - p^{-1})^{-1},\ 1 < p < \infty \\ &\qquad \leqq \sup_w |g(w)| \Delta_{n1}, \qquad p = 1, \end{aligned} \tag{4}$$

with $g(w) = h'(w)$ instead of (2) of [6] and with $g(w) = f''(w + \beta)$ in the bounding of $|t_{2i}|$ of [6]. Then note that

$$\left(\int_{-\infty}^{\infty} |g(w)|^q dw \right)^{1/q} \leqq (\sup_w |g(w)|)^{1/p} \left(\int_{-\infty}^{\infty} |g(w)| dw \right)^{1/q} \tag{5}$$

while simple calculations on the functions g used in the proof of [6] reveal that in each case

$$\sup_w |g(w)| \leqq \int_{-\infty}^{\infty} |g(w)| dw \tag{6}$$

so that, using (5) and (6) in (4),

$$\left| \int_{-\infty}^{\infty} g(w)\{P(W \leqq w) - \Phi(w)\} dw \right| \leqq \Delta_{np} \int_{-\infty}^{\infty} |g(w)| dw, \qquad 1 \leqq p \leqq \infty.$$

The result (2) then follows via the proof of [6] with the same universal constant C as obtained by Hall and Barbour for the case $p = \infty$.

The result (3), on the other hand, is a consequence of Theorem 2.2 (p. 25) and Theorem 2.3 (p. 44) of Hall [5] and this completes the proof of the theorem.

Proof of Theorem 1. Applying the results of Theorem 2 to the case of identically distributed random variables, we can put $Y_i = X_i/n^{\frac{1}{2}}$ so that $\sum_{i=1}^{n} \sigma_i^4 = n^{-1}$ and $\varepsilon_n = \lambda n^{-\frac{1}{4}}$ where

$$\lambda = \min[\mu\colon EX_1^2 I(|X_1| > \mu) \leqq \tfrac{1}{8}].$$

When $x^3 P(|X_1| > x) \to \infty$ as $x \to \infty$ we have from (2) that $n^{\frac{1}{2}} \Delta_{np} \to \infty$ as $n \to \infty$ since

$$n^{\frac{1}{2}} \delta_n \geqq n^{\frac{1}{2}} EX_1^2 I(|X_1| > n^{\frac{1}{2}}) \geqq n^{\frac{3}{2}} P(|X_1| > n^{\frac{1}{2}}) \to \infty$$

as $n \to \infty$. The results (2) and (3) then give asymptotic equivalence of the L_p metrics and $\Delta_{np} \asymp \delta_n$ as $n \to \infty$. This completes the proof for part (i) of the theorem.

Next suppose that $E|X_1|^3 < \infty$. We have from results due to Esseen [2], [3] (e.g. Ibragimov and Linnik [10], Theorem 5.3.2, p. 147 and Theorem 5.3.3, p. 150) that for $1 \leqq p \leqq \infty$,

$$\Delta_{np} = A_p |EX_1^3| n^{-\frac{1}{2}} + o(n^{-\frac{1}{2}})$$

where

$$A_\infty = 1/6(2\pi)^{\frac{1}{2}}$$

$$A_p = A_\infty \left(\int_{-\infty}^{\infty} |1 - x^2|^p e^{-\frac{1}{2}px^2} dx \right)^{1/p}, \quad 1 \leqq p < \infty,$$

if X_1 does not have a lattice distribution, while

$$\Delta_{np} = M_p n^{-\frac{1}{2}} + o(n^{-\frac{1}{2}})$$

where

$$M_\infty = (2\pi)^{-\frac{1}{2}} (\tfrac{1}{6} |EX_1^3| + \tfrac{1}{2} h)$$

$$M_p = (2\pi)^{-\frac{1}{2}} \left(\int_{-\infty}^{\infty} |Q_1(x) + \tfrac{1}{6} EX_1^3 (1 - x^2)|^p e^{-\frac{1}{2}px^2} dx \right)^{1/p}, \quad 1 \leqq p < \infty \tag{7}$$

with

$$Q_1(x) = hQ(h^{-1} n^{\frac{1}{2}} (x - \xi_n))$$

and

$$Q(x) = [x] - x + \tfrac{1}{2},$$

$[x]$ denoting the integer part of x and

$$\xi_n = h n^{-\frac{1}{2}} (n x_0 h^{-1} - [n x_0 h^{-1}]),$$

the lattice distribution being situated on the points $x_0 + vh$, $v = 0, \pm 1, \pm 2, \ldots$.

To obtain the result for part (ii) in the non-lattice case we first note that

$$\lim_{n \to \infty} \Delta_{np'}/\Delta_{np} = A_{p'}/A_p.$$

Furthermore, it is a straightforward exercise to show that $\Lambda_1=4e^{-\frac{1}{2}}\Lambda_\infty$ and $\Lambda_p\downarrow\Lambda_\infty$ as $p\uparrow\infty$. Thus, taking $p'>p$ without loss of generality we have

$$\tfrac{1}{4}e^{\frac{1}{2}}\leqq\lim_{n\to\infty}\Delta_{np'}/\Delta_{np}\leqq 1.$$

To deal with parts (ii) (in the lattice case) and (iii) of the theorem we need upper and lower bounds on M_p in (7). Using Minkowski's inequality we have

$$\begin{aligned}&\left(\int_{-\infty}^{\infty}|Q_1(x)+\tfrac{1}{6}EX_1^3(1-x^2)|^p e^{-\frac{1}{2}px^2}dx\right)^{1/p}\\&\quad\leqq\left(\int_{-\infty}^{\infty}|Q_1(x)|^p e^{-\frac{1}{2}px^2}dx\right)^{1/p}+ch\left(\int_{-\infty}^{\infty}|1-x^2|^p e^{-\frac{1}{2}px^2}dx\right)^{1/p}\end{aligned}\tag{8}$$

where $hc=\frac{1}{6}|EX_1^3|$. But, as noted previously, for $p\geqq 1$,

$$\left(\int_{-\infty}^{\infty}|1-x^2|^p e^{-\frac{1}{2}px^2}dx\right)^{1/p}\downarrow\quad\text{as } p\uparrow$$

while, since $|Q_1(x)|\leqq\frac{1}{2}h$,

$$\begin{aligned}&\left(\int_{-\infty}^{\infty}|Q_1(x)|^p e^{-\frac{1}{2}px^2}dx\right)^{1/p}\leqq\tfrac{1}{2}h\left(\int_{-\infty}^{\infty}e^{-\frac{1}{2}px^2}dx\right)^{1/p}\\&\quad=\tfrac{1}{2}h(2\pi/p)^{1/2p}\leqq\tfrac{1}{2}h(2\pi)^{\frac{1}{2}},\end{aligned}$$

and hence (8) gives

$$\begin{aligned}M_p&\leqq\tfrac{1}{2}h+hc(2\pi)^{-\frac{1}{2}}\int_{-\infty}^{\infty}|1-x^2|e^{-\frac{1}{2}x^2}dx\\&=\tfrac{1}{2}h+4hc(2\pi e)^{-\frac{1}{2}}\end{aligned}\tag{9}$$

after a straightforward calculation.

To obtain a lower bound on M_p we first note that

$$\begin{aligned}&\left(\int_{-\infty}^{\infty}|Q_1(x)+\tfrac{1}{6}EX_1^3(1-x^2)|^p e^{-\frac{1}{2}px^2}dx\right)^{1/p}\\&\quad\geqq\left(\int_{\xi_n}^{\frac{1}{2}+\xi_n}|hQ(h^{-1}n^{\frac{1}{2}}(x-\xi_n))+\tfrac{1}{6}EX_1^3(1-x^2)|^p e^{-\frac{1}{2}px^2}dx\right)^{1/p}.\end{aligned}$$

Now suppose for definiteness that $EX_1^3\geqq 0$; the argument for $EX_1^3\leqq 0$ is similar. We also note that $|\xi_n|\leqq h/n^{\frac{1}{2}}$ and we choose n so large that $h/n^{\frac{1}{2}}<\frac{1}{4}$. Then, we have

$$
\begin{aligned}
(2\pi)^{\frac{1}{2}} M_p &\geqq h\left(\int_0^{\frac{1}{2}} |Q(h^{-1}n^{\frac{1}{2}}x)+c(1-(x+\xi_n)^2)|^p e^{-9p/32}\,dx\right)^{1/p} \\
&\geqq h e^{-9/32}\left(\sum_{i=0}^{[\frac{1}{2}n^{\frac{1}{2}}h^{-1}]-1} \int_{hin^{-\frac{1}{2}}}^{h(i+\frac{1}{2})n^{-\frac{1}{2}}} |Q(h^{-1}n^{\frac{1}{2}}x)+\tfrac{7}{16}c|^p\,dx\right)^{1/p} \\
&= h e^{-9/32}\left(\sum_{i=0}^{[\frac{1}{2}n^{\frac{1}{2}}h^{-1}]-1} \int_{hin^{-\frac{1}{2}}}^{h(i+\frac{1}{2})n^{-\frac{1}{2}}} ([h^{-1}n^{\frac{1}{2}}x]-h^{-1}n^{\frac{1}{2}}x+\tfrac{1}{2}+\tfrac{7}{16}c)^p\,dx\right)^{1/p} \\
&= h e^{-9/32}\left(\sum_{i=0}^{[\frac{1}{2}n^{\frac{1}{2}}h^{-1}]-1} hn^{-\frac{1}{2}}\int_0^{\frac{1}{2}} (\tfrac{1}{2}-y+\tfrac{7}{16}c)^p\,dx\right)^{1/p} \\
&= h e^{-9/32}\left(\int_0^{\frac{1}{2}} (\tfrac{1}{2}-y+\tfrac{7}{16}c)^p\,dx\right)^{1/p} (hn^{-\frac{1}{2}}[\tfrac{1}{2}h^{-1}n^{\frac{1}{2}}])^{1/p} \\
&\geqq \tfrac{1}{4} h e^{-9/32}(p+1)^{-1/p}((\tfrac{1}{2}+\tfrac{7}{16}c)^{p+1}-(\tfrac{7}{16}c)^{p+1})^{1/p} \\
&\geqq \tfrac{1}{4} h e^{-9/32}(p+1)^{-1/p}2^{-1/p}(\tfrac{1}{2}+\tfrac{7}{16}c) \\
&\geqq \tfrac{1}{16} h e^{-9/32}(\tfrac{1}{2}+\tfrac{7}{16}c). \qquad (10)
\end{aligned}
$$

From (9) and (10) we obtain

$$R^{-1} \leqq \liminf_{n\to\infty} \Delta_{np'}/\Delta_{np} \leqq \limsup_{n\to\infty} \Delta_{np'}/\Delta_{np} \leqq R$$

where

$$R = 16 e^{9/32}(2\pi)^{\frac{1}{2}}\left(\tfrac{1}{2}+\frac{4c}{(2\pi e)^{\frac{1}{2}}}\right)\Big/(\tfrac{1}{2}+\tfrac{7}{16}c).$$

However, note that

$$1 \leqq \left(\tfrac{1}{2}+\frac{4c}{(2\pi c)^{\frac{1}{2}}}\right)\Big/(\tfrac{1}{2}+\tfrac{7}{16}c) \leqq 64/(7(2\pi e)^{\frac{1}{2}})$$

since $4/(2\pi e)^{\frac{1}{2}} > 7/16$. Consequently, whatever the value of c,

$$7(2^{-10})e^{7/32} \leqq \liminf_{n\to\infty} \Delta_{np'}/\Delta_{np} \leqq \limsup_{n\to\infty} \Delta_{np'}/\Delta_{np} \leqq \tfrac{1}{7}2^{10}e^{-7/32}$$

and the proof of part (ii) in the lattice case and part (iii) is complete.

Finally, suppose that $\limsup_{|t|\to\infty} |Ee^{itX_1}| < 1$ (often known as Cramér's condition (C)). We may, in view of the foregoing discussion, confine attention to the case $E|X_1|^3 < \infty$, $EX_1^3 = 0$. Then, if X_1 does not have the unit normal distribution there is a first (integer) moment $\alpha_{k+2} = EX_1^{k+2}$, $k \geqq 2$, which does not match that of the unit normal law. This holds since the normal law is characterized by its moments. By assumption $E|X_1|^{k+2} < \infty$.

Now we can approximate $F_n(x)$ by $\Phi(x) + R_{nk}(x)$ where

$$R_{nk}(x) = (2\pi)^{-\frac{1}{2}} e^{-\frac{1}{2}x^2} \sum_{s=1}^{k} P_s(x) n^{-\frac{1}{2}s}$$

is a portion of the Čebyšev-Cramér series corresponding to X_1 (see for example Gnedenko and Kolmogorov [4], Chap. 8), the $P_i(x)$ being polynomials of degree $3j-1$ whose coefficients depend on the first $j+2$ moments of X_1. Furthermore, $P_j(x)\equiv 0$ if the first $j+2$ moments of X_1 match those of the unit normal law.

Now use $\|u(x)\|_p$, $1\leqq p\leqq\infty$, to denote the norm

$$\|u(x)\|_p=\left(\int_{-\infty}^{\infty}|u_n(x)|^p dx\right)^{1/p}, \qquad 1\leqq p<\infty,$$
$$=\sup_x |u_u(x)|, \qquad p=\infty.$$

Then,

$$\Delta_{np}=\|F_n(x)-\Phi(x)\|_p\leqq\|R_{nk}(x)\|_p+\|F_n(x)-\Phi(x)-R_{nk}(x)\|_p \tag{11}$$

and

$$\|F_n(x)-\Phi(x)\|_p\geqq\|R_{nk}(x)\|_p-\|F_n(x)-\Phi(x)-R_{nk}(x)\|_p. \tag{12}$$

But, it follows from Theorem 1 of Erickson [1] for the case $1\leqq p<\infty$ and Theorem 1 of Ibragimov [9] for the case $p=\infty$ that

$$\|F_n(x)-\Phi(x)-R_{nk}(x)\|_p=o(n^{-k/2}) \tag{13}$$

as $n\to\infty$. On the other hand, since the moments EX_1^j, $1\leqq j\leqq k+1$, match those of the unit normal distribution, $P_j(x)\equiv 0$ for $1\leqq j\leqq k-1$ and

$$\|R_{nk}(x)\|_p=n^{-k/2}\,\|(2\pi)^{-\frac{1}{2}}e^{-\frac{1}{2}x^2}P_k(x)\|_p,$$

so that, using (11), (12) and (13),

$$\Delta_{np}\sim\|(2\pi)^{-\frac{1}{2}}e^{-\frac{1}{2}x^2}P_k(x)\|_p n^{-k/2}$$

as $n\to\infty$. But, the condition that the moments EX_1^j, $1\leqq j\leqq k+1$, match those of the unit normal law gives

$$P_k(x)=-(EX_1^{k+2}/(k+2)!)\,H_{k+1}(x)$$

where $H_j(x)$ denotes the jth Hermite-Chebyshev polynomial,

$$H_j(x)=(-1)^j e^{\frac{1}{2}x^2}\frac{d^j}{dx^j}e^{-\frac{1}{2}x^2}.$$

Thus,

$$\Delta_{np}\sim(2\pi)^{-\frac{1}{2}}EX_1^{k+2}((k+2)!)^{-1}\,\|e^{-\frac{1}{2}x^2}H_{k+1}(x)\|_p n^{-k/2}$$

and for $p'\neq p$,

$$\Delta_{np'}/\Delta_{np}\to\|e^{-\frac{1}{2}x^2}H_{k+1}(x)\|_{p'}/\|e^{-\frac{1}{2}x^2}H_{k+1}(x)\|_p$$

as $n\to\infty$. However, in view of the Lemma, this limit may be bounded between the universal constants U^{-1} and U (which may depend on k) where

$$U=\max_p\|e^{-\frac{1}{2}x^2}H_{k+1}(x)\|_p/\min_p\|e^{-\frac{1}{2}x^2}H_{k+1}(x)\|_p$$

and this completes the proof.

Proof of Theorem 3. When $x^3 P(|X_1|>x)\to\infty$ as $x\to\infty$, we have from Theorem 2 that there exist universal constants C_1 and C_2 such that

$$0<C_1\leqq\liminf_{n\to\infty}\delta_n^{-1}\Delta_{np}\leqq\limsup_{n\to\infty}\delta_n^{-1}\Delta_{np}\leqq C_2<\infty, \tag{14}$$

and hence the result of the theorem follows (with the aid of the Lemma) if there are universal constants C_3 and C_4 for which

$$0<C_3\leqq\liminf_{n\to\infty}\delta_n^{-1}\Delta_{np}(A)\leqq\limsup_{n\to\infty}\delta_n^{-1}\Delta_{np}(A)\leqq C_4<\infty \tag{15}$$

where $\Delta_{np}(A)=\Delta_{np}((-A,A))$. However,

$$\limsup_{n\to\infty}\delta_n^{-1}\Delta_{np}(A)\leqq\limsup_{n\to\infty}\delta_n^{-1}\Delta_{np}\leqq C_2<\infty,$$

while, from the Lemma,

$$\liminf_{n\to\infty}\delta_n^{-1}\Delta_{np}(A)\geqq\liminf_{n\to\infty}\delta_n^{-1}\Delta_{n1}(A),$$

and hence (15) is satisfied if $\delta_n^{-1}\Delta_{n1}(A)$ is universally bounded below (away from zero) as $n\to\infty$. This result, however, follows from a minor modification of the proof of Theorem 1.3 of [5] wherein it is necessary to replace (1.14) by

$$\left|\int_{-\infty}^{\infty}\{\alpha^n(t/n^{\frac{1}{2}})-e^{-\frac{1}{2}t}\}t^{-2}(1-\cos At)\,dt\right|\leqq\pi\Delta_{n1}(A)$$

and to observe that each constant being used can be made universal.

3. Concluding Remarks

It is clearly a plausible conjecture that all the metrics Δ_{np}, $1\leqq p\leqq\infty$, have the same asymptotic behaviour in general under $EX_1=0$, $EX_1^2=1$ without the constraints imposed in Theorem 1. The principal difficulty in establishing this result is in dealing with the class of discrete non-lattice distributions. If $\limsup_{|t|\to\infty}|Ee^{itX_1}|=1$ and $E|X_1|^r<\infty$ for some integer $r>3$ it is necessary to augment the Chebyshev-Cramér approximation to $F_n(x)$ with discontinuous terms and these are crucial in determining the asymptotic behaviour of Δ_{np}. In the non-lattice case there appears to be no general and systematic way of accomplishing this aim.

The other difficulty in extending Theorem 1 is in coping with cases such as $\delta_n\asymp n^{-\frac{1}{2}}$ or $n^{-\frac{1}{2}}\delta_n\to 0$ as $n\to\infty$ but $E|X_1|^3=\infty$. Here the problems are mostly those of establishing universality of bounds for asymptotic equivalence. Non-universal bounds on the ratios are mostly obtainable.

Finally, a small improvement to the result of Theorem 1 may be achievable, by using the ideas of Osipov [11], to extend the conditions (iii) by removing the requirement that $E|X_1|^3<\infty$ and the condition (iv) to include the case where $EX_1^k=\int_{-\infty}^{\infty}x^k\,d\Phi(x)$, $k=1,2,\dots,r-1$, $E|X_1|^r=\infty$.

Acknowledgement. This paper has benefited from a particularly careful refereeing.

References

1. Erickson, R.V.: On L_p Čebyšev-Cramér asymptotic expansions. Ann. Probability **1**, 355–361 (1973)
2. Esseen, C.-G.: Fourier analysis of distribution functions. Acta Math. **77**, 1-125 (1945)
3. Esseen, C.-G.: On mean central limit theorems. Trans. Roy. Inst. Tech., Stockholm No. 121 (1958)
4. Gnedenko, B.V., Kolmogorov, A.N.: Limit distributions for sums of independent random variables. Reading, Mass.: Addison-Wesley 1954
5. Hall, P.: Rates of convergence in the central limit theorem. London: Pitman 1982
6. Hall, P., Barbour, A.D.: Reversing the Berry-Esseen inequality. Submitted for publication in Proc. Amer. Math. Soc. (1983)
7. Heyde, C.C.: A nonuniform bound on convergence to normality. Ann. Probability **3**, 903-907 (1975)
8. Ibragimov, I.A.: On the accuracy of the Gaussian approximation to the distribution functions of sums of independent random variables. Theor. Probability Appl. **11**, 559-579 (1966)
9. Ibragimov, I.A.: On the Chebyshev-Cramér asymptotic expansions. Theor. Probability Appl. **12**, 455-469 (1967)
10. Ibragimov, I.A., Linnik, Yu.V.: Independent and stationary sequences of random variables. Groningen: Wolters-Noordhoff 1971
11. Osipov, L.V.: On asymptotic expansions for the distributions of sums of independent random variables. Theor. Probability Appl. **16**, 333-343 (1971)
12. Rozovskii, L.V.: On the precision of an estimate of the remainder term in the central limit theorem. Theor. Probability Appl. **23**, 712-730 (1978)

Received March 29, 1983; in revised form March 9, 1984

International Statistical Review (1987), **55**, 3, pp. 231–244. Printed in Great Britain

Correspondent paper

Quasi-likelihood and Optimal Estimation

V.P. Godambe[1] and C.C. Heyde[2]

[1] *Department of Statistics and Actuarial Science, University of Waterloo, Waterloo, Ontario, Canada N2L 3G1.* [2] *Department of Statistics, IAS, Australian National University, GPO Box 4, Canberra, ACT 2601, Australia.*

Summary

Within the framework of estimating function theory, this paper provides a very general definition of quasi-likelihood estimating equations. Applications to stochastic processes are discussed. This work extends the previous results of Godambe (1985) and Hutton & Nelson (1986).

Key words: Asymptotic theory; Estimating functions; Generalized least-squares; Martingales; Maximum likelihood; Optimal estimators; Quasi-likelihood; Quasi-score function.

1 Historical introduction

Historically there are two distinct and sharply contrasting traditions in relation to estimation problems arising out of what are now known as classical linear models: (I) Gauss's relying exclusively on finite sample properties, and (II) Laplace's based on asymptotics.

Gauss in 1821, 1823 (Gauss, 1880) founded and substantially developed the methodology of what is usually called the Gauss–Markov theorem. Using only the 'first two moments' of the underlying distribution, the theorem, in terms of the modern concept of unbiasedness, asserts that in the class of linear unbiased estimates the variance is minimized for the 'least-squares estimate'. In this sense the least-squares estimates are *optimal.* This optimality is in relation to *finite samples.* Gauss did not investigate asymptotic properties of the least-squares estimates. However, he demonstrated (and emphasized) that in the location family of distributions *uniquely* for 'normal distributions', the 'maximum likelihood estimates' (using the present day terminology) *coincide* with the 'least-squares estimates', implying some kind of justification for the normal models. To clarify the point, we emphasize that the optimality of the least-squares estimate mentioned above depends on assumptions concerning the first two moments of the distribution but is otherwise *independent* of the distribution or distributional form. On the other hand, the maximum likelihood estimate generally depends on the entire form of the distribution. The observation of Gauss, that the methods of maximum likelihood and least-squares give identical estimates, if the distribution is normal, contain the germ of the modern theory of 'quasi-likelihood'. This will be clear from the subsequent discussion.

Gauss in his astronomical calculations apparently needed 'estimates' as numbers (point estimates) rather than 'intervals'. This is also suggested by the fact that he does *not* write of unbiased estimates but refers to estimates which equal 'true' values when the observations are without errors (Bertrand, 1889; Sprott, 1983).

Laplace, on the other hand, provided 'asymptotics'. He proved that in a certain class of symmetric confidence intervals (to use the terminology common now), the one based on the least-squares estimate is asymptotically shortest. For an account of these historical contributions see Heyde & Seneta (1977, Ch. 4).

In the following we, rather informally, state the Gauss–Markov theorem in the framework of estimating function theory. This will help establish the connection of the above-mentioned historical results with what, in this paper, we define as the *quasi-score function* and the *quasi-likelihood*.

Let $y_1, \ldots, y_n$ be independent real random variables jointly distributed with cumulative distribution function F where $F \in \mathscr{F}$. On $\mathscr{F}$ are defined real parameters θ and σ^2 such that for $i = 1, \ldots, n$

$$E_F(y_i - \theta(F)) = 0, \quad E_F(y_i - \theta(F))^2 = \sigma^2(F) \quad (F \in \mathscr{F}),$$

where E_F denotes expectation under F. Now, for any specified numbers α_i $(i = 1, \ldots, n)$, we call

$$g = \sum_{i=1}^{n} (y_i - \theta)\alpha_i \tag{1}$$

a *linear unbiased estimating function*; 'linear' for g is 'linear in $(y_i - \theta)$' $(i = 1, \ldots, n)$, 'unbiased' for $E_F(g) = E_F\{\sum (y_i - \theta(F))\alpha_i\} = 0$, where the sum is over $i = 1, \ldots, n$, $F \in \mathscr{F}$, and 'estimating function' for the equation $g = 0$, when solved for θ, provides an estimate $\hat{\theta}$ for θ. Let $\mathscr{G}$ be the class of all linear unbiased estimating functions subject to the condition $\sum \alpha_i = c$, a constant, where the sum is over $i = 1, \ldots, n$. Now, in this context, an estimating function g^* is said to be *optimal* in $\mathscr{G}$ if $g^* \in \mathscr{G}$ and if, for all $g \in \mathscr{G}$,

$$E_F\{(g^*)^2\} \leqslant E_F(g^2) \quad (F \in \mathscr{F}). \tag{2}$$

Here this means that the unbiased estimator obtained from g^* has minimal variance among the estimators obtained from estimating functions in $\mathscr{G}$. It is easy to see that *up to a constant multiple* the estimating function g^* in $\mathscr{G}$ satisfying (2) is given by

$$g^* = \bar{y} - \theta, \tag{3}$$

where $\bar{y} = \sum y_i/n$, with the sum over $i = 1, \ldots, n$.

Now suppose that the class of underlying distributions $\mathscr{F} = \{F\}$ is such that it can be written as

$$\mathscr{F} = \{f\} \times \{\theta\}, \tag{4}$$

where 'f' denotes the 'form' of the distribution, and 'θ' the 'parameter'. That is, $\mathscr{F}$ is a 'union' of 'families' of parametric distributions, each family being indexed by the 'same' parameter. In other words, in (4) for each 'f' the 'parameter range' is the same, namely $\{\theta\}$. This covers examples like models with a nuisance parameter or semiparametric models. More conventionally, we may write

$$dF = f_\theta \, d\mu, \tag{5}$$

where f_θ is the density with respect to the measure μ on $\mathbb{R}^n$.

Now we assume, in relation to (5), the existence of the *score function*, namely $\partial \log f_\theta / \partial\theta$, for all θ and f in (4). It can be shown that the 'optimality' of the estimating function g^*, that is g^* in $\mathscr{G}$ satisfying the property (2), is *equivalent* to each of:

(i) $E_F(g^* - \partial \log f_\theta/\partial\theta)^2 \leqslant E_F(g - \partial \log f_\theta/\partial\theta)^2$,
(ii) $\text{corr}\,(g^*, \partial \log f_\theta/\partial\theta) \geqslant \text{corr}\,(g, \partial \log f_\theta/\partial\theta)$,

for all $g \in \mathscr{G}$ and $F \in \mathscr{F}$, where corr denotes correlation (Godambe, 1985; Godambe &

Thompson, 1985). Here $\partial \log f_\theta / \partial \theta$ is not known, in the sense that it is not known which element of $\{f\}$ is the true one. Further, as an implication of a result proved in this paper, under suitable conditions,

(iii) for large samples the confidence interval for θ associated with g^* is smaller or equal to that associated with g for all $g \in \mathcal{G}$ and $F \in \mathcal{F}$.

Now, because of the properties (i), (ii) and (iii) above of the estimating function g^*, and noting that (iii) is a characteristic property of the score function in a parametric family of distributions (Wilks, 1938), we *define* initially,

$$\text{'}\textit{the quasi-score function}\text{'} \equiv \text{'}g^*\text{'}, \tag{6}$$

(noting that g^* is defined only up to a constant multiple). The *quasi-likelihood* is obtained by the appropriate integration of (6).

Gauss's justification for the 'normal model', discussed above, would seem to be fairly in line with our definition of 'quasi-likelihood' if we put, as in (3), $g^* = \bar{y} - \theta$ in (6).

In the following sections, the above concepts of quasi-score function and quasi-likelihood are cast in the general setting of stochastic processes. Within this general framework, it is shown that the confidence intervals obtained by the inversion of the appropriately standardized optimal estimating function (i.e. quasi-score function) are asymptotically smallest, generalizing the corresponding result of Laplace, in relation to the least-squares estimator, mentioned above.

The present-day usage of the term 'quasi-likelihood' seems to have been initiated by Wedderburn (1974). Deeply intuitively, he observed that the expression $\bar{y} - \theta$ in (3), or a similar expression in a more general setting, possesses mathematical properties of a 'score function'. The statistical significance of this mathematical fact is indicated by the term 'quasi-likelihood'. The underlying idea comes from the scalar linear regression model with independent error term, whose distribution is of exponential family type. Here, as noted by Bradley (1973) and Wedderburn (1974), the true score function depends on the parameters only through the means and variances. These authors further noted that this score function can also be written as a weighted least-squares estimating function. To estimate the parameters, Wedderburn (1974) *suggested* using the exponential family score function, even when the 'error' distribution is *unspecified.* Clearly then, the estimation depends on the first two moments only, as in the Gauss–Markov theorem. But this suggestion raises the *question* of what happens when the true underlying distribution is *not* from the exponential family? Both the suggestion and the question can be better formulated, extended and answered by relating the 'quasi-likelihood' to the 'optimal estimating function' as in (6) in view of its properties (i), (ii) and (iii). This is the subject of the remainder of the paper.

2 The general framework

Let $\{\mathbf{X}_t, 0 \leqslant t \leqslant T\}$ be a sample in discrete or continuous time drawn from some process taking values in r-dimensional Euclidean space whose distribution depends on a parameter $\boldsymbol{\theta}$ taking values in an open subset Θ of p-dimensional Euclidean space. Suppose that the possible probability measures for $\{\mathbf{X}_t\}$ are $\mathcal{P} = \{\mathcal{P}_{\boldsymbol{\theta}}\}$, a *union* of parametric families each family being indexed by the same parameter $\boldsymbol{\theta}$, and that each $(\Omega, \mathcal{B}, P_{\boldsymbol{\theta}})$ is a complete probability space. Let $\{\mathcal{B}_t, t \geqslant 0\}$ denote a standard filtration, $\mathcal{B}_s \subseteq \mathcal{B}_t \subseteq \mathcal{B}$ for $s \leqslant t$; $\mathcal{B}_0$ is augmented by sets of measure zero of $\mathcal{B}$ and $\mathcal{B}_t = \mathcal{B}_{t+}$, where $\mathcal{B}_{t+} = \bigcap_{s>t} \mathcal{B}_s$.

We shall focus attention on the class $\mathcal{G}$ of zero mean, square integrable estimating functions $\mathbf{G}_T = \mathbf{G}_T(\{\mathbf{X}_t, 0 \leq t \leq T\}, \boldsymbol{\theta})$ for which $E\mathbf{G}_T(\boldsymbol{\theta}) = \mathbf{0}$ for each $P_{\boldsymbol{\theta}} \in \mathcal{P}$ with index $\boldsymbol{\theta}$ and on the class $\mathcal{M} \subset \mathcal{G}$ of martingale estimating functions $\{\mathbf{G}_T, \mathcal{B}_T\}$ which are martingales for each $P_{\boldsymbol{\theta}} \in \mathcal{P}$ with index $\boldsymbol{\theta}$. Here $\mathbf{G}_T$ is a vector of dimension p. Estimators $\boldsymbol{\theta}^*$ are found by solving the estimating equation $\mathbf{G}_T(\boldsymbol{\theta}^*) = \mathbf{0}$.

We should emphasize that this framework includes the standard methods of estimation: maximum likelihood, least-squares, conditional least-squares, minimum chi-squared, etc., under minor regularity conditions. Indeed, it is essentially the class of martingale M-estimators, albeit in a more general setting than the usual one; see for example, Basawa (1985).

The estimating function approach focuses on the function rather than the estimator derived from it and optimality properties are for the function. There is often loss of information about the parameter in moving from estimating function to estimator. For example, it is well known that, under minor regularity conditions, the score function provides a minimal sufficient partitioning of the sample space. However, there is often no single sufficient statistic.

The properties of estimators derived from optimal estimating functions can be investigated in detail for large samples but little explicit information can be given in general about estimators from small samples. This is a manifestation in a broader setting of the fact that, outside the field of sufficient statistics, the optimal properties of maximum likelihood (ML) estimators are asymptotic ones.

For a matrix $\mathbf{A}$ we shall write $\mathbf{A}'$ for the transpose and, if $\mathbf{A}$ is square, $|\mathbf{A}|$ for the determinant. If $\mathbf{A}$ is symmetric and nonnegative-definite its unique nonnegative-definite square root will be written as $\mathbf{A}^{\frac{1}{2}}$, while $\mathbf{A}^+$ denotes its Moore–Penrose pseudo inverse, namely the unique matrix possessing the properties $\mathbf{AA}^+\mathbf{A} = \mathbf{A}$, $\mathbf{A}^+\mathbf{AA}^+ = \mathbf{A}^+$, $\mathbf{A}^+\mathbf{A} = \mathbf{AA}^+$.

For $n \times 1$ vector valued martingales $\mathbf{M}_T$ and $\mathbf{N}_T$, the $n \times n$ process $\langle \mathbf{M}, \mathbf{N}' \rangle_T$ is the *mutual quadratic characteristic*, a predictable increasing process such that $\mathbf{M}_T\mathbf{N}_T' - \langle \mathbf{M}, \mathbf{N}' \rangle_T$ is an $n \times n$ martingale. We shall write $\langle \mathbf{M} \rangle_T$ for $\langle \mathbf{M}, \mathbf{M}' \rangle_T$, the *quadratic characteristic* of $\mathbf{M}_T$. A convenient sketch of these concepts is given by Shiryaev (1981).

3 Fixed sample criteria

Here we shall confine attention to the subclass $\mathcal{G}_1$ of $\mathcal{G}$ of estimating functions $\mathbf{G}_T$ which are almost surely differentiable with respect to the components of $\boldsymbol{\theta}$, and for which $E\dot{\mathbf{G}}_T(\boldsymbol{\theta}) = (E\, \partial G_{T,i}(\boldsymbol{\theta})/\partial\theta_j)$ and $E(\mathbf{G}_T(\boldsymbol{\theta})\mathbf{G}_T'(\boldsymbol{\theta}))$ are nonsingular. The expectations are always with respect to $P_{\boldsymbol{\theta}}$.

We shall usually suppose that $P_{\boldsymbol{\theta}}$ is absolutely continuous with respect to some σ-finite measure λ_t giving density $p_t(\boldsymbol{\theta})$. Then, we write $\mathbf{U}_t(\boldsymbol{\theta}) = p_t^{-1}(\boldsymbol{\theta})\dot{\mathbf{p}}_t(\boldsymbol{\theta})$ for the score function which we further suppose to be almost surely differentiable with respect to the components of $\boldsymbol{\theta}$. In addition we suppose that differentiation and integration can be interchanged in $E(\mathbf{G}_T\mathbf{U}_T')$ and $E(\mathbf{U}_T\mathbf{G}_T')$, for $\mathbf{G}_T \in \mathcal{G}_1$.

Here we note that corresponding to each parametric family in $\mathcal{P}$ is defined a score function. When the true underlying parametric family and hence the score function is unknown it may be argued that the best strategy is to choose an estimating function $\mathbf{G}_T$ which has minimum distance, in an appropriate sense, from $\mathbf{U}_T$ or maximum vector correlation with $\mathbf{U}_T$. These ideas are formalized in the following equivalent properties of an optimal estimating function $\mathbf{G}_T^*$.

Criterion 1. We have that

$$(E\dot{\mathbf{G}}_T)^{-1}(E\mathbf{G}_T\mathbf{G}_T')((E\dot{\mathbf{G}}_T)^{-1})' - (E\dot{\mathbf{G}}_T^*)^{-1}(E\mathbf{G}_T^*\mathbf{G}_T^{*\prime})((E\dot{\mathbf{G}}_T^*)^{-1})'$$

is nonnegative-definite for all $\mathbf{G}_T \in \mathscr{G}_1$, $\boldsymbol{\theta} \in \Theta$ and $P_{\boldsymbol{\theta}} \in \mathscr{P}$.

Criterion 2. For some fixed matrix function $\boldsymbol{\alpha}$ depending on $\boldsymbol{\theta}$ and T,

$$E((\boldsymbol{\alpha}_T \mathbf{U}_T - \mathbf{G}_T^*)\mathbf{G}_T') = E(\mathbf{G}_T(\boldsymbol{\alpha}_T \mathbf{U}_T - \mathbf{G}_T^*)') = \mathbf{0}$$

for every $\mathbf{G}_T \in \mathscr{G}_1$ for all $\boldsymbol{\theta} \in \Theta$ and $P_{\boldsymbol{\theta}} \in \mathscr{P}$.

Criterion 3. For some fixed matrix function $\boldsymbol{\alpha}$ depending on $\boldsymbol{\theta}$ and T,

$$E(\boldsymbol{\alpha}_T \mathbf{U}_T - \mathbf{G}_T)(\boldsymbol{\alpha}_T \mathbf{U}_T - \mathbf{G}_T)' - E(\boldsymbol{\alpha}_T \mathbf{U}_T - \mathbf{G}_T^*)(\boldsymbol{\alpha}_T \mathbf{U}_T - \mathbf{G}_T^*)'$$

is nonnegative-definite for all $\mathbf{G}_T \in \mathscr{G}_1$, $\boldsymbol{\theta} \in \Theta$ and $P_{\boldsymbol{\theta}} \in \mathscr{P}$.

The proof of the equivalence of these criteria, which extends the results of Godambe (1985), Godambe & Thompson (1985) and Thavaneswaran & Thompson (1986) to the vector case, is given in the Appendix. Note from Criterion 1 that an optimal estimating function is defined only up to a constant (matrix) multiplier, while Criterion 3 can be interpreted as $\mathbf{G}_T^*$ having minimum dispersion distance from $\boldsymbol{\alpha}_T \mathbf{U}_T$. Criterion 1 can also be inverted to give an equivalent form as follows.

Criterion 1*. We have that

$$(E\dot{\mathbf{G}}_T^*)'(E\mathbf{G}_T^*\mathbf{G}_T^{*\prime})^{-1}(E\dot{\mathbf{G}}_T^*) - (E\dot{\mathbf{G}}_T)'(E\mathbf{G}_T\mathbf{G}_T')^{-1}(E\dot{\mathbf{G}}_T)$$

is nonnegative-definite for all $\mathbf{G}_T \in \mathscr{G}_1$, $\boldsymbol{\theta} \in \Theta$ and $P_{\boldsymbol{\theta}} \in \mathscr{P}$.

We shall subsequently use this form. It should be noted that Criterion 1* (or 1) does not require the existence of the score function and this is taken as the general *defining* relationship for an optimal estimating function for fixed samples, denoted by O_F.

In particular, if

$$(E\dot{\mathbf{G}}_T^*)'(E\mathbf{G}_T^*\mathbf{G}_T^{*\prime})^{-1}(E\dot{\mathbf{G}}_T^*) - (E\dot{\mathbf{G}}_T)'(E\mathbf{G}_T\mathbf{G}_T')^{-1}(E\dot{\mathbf{G}}_T)$$

is nonnegative-definite for all $\mathbf{G}_T \in \mathscr{G}_2 \subset \mathscr{G}_1$, $\boldsymbol{\theta} \in \Theta$ and $P_{\boldsymbol{\theta}} \in \mathscr{P}$, we will say that $\mathbf{G}_T^*$ is *O_F-optimal within* $\mathscr{G}_2$. We emphasize that Criteria 1, 2 and 3 are not necessarily equivalent if consideration is restricted to a subset $\mathscr{G}_2 \subset \mathscr{G}_1$.

Now the vector correlation which measures the association between $\mathbf{G}_T = (G_{T,1}, \ldots, G_{T,p})'$ and $\mathbf{U}_T = (U_{T,1}, \ldots, U_{T,p})'$, defined, for example, by Hotelling (1936), is

$$\rho^2 = \frac{|E(\mathbf{G}_T\mathbf{U}_T')|^2}{|E(\mathbf{G}_T\mathbf{G}_T')|\,|E(\mathbf{U}_T\mathbf{U}_T')|}.$$

However, under the regularity conditions that have been imposed, $E\dot{\mathbf{G}}_T = -E(\mathbf{G}_T\mathbf{U}_T')$, so a maximal correlation requirement is to maximize

$$|E\dot{\mathbf{G}}_T|^2/|E\mathbf{G}_T\mathbf{G}_T'|, \tag{7}$$

which can be achieved by maximizing

$$(E\dot{\mathbf{G}}_T)'(E\mathbf{G}_T\mathbf{G}_T')^{-1}(E\dot{\mathbf{G}}_T) \tag{8}$$

in the (partial) order of nonnegative-definite matrices. This corresponds to Criterion 1*.

The single-parameter optimality criterion, obtained from (7), is originally due to Godambe (1960). Multiparametric optimality criteria corresponding to (7) and (8) above were previously proposed by Durbin (1960), Bhapkar (1972), Morton (1981), Ferreira (1982), Chandrasekar & Kale (1984) and Godambe & Thompson (1986).

Following discussion in § 1, we call the optimal estimating function $\mathbf{G}_T^*$ the *quasi-score*

(QS) *function* and the corresponding equation $\mathbf{G}_T^* = \mathbf{0}$ the *quasi-likelihood* (QL) *equation*. A solution of the QL equation would be called a *maximum quasi-likelihood estimator* (MQLE).

Optimal estimating functions in the sense of Criterion 1* do not necessarily exist for the full class $\mathscr{G}_1$; see, for example, Godambe & Thompson (1985). However, by restricting consideration to a statistically meaningful subclass of $\mathscr{G}_1$, such as in (13) of § 5 below, an optimum may be obtained.

4 An asymptotic criterion

Let $\mathscr{M}_1$ denote the subset of $\mathscr{G}_1$ which are square integrable martingales. For $\{\mathbf{G}_T, \mathscr{B}_T\} \in \mathscr{M}_1$ there is, under quite broad conditions, a multivariate central limit result

$$\langle \mathbf{G} \rangle_T^{-\frac{1}{2}} \mathbf{G}_T \to MVN(\mathbf{0}, \mathbf{I}_p) \tag{9}$$

in distribution, as $T \to \infty$; see, for example, Feigin (1985), Hall & Heyde (1980, Ch. 3), with the Cramér–Wold device in mind to obtain multivariate versions, and Hutton & Nelson (1984). Let $\mathscr{M}_2 \subset \mathscr{M}_1$ be the subclass for which (9) obtains.

Next, with $\mathbf{G}_T \in \mathscr{M}_2$ let $\boldsymbol{\theta}^*$ be a solution of $\mathbf{G}_T(\boldsymbol{\theta}) = \mathbf{0}$ and use Taylor's expansion to obtain

$$\mathbf{0} = \mathbf{G}_T(\boldsymbol{\theta}^*) = \mathbf{G}_T(\boldsymbol{\theta}) + \dot{\mathbf{G}}_T(\boldsymbol{\theta}\dagger)(\boldsymbol{\theta}^* - \boldsymbol{\theta}), \tag{10}$$

where $\|\boldsymbol{\theta} - \boldsymbol{\theta}\dagger\| < \|\boldsymbol{\theta} - \boldsymbol{\theta}^*\|$, the norm denoting sum of squares of elements. Then, if $\dot{\mathbf{G}}_T(\boldsymbol{\theta})$ is nonsingular for $\boldsymbol{\theta}$ in a suitable neighbourhood and $(\dot{\mathbf{G}}_T(\boldsymbol{\theta}\dagger))^{-1}\dot{\mathbf{G}}_T(\boldsymbol{\theta}) \to \mathbf{I}_p$ in probability as $T \to \infty$, expressions (9) and (10) lead to

$$\langle \mathbf{G}(\boldsymbol{\theta}) \rangle_T^{-\frac{1}{2}} \dot{\mathbf{G}}_T(\boldsymbol{\theta})(\boldsymbol{\theta}^* - \boldsymbol{\theta}) \to MVN(\mathbf{0}, \mathbf{I}_p)$$

in distribution.

Now define the predictable process

$$\bar{\mathbf{G}}_t(\boldsymbol{\theta}) = \int_0^t E(d\dot{\mathbf{G}}_s(\boldsymbol{\theta}) \mid \mathscr{B}_{s-}),$$

$\mathscr{B}_{s-}$ being the σ-field generated by $\bigcup_{r<s} \mathscr{B}_r$, and assume that $\dot{\mathbf{G}}_T(\boldsymbol{\theta})$ admits a Doob–Meyer type decomposition

$$\dot{\mathbf{G}}_T(\boldsymbol{\theta}) = \mathbf{M}_{\mathbf{G},T}(\boldsymbol{\theta}) + \bar{\mathbf{G}}_T(\boldsymbol{\theta}),$$

$\mathbf{M}_{\mathbf{G},T}(\boldsymbol{\theta})$ being a martingale. Then, under modest conditions, for example if $\dot{\mathbf{G}}_T(\boldsymbol{\theta}) \to -\infty$ almost surely as $T \to \infty$,

$$|\mathbf{M}_{\mathbf{G},T}(\boldsymbol{\theta})| = o_p(|\bar{\mathbf{G}}_T(\boldsymbol{\theta})|)$$

as $T \to \infty$, o_p denoting small order in probability.

Thus, considerations which can be formalized under appropriate regularity conditions indicate that, for $\mathbf{G}_T$ belonging to some $\mathscr{M}_3 \subset \mathscr{M}_2$,

$$\langle \mathbf{G}(\boldsymbol{\theta}) \rangle_T^{-\frac{1}{2}} \bar{\mathbf{G}}_T(\boldsymbol{\theta})(\boldsymbol{\theta}^* - \boldsymbol{\theta}) \to MVN(\mathbf{0}, \mathbf{I}_p)$$

in distribution, and hence

$$(\boldsymbol{\theta}^* - \boldsymbol{\theta})' \bar{\mathbf{G}}_T'(\boldsymbol{\theta}) \langle \mathbf{G}(\boldsymbol{\theta}) \rangle_T^{-1} \bar{\mathbf{G}}_T(\boldsymbol{\theta})(\boldsymbol{\theta}^* - \boldsymbol{\theta}) \to \chi_p^2 \tag{11}$$

in distribution. Best (asymptotic) estimation within a class $\mathscr{M}_4 \subseteq \mathscr{M}_3$ of estimating functions is then achieved by choosing $\mathbf{G}_T^* \in \mathscr{M}_4$ so that Criterion 4 is satisfied.

Criterion 4. We have that

$$\bar{\mathbf{G}}_T^{*\prime}(\boldsymbol{\theta})\langle \mathbf{G}^*(\boldsymbol{\theta})\rangle_T^{-1}\bar{\mathbf{G}}_T^*(\boldsymbol{\theta}) - \bar{\mathbf{G}}_T'(\boldsymbol{\theta})\langle \mathbf{G}(\boldsymbol{\theta})\rangle_T^{-1}\bar{\mathbf{G}}_T(\boldsymbol{\theta})$$

is nonnegative-definite for all $\mathbf{G}_T \in \mathcal{M}_4$, $\boldsymbol{\theta} \in \Theta$, $P_{\boldsymbol{\theta}} \in \mathcal{P}$ and $T > 0$.

This follows because maximizing

$$\mathbf{C}_T(\boldsymbol{\theta}) = \bar{\mathbf{G}}_T'(\boldsymbol{\theta})\langle \mathbf{G}(\boldsymbol{\theta})\rangle_T^{-1}\bar{\mathbf{G}}_T(\boldsymbol{\theta})$$

leads to (asymptotic) confidence regions centred on $\boldsymbol{\theta}^*$ of minimum size, for example Rao (1965, § 4b.2). Note that $\mathbf{C}_T(\boldsymbol{\theta})$ can be replaced by $\mathbf{C}_T(\boldsymbol{\theta}^*)$ in (11). If $\mathbf{G}_T^*$ satisfies Criterion 4 we will say that it is *O_A-optimal within $\mathcal{M}_4$*, O_A meaning optimal in the asymptotic sense.

The relation between O_F and O_A optimality, both restricted to the same class of estimating functions, is very close and it should be noted that

$$E\bar{\mathbf{G}}_T(\boldsymbol{\theta}) = E\dot{\mathbf{G}}_T(\boldsymbol{\theta}), \quad E\langle \mathbf{G}(\boldsymbol{\theta})\rangle_T = E\mathbf{G}_T(\boldsymbol{\theta})\mathbf{G}_T'(\boldsymbol{\theta}).$$

Both criteria are satisfied by the same (quasi-score) estimating function under broad conditions as we shall indicate below.

5 A particular QL solution

We shall illustrate the use of the criteria for O_F and O_A optimality by supposing that the observed process $\{\mathbf{X}_t, t \geqslant 0\}$ is (a semimartingale) representable for all T in the form

$$\mathbf{X}_T = \int_0^T \mathbf{f}_t(\boldsymbol{\theta})\, d\lambda_t + \mathbf{m}_T(\boldsymbol{\theta}). \tag{12}$$

Here $\{\lambda_t\}$ is a real, monotone increasing, right continuous process with $\lambda_0 = 0$, while $\{\mathbf{m}_T(\boldsymbol{\theta}), \mathcal{B}_T\}$ is a cadlag, square integrable $r \times 1$ vector martingale with characteristic $\langle \mathbf{m}(\boldsymbol{\theta})\rangle_T$ given for all T by

$$\langle \mathbf{m}(\boldsymbol{\theta})\rangle_T = \int_0^T \mathbf{a}_t(\boldsymbol{\theta})\, d\lambda_t,$$

the processes $\{\lambda_t\}$, $\{\mathbf{a}_t(\boldsymbol{\theta})\}$ and $\{\mathbf{f}_t(\boldsymbol{\theta})\}$ being predictable and the elements of $\mathbf{f}_t(\boldsymbol{\theta})$ being almost surely continuously differentiable with respect to the elements of $\boldsymbol{\theta}$. This is the setting used by Hutton & Nelson (1986) and it is very widely applicable. It covers many continuous time stochastic models of the diffusion, counting, queueing or population process type as well as most discrete time models via the use of $\lambda_t = [t]$, the greatest integer less than or equal to t.

Now in the context of (12), utilizing the motivation provided by Godambe (1985, § 4), we shall confine attention to estimating functions of the form

$$\mathbf{G}_T(\boldsymbol{\theta}) = \int_0^T \boldsymbol{\alpha}_s(\boldsymbol{\theta})\, d\mathbf{m}_s(\boldsymbol{\theta}), \tag{13}$$

where $\{\boldsymbol{\alpha}_s(\boldsymbol{\theta})\}$ is a $p \times r$ predictable process whose elements are almost surely continuously differentiable with respect to the components of $\boldsymbol{\theta}$. We shall proceed to find an estimator which is optimal in both the O_F and O_A senses within this class. It should be remarked, however, that this discussion is illustrative and cases in which both O_F and O_A optimality are satisfied are certainly not peculiar to the semimartingale setting. This point is discussed further in § 6.

We have, under appropriate regularity conditions, that

$$\dot{\mathbf{G}}_T(\boldsymbol{\theta}) = \int_0^T \dot{\boldsymbol{\alpha}}_s(\boldsymbol{\theta})\, d\mathbf{m}_s(\boldsymbol{\theta}) + \int_0^T \boldsymbol{\alpha}_s(\boldsymbol{\theta})\, d\dot{\mathbf{m}}_s(\boldsymbol{\theta})$$

and, since $\dot{\mathbf{m}}_s(\theta)$ is $\mathscr{B}_{s-}$-measurable from (12),

$$\bar{\mathbf{G}}_T(\boldsymbol{\theta}) = \int_0^T \boldsymbol{\alpha}_s(\boldsymbol{\theta})\, d\dot{\mathbf{m}}_s(\boldsymbol{\theta}) = -\int_0^T \boldsymbol{\alpha}_s(\boldsymbol{\theta})\dot{\mathbf{f}}_s(\boldsymbol{\theta})\, d\lambda_s.$$

Furthermore,

$$\langle \mathbf{G}(\boldsymbol{\theta})\rangle_T = \int_0^T \boldsymbol{\alpha}_s(\boldsymbol{\theta}) \frac{d\langle \mathbf{m}(\boldsymbol{\theta})\rangle_s}{d\lambda_s} \boldsymbol{\alpha}_s'(\boldsymbol{\theta})\, d\lambda_s,$$

and, more generally, if $\mathbf{H}_T(\boldsymbol{\theta}) = \int \boldsymbol{\beta}_s(\boldsymbol{\theta})\, d\mathbf{m}_s(\boldsymbol{\theta})$, where the integral is over $(0, T)$, and where $\{\boldsymbol{\beta}_s(\boldsymbol{\theta})\}$ is predictable, then

$$\langle \mathbf{G}(\boldsymbol{\theta}), \mathbf{H}'(\boldsymbol{\theta})\rangle_T = \int_0^T \boldsymbol{\alpha}_s(\boldsymbol{\theta}) \frac{d\langle \mathbf{m}(\boldsymbol{\theta})\rangle_s}{d\lambda_s} \boldsymbol{\beta}_s'(\boldsymbol{\theta})\, d\lambda_s.$$

Of course

$$E\langle \mathbf{G}(\boldsymbol{\theta}), \mathbf{H}'(\boldsymbol{\theta})\rangle_T = E\mathbf{G}_T(\boldsymbol{\theta})\mathbf{H}_T'(\boldsymbol{\theta}), \quad E\bar{\mathbf{G}}_T(\boldsymbol{\theta}) = E\dot{\mathbf{G}}_T(\boldsymbol{\theta}).$$

We shall suppose that $\langle \mathbf{G}(\boldsymbol{\theta})\rangle_T$ and $\bar{\mathbf{G}}_T(\boldsymbol{\theta})$ are almost surely nonsingular and that their expected values are nonsingular.

Now we introduce the estimating function $\mathbf{Q}_T$ defined by

$$\mathbf{Q}_T(\boldsymbol{\theta}) = \int_0^T \dot{\mathbf{f}}_s'(\boldsymbol{\theta})\mathbf{a}_s^+(\boldsymbol{\theta})\, d\mathbf{m}_s(\boldsymbol{\theta}). \tag{14}$$

We shall show that this estimating function is optimal in the sense of the criteria for both O_F and O_A optimality within the class of estimating functions of the form (13). This justifies the use of the terminology *quasi-score* function for $\mathbf{Q}_T$. It is useful, where possible, to interpret the quasi-score function as the derivative of an underlying log quasi-likelihood whose maximum provides the quasi-likelihood estimator. In fact, it is frequently, but not always, possible to obtain (14) as the true score function for members of a certain exponential family. This important topic will be taken up elsewhere.

Note that, if $\mathbf{a}_s(\boldsymbol{\theta})$ is positive-definite,

$$\langle \mathbf{G}(\boldsymbol{\theta}), (\mathbf{Q}(\boldsymbol{\theta}))'\rangle_T = -\bar{\mathbf{G}}_T(\boldsymbol{\theta}),$$

while

$$\langle \mathbf{Q}(\boldsymbol{\theta})\rangle_T = \int_0^T \dot{\mathbf{f}}_s'(\boldsymbol{\theta})\mathbf{a}_s^+(\boldsymbol{\theta})\dot{\mathbf{f}}_s(\boldsymbol{\theta})\, d\lambda_s = -\bar{\mathbf{Q}}_T(\boldsymbol{\theta}).$$

Next, write $\mathbf{G}_T = (G_{1,T}, \ldots, G_{p,T})'$, $\mathbf{Q}_T = (Q_{1,T}, \ldots, Q_{p,T})'$. The $2p \times 2p$ dispersion and mutual quadratic characteristic matrices of the set of variables

$$(G_{1,T}, \ldots, G_{p,T}, Q_{1,T}, \ldots, Q_{p,T})$$

may be written in partitioned matrix form as

$$\mathbf{D} = \begin{pmatrix} E\mathbf{G}_T\mathbf{G}_T' & E\mathbf{G}_T\mathbf{Q}_T' \\ (E\mathbf{G}_T\mathbf{Q}_T')' & E\mathbf{Q}_T\mathbf{Q}_T' \end{pmatrix} = \begin{pmatrix} E\mathbf{G}_T\mathbf{G}_T' & -E\dot{\mathbf{G}}_T \\ (-E\dot{\mathbf{G}}_T)' & E\mathbf{Q}_T\mathbf{Q}_T' \end{pmatrix},$$

$$\mathbf{C} = \begin{pmatrix} \langle \mathbf{G}\rangle_T & \langle \mathbf{G}, \mathbf{Q}'\rangle_T \\ \langle \mathbf{G}, \mathbf{Q}'\rangle_T' & \langle \mathbf{Q}\rangle_T \end{pmatrix} = \begin{pmatrix} \langle \mathbf{G}\rangle_T & -\bar{\mathbf{G}}_T \\ -\bar{\mathbf{G}}_T' & \langle \mathbf{Q}\rangle_T \end{pmatrix},$$

respectively, the former matrix being the expected value of the latter.

Now **C** and **D** are nonnegative-definite since, if

$$\mathbf{Z}' = (G_{1,T}, \ldots, G_{p,T}, Q_{1,T}, \ldots, Q_{p,T})$$

and **u** is an arbitrary $2p \times 1$ vector,

$$\mathbf{u}'\mathbf{D}\mathbf{u} = \mathbf{u}'E(\mathbf{Z}\mathbf{Z}')\mathbf{u} = E(\mathbf{u}'\mathbf{Z}\mathbf{Z}'\mathbf{u}) = E(\mathbf{Z}'\mathbf{u})^2 \geqslant 0,$$

while, almost surely,

$$\mathbf{u}'\mathbf{C}\mathbf{u} = \mathbf{u}'\langle \mathbf{Z}, \mathbf{Z}'\rangle_T \mathbf{u} = \langle \mathbf{u}'\mathbf{Z}, \mathbf{Z}'\mathbf{u}\rangle_T = \langle \mathbf{u}'\mathbf{Z}\rangle_T \geqslant 0,$$

and the method of Rao (1965, p. 266) gives the nonnegative-definiteness of

$$E\mathbf{G}_T\mathbf{G}_T' - (E\dot{\mathbf{G}}_T)(E\mathbf{Q}_T\mathbf{Q}_T')^{-1}(E\dot{\mathbf{G}}_T)',$$

and, almost surely, $\langle \mathbf{G}\rangle_T - \bar{\mathbf{G}}_T\langle Q\rangle_T^{-1}\bar{\mathbf{G}}_T'$. Furthermore, since all the matrices involved in these expressions are almost surely nonsingular, we obtain by inversion that

$$E\mathbf{Q}_T\mathbf{Q}_T' - (E\dot{\mathbf{G}}_T)'(E\mathbf{G}_T\mathbf{G}_T')^{-1}(E\dot{\mathbf{G}}_T),$$

and, almost surely, $\langle \mathbf{Q}\rangle_T - \bar{\mathbf{G}}_T'\langle \mathbf{G}\rangle_T^{-1}\bar{\mathbf{G}}_T$ are nonnegative-definite.

Then, since

$$E\mathbf{Q}_T\mathbf{Q}_T' = -E\dot{\mathbf{Q}}_T, \quad \langle \mathbf{Q}\rangle_T = -\bar{\mathbf{Q}}_T,$$

both $(E\dot{\mathbf{G}}_T)'(E\mathbf{G}_T\mathbf{G}_T')^{-1}(E\dot{\mathbf{G}}_T)$ and $\bar{\mathbf{G}}_T'\langle \mathbf{G}\rangle_T^{-1}\bar{\mathbf{G}}_T$ are maximized in the partial order of nonnegative-definite matrices by choosing $\mathbf{G}_T = \mathbf{Q}_T$. Thus, the estimating function defined by (14) is optimal in both the senses of O_F and O_A optimality as required.

The maximum quasi-likelihood estimator (MQLE) is a solution of the QL estimating equation and general properties of the MQLE can be established along similar lines to those for the MLE which they closely resemble. Sufficient conditions for strong consistency and asymptotic normality are given in Theorems 3.1 and 4.1, respectively, of Hutton & Nelson (1986).

The framework discussed above specializes to what has been described as quasi-likelihood in the context of the general linear model, for example, McCullagh (1983) and McCullagh & Nelder (1983, Ch. 8) provided the $\mathbf{X}_t$ are sums of independent random vectors $\mathbf{Y}_s$, for $1 \leqslant s \leqslant t$. Then

$$\mathbf{m}_T(\boldsymbol{\theta}) = \sum_{t=1}^{T} (\mathbf{Y}_t - \boldsymbol{\mu}_t(\boldsymbol{\theta})),$$

where $\boldsymbol{\mu}_t(\boldsymbol{\theta}) = E\mathbf{Y}_t$, and, if

$$E(\mathbf{Y}_t - \boldsymbol{\mu}_t(\boldsymbol{\theta}))(\mathbf{Y}_t - \boldsymbol{\mu}_t(\boldsymbol{\theta}))' = \mathbf{v}_t(\boldsymbol{\theta}) \quad (t = 1, 2, \ldots, T),$$

then the QL estimating function given by (14) is

$$\mathbf{Q}_T(\boldsymbol{\theta}) = \sum_{t=1}^{T} \dot{\boldsymbol{\mu}}_t'(\boldsymbol{\theta})\mathbf{v}_t^+(\boldsymbol{\theta})(\mathbf{Y}_t - \boldsymbol{\mu}_t(\boldsymbol{\theta})) = \dot{\boldsymbol{\mu}}'\mathbf{V}^+(\mathbf{Y} - \boldsymbol{\mu}),$$

where

$$\boldsymbol{\mu}' = (\boldsymbol{\mu}_1'(\boldsymbol{\theta}) \vdots \ldots \vdots \boldsymbol{\mu}_T'(\boldsymbol{\theta})), \quad \mathbf{Y}' = (\mathbf{Y}_1' \vdots \ldots \vdots \mathbf{Y}_T'), \quad \mathbf{V}^+ = \text{diag}\,(\mathbf{v}_1^+(\boldsymbol{\theta}) \vdots \ldots \vdots \mathbf{v}_T^+(\boldsymbol{\theta})).$$

This agrees with the usual definition but the extensions that have been used for the generalized linear model cannot be used directly to deal with most cases of dependent variables.

A simple example of dependent variables for which the QL formalism discussed herein is essential is given by the estimation of the mean of the offspring distribution in a single type (Bienaymé-) Galton–Watson branching process $\{Z_0 = 1, Z_1, \ldots, Z_T\}$. Here we have $EZ_1 = \theta$ and we assume var $Z_1 = \sigma^2 < \infty$.

We take $X_T = \sum Z_t$, $m_T(\theta) = \sum (Z_t - \theta Z_{t-1})$, and then $f_t(\theta) = \theta Z_{t-1}$ and $\langle m \rangle_T = \sigma^2 \sum Z_{t-1}$, where the sums are all over $t = 1, \ldots, T$, so that $a_t(\theta) = \sigma^2 Z_{t-1}$. Here the quasi-score function given by (14) is $\sigma^{-2} \sum (Z_t - \theta Z_{t-1})$, with the sum over $t = 1, \ldots, T$. The maximum quasi-likelihood estimating equation is then

$$\sum_{t=1}^{T} (Z_t - \hat{\theta} Z_{t-1}) = 0.$$

The estimator $\hat{\theta}$ so obtained is easily seen to be the maximum likelihood estimator (MLE) for the power series family of offspring distributions:

$$P(Z_1 = j) = A(j) \frac{(a(\theta))^j}{F(\theta)} \quad (j = 0, 1, 2, \ldots),$$

where $F(\theta) = \sum A(j)(a(\theta))^j$, with the sum over $j = 0, \ldots, \infty$, for example, Heyde (1975). These form the discrete linear exponential family for this context. The estimator $\hat{\theta}$ can also be obtained quite generally as a nonparametric MLE, (Feigin, 1977).

As another example, now in continuous time, we consider the multivariate counting process $\mathbf{X}_T = (X_{1,T}, \ldots, X_{p,T})'$ such that each $X_{i,T}$ is of the form

$$X_{i,T} = \theta_i \int_0^T J_i(s)\, ds + M_{i,T}$$

with multiplicative intensity $\Lambda_i(t) = \theta_i J_i(t)$, $J_i(t) > 0$ almost surely being a predictable process, and $M_{i,t}$ a square integrable martingale. This is a special case of the framework considered by Aalen (1978), see also Andersen et al. (1982), and it covers a variety of contexts for processes such as those of birth and death type. The case $p = 1$ has been discussed by Thavaneswaran & Thompson (1986).

Now for counting processes,

$$\langle M_i \rangle_T = A_{i,T} = \theta_i \int_0^T J_i(s)\, ds,$$

$A_{i,T}$ being a (nondecreasing) compensator and $\langle M_i, M_j \rangle_T = 0$ whenever $i \neq j$ (Aalen, 1978, Th. 3.2). Thus we can write

$$\mathbf{X}_T = \left(\int_0^T \mathbf{J}(s)\, ds \right) \boldsymbol{\theta} + \mathbf{M}_T,$$

where

$$\mathbf{J}(s) = \operatorname{diag}(J_1(s), \ldots, J_p(s)), \quad \boldsymbol{\theta}' = (\theta_1, \ldots, \theta_p), \quad \mathbf{M}_T = (M_{1,T}, \ldots, M_{p,T})',$$

so that $\mathbf{f}_t(\boldsymbol{\theta}) = \mathbf{J}(t)\boldsymbol{\theta}$ and, since $\langle \mathbf{M} \rangle_T = \operatorname{diag}(A_{1,T}, \ldots, A_{p,T})$,

$$a_t = \operatorname{diag}(\theta_1 J_1(t), \ldots, \theta_p J_p(t)).$$

The MQLE is then given by $\mathbf{X}_T = \hat{\boldsymbol{\theta}} \int \mathbf{J}(s)\, ds$, where the integral is over $(0, T)$. That this $\hat{\boldsymbol{\theta}}$ is also the MLE follows from § 3.3 of Aalen (1978). The simplest particular case is where each $X_{i,t}$ is a Poisson process with parameter θ_i.

6 Extensions

The results of § 5 are focused on the case where the observed process $\{X_t, t \geq 0\}$ is a semimartingale of the form (12) but it is not essential to restrict consideration to such a setting. All that is required is the choice of a basic martingale $\{\mathbf{m}_T(\boldsymbol{\theta}), \mathscr{B}_T\}$ and then a quasi-score estimating function can be chosen from amongst the competitors of the form $\int \boldsymbol{\alpha}_s(\boldsymbol{\theta})\, d\mathbf{m}_s(\boldsymbol{\theta})$, where the integral is over $(0, T)$, and where $\{\boldsymbol{\alpha}_s(\boldsymbol{\theta})\}$ is a predictable process. Here the QS estimating function is

$$\int_0^T (d\bar{\mathbf{m}}_s(\boldsymbol{\theta}))'(d\langle \mathbf{m}(\boldsymbol{\theta})\rangle_s)^+ \, d\mathbf{m}_s(\boldsymbol{\theta}),$$

where $d\bar{\mathbf{m}}_t(\boldsymbol{\theta}) = E(d\dot{\mathbf{m}}_t(\boldsymbol{\theta}) \mid \mathscr{B}_{t-})$, and it is easily seen that this reduces to (14) in the particular case of the model (12).

Of course the martingale $\{\mathbf{m}_t(\boldsymbol{\theta}), \mathscr{B}_t\}$ can always be chosen in a variety of ways, for example with robustness requirements in mind as in the M-estimation context. A family of competing quasi-score estimating functions can then be envisaged based on the different possible choices of the basic martingale. The comparison of these and their combination, when this is advantageous, is treated by Heyde (1987). For example, referring to the (Bienaymé–) Galton–Watson branching process $\{Z_0 = 1, Z_1, \ldots, Z_T\}$ mentioned in § 5, the quasi-score estimating functions

$$G_{1,T}^* = \sum_{t=1}^{T} (Z_t - \theta Z_{t-1}), \quad G_{2,T}^* = \sum_{t=1}^{T} \{(Z_t - \theta Z_{t-1})^2 - \sigma^2 Z_{t-1})\}/(C + 2Z_{t-1}),$$

$C = \sigma^{-4}E(Z_1 - \theta)^4 - 3$ being the kurtosis of the distribution of Z_1, are *jointly* optimal for estimating the mean $EZ_1 = \theta$ and the variance $E(Z_1 - \theta)^2 = \sigma^2$ provided that $E(Z_1 - \theta)^3 = 0$. This result is also given by Godambe (1986). On the other hand, without this third moment condition, the *separate* optimality of $G_{1,T}^*$ and $G_{2,T}^*$, for estimating θ and σ^2 respectively, holds (Godambe, 1985). Godambe (1987) has initiated a theory of estimation which is aimed at spatial processes. This is based on a generalization of the martingale structure and the standard filtration of § 2.

Acknowledgment

Particular thanks are due to a referee for a very detailed set of comments.

Appendix: Equivalence of optimality criteria from § 3

The proof that Criteria 1, 2 and 3 are equivalent follows the essential lines of that of Godambe & Thompson (1985) for the scalar case. We shall drop the subscript T for convenience.

Let $\mathbf{G}_f = (E\dot{\mathbf{G}})^{-1}\mathbf{G}$. Then, note that

$$E\mathbf{G}_f\mathbf{U}' = (E\dot{\mathbf{G}})^{-1}E\mathbf{G}\mathbf{U}' = -\mathbf{I}_p, \quad E\mathbf{U}\mathbf{G}_f' = E\mathbf{U}\mathbf{G}'((E\dot{\mathbf{G}})^{-1})' = -\mathbf{I}_p.$$

We shall first show that Criteria 2 and 3 are equivalent and then that Criteria 1 and 2 are equivalent.

Suppose that Criterion 2 holds. Then

$$\begin{aligned} &E(\boldsymbol{\alpha}\mathbf{U} - \mathbf{G})(\boldsymbol{\alpha}\mathbf{U} - \mathbf{G})' - E(\boldsymbol{\alpha}\mathbf{U} - \mathbf{G}^*)(\boldsymbol{\alpha}\mathbf{U} - \mathbf{G}^*)' \\ &\qquad = -\boldsymbol{\alpha}E(\mathbf{U}\mathbf{G}') - E(\mathbf{G}\mathbf{U}')\boldsymbol{\alpha}' + E(\mathbf{G}\mathbf{G}') \\ &\qquad\quad + \boldsymbol{\alpha}E(\mathbf{U}\mathbf{G}^{*\prime}) + E(\mathbf{G}^*\mathbf{U}')\boldsymbol{\alpha}' - E(\mathbf{G}^*\mathbf{G}^{*\prime}) \\ &\qquad = -E(\mathbf{G}^*\mathbf{G}') - E(\mathbf{G}\mathbf{G}^{*\prime}) + E(\mathbf{G}\mathbf{G}') + E(\mathbf{G}^*\mathbf{G}^{*\prime}) \\ &\qquad = E(\mathbf{G} - \mathbf{G}^*)(\mathbf{G} - \mathbf{G}^*)', \end{aligned}$$

which is nonnegative-definite, being a dispersion matrix. This gives Criterion 3.

Now suppose that Criterion 3 holds. Then for all scalar β and $\mathbf{G} \in \mathcal{G}_1$,

$$E(\boldsymbol{\alpha}\mathbf{U} - \mathbf{G}^* - \beta\mathbf{G})(\boldsymbol{\alpha}\mathbf{U} - \mathbf{G} - \beta\mathbf{G})' - E(\boldsymbol{\alpha}\mathbf{U} - \mathbf{G}^*)(\boldsymbol{\alpha}\mathbf{U} - \mathbf{G}^*)'$$

is nonnegative-definite which gives, after some algebra, that

$$\beta^2 E\mathbf{G}\mathbf{G}' - \beta E((\boldsymbol{\alpha}\mathbf{U} - \mathbf{G}^*)\mathbf{G}' + \mathbf{G}(\boldsymbol{\alpha}\mathbf{U} - \mathbf{G}^*)')$$

is nonnegative-definite. This is of the form $\beta^2\mathbf{A} - \beta\mathbf{B}$, where $\mathbf{A}$ and $\mathbf{B}$ are symmetric and $\mathbf{A}$ is positive-definite. Then, using the results of Rao (1965, 1c.3(ii), p. 37), there exists a nonsingular matrix $\mathbf{R}$ such that

$$\mathbf{A} = (\mathbf{R}^{-1})'\mathbf{R}^{-1}, \quad \mathbf{B} = (\mathbf{R}^{-1})'\boldsymbol{\Lambda}\mathbf{R}^{-1},$$

where $\boldsymbol{\Lambda}$ is a diagonal matrix. Thus $(\mathbf{R}^{-1})'[\beta^2\mathbf{I_p} - \beta\boldsymbol{\Lambda}]\mathbf{R}^{-1}$ is nonnegative-definite, and hence $\mathbf{I}_p - \beta^{-1}\boldsymbol{\Lambda}$ is nonnegative-definite for all β which forces $\boldsymbol{\Lambda} = \mathbf{0}$ and hence $\mathbf{B} = \mathbf{0}$. But, this holds for each $\mathbf{G} \in \mathcal{G}_1$ so it is possible to replace $\mathbf{G}$ by $\mathbf{DG}$ where $\mathbf{D} = \operatorname{diag}(\lambda_1, \ldots, \lambda_p)$ is an arbitrary constant matrix. Then

$$E\{(\boldsymbol{\alpha}\mathbf{U} - \mathbf{G}^*)\mathbf{G}'\mathbf{D} + \mathbf{DG}(\boldsymbol{\alpha}\mathbf{U} - \mathbf{G}^*)'\} = \mathbf{0},$$

which yields, in obvious notation,

$$\lambda_j E((\boldsymbol{\alpha}\mathbf{U} - \mathbf{G}^*)_i\mathbf{G}_j) + \lambda_i E((\boldsymbol{\alpha}\mathbf{U} - \mathbf{G}^*)_j\mathbf{G}_i) = 0 \quad (i, j = 1, 2, \ldots, p).$$

Then, first taking $\lambda_i = \lambda_j = 1$, we find

$$E((\boldsymbol{\alpha}\mathbf{U} - \mathbf{G}^*)_i\mathbf{G}_j) = -E((\boldsymbol{\alpha}\mathbf{U} - \mathbf{G}^*)_j\mathbf{G}_i)$$

and hence $(\lambda_i - \lambda_j)E((\boldsymbol{\alpha}\mathbf{U} - \mathbf{G}^*)_j\mathbf{G}_i) = 0$, which forces, for all i, j,

$$E((\boldsymbol{\alpha}\mathbf{U} - \mathbf{G}^*)_j\mathbf{G}_i) = 0,$$

and hence Criterion 2 holds. Thus Criteria 2 and 3 are equivalent.

Next suppose that Criterion 2 holds and note that this gives

$$E(\mathbf{G}_f^*(\mathbf{G}_f - \mathbf{G}_f^*)') = \mathbf{0} = E((\mathbf{G}_f - \mathbf{G}_f^*)\mathbf{G}_f^{*\prime}). \tag{A1}$$

Then,

$$\begin{aligned}(E\dot{\mathbf{G}})^{-1}E\mathbf{G}\mathbf{G}'((E\dot{\mathbf{G}})^{-1})' &- (E\dot{\mathbf{G}}^*)^{-1}E\mathbf{G}^*\mathbf{G}^{*\prime}((E\dot{\mathbf{G}}^*)^{-1})' \\ &= E\mathbf{G}_f\mathbf{G}_f' - E\mathbf{G}_f^*\mathbf{G}_f^{*\prime} \\ &= E(\mathbf{G}_f - \mathbf{G}_f^*)(\mathbf{G}_f - \mathbf{G}_f^*)' + E(\mathbf{G}_f^*(\mathbf{G}_f - \mathbf{G}_f^*)') + E((\mathbf{G}_f - \mathbf{G}_f^*)\mathbf{G}_f^{*\prime}) \\ &= E(\mathbf{G}_f - \mathbf{G}_f^*)(\mathbf{G}_f - \mathbf{G}_f^*)'\end{aligned}$$

in view of (A1), and this dispersion matrix is nonnegative-definite giving Criterion 1.

Finally, suppose that Assumption 1 holds. Let $\tilde{\mathbf{G}} = \mathbf{G}_f^* + \mathbf{B}\mathbf{G}_f$, where $\mathbf{B}$ is a constant diagonal matrix. Then,

$$E d\tilde{G}/d\theta = E\dot{\mathbf{G}}_f^* + \mathbf{B}E\dot{\mathbf{G}}_f = \mathbf{I}_p + \mathbf{B},$$

and Criterion 1 implies that

$$(\mathbf{I}_p + \mathbf{B})^{-1}E\tilde{\mathbf{G}}\tilde{\mathbf{G}}'(\mathbf{I}_p + \mathbf{B})^{-1} - E\mathbf{G}_f^*\mathbf{G}_f^{*\prime}$$

is nonnegative-definite. This gives, after some algebra, that

$$\mathbf{B}(E\mathbf{G}_f\mathbf{G}_f' - E\mathbf{G}_f^*\mathbf{G}_f^{*\prime})\mathbf{B} - \mathbf{B}(E(\mathbf{G}_f^* - \mathbf{G}_f)\mathbf{G}_f^{*\prime} - E\mathbf{G}_f^*(\mathbf{G}_f^* - \mathbf{G}_f)')\mathbf{B}$$

is nonnegative-definite for all $\mathbf{B}$.

Now put $\mathbf{B} = \beta\mathbf{B}_1$ where β is a scalar. A similar argument to that used to show that

Criterion 3 implies Criterion 2 first gives

$$\mathbf{B}(E(\mathbf{G}_f^* - \mathbf{G}_f)\mathbf{G}_f^{*\prime} + E\mathbf{G}_f^*(\mathbf{G}_f^* - \mathbf{G}_f)')\mathbf{B} = \mathbf{0},$$

and then

$$E((\mathbf{G}_f^* - \mathbf{G}_f)\mathbf{G}_f^{*\prime}) = E(\mathbf{G}_f^*(\mathbf{G}_f^* - \mathbf{G}_f)') = \mathbf{0}. \tag{A2}$$

But, if $\boldsymbol{\alpha} = -(E\dot{\mathbf{G}}^*)E\mathbf{G}_f^*\mathbf{G}_f^{*\prime}$, then

$$E((\mathbf{G}_f^* - \mathbf{G}_f)\mathbf{G}_f^{*\prime}) = E\mathbf{G}_f\mathbf{U}'\boldsymbol{\alpha}'((E\dot{\mathbf{G}}^*)^{-1})' - E\mathbf{G}_f\mathbf{G}_f^{*\prime} = E(\mathbf{G}_f(\boldsymbol{\alpha}\mathbf{U} - \mathbf{G}^*)')((E\dot{\mathbf{G}}^*)^{-1})'$$

and hence (A2) yields Criterion 2. Thus Criteria 1 and 2 are equivalent and the proof is complete.

References

Aalen, O. (1978). Nonparametric inference for a family of counting processes. *Ann. Statist.* **6,** 701–726.

Andersen, P.K., Borgan, O., Gill, R. & Keiding, N. (1982). Linear nonparametric tests for comparison of counting processes, with applications to censored survival data. *Int. Statist. Rev.* **50,** 219–258.

Basawa, I.V. (1985). Neyman–Le Cam tests based on estimating functions. In *Proceedings of the Berkeley Conference in Honor of Jerzy Neyman and Jack Kiefer,* **2,** Ed. L. Le Cam and R.A. Olshen, pp. 811–825. Belmont, CA.: Wadsworth.

Bertrand, J. (1889). *Calcul des Probabilités.* Paris: Gauthier–Villars; 2nd ed. 1907. Reprinted (1973), New York: Chelsea.

Bhapkar, V.P. (1972). On a measure of efficiency in an estimating equation. *Sankhyā* A **34,** 467–472.

Bradley, E.L. (1973). The equivalence of maximum likelihood and weighted least squares estimates in the exponential family. *J. Am. Statist. Assoc.* **68,** 199–200.

Chandrasekar, B. & Kale, B.K. (1984). Unbiased statistical estimation functions in presence of nuisance parameters. *J. Statist. Plan. Inf.* **9,** 45–54.

Durbin, J. (1960). Estimation of parameters in time-series regression models. *J. R. Statist. Soc.* B **22,** 139–153.

Feigin, P.D. (1977). A note on maximum likelihood estimation for simple branching processes. *Aust. J. Statist.* **19,** 152–154.

Feigin, P.D. (1985). Stable convergence of semimartingales. *Stoch. Processes Applic.* **19,** 125–134.

Ferreira, P.E. (1982). Multiparametric estimating equations. *Ann. Inst. Statist. Math.* A **34,** 423–431.

Gauss, C.F. (1880). Theoria Combinationis Observationum Erroribus Minimus Obnaxiae. Part 1, 1821; Part 2, 1823; Suppl., 1826. In *Werke* **4,** 1–108. Göttingen.

Godambe, V.P. (1960). An optimum property of regular maximum likelihood estimation. *Ann. Math. Statist.* **31,** 1208–1212.

Godambe, V.P. (1985). The foundations of finite sample estimation in stochastic processes. *Biometrika* **72,** 419–428.

Godambe, V.P. (1987). The foundations of finite sample estimation in stochastic processes—II. In *Proc. First World Congress of Bernoulli Soc., Tashkent, 1986.* To appear.

Godambe, V.P. & Thompson, M.E. (1985). Logic of least squares revisited. Preprint.

Godambe, V.P. & Thompson, M.E. (1986). Parameters of superpopulation and survey population: their relationship and estimation. *Int. Statist. Rev.* **54,** 127–138.

Hall, P.G. & Heyde, C.C. (1980). *Martingale Limit Theory and its Application.* New York: Academic Press.

Heyde, C.C. (1975). Remarks on efficiency in estimation for branching processes. *Biometrika* **62,** 49–55.

Heyde, C.C. (1987). On combining quasi-likelihood estimating functions. *Stoch. Processes Applic.* **25.** To appear.

Heyde, C.C. & Seneta, E. (1977). *I.J. Bienaymé: Statistical Theory Anticipated.* New York: Springer.

Hotelling, H. (1936). Relations between two sets of variables. *Biometrika* **28,** 321–377.

Hutton, J.E. & Nelson, P.I. (1984). A mixing and stable central limit theorem for continuous time martingales. Preprint.

Hutton, J.E. & Nelson, P.I. (1986). Quasi-likelihood estimation for semimartingales. *Stoch. Processes Applic.* **22,** 245–257.

McCullagh, P. (1983). Quasi-likelihood functions. *Ann. Statist.* **11,** 59–67.

McCullagh, P. & Nelder, J.A. (1983). *Generalized Linear Models.* London: Chapman and Hall.

Morton, R. (1981). Efficiency of estimating equations and the use of pivots. *Biometrika* **68,** 227–233.

Rao, C.R. (1965). *Linear Statistical Inference and its Applications.* New York: Wiley.

Thavaneswaran, A. & Thompson, M.E. (1986). Optimal estimation for semimartingales. *J. Appl. Prob.* **23,** 409–417.

Shiryaev, A.N. (1981). Martingales: recent developments, results and applications. *Int. Statist. Rev.* **49,** 199–233.

Sprott, D.A. (1983). Gauss, Carl Frederich. In *Encyclopedia of Statistical Sciences,* **3,** Ed. S. Kotz and N.L. Johnson, pp. 305–308. New York: Wiley.

Wedderburn, R.W.M. (1974). Quasi-likelihood functions, generalized linear models, and the Gauss–Newton method. *Biometrika* **61,** 439–447.
Wilks, S.S. (1938). Shortest average confidence intervals from large samples. *Ann. Math. Statist.* **9,** 166–175.

Résumé

Cet article étudie l'estimation des paramètres optimaux pour les processus stochastiques. Les équations d'estimation sont utilisées. Une définition très générale est présentée pour l'estimateur de quasi-vraisemblance. Ce travail généralise les résultats de Godambe (1985) et de Hutton & Nelson (1986).

[*Received November* 1986, *revised May* 1987]

FISHER LECTURE

QUASI-LIKELIHOOD AND OPTIMALITY FOR ESTIMATING FUNCTIONS: SOME CURRENT UNIFYING THEMES

C.C. Heyde

Department of Statistics, IAS
The Australian National University
GPO Box 4, Canberra, ACT 2601
Australia

Key words: Optimality in estimation, quasi-likelihood, efficiency, information, minimal confidence zones, estimating functions, least squares, maximum likelihood

1. INTRODUCTION

This paper is concerned with recent progress towards a general unification of two of the principal themes in statistical estimation theory, those of least squares, which is founded on finite sample considerations, and maximum likelihood, whose justification is primarily asymptotic. The gestation period for this progress has been long and very many authors have contributed. Many key references are given in Godambe and Heyde (1987).

The methods underlying the abovementioned unification are based on extensions of the concept of Fisher information and it is especially appropriate to remember Fisher at the 47th Session of the ISI, the one which is closest to the centenary of his birth (February 17, 1890 in London). His introduction of maximum likelihood (although not yet under that name) in Fisher (1912) and his seminal papers Fisher (1920), (1922), (1925) introducing information, consistency, sufficiency, ancillarity, etc., provided a basis on which all current discussions are built.

R. Maller et al. (eds.), *Selected Works of C.C. Heyde*, Selected Works in Probability and Statistics,
DOI 10.1007/978-1-4419-5823-5_50, © Springer Science+Business Media, LLC 2010

It is through the use of estimating functions (functions of both the data and the parameter) rather than the estimators themselves that it is possible to incorporate the principal advantages of the methods of least squares and of maximum likelihood. The virtue of this indirect approach of estimating an unknown function which has the true value of the parameter as a root rather than estimating the parameter directly may be less than transparent at first sight. This is despite the long history of the subject, which dates back at least to K. Pearson's (1894) introduction of the method of moments, and the fact that the standard methods of estimation: maximum likelihood, least-squares, conditional least-squares, minimum chi-squared, M-estimation, etc., are included under minor regularity conditions. Also, Fisher's information is an estimating function property, namely of the score function (the derivative of the log-likelihood with respect to the parameter) rather than of the maximum likelihood estimator. Indeed, the rationale lies in the more fundamental character of the estimating function than that of an estimator derived therefrom. For example, under minor regularity conditions, the score function provides a minimal sufficient partitioning of the sample space. However, there is often no single sufficient statistic. Also, the asymptotic properties of an estimator are almost invariably obtained, as in the case of maximum likelihood, via the asymptotics of the estimating function and then transferred to the parameter space via local linearity. For details concerning the derivation of the asymptotic properties of the MLE from those of the score function together with a further justification for use of estimating functions in terms of marginalization for incomplete data problems, see Chapter 1 of McLeish and Small (1988). To these considerations we shall add another, namely the ready capacity to combine separate estimating function, each with information to offer about the unknown parameter (Heyde (1987), (1989)).

Within broad classes of unbiased estimating functions it is possible to develop a theory of optimality, founded on information based ideas, which encapsulates the virtues of least squares and maximum likelihood. This carries with it, as simple corollaries, such diverse results as the Gauss-Markov Theorem, the Cramér-Rao Inequality and the minimum size asymptotic confidence zones property possessed by the maximum likelihood estimator (subject to the usual regularity conditions). The estimating functions so produced can be interpreted as quasi-score estimating functions and the estimators derived therefrom as quasi-likelihood estimators.

2. THE SETTING

Let $\{X_t, 0 \leq t \leq T\}$ be a sample in discrete or continuous time drawn from a process taking values in r-dimensional Euclidean space whose distribution involves a "parameter" θ taking values in an open subset Θ of p-dimensional Euclidean space. The setting may be parametric or nonparametric; θ could be, for example, the mean of stationary process. The true value of the "parameter" is θ_0 and this is to be estimated.

Suppose that the possible probability measures for $\{X_t\}$ are $\{P_\theta\}$ and that each $(\Omega, \mathcal{F}, P_\theta)$ is a complete probability space.

We shall confine attention to the class $\mathcal{G}$ of zero mean square integrable estimating functions $G_T = G_T(\{X_t, 0 \leq t \leq T\}, \theta)$ for which $EG_T(\theta) = 0$ for each P_θ. Here G_T is a vector of dimension p. Estimators $\hat{\theta}$ are found by solving the estimating equation $G_T(\hat{\theta}) = 0$.

We consider a class of $\mathcal{H} \in \mathcal{G}$ of estimating functions G_T which are a.s. differentiable with respect to the components of θ and such that

$$E\dot{G}_T = (E\partial G_{T,i}/\partial\theta_j)$$

and EG_TG_T' are nonsingular, the prime denoting transpose.

Recent studies on the performance of estimating functions G_T involve focusing on the quantity

$$\mathcal{E}(G_T) = (E\dot{G}_T)' (EG_TG_T')^{-1} (E\dot{G}_T)$$

which may be thought of as an information matrix (and is indeed the Fisher information matrix when $G_T = U_T$, the score function). Our object is to maximize this estimating function information within specified classes of useful estimating functions and to this end we adopt the following definition whose origins date back to Godambe (1960) and Durbin (1960).

Definition 1. *Suppose that* $G_T^* \in \mathcal{G}_1 \subseteq \mathcal{G}$. *If*

$$\mathcal{E}(G_T^*) - \mathcal{E}(G_T)$$

is nonnegative definite for all $G_T \in \mathcal{G}_1$, *we say that* G_T^* *is* O_F*–optimal within* $\mathcal{G}_1$.

The terminology O_F–optimal refers to finite sample optimality. There is also a closely related concept of O_A–optimality (asymptotic optimality) (see Godambe and Heyde (1987), Heyde (1988)) and both hold under broad conditions. We shall confine attention to O_F–optimality in this paper for clarity of exposition.

Note that the definition of O_F–optimality does not require the existence of the score function, and indeed the setting may be nonparametric. Nevertheless, to motivate the definition it is useful to suppose that the score function U_T exists and is a.s. differentiable with respect to the components of θ , and that differentiation and integration can be interchanged in $EG_T U_T'$ and $EU_T G_T'$ for all G_T under consideration.

Now, as noted above, the score function typically provides a minimal sufficient partitioning of the sample space so we should use this estimating function as a basis for inference if it is known. However, if the true underlying parametric family and hence the score function is unknown we can seek an estimating function G_T which has minimum dispersion distance from U_T . This is precisely what an O_F–optimal solution provides. That is, for some fixed matrix α depending on θ and T ,

$$E(\alpha_T U_T - G_T)(\alpha_T U_T G_T)' - E(\alpha_T U_T - G_T^*)(\alpha_T U_T - G_T^*)'$$

is nonnegative definite for all $G_T \in \mathcal{G}_1$. (Of course an optimal estimating function is defined only up to a constant (matrix) multiplier.) Furthermore, if the dimension of

θ, $p = 1$, another equivalent formulation is that G_T^* has maximum squared correlation with U_T. These are, of course, least squares principles built into the formulation. For details of the equivalences see Godambe and Heyde (1987).

Additionally, under a fairly broad range of conditions the covariance matrix of the estimator $\bar{\theta}_T$ given by $G_T(\bar{\theta}_T) = 0$ is asymptotically $(\mathcal{E}(G_T(\theta_0)))^{-1}$ and, indeed,

$$(EG_T(\theta_0)\, G_T'(\theta_0))^{-1/2}\, (E\dot{G}_T(\theta_0))\, (\bar{\theta}_T - \theta_0) \xrightarrow{d} N_p(0,I_p)\,. \tag{1}$$

where "d" denotes convergence in distribution to $N_p(0,I_p)$, the standard p–variate normal. Hence, asymptotic confidence zones of minimum size are associated with an O_F–optimal estimator for which $(\mathcal{E}(G_T(\theta_0)))^{-1}$ is minimized in the partial order of nonnegative definite matrices. This is of course, the characteristic property of maximum likelihood but here there is a restricted class of competitors. Nevertheless, the prime advantage of maximum likelihood is built into the formulation.

The heuristics are as follows. By Taylor expansion

$$0 = G_T(\bar{\theta}_T) = G_T(\theta_0) + \dot{G}_T(\theta_{1,T})\, (\bar{\theta}_T - \theta_0)$$

where $\|\theta_0 - \theta_{1,T}\| \le \|\theta_0 - \bar{\theta}_T\|$ and under certain conditions,

$$\dot{G}_T(\theta_{1,T})\, (E\dot{G}_T(\theta_0))^{-1} \xrightarrow{p} I_p\,,$$

$$(EG_T(\theta_0)\, G_T'(\theta_0))^{-1/2}\, G_T(\theta_0) \xrightarrow{d} N_p(0,I_p)$$

as $T \to \infty$ from which (1) readily follows.

The criterion of Definition 1 is often not of direct practical value but an equivalent from is given in the following theorem of Heyde (1988) which is easier to use and is preferabe to other equivalent terms such as those mentioned above.

Theorem 1. *Suppose that* $\mathcal{X}_1 \subseteq \mathcal{X}$ *is a convex set. Then,* $G_T^* \in \mathcal{X}_1$ *is* O_F*–optimal within* $\mathcal{X}_1$ *if and only if*

$$(E\dot{G}_T)^{-1} EG_T G_T^{*\prime} = (E\dot{G}_T^*)^{-1} EG_T^* G_T^{*\prime}$$

for all $G_T \in \mathcal{X}_1$.

3. SOME RAMIFICATIONS

An estimating function which is O_F-optimal can be regarded as a *quasi-score* estimating function and an estimator derived therefrom as a *quasi-likelihood estimator.* We shall henceforth employ this terminology which was introduced by Wedderburn (1974) and has been widely used in the context of the general linear model (e.g. McCullagh and Nelder (1983, Ch.8)) as it is a helpful description and underlines its useful properties. A quasi-score estimating function will indeed be a true score in a wide variety of situations, such as in an exponential family environment; for some general discussion of this issue see Sorensen (1989).

The classical quasi-likelihood setting concerns estimation of θ where the sample is of independent random vectors Y_t, $1 \le t \le T$, with means $\mu_t(\theta)$ and dispersions

$$E(Y_t - \mu_t(\theta))(Y_t - \mu_t(\theta))' = v_t(\theta) .$$

The quasi-score estimating function is defined as

$$Q_T = \sum_{t=1}^{T} \dot{\mu}_t' v_t^+ (Y_t - \mu_t) = \dot{\mu}' V^+ (Y - \mu)$$

where

$$\dot{\mu}' = (\dot{\mu}_1' \vdots \dots \vdots \dot{\mu}_T') , \quad Y' = (Y_1' \vdots \dots \vdots Y_T') , \quad V^+ = \text{diag}(v_1^+ \vdots \dots \vdots v_T^+)$$

and for a matrix A, A^+ denotes its Moore-Penrose pseudoinverse, the unique matrix possessing the properties $AA^+A = A$, $A^+AA^+ = A^+$, $A^+A = AA^+$. That Q_T is O_F-optimal within the relevant class of weighted sums of $(Y_t - \mu_t)$'s follows immediately from Theorem 1.

As another application of Theorem 1 we note that under the regularity conditions which are imposed,

$$E\dot{G}_T = -EG_T U_T'$$

24

and if $U_T \in \mathcal{X}_1$, then

$$(E\dot{G}_T)^{-1} EG_T U_T' = -I_p .$$

Thus, U_T is O_F-optimal within $\mathcal{X}_1$ via Theorem 1. This means that for $G_T \in \mathcal{X}_1$,

$$(E\dot{G}_T)^{-1} (EG_T G_T') ((E\dot{G}_T)^{-1})' - (EU_T U_T')^{-1} \tag{2}$$

is nonnegative definite which encapsulates the multivariate *Cramer–Rao Inequality*. This is usually presented in the case where G_T is of the form $G_T = S_T - \theta$ when S_T is a statistic with $ES_T = \theta$, and then (2) gives the nonnegative definiteness of

$$E(S_T - \theta)(S_T - \theta)' - (EU_T U_T')^{-1} .$$

For random sampling $EU_T U_T' = n\,\Sigma$, say, where Σ is the Fisher information contained in a single observation.

A final application of Theorem 1 treats the *Gauss–Markov Theorem*. This has traditionally been set in the context of a multiparameter linear regression model, for example

$$Y = X\theta + \varepsilon$$

where $Y = (y_1,...,y_n)'$, X is an $n \times p$ design matrix of rank p, $\theta = (\theta_1,...,\theta_p)'$ is a parameter vector and ε is a zero mean error vector with independent components and covariance matrix V.

Unbiased linear estimates of θ are of the form AY where A is a $p \times n$ matrix such that $AX = I_p$, and we may think of these as coming from a corresponding linear unbiased estimating function $A(Y - X\theta)$ which, when set to zero, yields the estimator AY.

The standard approach via least squares is to minimize the covariance matrix

$$E(AY - \theta)(AY - \theta)' = AVA' ,$$

subject to $AX = I_p$, in the partial order of nonnegative definite matrices. This leads to

25

the solution

$$A^* = (X'V^{-1}X)^{-1} X'V^{-1} , \qquad (3)$$

and estimator A^*Y for θ , provided V is nonsingular.

Now an alternative approach is to seek the O_F-optimal estimating function within the family of estimating functions

$$\mathcal{X} = \{A(Y - X\theta) , \quad AX = I_p\} .$$

Then, writing

$$G = A(Y - X\theta) , \quad G^* = A^*(Y - X\theta)$$

with A^* as in (3), we have, via Theorem 1, that G^* provides the desired solution if

$$(E\dot{G})^{-1} EGG^{*\prime}$$

is a constant matrix. This is, furthermore, easily checked since

$$E\dot{G} = -I$$

and

$$EGG^{*\prime} = AE(\varepsilon\varepsilon')A^{*\prime} = AVA^{*\prime} = (X'V^{-1}X)^{-1} .$$

This theory can be put in a more general setting of best linear unbiased estimators (BLUE's) for stochastic processes, as in Grenander (1981, Ch.4) with minimal change to the mathematical details. Here it is assumed that $Y(t)$ has mean $m(t)$ where $m(t)$ is expressible in the form

$$m(t) = \sum_{\nu=1}^{p} a_\nu \psi_\nu(t) = \alpha'\Psi$$

where the coefficients $(\alpha_1,\ldots,\alpha_p)' = \alpha$ are real or complex unknown constants and the $(\psi_1,\ldots,\psi_p)' = \Psi$ are given functions. Similar considerations to the above lead to Theorem 2, Chapter 4 of Grenander (1981) and we can interpret the BLUE as coming from the O_F-optimal estimating function within the class

$$\{cY, c'\Psi = I_p\} .$$

4. FURTHER CONSIDERATIONS AND EXTENSIONS

In this concluding section we mention a number of related issues of importance.

Many estimation problems are based in a natural way on a semimartingale model which thereby leads to consideration of martingale estimating functions. For such models, O_A-optimality, which is briefly mentioned above and is very closely related to O_F-optimality, is generally a more pertinent concept. For the beginnings of a comparison of the different criteria see Heyde (1988).

O_F and O_A-optimality are exact properties and in some applications optimality holds only in an asymptotic sense. For a treatment of asymptotic quasi-likelihood see Heyde and Gay (1989).

Quasi-score estimating functions based on different classes of estimating functions can be derived in many contexts and it is important to be able to compare them and to combine them if this should be advantageous. Furthermore, various problems present a collection of conditional or marginal processes leading to separate quasi-score or score functions which can potentially be combined. This may occur, for example, when likelihood methods are very difficult to use or when it is only part of the model that can be treated satisfactorily. In either case, one has to consider a set of quasi-score functions $Q_{i,T}$, $1 \leq i \leq k$, and the maximizing of information over the class of estimating functions $\{\Sigma_{i=1}^{k} w_i Q_{i,T}\}$. New results on this problem are given in Heyde (1989) building upon discussions of the case $k = 2$ in Heyde (1987) and of the combinations of true score functions in Lindsay (1988).

Much more remains to be said on the subject of optimal inference in problems where there are nuisance parameters. For some discussion and references see McLeish and small (1988). Their approach to optimal inference is via concepts of E-ancillarity and E-sufficiency which permit inferential reduction analogous to the usual sufficiency and ancillarity reductions but within the class of unbiased estimating functions.

Finally, rather little theory has yet been developed for extensions of optimal

27

estimation beyond the case of finite dimensional parameters. Contributions include the thesis of Thavaneswaran (1986) and treatment of the special case of linear inference (Grenander (1981)). Many important applications involving estimating of functions await this theory and *ad hoc* methods abound.

REFERENCES

Durbin, J. (1960). Estimation of parameters in time-series regression models. *J.R. Statist. Soc.* B 22, 139-153.

Fisher, R.A. (1912). On an absolute criterion for fitting frequency curves. *Messeng. Math.* 41, 155-160.

Fisher, R.A. (1920). A mathematical examination of the methods of determining the accuracy of an observation by the mean error, and by the mean square error. *Mon. Not. Roy. Astron. Soc.* 80, 758-770.

Fisher, R.A. (1922). On the mathematical foundations of theoretical statistics. *Phil. Trans.* A 222, 309-368.

Fisher, R.A. (1925). Theory of statistical estimation. *Proc. Camb. Phil. Soc.* 22, 700-725.

Godambe, V.P. (1960). An optimal property of regular maximum likelihood estimation. *Ann. Math. Statist.* 31, 1208-1212.

Godambe, V.P. and Heyde, C.C. (1987). Quasi-likelihood and optimal estimation. *Int. Statist. Rev.* 55, 231-244.

Grenander, U. (1981). *Abstract Inference*. Wiley, New York.

Heyde, C.C. (1987). On combining quasi-likelihood estimating functions. *Stoch. Processes Applic.* 25, 281-287.

Heyde, C.C. (1988). Fixed sample and asymptotic optimality for classes of estimating functions. *Contemporary Mathematics* 80, 241-247.

Heyde, C.C. (1989). On efficiency for quasi-likelihood and composite quasi-likelihood methods. In Proceedings of International Conference on Recent Developments in Statistical Data Analysis and Inference, Neuchâtel, August 1989. Elsevier, Amsterdam, to appear.

28

Heyde, C.C. and Gay, R. (1989). On asymptotic quasi-likelihood estimation. *Stoch. Processes. Applic.* **31**, in press.

Lindsay, B.G. (1988). Composite likelihood methods. *Contemporary Mathematics* 80, 221-239.

McCullagh, P. and Nelder, J.A. (1983). *Generalized Linear Models.* Chapman and Hall, London.

McLeish, D.L. and Small, C.G. (1988). *The Theory and Application of Statistical Inference Functions.* Springer Lecture Notes in Statistics No. 44, Springer, New York.

Pearson, K. (1894). Contributions to the mathematical theory of evolution. *Phil. Trans. Roy. Soc.* **A 185**, 71-110.

Sorensen, M. (1989). On quasi-likelihood for semimartingales. *Stoch. Processes Applic.*, to appear.

Thavaneswaran, A. (1986). *Estimations of Semimartingales.* PhD thesis, University of Waterloo, Canada.

Wedderburn, R.W.M. (1974). Quasi-likelihood functions, generalized linear models, and the Gauss-Newton method. *Biometrika* **61**, 439-447.

SUMMARY

This paper outlines the recent development of a general theory of quasi-likelihood which embraces the principal features of the methods of least squares and maximum likelihood.

RESUMÉ

Cet article expose le developpement nouveau de une théorie générale de quasi-vraisemblance. par où on embrasse les méthodes de moindre carrés et de maximum de vraisemblance.

29

The Annals of Statistics
1992, Vol. 20, No. 1, 603–607

ON BEST ASYMPTOTIC CONFIDENCE INTERVALS FOR PARAMETERS OF STOCHASTIC PROCESSES

BY C. C. HEYDE

Australian National University

This paper is concerned with the size of confidence intervals for parameters of stochastic processes based on limit laws with two competing normalizations, one producing asymptotic normality and the other asymptotic mixed normality. It is shown that, in a certain sense, the interval based on asymptotic normality is preferable on average. Applications to estimation of parameters in nonergodic stochastic processes and to estimation of steady-state parameters in a simulation are given to illustrate the theory.

1. Introduction and results. Much asymptotic inference for stochastic processes is based on use of some obvious consistent estimator but involves a choice between competing normalizations, one being of constants and the other of random variables. In this paper we study the common situation where either asymptotic normality or asymptotic mixed normality is achievable through suitable normalization. It is shown that there is a certain sense in which, on average, confidence intervals based on the asymptotic normality result are preferable to those based on asymptotic mixed normality.

THEOREM. *Let $\{\hat{\theta}_n\}$ be a sequence of estimators which is consistent for θ and $\{c_n\}, \{d_n\}$ be norming sequences, possibly random, such that*

$$c_n\left(\hat{\theta}_n - \theta\right) \to_d W, \tag{1}$$

$$d_n\left(\hat{\theta}_n - \theta\right) \to_d \eta^{-1}W \tag{2}$$

as $n \to \infty$, where W is standard normal and $\eta > 0$ a.s. is random, independent of W and such that

$$c_n d_n^{-1} \to_d \eta$$

as $n \to \infty$. Suppose that $L_n^{(i)}(\delta)$, $i = 1, 2$, is the minimum length of an $100(1 - \delta)\%$ confidence interval for θ based on an exact, approximate or assumed distribution of the pivot in (1) and (2), respectively, for which the specified convergence result holds. Then

$$\liminf_{n \to \infty} E\left(L_n^{(2)}(\delta)/L_n^{(1)}(\delta)\right) \geq 1.$$

Received January 1991; revised July 1991.
AMS 1980 *subject classifications*. Primary 62F11; secondary 62F25, 62M09.
Key words and phrases. Best confidence intervals, normalization, asymptotic normality, asymptotic mixed normality, nonergodic models.

R. Maller et al. (eds.), *Selected Works of C.C. Heyde*, Selected Works in Probability and Statistics, DOI 10.1007/978-1-4419-5823-5_51, © Springer Science+Business Media, LLC 2010

REMARK 1. The theorem gives a sense in which, on average, confidence intervals based on the pivot in (1) are to be preferred to those based on the pivot in (2) whether or not a random norming is required.

REMARK 2. A principal application of the theorem is in the context of inference for the nonergodic models that are common in stochastic process estimation [e.g., Basawa and Scott (1983), Hall and Heyde (1980), Chapter 6]. In this context we have $\{c_n\}$ as random and $\{d_n\}$ as constants so that $L_n^{(1)}(\delta)$ is a random variable and $L_n^{(2)}(\delta)$ is a constant.

One of the most important special cases is that of locally asymptotic mixed normal (LAMN) families. This is the situation in which the log-likelihood ratio has asymptotically a mixed normal distribution and results of type (2) hold with $\{d_n\}$ as constants and (1) with $\{c_n\}$ as random variables, while $Ec_n = d_n$ for each n. For LAMN families one has, under modest regularity conditions, a striking result known as Hajék's convolution theorem [e.g., Theorem 2, page 47, of Basawa and Scott (1983)] to the effect that if $\{T_n\}$ is any other sequence of consistent estimators of θ such that

$$d_n(T_n - \theta) \to_d U \tag{3}$$

for some nondegenerate U, then

$$U =_d V + \eta^{-1}W,$$

where V is independent of η and W.

It is readily shown that confidence intervals based on the pivot in (3) are wider than those for the corresponding result (2). That is, using (1) is better on average than (2), which is better than (3). If Φ is the standard normal distribution function, we have for any real a, β, α with $\beta > \alpha$,

$$\Phi((\beta - a)y) - \Phi((\alpha - a)y) \le \Phi(\tfrac{1}{2}(\beta - \alpha)y) - \Phi(-\tfrac{1}{2}(\beta - \alpha)y) \tag{4}$$

and hence, integrating with respect to $dP(\eta \le y)$,

$$P(\alpha < \eta^{-1}W + a < \beta) \le P(-\tfrac{1}{2}(\beta - \alpha) < \eta^{-1}W < \tfrac{1}{2}(\beta - \alpha)), \tag{5}$$

so that

$$\begin{aligned} P(\alpha < \eta^{-1}W + V < \beta) &= \int_{-\infty}^{\infty} P(\alpha < \eta^{-1}W + v < \beta)\, dP(V \le v) \\ &\le P(-\tfrac{1}{2}(\beta - \alpha) < \eta^{-1}W < \tfrac{1}{2}(\beta - \alpha)). \end{aligned}$$

REMARK 3. A particular case of the result of the theorem has been obtained by Glynn and Iglehart (1990) who studied the problem of finding minimum size asymptotic confidence intervals for steady state parameters of the simulation output process from a single simulation run. They contrasted the approach of consistently estimating the variance constant in the relevant central limit theorem with the standardized time series approach which avoids estimation of the variance in a manner reminiscent of the t-statistic and suggested that the former approach is preferable on average.

2. Proof of theorem. Suppose that Φ_n and Ψ_n are exact, approximate or assumed distribution functions of the pivots $c_n(\hat{\theta}_n - \theta)$ and $d_n(\hat{\theta}_n - \theta)$, respectively. We are given the complete convergence results

$$\Phi_n \to_c \Phi, \qquad \Psi_n \to_c \Psi$$

as $n \to \infty$, where

$$\Phi(x) = P(W \leq x) \quad \text{and} \quad \Psi(x) = P(\eta^{-1}W \leq x) = \int_0^\infty \Phi(xy)\, dG(y)$$

with $G(x) = P(\eta \leq x)$.

Now let a_n, b_n be any numbers for which

$$\Phi_n(b_n) - \Phi_n(a_n) \geq 1 - \delta.$$

This gives a confidence interval for θ of length

$$l_n^{(1)}(\delta) = (b_n - a_n)/c_n.$$

By passing to a subsequence, we may suppose that $a_n \to a$ and $b_n \to b$, where $-\infty \leq a \leq b \leq \infty$. Then, by complete convergence,

$$\Phi_n(a_n) - \Phi_n(b_n) \to \Phi(b) - \Phi(a),$$

and, in view of (4), $b - a \geq 2z_\delta$, where $2\Phi(z_\delta) = 2 - \delta$. Therefore, noting that $c_n l_n^{(1)}(\delta)$ is not random, we have

$$\liminf_{n\to\infty} c_n L_n^{(1)}(\delta) \geq 2z_\delta. \tag{6}$$

Next, let z_n be the smallest z for which

$$\Phi_n(z) - \Phi_n(-z) \geq 1 - \delta.$$

As above, we may suppose that $z_n \to z_\infty$ and, because Φ is continuous,

$$\Phi_n(z_n) - \Phi_n(-z_n) = 1 - \delta + o(1)$$

as $n \to \infty$. By complete convergence and using the result $\Phi(-x) = 1 - \Phi(x)$, $x > 0$, it follows that $2\Phi(z_\infty) = 2 - \delta$ and hence that $z_\infty = z_\delta$. Then, $L_n^{(1)}(\delta) \leq 2z_n/c_n$, so that

$$\liminf_{n\to\infty} c_n L_n^{(1)}(\delta) \leq 2z_\delta, \tag{7}$$

and (6) and (7) imply

$$\lim_{n\to\infty} c_n L_n^{(1)}(\delta) = 2z_\delta. \tag{8}$$

Similar reasoning to that which led to (8) also applies to show that

$$\lim_{n\to\infty} d_n L_n^{(2)}(\delta) = 2\zeta_\delta, \tag{9}$$

where $2\Psi(\zeta_\delta) = 2 - \delta$. We merely replace (4) by (5) to show that the symmetric confidence interval is best, while $\Psi(-x) = 1 - \Psi(x)$, $x > 0$, is evident from the definition and the corresponding result for Φ.

From (6) and (9), it then follows that, as $n \to \infty$,

$$d_n L_n^{(2)}(\delta)/c_n L_n^{(1)}(\delta) \to \zeta_\delta / z_\delta .$$

Since $c_n/d_n \to_d \eta$, Slutsky's theorem gives

$$L_n^{(2)}(\delta)/L_n^{(1)}(\delta) \to_d \eta\zeta_\delta/z_\delta$$

and then, using Billingsley [(1968), Theorem 5.3, page 32],

$$\liminf_{n\to\infty} E\big(L_n^{(2)}(\delta)/L_n^{(1)}(\delta)\big) \geq (E\eta)\zeta_\delta/z_\delta .$$

Thus, in order to complete the proof, it remains to show that

$$(E\eta)\zeta_\delta \geq z_\delta . \tag{10}$$

Now note that $\zeta_\delta = \zeta(\eta;\delta)$ solves the equation

$$\Psi(\zeta(\eta;\delta)) = P(\eta^{-1}W \leq \zeta(\eta;\delta)) = 1 - \delta/2.$$

Then, taking $b > 0$, we have

$$1 - \frac{\delta}{2} = P(W \leq \zeta(\eta;\delta)\eta) = P\left(W \leq \frac{1}{b}\zeta(\eta;\delta)b\eta\right) = P(W \leq \zeta(b\eta;\delta)b\eta)$$

so that continuity and strict monotonicity of Ψ imply that

$$\zeta(b\eta;\delta) = \frac{1}{b}\zeta(\eta;\delta)$$

and hence, if

$$\psi(\eta) = (E\eta)\zeta(\eta;\delta),$$

we have

$$\psi(b\eta) = \psi(\eta). \tag{11}$$

Now we see from (10) that it is required to show

$$\psi(\eta) \geq \Phi^{-1}(1-\delta/2)$$

and using (11) we may, without loss of generality, scale η so that

$$\zeta(\eta;\delta) = 1. \tag{12}$$

But (12) implies

$$\Psi(1) = 1 - \delta/2,$$

or equivalently,

$$\int_0^\infty \Phi(y)G(y) = E\Phi(\eta) = 1 - \delta/2.$$

Thus, we have to show that

$$\psi(\eta) = E\eta \geq \Phi^{-1}(1-\delta/2)$$

subject to

$$E\Phi(\eta) = 1 - \delta/2,$$

and since Φ is monotone, this holds if

$$\Phi(E\eta) \geq E\Phi(\eta). \tag{13}$$

However, since $1 - \Phi(x)$ is convex for $x > 0$, as is easily checked by differentiation, we have from Jensen's inequality that

$$1 - \Phi(E\eta) \leq E(1 - \Phi(\eta))$$

and (13) follows. This completes the proof.

FINAL REMARKS. It is worth noting that (10), which relates the average confidence intervals when (1) and (2) hold exactly rather than as limits, is a paraphrase of the fact that, on average, extraneous randomization does not improve confidence intervals for a normal mean. That is, if X is distributed as $N(\theta, 1)$, the usual symmetric confidence interval for θ based on $X - \theta$ is not improved, on average, if one instead uses the pivot $(x - \theta)/\eta$, where η is independent of θ. In particular, when $\bar{X}_n, s_n^2$ are the mean and variance of a random sample from the $N(\theta, \sigma^2)$ distribution where σ *is known*, using the t-intervals for θ based on $(\bar{X}_n - \theta)/s_n$ produces confidence intervals which are longer on average than z-intervals.

Acknowledgment. The perceptive comments of an Associate Editor which purged the theorem of a number of unnecessary restrictions are gratefully acknowledged.

REFERENCES

BASAWA, I. V. and SCOTT, D. J. (1983). *Asymptotic Optimal Inference for Non-ergodic Models. Lecture Notes in Statist.* **17**. Springer, New York.

BILLINGSLEY, P. (1968). *Convergence of Probability Measures*. Wiley, New York.

GLYNN, P. and IGLEHART, D. L. (1990). Simultaneous output analysis using standardized time series. *Math. Oper. Res.* **15** 1–16.

HALL, P. and HEYDE, C. C. (1980). *Martingale Limit Theory and Its Application*. Academic, New York.

STATISTICS RESEARCH SECTION
INSTITUTE OF ADVANCED STUDIES
AUSTRALIAN NATIONAL UNIVERSITY
GPO BOX 4, CANBERRA
ACT 2601
AUSTRALIA

A quasi-likelihood approach to estimating parameters in diffusion-type processes

C. C. HEYDE

Abstract

Estimation of parameters in diffusion models is usually handled by maximum likelihood and involves the calculation of a Radon–Nikodym derivative. This methodology is often not available when minor changes are made to the model. However, these complications can usually be avoided and results obtained under more general conditions using quasi-likelihood methods. The basic ideas are explained in this paper and are illustrated through discussion of the Cox–Ingersoll–Ross model and a modification of the Langevin model.

COX–INGERSOLL–ROSS MODEL; LANGEVIN MODEL; INFERENCE; PARAMETER ESTIMATION; DIFFUSION PROCESSES

AMS 1991 SUBJECT CLASSIFICATION: PRIMARY 60J60

1. Introduction

The standard method of estimation for parameters in the drift coefficient of a diffusion process involves calculation of a likelihood ratio (Radon–Nikodym derivative) and thence the maximum likelihood estimator(s). This is less than straightforward for more complicated models, and indeed it is often not available at all because of the non-existence of the Radon–Nikodym derivative. New methods of quasi-likelihood, however, allow estimators to be obtained straightforwardly under very general conditions. They can deal, in particular, with the situation in which the Brownian motion in a diffusion is replaced by a general square-integrable martingale. The approach, which is based on selection of an optimal estimating function from within a specified class of such functions, involves assumptions on only the first two conditional moments of the underlying process. Nevertheless, the quasi-likelihood estimators will ordinarily be true maximum likelihood estimators in a context where the Radon–Nikodym derivative is available. Furthermore, they will generally be consistent, asymptotically normally distributed, and can be used to construct minimum-size asymptotic confidence zones for the unknown parameters among estimators coming from the specified class. The purpose of this paper is to illustrate the value of the methodology through a general discussion and application to the

R. Maller et al. (eds.), *Selected Works of C.C. Heyde*, Selected Works in Probability and Statistics, DOI 10.1007/978-1-4419-5823-5_52, © Springer Science+Business Media, LLC 2010

Cox–Ingersoll–Ross model for interest rates and to a modification of the Langevin model for dynamical systems.

Since the general theory of quasi-likelihood is not widely known we shall begin with a sketch of the basic principles. For a detailed discussion of the general framework see Godambe and Heyde (1987). Various amplifications are provided in Heyde (1988) and Sørensen (1990). For a particularly accessible introduction see Heyde (1989).

It should be noted that the term 'quasi-likelihood' also appears in the literature with a rather different interpretation from that given here (e.g. Wedderburn (1974), McCullagh and Nelder (1989)). A discussion of the relationship between these interpretations is provided in Lin and Heyde (1993).

2. General principles

The general framework is as follows. We are given a sample $\{Z_t, t \in T\}$, say, T being discrete or continuous, taking values in r-dimensional Euclidean space, and the set of probability measures $\{\boldsymbol{P}_\theta\}$ for $\{Z_t\}$ is a union of families of models, each being indexed by a characteristic θ belonging to an open subset of p-dimensional Euclidean space. The object is the efficient estimation of θ in situations where a likelihood function may not be available.

Attention is focused on the class $\mathscr{G}$ of zero-mean square-integrable estimating functions $G_T(\theta) = G_T(\{Z_t, t \in T\}, \theta)$ which are vectors of dimension p for which $\boldsymbol{E}G_T(\theta) = 0$ for each $\boldsymbol{P}_\theta$ and for which the p-dimensional matrices $\boldsymbol{E}\dot{G}_T(\theta) = (\boldsymbol{E}\partial G_{T,i}(\theta)/\partial\theta_j)$ and $\boldsymbol{E}G_T(\theta)G'_T(\theta)$ are non-singular, the prime denoting transpose. The expectations are always with respect to $\boldsymbol{P}_\theta$. Within $\mathscr{G}$, estimators $\hat{\theta}_T$ are found by solving the estimating equation $G_T(\hat{\theta}_T) = 0$.

Quasi-likelihood theory is focused on suitably chosen subsets of $\mathscr{G}$ and involves choice of the estimating function G to maximize, in the partial order of non-negative definite (n.n.d.) matrices, the information criterion

$$\mathscr{E}(G_T) = (\boldsymbol{E}\dot{G}_T)'(\boldsymbol{E}G_TG'_T)^{-1}\boldsymbol{E}\dot{G}_T$$

which is a natural generalization of Fisher information (see e.g. Godambe and Heyde (1987)). (Here and below we omit the argument θ for simplicity.)

Definition. Suppose that $G^*_T \in \mathscr{H} \subseteq \mathscr{G}$. If

$$\mathscr{E}(G^*) - \mathscr{E}(G_T)$$

is n.n.d. for all $G_T \in \mathscr{H}$ we say that G^*_T is a quasi-score estimating function within $\mathscr{H}$.

The estimator θ^*_T obtained from $G^*_T(\theta^*_T) = 0$, which is termed a *quasi-likelihood estimator*, has, under broad conditions, certain minimum variance and minimum size asymptotic confidence zone properties for θ, at least within the class $\mathscr{H}$. Indeed, the basic properties are those of the maximum likelihood estimator but restricted to the class $\mathscr{H}$.

The theory does not require a parametric setting, let alone the existence of a likelihood score function U_T (derivative of the log-likelihood with respect to θ), but if U_T exists a judicious choice of $\mathscr{H}$ will usually produce it as the quasi-score estimating function. If $U_T \in \mathscr{H}$, as can ordinarily be arranged in exponential family problems, then U_T is the quasi-score estimating function within $\mathscr{H}$ and can easily be calculated. Godambe (1960) showed that U_T is the quasi-score estimating function within $\mathscr{G}$.

It is not usually practicable to find a quasi-score estimating function directly from the use of the definition. However, the following proposition (Heyde (1988)) provides for ready derivation.

Proposition 1. *Let $\mathscr{H} \subset \mathscr{G}$. Then $G_T^* \in \mathscr{H}$ is a quasi-score estimating function within $\mathscr{H}$ if*

$$(1) \qquad (\boldsymbol{E}\dot{G}_T)^{-1}\boldsymbol{E}G_T G_T^{*'} = C_T$$

for all $G_T \in \mathscr{H}$, where C_T is a fixed matrix. Conversely, if $\mathscr{H}$ is convex and G_T^* is a quasi-score estimating function then (1) *holds.*

3. The application

The models which we shall consider in this paper can all be written in the semimartingale form

$$(2) \qquad dX_t = dA_t(\theta) + dM_t(\theta)$$

where the finite variation process $\{A_t\}$ can be interpreted as the *signal* and the local martingale $\{M_t\}$ can be interpreted as the *noise*. The local martingale has a natural role to play in inference as it represents the residual after fitting of the signal. Then, the quasi-score estimating function based on the family of local martingale estimating functions

$$\mathscr{H} = \left\{\int_0^T a_t(\theta)dM_t(\theta), \{a_t\} \text{ non-anticipating}\right\}$$

is easily seen from Proposition 1 to be

$$(3) \qquad \int_0^T (\boldsymbol{E}(d\dot{M}_t(\theta)|\mathscr{F}_{t-}))'(d\langle M(\theta)\rangle_t)^+ dM_t(\theta)$$

where $\{\mathscr{F}_t\}$ is a filtration of past-history σ-fields, $\langle M(\theta)\rangle_t$ is the quadratic characteristic and the $+$ denotes the Moore–Penrose generalized inverse which is, for a matrix A, the unique matrix A^+ possessing the properties $AA^+A = A$, $A^+AA^+ = A^+$, $A^+A = AA^+$.

Now (3) can be rewritten as

$$\int_0^T (\boldsymbol{E}(d\dot{A}_t(\theta)|\mathscr{F}_{t-}))'(d\langle M(\theta)\rangle_t)^+(dA_t(\theta) - dX_t)$$

from which it is clear that the quasi-score estimating function is unaffected if the local martingale noise $\{M_t(\theta)\}$ is replaced by another whose quadratic characteristic is the same. The precise distributional form of the noise does not need to be known. In the commonly met situation where $M_t(\theta) = \sigma W_t$ with $\sigma > 0$ and W_t being standard Brownian motion, the results are robust to the extent that $\{M_t(\theta)\}$ could be replaced by any local martingale $\{Z_t(\theta)\}$, for example one with independent increments, for which $\langle Z\rangle_t = \sigma^2 t$ without changing the estimators.

In the particular case of a diffusion process the components on the right-hand side of the representation (2) can be written as

$$(4) \qquad dA_t(\theta) = a(t, X_t, \theta)dt, \qquad dM_t(\theta) = b^{1/2}(t, X_t)dW_t$$

where a and b are known vector and matrix functions respectively and $\{W_t\}$ is standard Brownian motion. Then, an appropriate Radon–Nikodym derivative of the measure induced by the process $\{X_t, 0 \leq t \leq T\}$ with parameter θ with respect to the corresponding measure for parameter θ_0 can be calculated and is given by

$$(5) \qquad \exp\left\{\int_0^T C(t, X_t)dX_t - \int_0^T D(t, X_t)dt\right\}$$

where

$$b(t,x)C(t,x) = a(t,x,\theta) - a(t,x,\theta_0)$$
$$D(t,x) = (a(t,x,\theta_0))'C(t,x) + \tfrac{1}{2}(C(t,x))'b(t,x)C(t,x)$$

(e.g. Basawa and Prakasa Rao (1980), p. 219).

From (5) it is easily checked that the likelihood score function (derivative of the logarithm of the Radon–Nikodym derivative with respect to θ) is given by (3). This means that the quasi-likelihood estimator is the maximum likelihood estimator for the model (4). However, as indicated above, the quasi-likelihood estimator is available much more generally.

The explanation for the quasi-score corresponding to the likelihood score for the model (4) is not hard to discern. A likelihood score is a martingale under modest regularity conditions and all square integrable martingales living on the same probability space as the Brownian motion in the noise term of the model (4) can be described as stochastic integrals with respect to the Brownian motion (see e.g. Theorem 5.17 of Lipster and Shiryaev (1977)). The likelihood score will be one such martingale and will therefore be included in the relevant family $\mathscr{H}$ over which optimization takes place and it solves the optimization problem.

As a concrete illustration of the methodology we shall discuss the stochastic differential equation

$$dX_t = \alpha(\beta - X_t)dt + \sigma\sqrt{X_t}dW_t \tag{6}$$

where $X_0 > 0, \alpha > 0, \beta > 0, \sigma > 0$. This form was proposed by Cox, Ingersoll and Ross (1985) as a model for interest rates and it has been widely used in finance.

In considering the model (6) we shall be concerned with the estimation of $\theta = (\alpha, \beta)'$. The parameter σ can be regarded as known whenever Brownian motion and continuous sampling are involved. Indeed, σ can be calculated with probability 1 on the basis of knowledge of a path of the process on any finite time interval. This follows from the definitions of the quadratic variation process and stochastic integrals with respect to Brownian motion (e.g. Rogers and Williams (1987), Chapter IV, Section 4) from which one obtains that, writing $t_i^{(n)} = \min(T, 2^{-n}i)$,

$$\lim_{n\to\infty} \sum_{i=0}^{\infty} (X_{t_{i+1}^{(n)}} - X_{t_i^{(n)}})^2 = \sigma^2 \int_0^T X_t dt \text{ a.s.}$$

and

$$\lim_{n\to\infty} \sum_{i=0}^{\infty} X_{t_i^{(n)}} (t_{i+1}^{(n)} - t_i^{(n)}) = \int_0^T X_t dt \text{ a.s.}$$

For the model (6) the Radon–Nikodym derivative of the measure based on $(\alpha, \beta)'$ with respect to that based on $(\alpha_0, \beta_0)'$ is easily seen from (5) to be

$$\exp\left\{\sigma^{-2} \int_0^T X_t^{-1}[\alpha(\beta - X_t) - \alpha_0(\beta_0 - X_t)]dX_t \right.$$
$$\left. - \frac{1}{2}\sigma^{-2} \int_0^T X_t^{-1}[\alpha^2(\beta - X_t)^2 - \alpha_0^2(\beta_0 - X_t)^2]dt \right\}$$

and differentiating the logarithm of this likelihood ratio with respect to $\theta = (\alpha, \beta)'$ gives the likelihood score

$$U_T = \sigma^{-2} \int_0^T \begin{pmatrix} \beta - X_t \\ \alpha \end{pmatrix} X_t^{-1/2} dW_t \tag{7}$$

which is also the quasi-score given by (2).

If we modify the model to the form

$$dX_t = \alpha(\beta - X_t)dt + \sigma(\alpha, \gamma)\sqrt{X_t}dW_t$$

where $\sigma(\alpha, \gamma)$ reflects a possibly rate-dependent noise, then the likelihood ratio does not exist in general. Indeed, when $\sigma(\alpha_1, \lambda) \neq \sigma(\alpha_2, \lambda)$ the supports of the distributions of the two processes are disjoint. The quasi-likelihood methodology, however, is unaffected by this change. The quasi-score estimating function continues to be given by (7) and the asymptotic properties of the estimators are also unaffected.

From (7), the maximum likelihood/quasi-likelihood estimators $\hat{\alpha}_T$, $\hat{\beta}_T$ are given by

$$\int_0^T (\hat{\beta}_T - X_t) X_T^{-1} [dX_t - \hat{\alpha}_T(\hat{\beta}_T - X_t)dt] = 0$$

$$\int_0^T X_t^{-1} [dX_t - \hat{\alpha}_T(\hat{\beta}_T - X_t)dt] = 0$$

and putting

$$I_T = \int_0^T X_t^{-1} dX_t, \qquad J_T = \int_0^T X_t^{-1} dt, \qquad K_T = \int_0^T X_t dt,$$

we find that

$$\hat{\alpha}_T = (I_T T - J_T(X_T - X_0))/(J_T K_T - T^2)$$

$$\hat{\beta}_T = (I_T K_T - T(X_T - X_0))/(I_T T - J_T(X_T - X_0)).$$

For the model with $2\alpha\beta \geq \sigma^2$ there is a strictly positive stationary ergodic solution to (6) at $T \to \infty$ whose distribution has gamma density $\Gamma(2\alpha\beta/\sigma^2, 2\alpha/\sigma^2)$ (see e.g. Kloeden and Platen (1992), p. 38). Suppose X_∞ has this density; then

$$\text{(8)} \qquad \boldsymbol{E}X_\infty = \beta, \qquad \boldsymbol{E}X_\infty^{-1} = 2\alpha/(2\alpha\beta - \sigma^2).$$

Using the ergodic theorem we obtain

$$\text{(9)} \qquad T^{-1} J_T = T^{-1} \int_0^T X_t^{-1} dt \xrightarrow{\text{a.s.}} \boldsymbol{E}X_\infty^{-1},$$

$$\text{(10)} \qquad T^{-1} K_T = T^{-1} \int_0^T X_t dt \xrightarrow{\text{a.s.}} \boldsymbol{E}X_\infty$$

and, since

$$I_T = \int_0^T X_t^{-1} dX_t = \log X_0^{-1} X_T + \tfrac{1}{2}\sigma^2 J_T$$

using Itô's formula,

$$T^{-1} I_T \xrightarrow{\text{a.s.}} \tfrac{1}{2}\sigma^2 \boldsymbol{E}X_\infty^{-1}$$

as $T \to \infty$. These results readily give the strong consistency of the estimators $\hat{\alpha}_T$, $\hat{\beta}_T$. Asymptotic normality of $T^{1/2}(\hat{\alpha}_T - \alpha, \hat{\beta}_T - \beta)'$ can be obtained by applying Theorem 2.1, p. 405 of Basawa and Prakasa Rao (1980).

All these results continue to hold under substantially weakened conditions on the noise component in the model. For example, they hold if $\{W_t\}$ is replaced by a square integrable martingale with stationary independent increments $\{Z_t\}$ for which $\langle Z\rangle_t \equiv t$. The details involve straightforward applications of the martingale strong law and central limit theorem and are omitted.

The formulation via (2) and (3) has to be used with care for problems with multiple sources of variation. Suppose for example that the Langevin stochastic differential equation (e.g. Kloeden and Platen (1992), pp. 104–105) is augmented with jumps coming from a Poisson process and becomes

(11) $$dX_t = \theta X_t dt + dW_t + dN_t,$$

N_t being a Poisson process with intensity λ. Then, the process may be written in semi-martingale form as

$$dX_t = (\theta X_t + \lambda)dt + dM_t$$

where $\{M_t\}$ is a martingale given by

(12) $$M_t = W_t + N_t - \lambda t.$$

Using (11) and (12) in (3), the quasi-score estimating function based on noise $\{M_t\}$ is

$$\int_0^T (X_{t-} 1)' dM_t,$$

leading to the estimating equations

$$\int_0^T X_{t-} dX_t = \hat{\theta}_T \int_0^T X_t^2 dt + \hat{\lambda}_T \int_0^T X_t dt$$

$$X_T = \hat{\theta}_T \int_0^T X_t^2 dt + \hat{\lambda}_T T.$$

These, however, are the maximum likelihood estimating equations for the model

$$dX_t = (\theta X_t + \lambda)dt + dW_t,$$

i.e. a version of (11) in which N_t has been replaced by its compensator λt. In this model the entire stochastic fluctuation is described by the Brownian process and this is only realistic if $\lambda \ll 1$.

This problem, first noted by Sørensen (1990), can be circumvented and the true maximum likelihood estimators for θ, λ obtained if we treat the sources of variation separately. We replace (11) by the equations

$$dX_t^c = \theta X_t dt + dW_t, \qquad X_t^c = X_t - N_t$$

$$dX_t - dX_t^c = dN_t = \lambda dt + (dN_t - \lambda t)$$

which focus on the continuous and discrete components respectively. Then, using separate quasi-score estimating functions based on the martingales $\{W_t\}$ and

$\{N_t - \lambda t\}$ respectively, we obtain estimating equations

$$\int_0^T X_{t-} dX_t^c = \tilde{\theta}_T \int_0^T X_t^2 dt$$

$$N_T = \tilde{\lambda}_T T$$

leading to maximum likelihood estimators $\tilde{\theta}_T$, $\tilde{\lambda}_T$.

Models of the above kind are quite common and the general message is to identify relevant (local) martingales which focus on the individual sources of variation. It is then possible to obtain quasi-score estimating functions for each, and to combine them, provided the appropriate sample information is available.

References

BASAWA, I. V. AND PRAKASA RAO, B. L. S. (1980) *Statistical Inference for Stochastic Processes.* Academic Press, London.

COX, J. S., INGERSOLL, J. E. AND ROSS, S. A. (1985) A theory of the term structure of interest rates. *Econometrica* **53**, 363–384.

GODAMBE, V. P. (1960) An optimum property of regular maximum likelihood estimation. *Ann. Math. Statist.* **31**, 1208–1211.

GODAMBE, V. P. AND HEYDE, C. C. (1987) Quasi-likelihood and optimal estimation. *Int. Statist. Rev.* **55**, 231–244.

HEYDE, C. C. (1988) Fixed sample and asymptotic optimality for classes of estimating functions. *Contemp. Math.* **80**, 241–247.

HEYDE, C. C. (1989) Quasi-likelihood and optimality for estimating functions: some current unifying themes. *Bull. Int. Statist. Inst.* **53**(1), 19–29.

KLOEDEN, P. E. AND PLATEN, E. (1992) *Numerical Solution of Stochastic Differential Equations.* Springer-Verlag, Berlin.

LIN, Y.-X. AND HEYDE, C. C. (1993) Optimal estimating functions and Wedderburn's quasi-likelihood. *Commun. Statist. Theory Meth.* **22**, 2341–2350.

LIPSTER, R. S. AND SHIRYAEV, A. N. (1977) *Statistics of Random Processes I. General Theory.* Springer-Verlag, New York.

MCCULLAGH, P. AND NELDER, J. A. (1989) *Generalized Linear Models*, 2nd edn. Chapman and Hall, London.

ROGERS, L. C. G. AND WILLIAMS, D. (1987) *Diffusions, Markov Processes and Martingales. Vol. 2 Itô Calculus.* Wiley, Chichester.

SØRENSEN, M. (1990) On quasi-likelihood for semimartingales. *Stoch. Proc. Appl.* **25**, 331–346.

WEDDERBURN, R. W. M. (1974) Quasi-likelihood functions, generalized linear models, and the Gauss–Newton method. *Biometrika* **61**, 439–447.

(origins or supply).

$$\sum_i x_{ij} = b_j, \quad j = 1, \dots, n$$

(destinations or demand), where $x_{ij} \geq 0$ and $\sum a_i = \sum b_j$, which is called the *balance condition*. The assignment problem arises when $m = n$ and all a_i and b_j are 1.

If all a_i and b_j in the transposed problem are integers, then there is an optimal solution for which all x_{ij} are integers (*Dantzig's theorem on integral solutions of the transport problem*).

In the assignment problem, for such a solution x_{ij} is either zero or one; $x_{ij} = 1$ means that person i is assigned to job j; the weight c_{ij} is the utility of person i assigned to job j.

The special structure of the transport problem and the assignment problem makes it possible to use algorithms that are more efficient than the **simplex method**. Some of these use the *Hungarian method* (see, e.g., [4], [5, Chapt. 7]), which is based on the König–Egervary theorem (see **König theorem**), the method of potentials (see [5], [1]), the *out-of-kilter algorithm* (see, e.g., [3]) or the transportation simplex method.

In turn, the transportation problem is a special case of the network optimization problem.

A totally different assignment problem is the **pole assignment problem** in control theory.

References

[1] FRISCH, R.: 'La résolution des problèmes de programme linéaire par la méthode du potentiel logarithmique', *Cahier Sém. Econom.* **4** (1956), 20–23.

[2] GRÖTSCHEL, M., LOVÁSZ, L., AND SCHRIJVER, A.: *Geometric algorithms and combinatorial optimization*, Springer, 1987.

[3] MURTZ, K.: *Linear and combinatorial programming*, Wiley, 1976.

[4] PAPADIMITRIOU, C.H., AND STEIGLITZ, K.: *Combinatorial optimization*, Prentice-Hall, 1982.

[5] YUDIN, D.B., AND GOL'SHTEIN, E.G.: *Linear programming*, Israel Program. Sci. Transl., 1965. (Translated from the Russian.)

M. Hazewinkel

MSC 1991: 90C05, 90B80, 90B06

ASYMPTOTIC INVARIANT OF A GROUP – A property of a **finitely-generated group** G which is a **quasi-isometry** invariant of the **metric space** (G, d_A), where d_A is the **word metric** associated to a finite generating set A of G (cf. also **Quasi-isometric spaces**). This definition does not depend on the choice of the set A, since if B is another finite set of generators of G, then the metric spaces (G, d_A) and (G, d_B) are quasi-isometric.

The theory of asymptotic invariants of finitely-generated groups has been recently brought to the foreground by M. Gromov (see, in particular, [2] and [3]). As Gromov says in [3, p. 8], 'one believes nowadays that the most essential invariants of an infinite group are asymptotic invariants'. For example, amenability (cf. **Invariant average**), hyperbolicity (in the sense of Gromov, cf. **Hyperbolic group**), the fact of being finitely presented (cf. **Finitely-presented group**), and the number of ends (cf. also **Absolute**) are all asymptotic invariants of finitely-generated groups. It is presently (1996) unknown whether the Kazhdan T-property is an asymptotic invariant. For an excellent survey on these matters, see [1].

A few examples of algebraic properties which are asymptotic invariants of finitely-generated groups are: being virtually nilpotent, being virtually Abelian, being virtually free.

References

[1] GHYS, E.: 'Les groupes hyperboliques', *Astérisque* **189-190** (1990), 203–238, Sém. Bourbaki Exp. 722.

[2] GROMOV, M.: 'Hyperbolic groups', in S.M. GERSTEN (ed.): *Essays in Group Theory*, Vol. 8 of *MSRI Publ.*, Springer, 1987, pp. 75–263.

[3] GROMOV, M.: 'Asymptotic invariants of infinite groups': *Proc. Symp. Sussex, 1991: II*, Vol. 182 of *London Math. Soc. Lecture Notes*, Cambridge Univ. Press, 1993, pp. 1–291.

A. Papadopoulos

MSC 1991: 20F32, 57M07

ASYMPTOTIC OPTIMALITY of estimating functions – Efficient estimation (cf. **Efficient estimator**) of parameters in stochastic models is most conveniently approached via properties of estimating functions, namely functions of the data and the parameter of interest, rather than estimators derived therefrom. For a detailed explanation see [3, Chapt. 1].

Let $\{X_t : 0 \leq t \leq T\}$ be a **sample** in discrete or continuous time from a stochastic system taking values in an r-dimensional Euclidean space. The distribution of X_t depends on a parameter of interest θ taking values in an open subset of a p-dimensional Euclidean space. The possible probability measures (cf. **Probability measure**) for X_t are $\{\mathsf{P}_\theta\}$, a union of families of models.

Consider the class $\mathcal{G}$ of zero-mean square-integrable estimating functions $G_T = G_T(\{X_t : 0 \leq t \leq T\}, \theta)$, which are vectors of dimension p and for which the matrices used below are non-singular.

Optimality in both the *fixed sample* and the *asymptotic* sense is considered. The former involves choice of an estimating function G_T to maximize, in the partial order of non-negative definite matrices, the information criterion

$$\mathcal{E}(G_T) = (\mathsf{E}\nabla G_T)'(\mathsf{E}G_T G_T')^{-1}(\mathsf{E}\nabla G_T),$$

which is a natural generalization of the **Fisher amount of information**. Here ∇G is the $(p \times p)$-matrix of derivatives of the elements of G with respect to those of θ and

R. Maller et al. (eds.), *Selected Works of C.C. Heyde*, Selected Works in Probability and Statistics, DOI 10.1007/978-1-4419-5823-5_53, © Springer Science+Business Media, LLC 2010

prime denotes transposition. If $\mathcal{H} \subset \mathcal{G}$ is a prespecified family of estimating functions, it is said that $G_T^* \in \mathcal{H}$ is *fixed sample optimal* in $\mathcal{H}$ if $\mathcal{E}(G_T^*) - \mathcal{E}(G_T)$ is nonnegative definite for all $G_T \in \mathcal{H}$, θ and P_θ. Then, G_T^* is the element of $\mathcal{H}$ whose dispersion distance from the maximum information estimating function in $\mathcal{G}$ (often the likelihood score) is least.

A focus on asymptotic properties can be made by confining attention to the subset $\mathcal{M} \subset \mathcal{G}$ of estimating functions which are martingales (cf. **Martingale**). Here one considers T ranging over the positive real numbers and for $\{G_T\} \in \mathcal{M}$ one writes $\{\langle G\rangle_T\}$ for the *quadratic characteristic*, the predictable increasing process for which $\{G_T G_T' - \langle G\rangle_T\}$ is a martingale. Also, write $\{\overline{G}_T\}$ for the predictable process for which $\{\nabla G_T - \overline{G}_T\}$ is a martingale. Then, $G_T^* \in \mathcal{M}_1 \subset \mathcal{M}$ is *asymptotically optimal* in $\mathcal{M}_1$ if $\overline{\mathcal{E}}(G_T^*) - \overline{\mathcal{E}}(G_T)$ is almost surely nonnegative definite for all $G_T \in \mathcal{M}_1$, θ, P_θ, and $T > 0$, where

$$\overline{\mathcal{E}}(G_T) = \overline{G}_T' \langle G\rangle_T^{-1} \overline{G}_T.$$

Under suitable regularity conditions, asymptotically optimal estimating functions produce estimators for θ which are consistent (cf. **Consistent estimator**), asymptotically unbiased (cf. **Unbiased estimator**) and asymptotically normally distributed (cf. **Normal distribution**) with minimum size asymptotic confidence zones (cf. **Confidence estimation**). For further details see [1], [2].

References

[1] GODAMBE, V.P., AND HEYDE, C.C.: 'Quasi-likelihood and optimal estimation', *Internat. Statist. Rev.* **55** (1987), 231–244.

[2] HEYDE, C.C.: *Quasi-likelihood and its application. A general approach to optimal parameter estimation*, Springer, 1997.

[3] MCLEISH, D.L., AND SMALL, C.G.: *The theory and applications of statistical inference functions*, Lecture Notes in Statistics. Springer, 1988.

C.C. Heyde

MSC 1991: 62F12, 62G20

J. Appl. Prob. **34**, 939–944 (1997)
Printed in Israel

ON DEFINING LONG-RANGE DEPENDENCE

C. C. HEYDE,* *Australian National University and Columbia University*
Y. YANG,** *Columbia University*

Abstract

Long-range dependence has usually been defined in terms of covariance properties relevant only to second-order stationary processes. Here we provide new definitions, almost equivalent to the original ones in that domain of applicability, which are useful for processes which may not be second-order stationary, or indeed have infinite variances. The ready applicability of this formulation for categorizing the behaviour for various infinite variance models is shown.

LONG-RANGE DEPENDENCE; PERSISTENCE; FRACTAL MODELS; ALLEN VARIANCE; INFINITE VARIANCE MODELS; FRACTIONAL STABLE NOISE

AMS 1991 SUBJECT CLASSIFICATION: PRIMARY 60G10; 60G18
SECONDARY 62M10

1. Introduction

The usual definition of long-range dependence is in terms of a second-order stationary process. Let $\{X_t, t=1, 2, \cdots\}$ be a zero mean second-order stationary process with covariances

$$\gamma_k = \mathrm{cov}(X_t, X_{t+k})$$

and spectral density

$$f(\omega) = \frac{1}{2\pi} \sum_{k=-\infty}^{\infty} \gamma_k \, e^{-ik\omega}, \qquad -\pi < \omega < \pi.$$

Then, long-range dependence holds if the spectral density is infinite at frequency $\omega = 0$, while short-range dependence holds if $f(0)$ is finite. We shall write LRD(SD) and SRD(SD) for these definitions, the SD referring to spectral density.

The LRD(SD) definition is inflexible in providing for extensions that allow for departure from stationarity and, indeed, infinite variances. Long-range dependence is a phenomenon which is widely observed in nature and it is characterized by sample paths displaying apparent trends and cycles, although these are ephemeral and are eventually replaced by equally convincing looking ones of quite different apparent direction and

Received 5 November 1996; revision received 11 March 1997.
* Postal address: Stochastic Analysis Group, School of Mathematical Sciences, The Australian National University, Canberra, ACT 0200, Australia and Department of Statistics, Columbia University, New York, NY 10027, USA.
** Postal address: Department of Statistics, Columbia University, New York, NY 10027, USA.

R. Maller et al. (eds.), *Selected Works of C.C. Heyde*, Selected Works in Probability and Statistics, DOI 10.1007/978-1-4419-5823-5_54, © Springer Science+Business Media, LLC 2010

period. Models in which variances may be infinite are needed for the analysis of a variety of such phenomena (e.g. in finance [5], geology [6], [7], [8] and hydrology [11]). For example, in [11] (e.g. Table 1 therein) the one-dimensional distributions required are of type $\boldsymbol{P}(X>x)\sim Cx^{-\beta}$ for large values of x, where β ranges from 1.10 to 3.22, while the self-similarity parameter of the process is close to 2/3. In [6], [7], [8], the permeability and porosity of sedimentary rock formations are analyzed and strong evidence is provided for fractional Lévy stable models with index α, $1<\alpha<2$, and self-similarity parameter $H<1/\alpha$.

Now consider the block means process $\{X_t^{(m)}\}$ given by

$$X_t^{(m)}=(X_{tm-m+1}+\cdots+X_{tm})/m$$

and the Allen variance

$$V_m=\operatorname{var}(X_t^{(m)}).$$

Then, as has been noted by various authors (e.g. [2]), a nearly equivalent formulation of the dichotomy between SRD(SD) and LRD(SD) is given by $mV_m\to$ finite constant as $m\to\infty$ (SRD(AV)) and $mV_m\to\infty$ as $m\to\infty$ (LRD(AV)). Here (AV) refers to definitions with respect to the Allen variance. Indeed, we see that

$$mV_m=\sum_{|j|\le m}\left(1-\frac{|j|}{m}\right)\gamma_j,$$

so that, using the Kronecker lemma,

$$mV_m\to\sum_{j=-\infty}^{\infty}\gamma_j=\gamma_0+2\sum_{j=1}^{\infty}\gamma_j$$

as $m\to\infty$ if the right-hand side is finite. Also, if $\Sigma_{|j|\le m}\gamma_j$ tends to $\pm\infty$ as $m\to\infty$ then so does mV_m. However, it is not always assured that $mV_m\to\sigma^2$, say, $0<\sigma^2<\infty$, will imply that $f(0)$ is finite.

2. Extensions

It turns out that the LRD(AV) definition is readily amenable to extension whereas that for the spectral density lacks similar flexibility. To deal with the possibility of non-stationarity we define a new block means process

$$Y_t^{(m)}=\frac{\Sigma_{tm-m+1}^{tm}X_j}{\Sigma_{tm-m+1}^{tm}\boldsymbol{E}X_j^2},$$

for zero mean variables, and we shall say that the process is LRD(AV) if

$$\left(\sum_{j=tm-m+1}^{tm}\boldsymbol{E}X_j^2\right)\operatorname{var}Y_t^{(m)}\to\infty$$

as $m\to\infty$. This conveniently allows for departure from stationarity, certainly including the periodic effects associated with seasonality, and it reduces to the original LRD(AV) definition in the stationary case.

An example of a non-stationary process which is LRD(AV) is the random walk $\{X_j = \Sigma_{k=1}^{j} Z_k\}$, the increments $\{Z_k\}$ being independent and identically distributed with zero mean and finite variance. This seems an entirely appropriate interpretation as the random walk is well known to produce long leads and rare sign changes despite being 'fair'.

To deal with the situation in which variances may be infinite we suggest replacing the LRD(AV) definition by an empirical version. Suppose that the X_j are centered to have zero mean when the mean is finite. We shall say that the process is LRD(SAV), SAV denoting sample Allen variance, if

$$\frac{(\Sigma_{j=tm-m+1}^{tm} X_j)^2}{\Sigma_{j=tm-m+1}^{tm} X_j^2} \xrightarrow{\mathrm{P}} \infty$$

as $m \to \infty$. The particular value of this criterion is that it provides a ready basis for statistical testing without *a priori* knowledge about the variances.

It should be noted that if LRD(SAV) holds for X_t stationary with finite variance then

$$\frac{(\Sigma_{j=tm-m+1}^{tm} X_j)^2}{\Sigma_{j=tm-m+1}^{tm} X_j^2} \xrightarrow{\mathrm{P}} \infty \Leftrightarrow m(X_t^{(m)})^2 \xrightarrow{\mathrm{P}} \infty \Rightarrow mV_m \to \infty,$$

so that LRD(AV) holds. Also, LRD(SAV) holds for the random walk example described above, as can be seen by applying Donsker's invariance principle for the central limit theorem (e.g. [1], pp. 137, 142) separately to numerator and denominator in the definition.

To underscore the value of the LRD(SAV) definition for dealing with infinite variance problems we illustrate by considering the collection of fractional stable noise models discussed by Samorodnitsky and Taqqu [10], Section 7.10. Properties such as covariation ([10], Section 2.7) and codifference ([10], Section 2.10), which are surrogates for the covariance when variances are not finite, do not provide a satisfactory basis for an extension of the standard definition of LRD, although an asymptotic analysis of the codifference for linear fractional stable noise is provided in [10]. In Section 7.10 the authors resort to analogy with the finite variance case to call the process LRD if $H > 1/\alpha$, $0 < \alpha < 2$. They note that the codifference does not even tend to zero as the lag increases for the other models described, namely increments of the real harmonizable stable noise process and the sub-Gaussian fractional stable noise. For fractional stable Lévy noise, Painter [7] refers to the cases $H > 1/\alpha$, $H < 1/\alpha$ as persistent and anti-persistent (descriptions typically associated with LRD and SRD respectively), again by analogy with the fractional Brownian case.

The models discussed in [10], which essentially form the natural paradigm models for long-range dependence in the infinite variance case are as follows.

(a) The linear fractional stable noise process has increments

$$X_j = \int_{-\infty}^{\infty} (a[(j+1-x)_+^{H-1/\alpha} - (j-x)_+^{H-1/\alpha}]$$
$$+ b[(j+1-x)_-^{H-1/\alpha} - (j-x)_-^{H-1/\alpha}])M(dx),$$

where a, b are real constants, $|a|+|b|>0$, $0<\alpha<2$, $0<H<1$, $H\neq 1/\alpha$ and M is an α-stable random measure with Lebesgue control measure. Here $a_+=\max(0, a)$, $a_-=-\min(0, a)$.

(b) The real harmonizable fractional stable noise process has increments

$$X_j = \mathrm{Re}\int_{-\infty}^{\infty} \mathrm{e}^{\mathrm{i}xj}\,\frac{\mathrm{e}^{\mathrm{i}x}-1}{\mathrm{i}x}\,|x|^{-H+1-1/\alpha}\tilde{M}(dx),$$

where $0<\alpha\leqq 2$, $0<H<1$ and where $\tilde{M}$ is a complete isotropic symmetric α-stable random measure with Lebesgue control measure.

(c) The sub-Gaussian fractional stable noise process has increments

$$X_j = A^{1/2}(B_H(j+1)-B_H(j)),$$

where B_H is fractional Brownian motion of index H, $0<H<1$, and A is an independent $\alpha/2$ stable random variable with Laplace transform $E\mathrm{e}^{-tA}=\mathrm{e}^{-t^{\alpha/2}}$, $t>0$. We shall discuss the use of the LRD(SAV) criterion on these models.

In each case it is easily checked that we have self-similarity with parameter H, so that

$$\sum_{tm-m+1}^{tm} X_j \stackrel{\mathrm{d}}{=} m^H X_1. \tag{1}$$

We shall henceforth confine attention to processes X_j which have this self-similarity property.

Our basic result on LRD is given in the following theorem.

Theorem 1. *Let X_j be a stationary process for which block sums are self-similar with parameter H (i.e. satisfy* (1)) *and suppose that $E|X_1|^p<\infty$ for some $0<p<2$. Then LRD(SAV) holds if $H>1/p$. In particular, if $P(|X_1|>x)\sim Cx^{-\alpha}$ as $x\to\infty$, $0<\alpha<2$, then LRD(SAV) holds if $H>1/\alpha$.*

The result of this theorem follows from the use of the relationship (1) to deal with the numerator in the LRD(SAV) criterion and the stability theorem E of Loève [4], p. 387, to deal with the denominator. The stability theorem gives $n^{-2/p}\sum_{j=1}^{n} X_j^2 \stackrel{\text{a.s.}}{\to} 0$ as $n\to\infty$. The particular case is established by noting that $E|X_1|^p<\infty$ for all $p<\alpha$.

To explore the LRD/SRD relationship further, we need to examine L_p variants of the SAV criterion. Write

$$F_m(p) = \frac{|\sum_{tm-m+1}^{tm} X_j|^p}{\sum_{tm-m+1}^{tm} |X_j|^p}, \qquad 0<p\leqq 2.$$

Suppose that X_j is stationary, ergodic, satisfies the self-similarity condition (1) and $E|X_1|^p<\infty$ for $0<p<\alpha<2$. Then, we have that $F_m(p)\stackrel{\text{a.s.}}{\to}\infty$ as $m\to\infty$ if $pH>1$, $p<\alpha$, i.e. $1/H<p<\alpha$. This follows from the use of self-similarity on the numerator and the ergodic theorem on the denominator of $F_m(p)$. The result corresponds to the case of long-range dependence as identified by the LRD(SAV) criterion, namely as holding for $H<1/\alpha$.

On the other hand, if $E|X_1|^p = \infty$ and if $\alpha < p < 2, pH < 1$, so that $\alpha < p < \min(2, 1/H)$, we have $F_m(p) \xrightarrow{\text{a.s.}} 0$ as $m \to \infty$. To obtain this result we again use self-similarity on the numerator of $F_m(p)$ and, for dealing with the denominator we note that, using the ergodic theorem with $I(\cdot)$ for the indicator function,

$$n^{-1} \sum_{j=1}^{n} |X_j|^p \geqq n^{-1} \sum_{j=1}^{n} |X_j|^p I(|X_j| \leqq k) \xrightarrow{\text{a.s.}} E|X_j|^p I(|X_j| \leqq k) \to \infty$$

as $k \to \infty$. The result here corresponds to short-range dependence.

Note that $\alpha < \min(2, 1/H)$ implies $H < 1/\alpha$ (since for $H \geqq 1/2$, $\alpha < \min(2, 1/H) = 1/H$, i.e. $H < 1/\alpha$, while for $H < 1/2$, $\alpha < \min(2, 1/H) = 2$ and hence $1/\alpha > 1/2 > H$). This clarifies the SRD ($H < 1/\alpha$) and LRD ($H > 1/\alpha$) dichotomy. We have chosen, however, not to make the $F_m(p)$, $p < 2$, the basis of a formal definition of LRD/SRD in order to avoid the uncertainty associated with p values which would generally be unknown in practice.

It should be observed that, for the sub-Gaussian fractional stable noise process mentioned above, the dichotomy is $H > 1/2$ (LRD) and $H < 1/2$ (SRD). The model here is not ergodic. A single history of the process appears to be from a fractional Brownian motion process and therefore to reflect, as this does, LRD or SRD according to $H > 1/2$ or $H < 1/2$. This can easily be seen via the AV or SAV criteria for LRD/SRD. Note that $H > 1/\alpha$ ensures $H > 1/2$ and the subtlety of ergodic or non-ergodic behaviour does not enter into the LRD consideration of Theorem 1.

Although the formal definition of LRD(SAV) can be used as a practical criterion for classifying behaviour, it must be used with care as it is based on the assumption of zero means, when these are finite. As an exercise we have investigated the time series of daily returns for the S&P 500 market index for the period 2 July 1962 to 31 December 1991, the data set having been supplied by a referee. Daily, weekly and monthly series were examined and there is no evidence of LRD. This assessment contrasts with that of Peters [9], p. 113, who used *R/S* analysis on the returns from the S&P 500 index for the period January 1950 to July 1988 and concluded that the value of the Hurst index H rises steadily from 0.59 for one-day increments to 0.78 for 30-day increments. *R/S* analysis is, however, well-known to be highly sensitive to departures from stationarity (e.g. Heyde and Dai [3]) of the kind that might be expected for financial series.

Acknowledgements

Helpful comments from a referee and from Huang Dawei are gratefully acknowledged.

References

[1] Billingsley, P. (1968) *Convergence of Probability Measures.* Wiley, New York.

[2] Cox, D. R. (1984) Long-range dependence: a review. In *Statistics: an Appraisal.* ed. H. A. David and H. T. David. Iowa State University Press, Ames. pp. 55–74.

[3] Heyde, C. C. and Dai, W. (1996) On the robustness to small trends of estimation based on the smoothed periodogram. *J. Time Series Anal.* **17**, 141–150.

[4] Loève, M. (1963) *Probability Theory.* 3rd edn. Van Nostrand, Princeton, NJ.

[5] Mittnik, S. and Rachev, S. T. (1997) *Modeling Financial Assets with Alternative Stable Models.* Wiley, New York.

[6] Painter, S. (1995) Random fractal models of heterogeneity: the Lévy-stable approach. *Math. Geol.* **27**, 813–830.

[7] Painter, S. (1996) Existence of non-Gaussian scaling behaviour in heterogeneous sedimentary formations. *Water Resources Res.* **32**, 1183–1195.

[8] Painter, S. (1996) Stochastic interpolation of aquifer properties using fractional Lévy motion. *Water Resouces Res.* **32**, 1323–1332.

[9] Peters, E. E. (1991) *Chaos and Order in the Capital Markets.* Wiley, New York.

[10] Samorodnitsky, G. and Taqqu, M. S. (1994) *Stable Non-Gaussian Random Processes. Stochastic Models with Infinite Variance.* Chapman and Hall, New York.

[11] Turcotte, D. L. (1994) Fractal theory and the estimation of extreme floods. *J. Res. Nat. Inst. Stand. Tech.* **99**, 377–389.

J. Appl. Prob. **36,** 1234–1239 (1999)
Printed in Israel

A RISKY ASSET MODEL WITH STRONG DEPENDENCE THROUGH FRACTAL ACTIVITY TIME

C. C. HEYDE,* *Australian National University and Columbia University*

Abstract

The geometric Brownian motion (Black–Scholes) model for the price of a risky asset stipulates that the log returns are i.i.d. Gaussian. However, typical log returns data shows a leptokurtic distribution (much higher peak and heavier tails than the Gaussian) as well as evidence of strong dependence. In this paper a subordinator model based on fractal activity time is proposed which simply explains these observed features in the data, and whose scaling properties check out well on various data sets.

Keywords: Risky asset model; Black–Scholes model; heavy tails; long-range dependence; fractal activity time; self-similarity

AMS 1991 Subject Classification: Primary 90A09
Secondary 62M10; 60G18

1. Introduction: the basic evidence

Over the past thirty years many risky asset models have been proposed, but these are generally deficient in not adequately explaining key features from the statistical evidence related to the issues of heavy tails and long-range dependence in financial time series. In the case of the latter, there have been problems even with understanding the nature of the characteristic, let alone its assessment.

In this note we provide a simple model which incorporates the statistical features as they are now understood. We outline the statistical evidence in Section 1. The model is given in Section 2 and its relation to the considerable previous literature is outlined in Section 3.

The paradigm model in mathematical finance is the geometric Brownian motion and this is a natural point of reference. Under this model the price P_t at time t of a risky asset is

$$P_t = P_0 \exp[\mu t + \sigma W(t)],$$

where $\mu, \sigma^2 > 0$ are fixed constants and $W(t)$ is a standard Brownian motion. Then the corresponding log returns,

$$X_t = \log P_t - \log P_{t-1} = \mu + \sigma(W(t) - W(t-1)),$$

are i.i.d. Gaussian with mean μ and variance σ^2.

In contrast, the typical log returns data shows:

- a pronounced leptokurtic distribution (much higher peaks and heavier tails than Gaussian),

Received 12 January 1998; revision received 28 January 1998.
* Postal address: School of Mathematical Sciences, Australian National University, Canberra, ACT 0200, Australia.
Email address: chris@maths.anu.edu.au

R. Maller et al. (eds.), *Selected Works of C.C. Heyde*, Selected Works in Probability and Statistics,
DOI 10.1007/978-1-4419-5823-5_55, © Springer Science+Business Media, LLC 2010

- a time series with high volatility and intermittency quite unlike white noise,
- strong dependence.

It is now widely appreciated that heavier tails than the normal are a necessary model feature for marginal distributions of returns. However, models with finite moments of all orders are still commonly advocated. Detailed comparisons using market index data have been given in Hurst *et al.* (1997) and Hurst and Platen (1997) and they favour a t-distribution with degrees of freedom ν typically in the range 3–5. This, of course, implies an infinite kth moment for $k \geq \nu$.

Most approaches to the marginal distribution of returns have involved the use of parametric families and the choice of a best fit within a family. Few authors have investigated general questions, such as whether particular moments are finite. Part of the difficulty has been a general supposition that tail behaviour would be a power function, at least asymptotically, and that estimating the power index is problematical. Indeed, the usual method, based on the Hill estimator, is founded on order statistic theory for i.i.d. samples and it can suffer from major shortcomings outside that context. See, for example, the 'Hill Horror Plot' in Resnick (1997) and the attendant discussion.

It is, however, quite possible to test for finiteness of moments based on asymptotics obtained via the use of the ergodic theorem if the assumption of stationarity of the returns is retained. In particular, if $\sum_{i=1}^{n} |X_i|^p/n$ converges as $n \to \infty$ then $E|X|^p < \infty$. If, on the other hand, $\max_{1 \leq i \leq n} |X_i|^p / \sum_{i=1}^{n} |X_i|^p$ stays away from zero as $n \to \infty$ then $E|X|^p = \infty$. Plots as n increases for large data sets should give a clear picture. For example, multiple tests on daily data from the well-known S&P 500 index lead to the conclusion that the 4th moment is infinite (e.g. Mikosch (1997); Yang (1998)).

The next issue for modelling is that of strong dependence in the data. Log returns always display a clustering of extremes (intermittency). Sample autocorrelations of the log returns die away quickly but this is not the case for their absolute values or squares, which have non-negligible values for large lags. This is convincingly illustrated in Ding and Granger (1996) and Granger and Ding (1996) with daily returns data from the S&P 500 index, the Nikkei index, the foreign exchange rate DM/US$, and Chevron stock. In the case of the absolute values from the S&P series, for example, the lagged autocorrelations are significantly different from zero for lags up to 2700, which is more than 10 years. The absolute values and squares of the returns exhibit long-range dependence (LRD), classically defined (for stationary finite variance processes) as holding if $\sum_{k=0}^{\infty} \gamma_k$ diverges, γ_k being the autocorrelation at lag k. For a discussion of the definition and its extensions see Heyde and Yang (1997). For a comprehensive overview of the subject of LRD see Beran (1994).

Claims of long-range dependence in the log returns themselves have sometimes been made (e.g. Peters (1991, 1994) for the S&P index) but this has generally been done on the basis of R/S analysis, which is known to be highly sensitive to departures from stationarity (see for example Heyde and Dai (1996)). However, the use of other methods which are robust against departures from stationarity does not support this conclusion (Heyde and Yang (1997)). But the debate still continues. Lo (1991) developed a modified R/S statistic and on the basis of its use rejected the hypothesis of LRD in the log returns, but weaknesses in his procedure have been pointed out by Willinger *et al.* (1999), who dispute his assessment, albeit without compelling evidence. Nevertheless, it remains an inescapable fact that the strong dependence in the data is principally reflected in the absolute values and squares of the log returns and not in the log returns themselves.

None of the existing models capture all these features satisfactorily and we now proceed to outline how this can be done straightforwardly.

2. Modelling to retain the essential features of the geometric Brownian motion (GBM) model

2.1. Subordinator model (Stage 1)

The $\{\log P_t\}$ process should have stationary differences and the absolute values and squares of the differences should exhibit LRD. We suppose that

$$P_t = P_0 \exp[\mu t + \sigma W(T_t)],$$

where $\{T_t\}$ is a positive increasing random process with stationary differences which is independent of the Brownian motion $\{W(t)\}$ and the differences of the $\{T_t\}$ process are LRD and have heavy tails.

The random time changed version of GBM has all the features suggested by the data. Clock time just does not correspond to activity time (also called trading time or intrinsic market time).

Without loss of generality, choose $T_t \sim t$ almost surely as $t \to \infty$.

Then

$$\begin{aligned} X_t &= \log P_t - \log P_{t-1} \\ &= \mu + \sigma(W(T_t) - W(T_{t-1})) \\ &\stackrel{\mathrm{d}}{=} \mu + \sigma(T_t - T_{t-1})^{1/2} W(1). \end{aligned}$$

The heavy tails of the X_t come from those of $\tau_t = T_t - T_{t-1}$.

Also, for $k = 1, 2, \ldots$ and centred variables,

$$\begin{aligned} \operatorname{cov}(X_t, X_{t+k}) &= 0, \\ \operatorname{cov}(|X_t|, |X_{t+k}|) &= \sigma^2 \operatorname{cov}(\tau_t^{1/2}, \tau_{t+k}^{1/2}), \\ \operatorname{cov}(X_t^2, X_{t+k}^2) &= 3\sigma^4 \operatorname{cov}(\tau_t, \tau_{t+k}), \end{aligned}$$

the last if $E\tau_t{}^2 < \infty$. Thus, LRD of $\{|X_t|\}$ and $\{X_t{}^2\}$ follows from that of $\{\tau_t^{1/2}\}$ and $\{\tau_t\}$ respectively.

Other features of the model are conditional heteroscedasity,

$$\operatorname{var}(X_t \mid \mathcal{F}_{t-1}) = \sigma^2 E(\tau_t \mid \mathcal{F}_{t-1}),$$

and leptokurtosis

$$\operatorname{kurtosis}(X_t) = 3(1 + \operatorname{var} \tau_t) > 3$$

(if finite).

The model provides a coherent formulation for all time scales, in contrast to ARCH, GARCH-type models which focus on a single time scale.

2.2. Subordinator model (Stage 2)

Now we consider the activity time process in more detail, noting that it should exhibit LRD, and that LRD is often suggestive of self-similar or multifractal behaviour.

Fortunately, the activity time process $\{T_t\}$ can be empirically constructed. Note that the use of Itô's formula gives

$$\mathrm{d}\log P_t - \frac{\mathrm{d}P_t}{P_t} = \tfrac{1}{2}\sigma^2\, \mathrm{d}T_t,$$

from which the increments of the $\{T_t\}$ process may be obtained in a discretized approximation. This allows the process to be checked for self-similar or multifractal behaviour.

The kind of multifractal behaviour which might be expected is a relationship of the form

$$T_{ct} - ct \stackrel{\mathrm{d}}{=} M(c)(T_t - t)$$

for positive c, where M and T are independent random functions and $\stackrel{\mathrm{d}}{=}$ denotes equality of finite dimensional distributions. Self-similar behaviour occurs if $M(c)$ is of the form c^K, $0 < K < 1$.

Approximate self-similarity can easily be checked via crude estimation of K. Suppose that the price P_t is observed over a fixed time interval $[0, \mathcal{T}]$ and let

$$S_q(\delta) = \sum_{i=1}^{N} |T_{i\delta} - T_{(i-1)\delta} - \delta|^q \,/\, N,$$

where $\mathcal{T} = N\delta$ and $0 < q \le p$ with $ET_1^p < \infty$. If $\{T_t - t\}$ is self-similar, then

$$\log ES_q(\delta) = Kq \log\delta + \log E|T_1 - 1|^q$$

for $0 < q \le p$. Now the ergodic theorem ensures that $S_q(\delta) \stackrel{\text{a.s.}}{\to} ES_q(\delta)$ as $N \to \infty$, so one checks whether $\log S_q(\delta)$ scales linearly with both $\log\delta$ and q. Linearity with respect to just $\log\delta$ and not q signifies multifractal behaviour.

Preliminary examination of various examples suggests that a self-similar model is usually satisfactory. For the S&P daily returns, for example, we find $K \approx 0.8$. A comprehensive study is being undertaken.

It should be noted that the increments of the process $\{T_t - t\}$ can easily be shown to be LRD (SAV) for $K > 1/p$ if $E|T_1 - 1|^p < \infty$ for some $0 < p \le 2$ (Heyde and Yang (1997)). This nicely explains the empirically observed strong dependence. The LRD (SAV) refers to a sample Allen variance based definition of LRD which copes with processes having possible infinite variance, such as is the case here, but essentially reduces to the classical definition in the finite variance case.

For some economic discussion, based on ideas of complexity and self-organized criticality, which suggest why the activity time process (standardized to have zero mean) could have a self-similar scaling, see Arthur (1995) and Bak (1996) (particularly Chapter 11).

The principle of Occam's razor is being used here. We have sought the simplest model which reasonably represents the data, while retaining the essential features of the original GBM model.

3. History

A subordinated process (mixture) model formulation dates back to Mandelbrot and Taylor (1967). They used a (one-sided) stable law of index $\alpha/2$ (where $\alpha < 2$) for the distribution of τ_t, which gives a symmetric stable distribution of index α for $W(T_t)$. This is theoretically attractive but does not match the data. It has infinite second moment and no LRD. Mandelbrot and Taylor conceived of the $\{T_t\}$ process as reflecting transaction time.

Perhaps the next was Clark (1973), who used the lognormal. This does not get the moments right and again there is no LRD. Clark related the $\{T_t\}$ process to trading volume.

Many others have subsequently tried other marginal distributions. A detailed assessment of these is given in Hurst and Platen (1997). They favour the t-distribution with degrees of freedom in the range 3–5 on the basis of an empirical analysis of market indices for the USA, Japan, Germany, Switzerland and Australia.

Various other interpretations of the $\{T_t\}$ process have also been given. For example, Andersen (1996) took for his τ_t what he called the 'intensity of information arrivals', an unobserved process, but one which also drives the volume process.

Even though an entirely satisfactory economic model for the $\{T_t\}$ process may not yet be available, the characteristics of the process are fairly clear from the available statistical information.

Recent contributions containing subordinator models along similar lines to those of the present paper are Hurst *et al.* (1997) and Mandelbrot *et al.* (1997), which also contain many references to earlier literature. In the former paper the activity time process $\{T_t\}$ is assumed to have (stationary) independent increments. This, however, results in independent returns and does not reflect the strong dependence in observed series.

In Mandelbrot *et al.* (1997) it is assumed that

$$P_t = P_0 \exp[B_H(\theta(t))],$$

where $B_H(t)$ is a fractional Brownian motion process with Hurst index H and the activity time $\theta(t)$ is a multifractal process with non-decreasing paths and stationary increments. This is required to satisfy a local scaling rule of the form

$$\theta(t + c\Delta t) - \theta(t) \stackrel{\mathrm{d}}{=} M(c)(\theta(t + \Delta t) - \theta(t)),$$

for all positive c, where $M(c)$ is a random variable. Now if $H = \frac{1}{2}$, $B_H(t) = W(t)$, standard Brownian motion, as used in this paper, but if $H \neq \frac{1}{2}$ the process is not a semimartingale, the standard stochastic calculus is lost, and arbitrage opportunities exist (e.g. Rogers (1997)). This makes the model unattractive for finance applications. Furthermore, if $H > \frac{1}{2}$, the returns process itself will exhibit LRD, which does not appear to be the case in practice. Also, since the increments of the positive process $\{\theta(t)\}$ are stationary, it is necessary, if the returns are to have finite first moment, that $\theta(t) \sim \alpha t$ almost surely as $t \to \infty$ for some $\alpha > 0$. Hence $\theta(t)$ itself can only have a trivial scaling. Finally, in fitting the model to DM/US$ exchange rates, Fisher *et al.* (1997) estimate the H in their fractional Brownian motion to be 0.53, which is close enough to 0.5 to confirm the adequacy of ordinary Brownian motion as the carrier process. They also seek multifractal behaviour for $\{X_t\}$, while it appears to this author that this is a property of the activity time process and not the log returns process itself.

The simple model of the present paper, which could be designated FATGBM (geometric Brownian motion with fractal activity time) captures enough of the reality of the actual processes to warrant detailed examination without obscuring the picture with additional sources of possible LRD.

References

ANDERSEN, T. (1996). Return volatility and trading volume. An information flow interpretation of stochastic volatility. *J. Finance* **51**, 169–204.

ARTHUR, W. B. (1995). Complexity in economic and financial markets. *Complexity* **1**, 20–25.

BAK, P. (1996). *How Nature Works.* Copernicus. Springer, New York.

BERAN, J. (1994). *Statistics for Long-Memory Processes.* Chapman and Hall, New York.

CLARK, P. K. (1973). A subordinated stochastic process model with finite variance for speculative prices. *Econometrica* **41**, 135–155.
DING, Z. AND GRANGER, C. W. J. (1996). Modeling volatility persistence of speculative returns: a new approach. *J. Econometrics* **73**, 185–215.
FISHER, A., CALVET, L. AND MANDELBROT, B. (1997). Multifractality of the Deutschmark/US Dollar exchange rates. Cowles Foundation Discussion Paper No. 1165.
GRANGER, C. W. J. AND DING, Z. (1996). Varieties of long memory models. *J. Econometrics* **73**, 61–77.
HEYDE, C. C. (1997). *Quasi-Likelihood and its Application. General Theory of Optimal Parameter Estimation.* Springer, New York.
HEYDE, C. C. AND DAI, W. (1996). On the robustness to small trends of estimation based on the smoothed periodogram. *J. Time Series Anal.* **17**, 141–150.
HEYDE, C. C. AND YANG, Y. (1997). On defining long-range dependence. *J. Appl. Prob.* **34**, 939–944.
HURST, S. R. AND PLATEN, E. (1997). The marginal distribution of returns and volatility. In *L_1-Statistical Procedures and Related Topics*, ed. Y. Dodge (IMS Lecture Notes-Monograph Series **31**). IMS, Hayward, CA, pp. 301–314.
HURST, S. R., PLATEN, E. AND RACHEV, S. R. (1997). Subordinated Markov models: a comparison. *Fin. Eng. Japanese Markets* **4**, 97–124.
LO, A. W. (1991). Long-term memory in stock market prices. *Econometrika* **59**, 1279–1313.
MANDELBROT, B. AND TAYLOR, H. (1967). On the distribution of stock price differences. *Operat. Res.* **15**, 1057–1062.
MANDELBROT, B., FISHER, A. AND CALVET, L. (1997). A multifractal model of asset returns. Cowles Foundation Discussion Paper No. 1164.
MIKOSCH, T. (1997). Heavy tails in finance. *Bull. Int. Statist. Inst.* **57**, Book 2, 109–112.
PETERS, E. E. (1991). *Chaos and Order in the Capital Markets.* John Wiley, New York.
PETERS, E. E. (1994). *Fractal Market Analysis.* John Wiley, New York.
RESNICK, S. (1997). Heavy tail modeling and teletraffic data. *Ann. Statist.* **25**, 1805–1869.
ROGERS, L. C. G. (1997). Arbitrage from fractional Brownian motion. *Math. Finance* **7**, 95–105.
WILLINGER, W., TAQQU, M. AND TEVEROVSKY, V. (1999). Stock market prices and long-range dependence. *Finance and Stochastics* **3**, 1–13.
YANG, Y. (1998). Long range dependence in the study of time series with finite or infinite variances. Ph.D. Dissertation, Columbia University, New York, NY.

ELSEVIER Stochastic Processes and their Applications 99 (2002) 295–321

stochastic processes and their applications

www.elsevier.com/locate/spa

Statistical estimation of nonstationary Gaussian processes with long-range dependence and intermittency

Jiti Gao[a,*], Vo Anh[b], Chris Heyde[c,d]

[a] *The University of Western Australia, Perth, Australia*
[b] *Queensland University of Technology, Brisbane, Australia*
[c] *Columbia University, New York, USA*
[d] *The Australian National University, Canberra, Australia*

Received 13 December 1999; received in revised form 23 October 2001; accepted 31 January 2002

Abstract

This paper considers statistical inference for nonstationary Gaussian processes with long-range dependence and intermittency. The existence of such a process has been established by Anh et al. (J. Statist. Plann. Inference 80 (1999) 95–110). We systematically consider the case where the spectral density of nonstationary Gaussian processes with stationary increments is of a general and flexible form. The spectral density function of fRBm is thus a special case of this general form. A continuous version of the Gauss–Whittle objective function is proposed. Estimation procedures for the parameters involved in the spectral density function are then investigated. Both the consistency and the asymptotic normality of the estimators of the parameters are established. In addition, a real example is presented to demonstrate the applicability of the estimation procedures.
© 2002 Published by Elsevier Science B.V.

Keywords: Asymptotic theory; Fractional Riesz–Bessel motion; Nonstationary process; Long-range dependence; Statistical estimation

1. Introduction

Many recent studies have indicated that data in a large number of fields display long-range dependence (LRD) (Beran, 1994; Baillie and King, 1996; Anh and Heyde, 1999). A basic process which has commonly been used to model LRD is fractional

* Corresponding author. School of Mathematics and Statistics, The University of Western Australia, 6009 Crawley, WA, Australia. Tel.: +618-9380-3354; fax: +618-9380-1028.
E-mail address: jiti@maths.uwa.edu.au (J. Gao).

0304-4149/02/$ - see front matter © 2002 Published by Elsevier Science B.V.
PII: S0304-4149(02)00092-3

Brownian motion (fBm) B_H with Hurst index $\frac{1}{2} < H < 1$. This is a Gaussian process which has stationary increments and spectral density of the form

$$\phi_1(\omega) = \frac{c}{|\omega|^{2H+1}}, \quad c > 0, \ 0 < H < 1, \quad \omega \in R^1. \tag{1.1}$$

In (1.1), the "spectral density" $\phi_1(\omega)$ must be understood in some specific sense (e.g. in the sense of time-scale analysis (Flandrin, 1989) or in a limiting sense (Solo, 1992)) since B_H is a nonstationary process.

Anh et al. (1999a,b) introduced fractional Riesz–Bessel motion (fRBm), which is a Gaussian process with stationary increments and the spectral density of the increments is given by

$$\phi_2(\omega) = \frac{\eta}{|\omega|^{2\beta}(1+\omega^2)^{\alpha}}, \quad \omega \in R^1, \tag{1.2}$$

where $\eta > 0$, $-\frac{1}{2} < \beta < \frac{1}{2}$ and $\alpha \geqslant 0$ are parameters. It is noted that model (1.2) is well defined and that the spectral density of the increments of fBm is given by (1.2) with $\alpha = 0$.

The significance of fRBm (via Eq. (1.2)) is in its behaviour when $|\omega| \to \infty$. It is noted that the $\phi_2(\omega)$ of (1.2) is well defined as $|\omega| \to \infty$ due to the presence of the component $(1+\omega^2)^{-\alpha}$, $\alpha > 0$, which is the Fourier transform of the Bessel potential. As a result, the covariances $R(t)$ of the increments of fRBm are strong for small $|t|$. That is, large (resp. small) values of the increments tend to be followed by large (resp. small) values with probability sufficiently close to one. This is the clustering phenomenon observed in stochastic finance (e.g. Shiryaev, 1999, p. 365). This phenomenon is referred to as (second-order) intermittency in the turbulence literature (e.g. Frisch, 1995). Summarising, the increments of fRBm display both LRD and intermittency, while those of fBm have LRD but no intermittency.

In this paper, we shall consider the case where the spectral density of nonstationary Gaussian processes with stationary increments is of a general and flexible form. The spectral density function of fRBm is thus a special case of this general form. A continuous version of the Gauss–Whittle objective function is proposed. Estimation procedures for the parameters involved in the spectral density function are then investigated. Both the consistency and the asymptotic normality of the estimators of the parameters are established. In addition, a real example is presented to demonstrate the applicability of the estimation procedures.

The organization of this paper is as follows. Section 2 proposes the estimation procedures and the corresponding asymptotic properties. A real example is given in Section 3. Mathematical proofs are given in the appendix.

2. The estimation procedures

Let $\{X(t): t \in R^1 = (-\infty, \infty)\}$ be a zero-mean Gaussian process. Denote the increments of $X(t)$ by $\Delta X(t,\tau) = X(t+\tau) - X(t)$ for $t, \tau \in R^1$. The Gaussian process $X(t)$

is said to have second-order stationary increments if

$$D(t,\tau_1,\tau_2)=E\{[\Delta X(t+s,\tau_1)][\Delta X(s,\tau_2)]\}$$

is independent of s for all $s,t,\tau_1,\tau_2\in R^1$. The function $D(t,\tau_1,\tau_2)$ then has the spectral representation

$$D(t,\tau_1,\tau_2)=\int_{-\infty}^{\infty} \mathrm{e}^{\mathrm{i}t\omega}(1-\mathrm{e}^{-\mathrm{i}\tau_1\omega})(1-\mathrm{e}^{\mathrm{i}\tau_2\omega})\frac{1+\omega^2}{\omega^2}F(\mathrm{d}\omega), \tag{2.1}$$

where $F(\mathrm{d}\omega)$ is a nonnegative measure on R^1 such that $\int_{-\infty}^{\infty}F(\mathrm{d}\omega)<\infty$ (e.g. Yaglom, 1986).

We shall assume that $F(\mathrm{d}\omega)$ is absolutely continuous with derivative of the form

$$f(\omega)=f(\omega,\mu)=\frac{\eta}{|\omega|^{2\gamma}(1+\omega^2)^{\alpha}}\frac{\omega^2}{1+\omega^2}\pi(\omega;\alpha,\gamma),\quad \omega\in R^1, \tag{2.2}$$

where $\mu=(\alpha,\gamma,\eta)\in\Omega=[0,\infty)\times(\frac{1}{2},\frac{3}{2})\times(0,\infty)$, and $\pi(\omega)=\pi(\omega;\alpha,\gamma)$ is a positive and continuous function to be specified in Condition 2.1. When $\pi(\omega)\equiv 1$, model (2.2) corresponds to the fractional Riesz–Bessel motion case.

According to (2.1) and (2.2), we can define the spectral density of $X(t)$ by

$$g(\omega)=g(\omega,\mu)=\frac{1+\omega^2}{\omega^2}f(\omega)=\frac{\eta}{|\omega|^{2\gamma}(1+\omega^2)^{\alpha}}\pi(\omega),\quad \omega\in R^1.$$

Let μ_0 be the true value of μ. We assume that μ_0 is in the interior of Ω_0, a compact subset of Ω.

We now define the estimator of $g(\omega)$ by

$$I_N^X(\omega)=\frac{1}{2\pi N}\left|\int_0^N \mathrm{e}^{-\mathrm{i}\omega t}X(t)\,\mathrm{d}t\right|^2,$$

where $N>0$ is the upper bound of the interval $[0,N]$, on which each $X(t)$ is observed. Throughout this paper, the stochastic integrals are limits in mean square of appropriate Riemann sums. We assume that $X(0)=0$. It can be easily shown that the conclusions of Theorems 2.1 and 2.2 are not affected when $X(t)$ is replaced by $X_-(t)=X(t)-X(0)$ if $X(0)\neq 0$.

Note that the question of whether $I_N^X(\omega)$ is a consistent estimate of $g(\omega)$ has not been answered in the literature.

We need to introduce the following condition.

Condition 2.1. (i) Assume that $\pi(\omega;\alpha,\gamma)$ is a positive and continuous function in both ω and (α,γ), bounded away from zero and chosen to satisfy

$$\int_{-\infty}^{\infty} f(\omega,\mu)\,\mathrm{d}\omega<\infty\quad \text{for } \mu\in\Omega.$$

In addition, $\pi(\omega;\alpha,\gamma)$ is a symmetric function in ω satisfying $0<\lim_{\omega\to 0}\pi(\omega;\alpha,\gamma)<\infty$ and $0<\lim_{\omega\to\infty}\pi(\omega;\alpha,\gamma)<\infty$ uniformly in $(\alpha,\gamma)\in[0,\infty)\times(\frac{1}{2},\frac{3}{2})$.

(ii) $\lim_{\omega\to 0}\omega^2 g(\omega,\mu_0)=0$.

(iii) For $\varepsilon \leqslant \|\mu - \mu_0\| < \frac{1}{4}$ with any small $\varepsilon > 0$,

$$\int_{-\infty}^{\infty} \frac{g(\omega, \mu_0)}{g(\omega, \mu)} \frac{1}{1+\omega^2} \, d\omega < \infty.$$

Now we state the following result and its proof is relegated to the appendix.

Theorem 2.1. *Assume that Condition* 2.1 *holds. In addition, if*

$$\tfrac{1}{2} < \gamma < 1 \quad \text{and} \quad \alpha \geqslant 0, \tag{2.3}$$

then we have as $N \to \infty$

$$E[I_N^X(\omega)] \to g(\omega, \mu_0), \tag{2.4}$$

$$E\{I_N^X(\omega) - E[I_N^X(\omega)]\}^2 = g^2(\omega, \mu_0) + o(1) \tag{2.5}$$

and

$$E\{(I_N^X(\omega) - E[I_N^X(\omega)])(I_N^X(\lambda)) - E[I_N^X(\lambda)])\} = o(1) \tag{2.6}$$

for all $\omega \neq 0$, $\lambda \neq 0$ *and* $\omega \neq \lambda$.

Remark 2.1. (i) The conditions of Theorem 2.1 are quite natural and mild. Condition 2.1(i) allows the dependence of $\pi(\omega)$ on α and γ in model (2.2). This provides more flexible models than the following form:

$$f(\omega) = f(\omega, \mu) = \frac{f^*(\omega)}{|\omega|^{2\gamma}(1+\omega^2)^{\alpha}},$$

where $f^*(\omega)$ is just a positive and continuous function of ω. A similar model for time series with only long-range dependence has already been discussed. See model (1) of Hurvich et al. (1998). Both Conditions 2.1(ii) and 2.1(iii) hold automatically when $\pi(\omega)$ is just a function of ω.

(ii) It is possible to further generalize model (2.2) to a form similar to model (1.4) of Robinson (1997). For this case, one needs to impose some more detailed conditions on $\pi(\omega)$. As the main objective of this paper is to estimate the parameters η, α and γ, we shall use only the decomposed form (2.2).

(iii) For the process $X(t)$, Theorem 2.1 shows that $I_N^X(\omega)$ is asymptotically unbiased while $I_N^X(\omega)$ is not a weakly consistent estimator of $g(\omega)$. In addition, Eq. (2.4) justifies the fact that $g(\omega)$ can be defined as a spectral density of $X(t)$. Recently, Solo (1992) established (2.4) for the case where

$$g(\omega) = |\omega|^{-2\gamma} \quad \text{for } \tfrac{1}{2} < \gamma \leqslant 1.$$

This part of the paper considers estimating the parameters involved in the spectral density of the nonstationary process $X(t)$. For this case, we define the following continuous version of the Gauss–Whittle objective function:

$$L_N^X(\mu) = \frac{1}{4\pi}\int_{-\infty}^{\infty}\left\{\log(g(\omega,\mu)) + \frac{I_N^X(\omega)}{g(\omega,\mu)}\right\}\frac{d\omega}{1+\omega^2} \tag{2.7}$$

and the minimum contrast estimator of μ by

$$\hat{\mu}_N = \arg\min_{\mu\in\Omega_0} L_N^X(\mu),$$

where Ω_0 is a compact subset of Ω.

Remark 2.2. This paper considers using the continuous version of the Gauss–Whittle contrast function to estimate the spectral density $g(\omega)$ with its frequency defined on $(-\infty,\infty)$. For the continuous case, due to the slow decay of the spectral density at ∞, the weight function $1/(1+\omega^2)$ must be used in (2.7) to ensure that Eq. (2.7) is well defined. Eq. (2.7) can also be justified by applying the entropy theory discussed in Dym and McKean (1976).

The weak consistency is given below and its proof is postponed to the appendix.

Theorem 2.2. *Assume that the conditions of Theorem* 2.1 *hold. In addition, if the true value parameter* μ_0 *of* μ *is in the interior of* Ω_0, *then*

$$\hat{\mu}_N \to_{\mathrm{p}} \mu_0 \quad \text{as } N\to\infty.$$

Remark 2.3. It is worthwhile to point out that condition (2.3) covers the important case where $\frac{1}{2}<\gamma<1$ and $0<\alpha<1$. More recently, Gao et al. (2001) consider another important case where $X(t)$ is a stationary Gaussian process and the parameters involved in its spectral density $g(\omega)$ satisfy $0<\gamma<\frac{1}{2}$ and $\alpha\geqslant\frac{1}{2}$. The latter case allows a stationary process to display both LRD and second-order intermittency.

As pointed out in Anh et al. (1999a,b), it is more useful to establish a consistent estimator for μ for the case where

$$1\leqslant\gamma<\tfrac{3}{2} \quad\text{and}\quad \alpha\geqslant 0. \tag{2.8}$$

Due to the fact that (2.4) and (2.5) are only true under condition (2.3), we can only establish a weakly consistent estimator for μ under condition (2.3). Therefore, in order to construct a consistent estimation procedure for the case of (2.8), we need to transform the nonstationary process $X(t)$ into a stationary process. This paper suggests using the following transform.

Define the Haar wavelet function

$$h(u) = \begin{cases} 1, & 0\leqslant u<\frac{1}{2},\\ -1, & \frac{1}{2}\leqslant u<1,\\ 0, & \text{otherwise.}\end{cases}$$

We have its Fourier transform $\hat{\psi}$ given by

$$\hat{h}(t) = \frac{\mathrm{i}t}{4}\,\mathrm{e}^{-\mathrm{i}t/2}\left(\frac{\sin(t/4)}{t/4}\right)^2.$$

We now define a new process

$$Y(t) = \int_{-\infty}^{\infty} X(s)h(s-t)\,\mathrm{d}s = \left\{\int_0^{1/2} X(t+u)\,\mathrm{d}u - \int_{1/2}^{1} X(t+u)\,\mathrm{d}u\right\}, \quad t \in R^1.$$

Then, it follows from (2.1) and (2.2) that

$$\begin{aligned}
&E[Y(t+\tau)Y(t)]\\
&= \int_{-\infty}^{\infty}\int_{-\infty}^{\infty} E[X(u+t+\tau)X(v+t)]h(u)h(v)\,\mathrm{d}u\,\mathrm{d}v\\
&= \int_{-\infty}^{\infty}\int_{-\infty}^{\infty}\left(\int_{-\infty}^{\infty}(1-\mathrm{e}^{-\mathrm{i}(u+\tau)\omega})(1-\mathrm{e}^{\mathrm{i}v\omega})\frac{1+\omega^2}{\omega^2}f(\omega)\,\mathrm{d}\omega\right)h(u)h(v)\,\mathrm{d}u\,\mathrm{d}v\\
&= \int_{-\infty}^{\infty}\left(\int_{-\infty}^{\infty}(1-\mathrm{e}^{-\mathrm{i}(u+\tau)\omega})h(u)\,\mathrm{d}u\right.\\
&\qquad \left.\int_{-\infty}^{\infty}(1-\mathrm{e}^{\mathrm{i}v\omega})h(v)\,\mathrm{d}v\right)\frac{1+\omega^2}{\omega^2}f(\omega)\,\mathrm{d}\omega\\
&= \int_{-\infty}^{\infty}\mathrm{e}^{-\mathrm{i}\tau x}|\hat{h}(\omega)|^2\frac{1+\omega^2}{\omega^2}f(\omega)\,\mathrm{d}\omega,
\end{aligned}$$

which is independent of t. Consequently, the process $Y(t)$ is a stationary Gaussian process with the following spectral density:

$$\phi(\omega) = \phi(\omega,\theta) = |\hat{h}(\omega)|^2\,\frac{1+\omega^2}{\omega^2}\,f(\omega) = \frac{1}{16}\,\frac{\eta}{|\omega|^{2\beta}}\,\frac{\pi(\omega;\alpha,\beta)}{(1+\omega^2)^{\alpha}}\left(\frac{\sin(\omega/4)}{\omega/4}\right)^4,$$

where $\omega \in R^1$, $\beta = \gamma - 1$ and $\theta = (\alpha,\beta,\eta) \in \Theta = [0,\infty)\times[0,1/2)\times(0,\infty)$. For convenience, we use $\pi(\omega) = \pi(\omega;\alpha,\beta) = \pi(\omega;\alpha,\gamma)$ throughout the rest of the paper. Let θ_0 be the true value of θ. We assume that θ_0 is in the interior of Θ_0, a compact subset of Θ.

Remark 2.4. (i) We believe that both the wavelet transform and the resulting estimation procedure provide a general approach to the estimation of (locally) self-similar processes. In addition, as pointed out in Vergassola and Frish (1991), the wavelet transform is continuous and linear, and the process $Y(t)$ is itself Gaussian when the process $X(t)$ is Gaussian. By contrast, the usual increments do not necessarily retain the Gaussianity property.

(ii) We also need to point out that this paper adopts a simple form of the Haar wavelet function for the stationarity transform. When using a more complicated form, the structure of $f(\omega)$ will remain unchanged. The only change is the form of $\hat{h}(\cdot)$, which will not affect the estimation of the parameters. See Anh et al. (1999a,b). In addition, our experience suggests that the choice of the simple form of $h(\cdot)$ can provide stable simulation and numerical results.

(iii) The main assumption of this paper is that the nonstationary process $X(t)$ is Gaussian. If $X(t)$ is not Gaussian but the increments of $X(t)$ are assumed to be Gaussian, we can still estimate the parameter θ. One can use for example the first difference $Y(t)=X(t+1)-X(t)$ to obtain a stationary process. If $Y(t)$ is Gaussian, then its spectral density can be proved to be

$$\phi_Y(\omega) = \frac{\eta}{|\omega|^{2\beta}} \frac{\pi(\omega;\alpha,\beta)}{(1+\omega^2)^\alpha} \left(\frac{\sin(\omega/2)}{\omega/2}\right)^2,$$

where θ is as defined above. It can be seen that the conclusions of Theorems 2.4 and 2.5 below remain true.

For any given $\omega \in (-\infty,\infty)$, we define the estimator of $\phi(\omega)=\phi(\omega,\theta)$ by

$$I_N^Y(\omega) = \frac{1}{2\pi N}\left|\int_0^N e^{-i\omega t} Y(t)\,dt\right|^2.$$

For the periodogram $I_N^Y(\omega)$, we have the following consistency result and its proof is postponed to the appendix.

Theorem 2.3. *Assume that condition* (2.8) *holds. In addition,* $Y(0)=0$. *Then we have as* $N \to \infty$

$$E[I_N^Y(\omega)] - \phi(\omega,\theta_0) = O(1/N), \tag{2.9}$$

$$E\{I_N^Y(\omega)) - E[I_N^Y(\omega)]\}^2 = \phi^2(\omega,\theta_0) + O(1/N^2) \tag{2.10}$$

and

$$E\{(I_N^Y(\omega)) - E[I_N^Y(\omega)])(I_N^Y(\lambda)) - E[I_N^Y(\lambda)])\} = O(1/N^2) \tag{2.11}$$

for all $\omega \neq 0$, $\lambda \neq 0$ *and* $\omega \neq \lambda$.

Remark 2.5. Eq. (2.9) implies that $I_N^Y(\omega)$ is asymptotically unbiased while (2.10) shows that $I_N^Y(\omega)$ is not a consistent estimator of $\phi(\omega)$, although $Y(t)$ is now stationary. Results (2.9)–(2.11) are similar to those for the discrete time processes with short-range dependence. See for example, Chapter 10 of Brockwell and Davis (1990).

As mentioned earlier, the main objective of this paper concentrates on the estimation of θ. This paper then considers using an extended version of the Gauss–Whittle contrast function to estimate the spectral density $\phi(\omega)$ with its frequency defined on $(-\infty,\infty)$. When considering the case where the spectral density is defined on $(-\pi,\pi]$, the weight function $1/(1+\omega^2)$ is not required (see, for example, Heyde and Gay, 1993 for example).

For the stationary process $Y(t)$, we define the following objective function:

$$L_N^Y(\theta) = \frac{1}{4\pi}\int_{-\infty}^{\infty}\left\{\log(\phi(\omega,\theta)) + \frac{I_N^Y(\omega)}{\phi(\omega,\theta)}\right\}\frac{d\omega}{1+\omega^2}.$$

The minimum contrast estimator of θ is defined by

$$\hat{\theta}_N = \arg\min_{\theta\in\Theta_0} L_N^Y(\theta),$$

where Θ_0 is a compact subset of Θ.

In order to state the asymptotic normality of $\hat{\theta}_N$, we need to introduce the following conditions:

Condition 2.2. (i) Assume that the first two derivatives of $\log(\phi(\omega,\theta))$ in θ exist and are continuous in $\omega\in(-\infty,\infty)$ and $\theta\in\Theta$.

(ii) Assume that

$$\int_{-\infty}^{\infty} \frac{\phi^2(\omega,\theta_0)}{(1+\omega^2)^2}\left(\frac{\partial\log(\phi(\omega,\theta))}{\partial\theta}\right)^T\left(\frac{\partial\log(\phi(\omega,\theta))}{\partial\theta}\right)\Bigg|_{\theta=\theta_0} < \infty,$$

where

$$\frac{\partial\log(\phi(\omega,\theta))}{\partial\theta} = \left(\frac{\partial\log(\phi(\omega,\theta))}{\partial\alpha},\frac{\partial\log(\phi(\omega,\theta))}{\partial\beta},\frac{\partial\log(\phi(\omega,\theta))}{\partial\eta}\right)^\tau.$$

(iii) For any real function $l\in L^2(-\infty,\infty)$,

$$\int_{-\infty}^{\infty} \frac{\phi^2(\omega,\theta_0)}{(1+\omega^2)^2}\, l^2(\omega)\left(\frac{\partial\log(\phi(\omega,\theta))}{\partial\theta}\right)^T\left(\frac{\partial\log(\phi(\omega,\theta))}{\partial\theta}\right)\Bigg|_{\theta=\theta_0} < \infty.$$

(iv) Assume that $\sum_{k=-\infty}^{\infty}\phi(\omega-2k\pi,\theta)$ converges uniformly in $\omega\in(-\pi,\pi]\backslash\{0\}$ and $\theta\in\Theta_0$.

Condition 2.3. (i) For $\theta\in\Theta$,

$$\Gamma(\theta) = \frac{1}{4\pi}\int_{-\infty}^{\infty}\left(\frac{\partial\log(\phi(\omega,\theta))}{\partial\theta}\right)\left(\frac{\partial\log(\phi(\omega,\theta))}{\partial\theta}\right)^T\frac{1}{(1+\omega^2)^2}\,\mathrm{d}\omega < \infty.$$

(ii) The inverse matrix, $\Gamma^{-1}(\theta_0)$, of $\Gamma(\theta_0)$ does exist.

Remark 2.6. (i) Conditions 2.2 and 2.3 are similar to those for the discrete case. See for example, Condition (A2) of Heyde and Gay (1993).

(ii) Conditions 2.2 and 2.3 indirectly impose some restrictions on the form of $\pi(\omega)$. When $\pi(\omega)\equiv 1$, Conditions 2.2 and 2.3 hold automatically. The justification is similar to that of Lemma B.1 of Gao et al. (2001).

(iii) Condition 2.2(iii) is required for an application of a continuous-time central limit theorem to the proof of the asymptotic normality.

We now state the next result of this paper and its proof is postponed to the appendix.

Theorem 2.4. (i) *Assume that Condition* 2.1 *holds. In addition, if the true value* θ_0 *of* θ *is in the interior of* Θ_0, *then under condition* (2.8)

$$\hat{\theta}_N \to \theta_0 \quad \textit{with probability one}$$

as $N\to\infty$.

(ii) *Assume that Conditions* 2.1–2.3 *hold. In addition, if the true value* θ_0 *of* θ *is in the interior of* Θ_0, *then under condition* (2.8)

$$\sqrt{N}(\hat{\theta}_N - \theta_0) \to_D \mathrm{N}(0, \Gamma^{-1}(\theta_0)),$$

where $\Gamma^{-1}(\theta_0)$ *is as defined in Condition* 2.3(ii).

Remark 2.7. (i) Theorem 2.4 establishes both the strong consistency and the asymptotic normality of $\hat{\theta}_N$. Previously, we were unable to establish the asymptotic normality of $\hat{\theta}_N$. Instead, we established the asymptotic normality for a discrete approximation of $\hat{\theta}_N$. See for example, Theorem 2.2 of Gao et al. (2001). Similarly, we can now establish the asymptotic normality of θ_N of Gao et al. (2001). Thus, for both the nonstationary and stationary cases, we can establish an asymptotically normal estimator for the vector of parameters involved in the spectral density. This theory is now complete.

(ii) Theorem 2.4 extends and complements some existing results. For example, Dahlhaus (1989) established asymptotic normality for estimators based the Whittle approach for the discrete case. Giraitis and Leipus (1995) discussed asymptotic consistency for estimators based the Whittle approach for the discrete case. Viano et al. (1994) established some probabilistic and asymptotic results for continuous-time fractional ARMA processes. Viano et al. (1995) considered a class of extended fractional ARMA processes, and discussed the asymptotic behaviour of their correlations.

As $Y(t)$ is stationary and Gaussian, we now can study the asymptotic distribution of $I_N^Y(\omega)$ for any fixed ω. Similar results for the discrete case have been discussed extensively (see, for example, Künsch, 1986; Hurvich and Beltrao, 1993; Robinson, 1995).

We state the last result of this section and its proof is postponed to the appendix.

Theorem 2.5. *Assume that the conditions of Theorem* 2.3 *hold. Then the normalized periodogram* $I_N^Y(\omega)/\phi(\omega,\theta_0)$ *is asymptotically distributed as* $\frac{1}{2}\chi_2^2$.

Remark 2.8. For the discrete case, Theorem 6 of Hurvich and Beltrao (1993) shows that $I_N^Y(\omega)/\phi(\omega,\theta_0)$ at a given frequency is asymptotically distributed as a quadratic form, and is only asymptotically distributed as $\frac{1}{2}\chi_2^2$ for a specific case. By contrast, Theorem 2.5 shows that for the continuous case the asymptotic distribution of $I_N^Y(\omega)/\phi(\omega,\theta_0)$ for any fixed ω is $\frac{1}{2}\chi_2^2$.

3. Implementation and application

In many practical circumstances, however, observations on $Y(t)$ are made at discrete intervals of time, even though the underlying process may be continuous. In addition, it is computationally easier to find a consistent estimate of θ based on a sequence of discrete observations on $Y(t)$. This section considers the following discrete process:

$$Z_t = Y(t), \quad t = 1, 2, \ldots . \tag{3.1}$$

According to (3.1), the autocovariance function of Z_t is well defined by

$$r_\tau = \mathrm{cov}(Z_t, Z_{t+\tau}) = \mathrm{cov}\{Y(t), Y(t+\tau)\} = \int_{-\infty}^{\infty} \mathrm{e}^{\mathrm{i}\tau\omega}\phi(\omega, \theta)\,\mathrm{d}\omega$$

since $\int_{-\infty}^{\infty} |\phi(\omega, \theta)|\,\mathrm{d}\omega < \infty$.

Then it can be shown that

$$\begin{aligned} r_\tau &= \int_{-\infty}^{\infty} \mathrm{e}^{\mathrm{i}\tau\omega}\phi(\omega, \theta)\,\mathrm{d}\omega = \sum_{k=-\infty}^{\infty} \int_{(2k-1)\pi}^{(2k+1)\pi} \mathrm{e}^{\mathrm{i}\tau\omega}\phi(\omega, \theta)\,\mathrm{d}\omega \\ &= \sum_{k=-\infty}^{\infty} \int_{-\pi}^{\pi} \exp\{\mathrm{i}\tau(\omega - 2k\pi)\}\phi(\omega - 2k\pi, \theta)\,\mathrm{d}\omega \\ &= \int_{-\pi}^{\pi} \exp(\mathrm{i}\tau\omega) \left\{ \sum_{k=-\infty}^{\infty} \phi(\omega - 2k\pi, \theta) \right\} \mathrm{d}\omega \end{aligned}$$

since $\sum_{k=-\infty}^{\infty} \phi(\omega - 2k\pi, \theta)$ converges uniformly in $\omega \in (-\pi, \pi]\backslash\{0\}$ and $\theta \in \Theta_0$.

Therefore, the spectral density function of Z_t can be defined by

$$f_Z(\omega) = f(\omega, \theta) = \sum_{k=-\infty}^{\infty} \phi(\omega - 2k\pi, \theta). \tag{3.2}$$

We now define the following discrete version of the Gauss–Whittle contrast function:

$$W(\theta) = W_T(\theta) = \frac{1}{2T} \sum_{s=1}^{T-1} \left\{ \log(f(\omega_s, \theta)) + \frac{I_T^Z(\omega_s)}{f(\omega_s, \theta)} \right\},$$

where $T = [N] \leqslant N$ is the smallest integer part, $\omega_s = 2\pi s/T$, and

$$I_T^Z(\omega) = \frac{1}{2\pi T} \left| \sum_{t=1}^{T} \mathrm{e}^{\mathrm{i}\omega t} Z_t \right|^2 .$$

Now, the minimum contrast estimator of θ can be defined by

$$\tilde{\theta}_T = \arg\min_{\theta \in \Theta_0} W(\theta). \tag{3.3}$$

It can be shown that as $N \to \infty$

$$\tilde{\theta}_T - \hat{\theta}_N \to_{\mathrm{p}} 0. \tag{3.4}$$

The derivation of (3.2)–(3.4) is similar to (2.4)–(2.6) of Gao et al. (2001).

Assume that the solution of (3.3) satisfies

$$W'(\tilde{\theta}_T) = 0. \tag{3.5}$$

Similar to Appendix C of Gao et al. (2001), one can implement (3.5) in practice.

Example 3.1. In this example, we consider two air pollution data sets provided by the Environment Protection Authority of New South Wales (NSWEPA) of Australia: the Lidcombe NO_x time series (1) and the Lidcombe ozone time series (2). These are measurements of maximum daily concentrations and cover the period 2/1/1982–30/4/1994. Lidcombe is an urban suburb at the centre of Sydney. Lidcombe is chosen because, at this site of the NSWEPA monitoring network, long series of monitoring data are available in a fairly complete form. We apply our estimation procedure to determine the values of the parameters and to detect whether the data sets exhibit both long-range dependence and intermittency. If we assume that the two data sets are stationary and Gaussian, then the spectral density of the data set (i) can be defined by

$$\phi_i(\omega,\theta) = \frac{\eta_i}{|\omega|^{2\beta_i}(1+\omega^2)^{\alpha_i}} \left(\frac{\sin(\omega/4)}{\omega/4}\right)^4, \quad \omega \in (-\infty,\infty), \tag{3.6}$$

where $\alpha_i \in [0,\infty)$, $\beta_i \in (-1/2,1/2)$ and $0 < \eta_i < \infty$ for $i = 1,2$. Note that the form of $\phi_i(\omega,\theta)$ is the special case of $\phi(\omega,\theta)$ where $\pi(\omega) \equiv 16$.

The plots and the power spectra of the data sets are given in Figs. 1–4. We should note the singularity at frequency 0 and the power law form of the sample spectra, indicating the presence of long-range dependence in the data. Applying the estimation procedure presented in Section 2 and the computational formulae given in (3.1)–(3.5) to the data sets, we found the solution of (3.5), and the values of the unknown parameters in (3.6) were $\alpha_1 = 0.04474$, $\beta_1 = 0.2668$, $\eta_1 = 12.571$, $\alpha_2 = 0.07832$, $\beta_2 = 0.3396$ and $\eta_2 = 359.618$, respectively. These results confirm that the two data sets exhibit both long-range dependence ($\beta_i > 0$) and intermittency ($\alpha_i > 0$), although the intermittency is relatively less pronounced.

Remark 3.1. This example has described some basic features of the air pollution time series for a single monitoring station at Lidcombe. The identification of long-range dependence and intermittency of the time series has facilitated the use of modern regression analysis to model the dependence of the maximum daily ozone on some factors such as morning temperature and wind speed. Given the complexity of air pollution time series, we suggest using nonparametric time series regression smoothing techniques (e.g. Fan and Gijbels, 1996) or semiparametric time series regression methods (e.g. Härdle et al., 2000) to model the dependence and then to provide better forecasts.

Another issue is the selection of avenues, as there are many avenues for further investigations. The characteristics of air quality at the other sites of Sydney could be investigated in a similar manner. These characteristics can be used to identify the structural relationships between different monitoring stations and different subregions

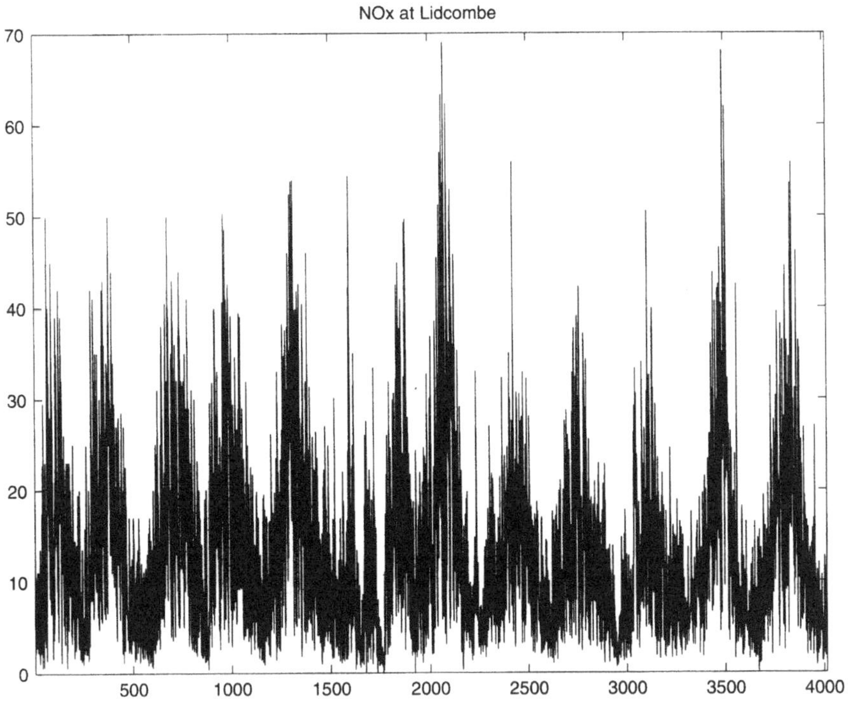
NOx at Lidcombe
70
60
50
40
30
20
10
0
500
1000
1500
2000
2500
3000
3500
4000

Fig. 1.

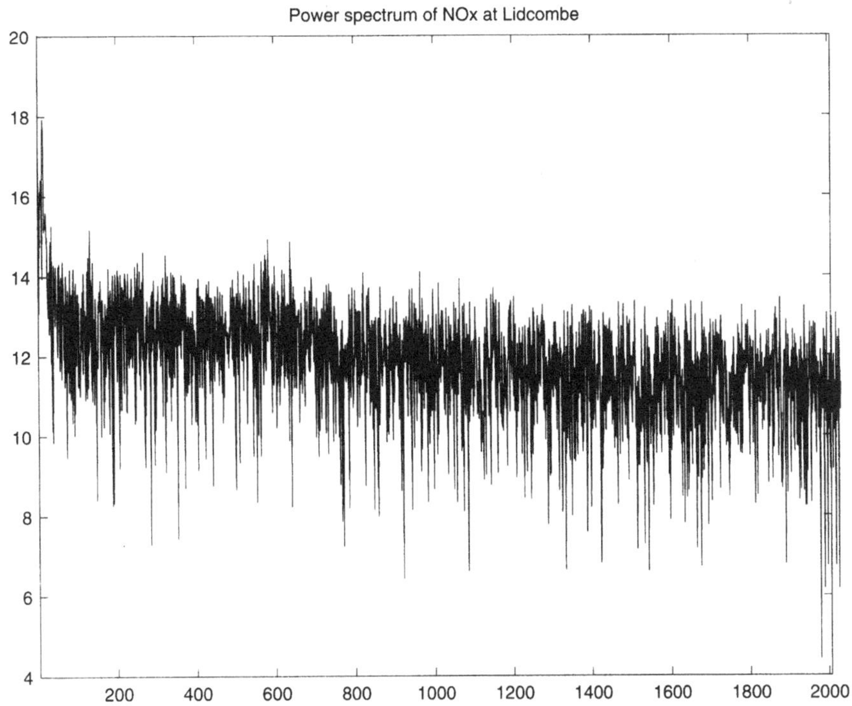
Power spectrum of NOx at Lidcombe
20
18
16
14
12
10
8
6
4
200
400
600
800
1000
1200
1400
1600
1800
2000

Fig. 2.

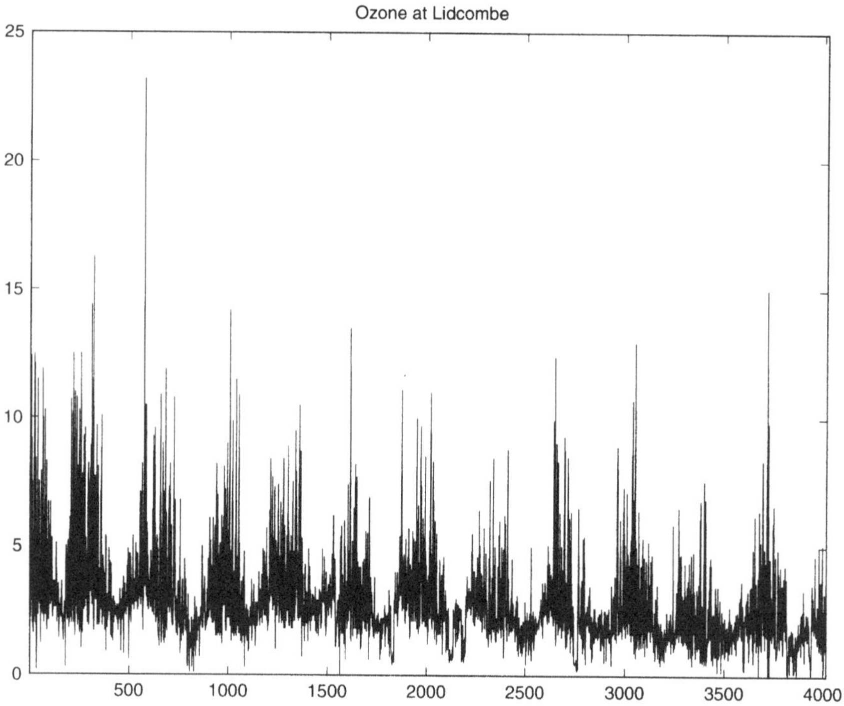
Ozone at Lidcombe
25
20
15
10
5
0
500
1000
1500
2000
2500
3000
3500
4000

Fig. 3.

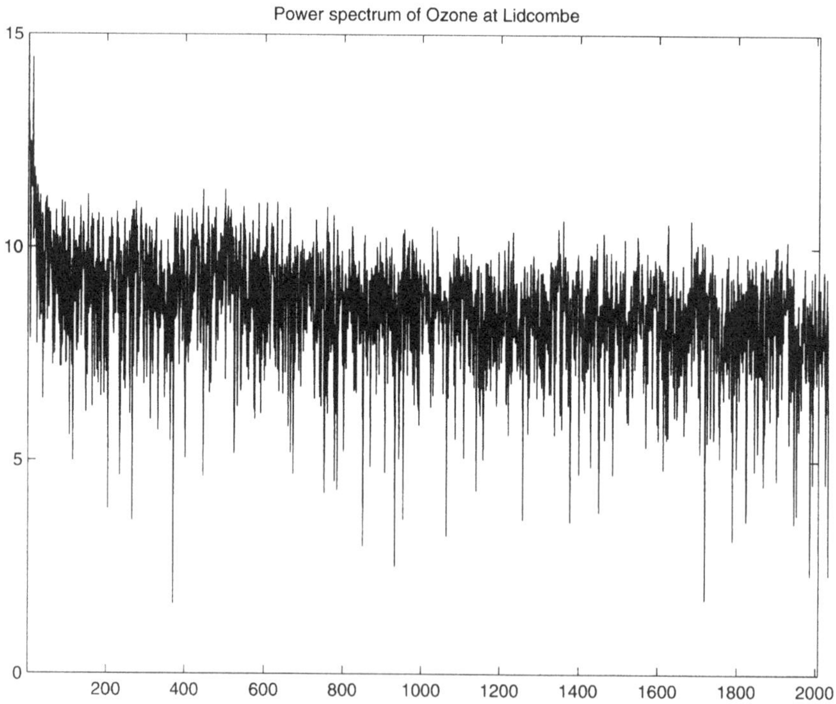
Power spectrum of Ozone at Lidcombe
15
10
5
0
200
400
600
800
1000
1200
1400
1600
1800
2000

Fig. 4.

of the Sydney airshed. For a more reliable forecasting system, it is worthwhile to investigate a much fuller set of information over all subregions.

Acknowledgements

The authors would like to thank the Editor, the Associate Editor and the referees for their constructive comments and suggestions. Thanks also go to the Australian Research Council for its financial support.

Appendix A

Proof of Theorem 2.1. It now follows from (2.1) and (2.2) that

$$E[X(s)X(t)] = \int_{-\infty}^{\infty} (1 - e^{it\lambda})(1 - e^{-is\lambda}) g_0(\lambda)\, d\lambda.$$

Thus

$$\begin{aligned} E[I_N^X(\omega)] &= \frac{1}{2\pi N} \int_{-\infty}^{\infty} \left(\int_0^N \int_0^N (1 - e^{it\lambda}) e^{-i\omega t} (1 - e^{-is\lambda}) e^{i\omega s}\, ds\, dt \right) g_0(\lambda)\, d\lambda \\ &= \frac{N}{2\pi} \int_{-\infty}^{\infty} |B(\omega N) - B((\omega - \lambda)N)|^2 g_0(\lambda)\, d\lambda, \end{aligned} \tag{A.1}$$

where $B(x) = (e^{ix} - 1)/ix$ and $g_0(\lambda) = g(\lambda, \mu_0)$.

Next, using the identity

$$B(\omega N) - B((\omega - \lambda)N) = \lambda(B((\omega - \lambda)N) - e^{i\omega N} B(\omega N))/\omega \tag{A.2}$$

we have

$$\begin{aligned} E[I_N^X(\omega)] &= \frac{N}{2\pi\omega^2} \int_{-\infty}^{\infty} \psi((\omega - \lambda)N)^2 g_0(\lambda) \lambda^2\, d\lambda \\ &\quad + \frac{N}{2\pi\omega^2} \int_{-\infty}^{\infty} \psi(\lambda N)^2 \lambda^2 g_0(\lambda)\, d\lambda \\ &\quad - \frac{2N}{2\pi\omega^2} \int_{-\infty}^{\infty} \psi((\omega - \lambda)N) \psi(\lambda N) \lambda^2 g_0(\lambda)\, d\lambda \equiv E_{1N} + E_{2N} - E_{12N}, \end{aligned}$$

where $\psi(\omega) = \sin(\omega/2)/(\omega/2)$, the symbol "$\equiv$" indicates that the terms on the left-hand side are represented correspondingly by those on the right-hand side.

Now, by Fejer's theorem (e.g. Goldberg, 1961, p. 17), we have as $N \to \infty$

$$E_{1N} \to g_0(\omega) = g(\omega, \mu_0), \quad E_{2N} \to \frac{1}{\omega^2} \lim_{\lambda \to 0} (g_0(\lambda)\lambda^2) = 0$$

and

$$\frac{\omega^2}{2\psi(\omega N)} E_{12N} \to \lim_{\lambda \to 0} (g_0(\lambda)\lambda^2) = 0 \tag{A.3}$$

using Condition 2.1(ii).

Therefore, the proof of (2.3) is completed.

Before proving (2.5), we introduce the following notation and identity:

$$r_X(s,t,u,v) = E[X(s)X(t)X(u)X(v)] \quad \text{and} \quad r_X(s,t) = E[X(s)X(t)],$$

$$r_X(s,t,u,v) - r_X(s,t)r_X(u,v) = r_X(s,u)r_X(t,v) + r_X(s,v)r_X(t,u) \tag{A.4}$$

since $X(t)$ is Gaussian and $X(0) = 0$.

It follows from (2.1) and (2.2) again that

$$\begin{aligned}
&4\pi^2 N^2 E\{I_N^X(\omega) - E[I_N^X(\omega)]\}^2 \\
&= \int_0^N \int_0^N \int_0^N \int_0^N \{[r_X(s,t,u,v) - r_X(s,t)r_X(u,v)]\} e^{i\omega(t-s)+i\omega(u-v)} \mathrm{d}s\, \mathrm{d}t\, \mathrm{d}u\, \mathrm{d}v \\
&= \int_0^N \int_0^N \int_0^N \int_0^N r_X(s,v)r_X(t,u) e^{i\omega(t+u)-i\omega(s+v)} \,\mathrm{d}t\, \mathrm{d}u\, \mathrm{d}s\, \mathrm{d}v \\
&\quad + \int_0^N \int_0^N \int_0^N \int_0^N r_X(s,u)r_X(t,v) e^{i\omega(t-v)+i\omega(u-s)} \,\mathrm{d}t\, \mathrm{d}v\, \mathrm{d}s\, \mathrm{d}u.
\end{aligned} \tag{A.5}$$

Similar to (A.1) and (A.2), we have

$$\begin{aligned}
&\frac{1}{N} \int_0^N \int_0^N r_X(t,u) e^{i\omega(t+u)} \,\mathrm{d}t\, \mathrm{d}u \\
&= -\frac{N}{\omega^2} \int_{-\infty}^{\infty} [B((\lambda+\omega)N) + e^{i\omega N} B(\lambda N)] \\
&\quad \times [B((\omega-\lambda)N) + e^{i\omega N} B(-\lambda N)] \lambda^2 g_0(\lambda) \,\mathrm{d}\lambda
\end{aligned}$$

and

$$\begin{aligned}
&\frac{1}{N} \int_0^N \int_0^N r_X(s,v) e^{-i\omega(s+v)} \,\mathrm{d}s\, \mathrm{d}v \\
&= -\frac{N}{\omega^2} \int_{-\infty}^{\infty} [B((\lambda-\omega)N) + e^{-i\omega N} B(\lambda N)] \\
&\quad \times [B(-(\lambda+\omega)N) + e^{-i\omega N} B(-\lambda N)] \lambda^2 g_0(\lambda) \,\mathrm{d}\lambda.
\end{aligned}$$

Observe that as $N \to \infty$

$$\frac{N}{\omega^2}\int_{-\infty}^{\infty} B((\lambda+\omega)N)B((\omega-\lambda)N)\lambda^2 g_0(\lambda)\,\mathrm{d}\lambda$$

$$=\frac{\mathrm{e}^{-\mathrm{i}\omega N}}{\omega^2}\int_{-\infty}^{\infty} \psi((\lambda+\omega)N)\psi((\omega-\lambda)N)\lambda^2 g_0(\lambda)\,\mathrm{d}(N\lambda)$$

$$=\mathrm{O}(\psi^2(\omega N)g_0(\omega))=\mathrm{O}(1/N), \tag{A.6}$$

$$\frac{N}{\omega^2}\int_{-\infty}^{\infty} |B(\lambda N)|^2\lambda^2 g_0(\lambda)\,\mathrm{d}\lambda \to \frac{1}{\omega^2}\lim_{\lambda\to 0}(\lambda^2 g_0(\lambda))=0 \tag{A.7}$$

and

$$\frac{N}{\omega^2}\int_{-\infty}^{\infty} B((\lambda+\omega)N)B(\lambda N)\lambda^2 g_0(\lambda)\,\mathrm{d}\lambda$$

$$=\mathrm{O}\left(\frac{\psi(\omega N)}{\omega^2}\lim_{\lambda\to 0}(\lambda^2 g_0(\lambda))\right)=\mathrm{o}(1/N) \tag{A.8}$$

for all $\omega \neq 0$.

Analogous to (A.6)–(A.8), we can show that as $N \to \infty$

$$\frac{1}{N^2}\int_0^N\int_0^N\int_0^N\int_0^N \{r_X(t,u)r_X(s,v)\}\mathrm{e}^{\mathrm{i}\omega(t-s)+\mathrm{i}\omega(u-v)}\,\mathrm{d}s\,\mathrm{d}t\,\mathrm{d}u\,\mathrm{d}v$$

$$=\mathrm{O}(1/N^2)+\mathrm{o}(1/N)+\mathrm{o}(1).$$

On the other hand, similar to (A.1)–(A.3) we have as $N \to \infty$

$$\frac{1}{N^2}\int_0^N\int_0^N\int_0^N\int_0^N \{r_X(t,v)r_X(s,u)\}\mathrm{e}^{\mathrm{i}\omega(t-v)+\mathrm{i}\omega(u-s)}\,\mathrm{d}s\,\mathrm{d}t\,\mathrm{d}u\,\mathrm{d}v$$

$$=\left(\frac{N}{2\pi}\int_{-\infty}^{\infty} |B(\omega N)-B((\omega-\lambda)N)|^2 g_0(\lambda)\,\mathrm{d}\lambda\right)$$

$$\times\left(\frac{N}{2\pi}\int_{-\infty}^{\infty} |B(\omega N)-B((\omega+\lambda)N)|^2 g_0(\lambda)\,\mathrm{d}\lambda\right)=g^2(\omega,\mu_0)+\mathrm{o}(1)$$

for all $\omega \neq 0$.

Thus, the proof of (2.5) follows from (A.5). Similar to (A.5)–(A.8), we can show that as $N \to \infty$

$$\frac{1}{N}\int_0^N\int_0^N r_X(t,u)\mathrm{e}^{\mathrm{i}xt+\mathrm{i}yu}\,\mathrm{d}t\,\mathrm{d}u$$

$$=\frac{-N}{2\pi xy}\int_{-\infty}^{\infty} \lambda^2(B((\lambda+y)N)+\mathrm{e}^{\mathrm{i}yN}B(\lambda N))(B((x-\lambda)N)$$

$$+\,\mathrm{e}^{\mathrm{i}xN}B(-\lambda N))g_0(\lambda)\,\mathrm{d}\lambda \to 0$$

and

$$\frac{1}{N}\int_0^N\int_0^N r_X(s,v)e^{-ixs-iyv}\,ds\,dv$$

$$=\frac{-N}{2\pi xy}\int_{-\infty}^{\infty}\lambda^2(B((\lambda-y)N)+e^{-iyN}B(\lambda N))(B((-x-\lambda)N)$$

$$+e^{-ixN}B(-\lambda N))g_0(\lambda)\,d\lambda\to 0 \tag{A.9}$$

for all $x\neq 0$ and $x\neq y$.

Analogously, we have as $N\to\infty$

$$\frac{1}{N}\int_0^N\int_0^N r_X(t,v)e^{ixt-iyv}\,dt\,dv$$

$$=\frac{N}{2\pi xy}\int_{-\infty}^{\infty}\lambda^2(B((\lambda-y)N)+e^{-iyN}B(\lambda N))(B((x-\lambda)N)$$

$$+e^{ixN}B(-\lambda N))g_0(\lambda)\,d\lambda\to 0$$

and

$$\frac{1}{N}\int_0^N\int_0^N r_X(s,u)e^{-ixs+iyu}\,ds\,du$$

$$=\frac{N}{2\pi xy}\int_{-\infty}^{\infty}\lambda^2(B((y-\lambda)N)+e^{iyN}B(-\lambda N))(B((\lambda-x)N)$$

$$+e^{-ixN}B(\lambda N))g_0(\lambda)\,d\lambda\to 0 \tag{A.10}$$

for all $x\neq 0$ and $x\neq y$.

Thus, (A.9) and (A.10) imply that as $N\to\infty$

$$E\{(I_N^X(x)-E[I_N^X(x)])(I_N^X(y)-E[I_N^X(y)])\}$$

$$=\frac{1}{N^2}\int_0^N\int_0^N\int_0^N\int_0^N r_X(t,u)r_X(s,v)e^{ixt+iyu}e^{-ixs-iyv}\,ds\,dt\,du\,dv$$

$$+\frac{1}{N^2}\int_0^N\int_0^N\int_0^N\int_0^N r_X(t,v)r_X(s,u)e^{ixt+iyu}e^{-ixs-iyv}\,ds\,dt\,du\,dv\to 0$$

for all $x\neq 0$ and $x\neq y$. Thus the proof of (2.6) is completed. □

Proof of Theorem 2.2. Before proving $\hat{\mu}_N\to\mu_0$ in probability, we show that as $N\to\infty$

$$L_N^X(\mu)\to L^X(\mu)=\frac{1}{4\pi}\int_{-\infty}^{\infty}\left\{\log(g(\omega,\mu))+\frac{g(\omega,\mu_0)}{g(\omega,\mu)}\right\}\frac{1}{1+\omega^2}\,d\omega \tag{A.11}$$

for $\varepsilon\leqslant\|\mu-\mu_0\|<\frac{1}{4}$ with any given $\varepsilon>0$.

In order to prove (A.11), it suffices to show that for

$$p(\omega) = p(\omega, \mu) = \frac{1}{1+\omega^2} g(\omega, \mu)^{-1} = \eta^{-1}\omega^{2\gamma}(1+\omega^2)^{\alpha-1}\pi(\omega)^{-1}$$

the following

$$\int_{-\infty}^{\infty} I_N^X(\omega) p(\omega, \mu)\, d\omega \to \int_{-\infty}^{\infty} g(\omega, \mu_0) p(\omega, \mu)\, d\omega \tag{A.12}$$

holds in probability as $N \to \infty$. It is easy to see that (A.12) follows from

$$\int_{-\infty}^{\infty} (E[I_N^X(\omega)] - g(\omega, \mu_0)) p(\omega, \mu)\, d\omega \to 0 \tag{A.13}$$

and

$$\int_{-\infty}^{\infty} (I_N^X(\omega) - E[I_N^X(\omega)]) p(\omega, \mu)\, d\omega \to 0 \tag{A.14}$$

in probability for $\varepsilon \leqslant \|\mu - \mu_0\| < \frac{1}{4}$ with any given $\varepsilon > 0$.

The proof of (A.13) follows from (A.3), Condition 2.1(iii) and

$$\int_{-\infty}^{\infty} g_0(\omega) p(\omega)\, d\omega = \int_{-\infty}^{\infty} \frac{g(\omega, \mu_0)}{g(\omega, \mu)} \frac{1}{1+\omega^2}\, d\omega < \infty \tag{A.15}$$

for $\varepsilon \leqslant \|\mu - \mu_0\| < \frac{1}{4}$, where $g_0(\omega) = g(\omega, \mu_0)$.

In order to prove (A.14), it suffices to show that as $N \to \infty$

$$\begin{aligned}
&4\pi^2 N^2 E\left\{\int_{-\infty}^{\infty} (I_N^X(\omega) - E\left[I_N^X(\omega)\right])\, p(\omega)\, d\omega\right\}^2 \\
&= 4\pi^2 N^2 \int_{-\infty}^{\infty}\int_{-\infty}^{\infty} E\{(I_N^X(x) - E[I_N^X(x)])(I_N^X(y) - E[I_N^X(y)])\} p(x) p(y)\, dx\, dy \\
&= \int_{-\infty}^{\infty}\int_{-\infty}^{\infty} \left(\int_0^N\int_0^N\int_0^N\int_0^N r_X(t,u) r_X(s,v) e^{ixt+iyu} e^{-ixs-iyv}\, ds\, dt\, du\, dv\right) p(x) p(y)\, dx\, dy \\
&\quad + \int_{-\infty}^{\infty}\int_{-\infty}^{\infty} \left(\int_0^N\int_0^N\int_0^N\int_0^N r_X(t,v) r_X(s,u) e^{ixt+iyu} e^{-ixs-iyv}\, ds\, dt\, du\, dv\right) p(x) p(y)\, dx\, dy \\
&= o(N^2).
\end{aligned} \tag{A.16}$$

Similar to (A.6)–(A.10), and using (A.15) we can show that (A.16) holds.

We now complete the proof of Theorem 2.2. It follows from (A.12) that in probability

$$L_N^X(\mu) - L_N^X(\mu_0) \to K(\mu, \mu_0),$$

where for $\mu \neq \mu_0$

$$K(\mu, \mu_0) = \frac{1}{4\pi} \int_{-\infty}^{\infty} \left\{\frac{g(\omega, \mu_0)}{g(\omega, \mu)} - 1 - \log\left(\frac{g(\omega, \mu_0)}{g(\omega, \mu)}\right)\right\} \frac{d\omega}{1+\omega^2} > 0$$

using the fact that for $x > 0$ and $x \neq 1$

$$x - 1 > \log(x).$$

Thus, for any given $\varepsilon > 0$

$$\liminf_{N\to\infty} \inf_{\|\mu-\mu_0\| \geqslant \varepsilon} (L_N^X(\mu) - L_N^X(\mu_0)) > 0$$

in probability. This implies that $\hat{\mu}_N \to \mu_0$ holds in probability. Thus the proof of Theorem 2.2 is completed. □

Proof of Theorem 2.3. It follows from

$$I_N^Y(\omega) = \frac{1}{2\pi N}\int_0^N \int_0^N Y(s)Y(t)\mathrm{e}^{\mathrm{i}(s-t)\omega}\,\mathrm{d}s\,\mathrm{d}t$$

that

$$\begin{aligned}
E[I_N^Y(\omega)] - \phi(\omega,\theta_0) &= \frac{1}{2\pi}\int_{-\infty}^{\infty} [\phi_0(\lambda) - \phi_0(\omega)]\psi^2((\lambda-\omega)N)\,\mathrm{d}(N\lambda)\\
&= \frac{1}{2\pi}\int_{-\infty}^{\infty} [\phi_0(\omega + u/N) - \phi_0(\omega)]\psi^2(u)\,\mathrm{d}u\\
&= \frac{1}{2\pi}\left\{\int_{-\infty}^{-N} + \int_{-N}^{N} + \int_{N}^{\infty}\right\}[\phi_0(\omega+u/N) - \phi_0(\omega)]\psi^2(u)\,\mathrm{d}u\\
&\equiv C_{1N} + C_{2N} + C_{3N},
\end{aligned}$$

where $\phi_0(\omega) = \phi(\omega,\theta_0)$.

It is obvious that for all $\omega \neq 0$

$$\begin{aligned}
C_{1N} &= \int_{-\infty}^{-N} [\phi_0(\omega + u/N) - \phi_0(\omega)]\psi^2(u)\,\mathrm{d}u = \mathrm{O}\left(\phi_0(\omega)\int_{-\infty}^{-N} u^{-2}\,\mathrm{d}u\right)\\
&= \mathrm{O}\left(\frac{\phi_0(\omega)}{N}\right)
\end{aligned}$$

and

$$\begin{aligned}
C_{3N} &= \int_{N}^{\infty} [\phi_0(\omega + u/N) - \phi_0(\omega)]\psi^2(u)\,\mathrm{d}u = \mathrm{O}\left(\phi_0(\omega)\int_{N}^{\infty} u^{-2}\,\mathrm{d}u\right)\\
&= \mathrm{O}\left(\frac{\phi_0(\omega)}{N}\right)
\end{aligned}$$

as $N \to \infty$.

For C_{2N}, a second-order Taylor expansion implies that as $N \to \infty$

$$\begin{aligned}
C_{2N} &= \int_{-N}^{N} [\phi_0(\omega + u/N) - \phi_0(\omega)]\psi^2(u)\,\mathrm{d}u\\
&= \int_{-N}^{N} \{\phi_0'(\omega)(u/N) + \phi_0''(\xi)(u^2/2N^2)\}\psi^2(u)\,\mathrm{d}u = \mathrm{O}\left(\frac{\phi_0''(\omega)}{N^2}\right)
\end{aligned}$$

for all $\omega \neq 0$, where ξ lies between ω and $\omega + u/N$ and $\phi''(\omega)$ is the second derivative of $\phi_0(\omega)$ with respect to ω.

This completes the proof of (2.9).

In order to prove (2.10), it suffices to show that

$$E\{I_N^Y(\omega) - E[I_N^Y(\omega)]\}^2 = \phi^2(\omega, \theta_0) + O(1/N^2).$$

Obviously,

$$\begin{aligned} &4\pi^2 N^2 E\{I_N^Y(\omega) - E[I_N^Y(\omega)]\}^2 \\ &= \int_0^N \int_0^N \int_0^N \int_0^N \{r_Y(s,t,u,v) - r_Y(s,t) r_Y(u,v)\} \mathrm{e}^{\mathrm{i}\omega(t-s)+\mathrm{i}\omega(u-v)} \,\mathrm{d}s\,\mathrm{d}t\,\mathrm{d}u\,\mathrm{d}v. \end{aligned}$$

Since $\{Y(s),\ -\infty < s < \infty\}$ is Gaussian with $Y(0) = 0$, we have

$$r_Y(s,t,u,v) - r_Y(s,t) r_Y(u,v) = r_Y(t,u) r_Y(s,v) + r_Y(t,v) r_Y(s,u). \tag{A.17}$$

Thus, it follows from (A.17) that

$$\begin{aligned} &4\pi^2 E\{I_N^Y(\omega) - E[I_N^Y(\omega)]\}^2 \\ &= \frac{1}{N^2} \int_0^N \int_0^N \int_0^N \int_0^N r_y(t,u) r_Y(s,v) \mathrm{e}^{\mathrm{i}(t+u)\omega} \mathrm{e}^{-\mathrm{i}(s+v)\omega} \,\mathrm{d}s\,\mathrm{d}t\,\mathrm{d}u\,\mathrm{d}v \\ &\quad + \frac{1}{N^2} \int_0^N \int_0^N \int_0^N \int_0^N r_Y(t,v) r_Y(s,u) \mathrm{e}^{\mathrm{i}(t-v)\omega} \mathrm{e}^{\mathrm{i}(u-s)\omega} \,\mathrm{d}s\,\mathrm{d}t\,\mathrm{d}u\,\mathrm{d}v. \end{aligned} \tag{A.18}$$

We now look at the first component of (A.18). For all $\omega \neq 0$, we have

$$\begin{aligned} &\frac{1}{N} \int_0^N \int_0^N r_Y(t,u) \mathrm{e}^{\mathrm{i}(t+u)\omega} \,\mathrm{d}t\,\mathrm{d}u \\ &= \frac{1}{N} \int_{-\infty}^{\infty} \int_0^N \int_0^N \mathrm{e}^{\mathrm{i}(t+u)\omega} \mathrm{e}^{\mathrm{i}(u-t)\lambda} \phi_0(\lambda) \,\mathrm{d}t\,\mathrm{d}u\,\mathrm{d}\lambda \\ &= \mathrm{e}^{-\mathrm{i}\omega N} \int_{-\infty}^{\infty} \psi(N(2\omega - z)) \psi(Nz) \phi_0(\omega - z) \,\mathrm{d}(Nz) \\ &= O(\psi(2\omega N) \phi_0(\omega)) = O(1/N) \end{aligned} \tag{A.19}$$

and

$$\begin{aligned} \frac{1}{N} \int_0^N \int_0^N r_Y(s,v) \mathrm{e}^{-\mathrm{i}(s+v)\omega} \,\mathrm{d}s\,\mathrm{d}v &= \mathrm{e}^{\mathrm{i}\omega N} \int_{-\infty}^{\infty} \psi(N(2\omega + x)) \psi(Nx) \phi_0(\omega + x) \,\mathrm{d}(Nx) \\ &= O(\psi(2\omega N) \phi_0(\omega)) = O(1/N) \end{aligned} \tag{A.20}$$

as $N \to \infty$.

Analogous to (A.19) and (A.20), we can estimate that the second component of (A.18) is

$$\frac{1}{N^2}\int_0^N\int_0^N\int_0^N\int_0^N r_Y(t,v)r_Y(s,u)\mathrm{e}^{\mathrm{i}(t-v)\omega}\mathrm{e}^{\mathrm{i}(u-s)\omega}\,\mathrm{d}s\,\mathrm{d}t\,\mathrm{d}u\,\mathrm{d}v$$

$$=\left(\int_{-\infty}^{\infty}\psi^2((\omega-\lambda)N)\phi_0(\lambda)\,\mathrm{d}\lambda\right)\left(\int_{-\infty}^{\infty}\psi^2((\omega+\lambda)N)\phi_0(\lambda)\,\mathrm{d}\lambda\right)\to\phi^2(\omega,\theta_0)$$

as $N\to\infty$.

This completes the proof of (2.10). Analogously, we can prove (2.11) and therefore we complete the proof of Theorem 2.3. □

Proof of Theorem 2.4. Before proving $\hat{\theta}_N\to\theta_0$ with probability one, we show that as $N\to\infty$

$$L_N^Y(\theta)\to L^Y(\theta)=\frac{1}{4\pi}\int_{-\infty}^{\infty}\left\{\log(\phi(\omega,\theta))+\frac{\phi(\omega,\theta_0)}{\phi(\omega,\theta)}\right\}\frac{\mathrm{d}\omega}{1+\omega^2} \tag{A.21}$$

holds with probability one for $\delta\leqslant\|\theta-\theta_0\|<\frac{1}{4}$ with any given $\delta>0$.

In order to prove (A.21), it suffices to show that for

$$q(\omega)=q(\omega,\theta)=\frac{1}{1+\omega^2}\phi(\omega)^{-1}=(16\eta)^{-1}\omega^{2\beta}(1+\omega^2)^{\alpha-1}\left(\frac{\sin(\omega/4)}{\omega/4}\right)^{-4}\pi^{-1}(\omega)$$

the following

$$\int_{-\infty}^{\infty}I_N^Y(\omega)q(\omega,\theta)\,\mathrm{d}\omega\to\int_{-\infty}^{\infty}\phi(\omega,\theta_0)q(\omega,\theta)\,\mathrm{d}\omega \tag{A.22}$$

holds with probability one as $N\to\infty$. It is easy to see that (A.22) follows from

$$\int_{-\infty}^{\infty}(E[I_N^Y(\omega)]-\phi(\omega,\theta_0))q(\omega,\theta)\,\mathrm{d}\omega\to 0 \tag{A.23}$$

and

$$\int_{-\infty}^{\infty}(I_N^Y(\omega)-E[I_N^Y(\omega)])q(\omega,\theta)\,\mathrm{d}\omega\to 0 \tag{A.24}$$

with probability one for $\delta\leqslant\|\theta-\theta_0\|<\frac{1}{4}$ with any given $\delta>0$. We first prove (A.23).

It follows from the definition of $I_N^Y(\omega)$ that as $N\to\infty$

$$\int_{-\infty}^{\infty}E[I_N^Y(\omega)]q(\omega,\theta)\,\mathrm{d}\omega\to\int_{-\infty}^{\infty}\phi(\omega,\theta_0)q(\omega,\theta)\,\mathrm{d}\omega$$

using for $\delta\leqslant\|\theta-\theta_0\|<\frac{1}{4}$,

$$\int_{-\infty}^{\infty}\phi_0(\omega)q(\omega)\,\mathrm{d}\omega<\infty \tag{A.25}$$

due to Condition 2.1(iii).

Hence, (A.25) implies (A.23). In order to prove (A.24), it suffices to show that as $N\to\infty$

$$E\left\{\int_{-\infty}^{\infty}\left(I_N^Y(\omega)-E\left[I_N^Y(\omega)\right]\right)q(\omega)\,\mathrm{d}\omega\right\}^2=\mathrm{O}\left(\frac{1}{N^2}\right). \tag{A.26}$$

The proof of (A.26) is similar to that of (A.6) of Gao et al. (2001).

The remainder of the proof of Theorem 2.4(i) is similar to that of Theorem 2.2. We shall not repeat the details here.

We now prove Theorem 2.4(ii). Note that

$$\left.\frac{\partial L_N^Y(\theta)}{\partial\theta}\right|_{\theta=\hat{\theta}_N} - \left.\frac{\partial L_N^Y(\theta)}{\partial\theta}\right|_{\theta=\theta_0} = \left[\left.\frac{\partial^2}{\partial\theta^2} L_N^Y(\theta)\right|_{\theta=\theta_N^*}\right](\hat{\theta}_N - \theta_0),$$

where $|\theta_N^* - \theta_0| \leqslant |\hat{\theta}_N - \theta_0|$. Since θ_0 is in the interior of Θ, Theorem 2.4(i) implies that $\hat{\theta}_N$ is in the interior of Θ for large N. Since $\hat{\theta}_N$ minimizes $L_N^Y(\theta)$, it follows that $\partial L_N^Y(\theta)/\partial\theta|_{\theta=\hat{\theta}_N} = 0$ for large N. Thus for large N

$$\left.\frac{\partial L_N^Y(\theta)}{\partial\theta}\right|_{\theta=\theta_0} = \left[-\left.\frac{\partial^2}{\partial\theta^2} L_N^Y(\theta)\right|_{\theta=\theta_N^*}\right](\hat{\theta}_N - \theta_0).$$

Thus, in order to prove that as $N \to \infty$

$$\sqrt{N}(\hat{\theta}_N - \theta_0) \to_{\mathrm{D}} \mathrm{N}(0, \Gamma^{-1}(\theta_0)),$$

it suffices to show that as $N \to \infty$

$$\sqrt{N}\int_{-\infty}^{\infty} \{I_N^Y(\omega) - E[I_N^Y(\omega)]\} D(\omega,\theta)\,\mathrm{d}\omega \to \mathrm{N}(0, 16\pi^2\Gamma(\theta_0)), \tag{A.27}$$

where

$$\Gamma(\theta) = \frac{1}{4\pi}\int_{-\infty}^{\infty} \frac{1}{(1+\omega^2)^2}\left\{\frac{\partial}{\partial\theta}\log(\phi(\omega,\theta))\right\}\left\{\frac{\partial}{\partial\theta}\log(\phi(\omega,\theta))\right\}^T \mathrm{d}\omega$$

and

$$\begin{aligned} D(\omega,\theta) &= -\frac{1}{1+\omega^2}\frac{\partial\phi^{-1}(\omega,\theta)}{\partial\theta} \\ &= -\frac{1}{1+\omega^2}\left(\frac{\partial\phi^{-1}(\omega,\theta)}{\partial\alpha}, \frac{\partial\phi^{-1}(\omega,\theta)}{\partial\beta}, \frac{\partial\phi^{-1}(\omega,\theta)}{\partial\eta}\right)^T \\ &\equiv (D_1(\omega,\theta), D_2(\omega,\theta), D_3(\omega,\theta))^T. \end{aligned}$$

To prove (A.27), it suffices to show that for any fixed $c = (c_1, c_2, c_3)^\tau$,

$$\sqrt{N}\int_{-\infty}^{\infty} \{I_N^Y(\omega) - E[I_N^Y(\omega)]\} H(\omega,\theta)\,\mathrm{d}\omega \to N(0, \sigma^2(\theta_0)), \tag{A.28}$$

as $N \to \infty$, where $H(\omega,\theta) = \sum_{j=1}^3 c_j D_j(\omega,\theta)$ and

$$\sigma^2(\theta) = 4\pi\int_{-\infty}^{\infty} \frac{1}{(1+\omega^2)^2}\left(\sum_{j=1}^3 c_j \frac{\partial}{\partial\theta_j}\log(\phi(\omega,\theta))\right)^2 \mathrm{d}\omega,$$

in which $\theta_1 = \alpha$, $\theta_2 = \beta$ and $\theta_3 = \gamma$.

Without loss of generality, we assume that there is a measurable function $k(\omega,\theta)$ such that $k(\omega,\theta)$ is symmetric in ω and

$$H(\omega,\theta)=\sum_{j=1}^{3}c_jD_j(\omega,\theta)=k^2(\omega,\theta)\geqslant 0.$$

Otherwise, one needs only to consider the positive and negative parts of $H(\omega,\theta)$ separately.

Let $k(\omega)=k(\omega,\theta)$,

$$G_N(\omega)=\frac{k(\omega)}{\sqrt{2\pi}}\int_0^N e^{i\omega t}Y(t)\,dt \text{ and } R_N(u,v)=E[G_N(u)G_N(-v)].$$

For any function $l\in L^2(-\infty,\infty)$, define the integral operator $\mathscr{L}_N\in L^2(-\infty,\infty)\to L^2(-\infty,\infty)$ by

$$\mathscr{L}_N l=\int_{-\infty}^{\infty}R_N(u,v)l(v)\,dv.$$

Let $\lambda_k^{(N)}$ be the eigenvalues of $\mathscr{L}_N$. Then it follows from the Karhunen–Loéve expansion (e.g. Yeh, 1973, Theorem 19.4) that

$$\int_{-\infty}^{\infty}|G_N(\omega)|^2\,d\omega=\sum_{i=1}^{\infty}\lambda_i^{(N)}(G_i^{(N)})^2, \tag{A.29}$$

where for any N, $\{G_i^{(N)}: i\geqslant 1\}$ is a sequence of independent $N(0,1)$ random variables.

It follows from (A.29) that

$$m_N=\int_{-\infty}^{\infty}E[|G_N(u)|^2]\,du=\int_{-\infty}^{\infty}R_N(u,u)\,du=\sum_{i=1}^{\infty}\lambda_i^{(N)}$$

and

$$\Delta_N^2=\mathrm{Var}\left(\int_{-\infty}^{\infty}|G_N(\omega)|^2\,d\omega\right)$$

$$=2\int_{-\infty}^{\infty}\int_{-\infty}^{\infty}R_N^2(u,v)\,du\,dv=2\sum_{i=1}^{\infty}(\lambda_i^{(N)})^2. \tag{A.30}$$

Note that

$$\sqrt{N}\int_{-\infty}^{\infty}\{I_N^Y(\omega)-E[I_N^Y(\omega)]\}H(\omega,\theta)\,d\omega$$

$$=\frac{1}{\sqrt{N}}\int_{-\infty}^{\infty}(|G_N(\omega)|^2-E[|G_N(\omega)|^2])\,d\omega. \tag{A.31}$$

In order to prove (A.28), in view of (A.29)–(A.31), it suffices to show that as $N\to\infty$

$$\frac{\max_k\lambda_k^{(N)}}{\sqrt{\sum_{k=1}^{\infty}(\lambda_k^{(N)})^2}}\to 0. \tag{A.32}$$

To prove (A.32), it suffices to show that as $N \to \infty$

$$\int_{-\infty}^{\infty}\int_{-\infty}^{\infty} R_N^2(u,v)\,\mathrm{d}u\,\mathrm{d}v \to \infty \tag{A.33}$$

and

$$\frac{\int_{-\infty}^{\infty}(\int_{-\infty}^{\infty} R_N(u,v)l(v)\,\mathrm{d}v)^2\,\mathrm{d}u}{\int_{-\infty}^{\infty}\int_{-\infty}^{\infty} R_N^2(u,v)\,\mathrm{d}u\,\mathrm{d}v} \to 0. \tag{A.34}$$

Similar to (A.19), it can be shown that as $N \to \infty$

$$\begin{aligned} R_N(u,v) &= E[G_N(u)G_N(-v)] \\ &= \frac{k(u)k(v)}{2\pi}\int_{-\infty}^{\infty}\left(\int_0^N\int_0^N \mathrm{e}^{\mathrm{i}t(v-\lambda)}\mathrm{e}^{-\mathrm{i}s(u-\lambda)}\,\mathrm{d}s\,\mathrm{d}t\right)\phi_0(\lambda)\,\mathrm{d}\lambda \\ &= \frac{k(u)k(v)N^2}{2\pi}\int_{-\infty}^{\infty} B(N(v-\lambda))B(-N(u-\lambda))\phi_0(\lambda)\,\mathrm{d}\lambda \\ &= N(1+\mathrm{o}(1))B(-N(u-v))\phi_0(v)k(u)k(v), \end{aligned}$$

where $B(x) = (\mathrm{e}^{\mathrm{i}x} - 1)/\mathrm{i}x$.

Thus, as $N \to \infty$

$$\begin{aligned} &\int_{-\infty}^{\infty}\int_{-\infty}^{\infty} R_N^2(u,v)\,\mathrm{d}u\,\mathrm{d}v \\ &= \frac{2\pi N^2(1+\mathrm{o}(1))}{2\pi}\int_{-\infty}^{\infty}\left(\int_{-\infty}^{\infty}|B(N(u-v))|^2\phi_0^2(v)k^2(v)\,\mathrm{d}v\right)k^2(u)\,\mathrm{d}u \\ &= 2\pi N(1+\mathrm{o}(1))\int_{-\infty}^{\infty}\phi_0^2(u)H^2(u)\,\mathrm{d}u \to \infty. \end{aligned} \tag{A.35}$$

This implies that (A.33) holds.

Similarly, as $N \to \infty$

$$\begin{aligned} \int_{-\infty}^{\infty} R_N(u,v)l(v)\,\mathrm{d}u\,\mathrm{d}v &= \frac{2\pi N^2(1+\mathrm{o}(1))}{2\pi}\int_{-\infty}^{\infty} B(-N(u-v))\phi_0(v)h(u)h(v)l(v)\,\mathrm{d}v \\ &= 2\pi\phi_0(u)l(u)H(u)(1+\mathrm{o}(1)). \end{aligned}$$

Therefore, as $N \to \infty$

$$\begin{aligned} &\int_{-\infty}^{\infty}\left(\int_{-\infty}^{\infty} R_N(u,v)l(v)\,\mathrm{d}u\,\mathrm{d}v\right)^2\mathrm{d}u \\ &= 4\pi^2(1+\mathrm{o}(1))\int_{-\infty}^{\infty}\phi_0^2(u)l^2(u)H^2(u)\,\mathrm{d}u < \infty. \end{aligned} \tag{A.36}$$

Eqs. (A.35) and (A.36) imply that (A.34) holds. Therefore, the asymptotic normality of Theorem 2.4(ii) is proved. □

Proof of Theorem 2.5. For any fixed ω, let

$$A(\omega)=\frac{1}{\sqrt{2\pi N}}\int_0^N \cos(\omega t)Y(t)\,\mathrm{d}t,\quad B(\omega)=\frac{1}{\sqrt{2\pi N}}\int_0^N \sin(\omega t)Y(t)\,\mathrm{d}t$$

and

$$C(\omega)=\frac{1}{\sqrt{2\pi N}}\int_0^N \mathrm{e}^{-\mathrm{i}\omega t}Y(t)\,\mathrm{d}t.$$

Similar to (A.19) and (A.20), we have for any fixed ω and λ

$$\begin{aligned}E[C(\omega)C(\lambda)]&=\frac{1}{2\pi N}\int_0^N\int_0^N \mathrm{e}^{-\mathrm{i}\omega s}\mathrm{e}^{-\mathrm{i}\lambda t}E[Y(s)Y(t)]\,\mathrm{d}s\,\mathrm{d}t\\&=\frac{1}{2\pi N}\int_0^N\int_0^N \mathrm{e}^{-\mathrm{i}\omega s}\mathrm{e}^{-\mathrm{i}\lambda t}\left(\int_{-\infty}^{\infty}\mathrm{e}^{\mathrm{i}x(t-s)}\phi(x)\,\mathrm{d}x\right)\mathrm{d}s\,\mathrm{d}t\\&=\frac{\mathrm{e}^{-(\lambda+\omega)N\mathrm{i}/2}}{2\pi}\int_{-\infty}^{\infty}\psi((\omega+x)N)\psi((\lambda-x)N)\phi(x)\,\mathrm{d}(Nx).\end{aligned}$$

Note that

$$A(\omega)A(\lambda)=\tfrac{1}{2}\operatorname{Re}(C(\omega)C(\lambda)+C(\omega)C(-\lambda)),$$

$$B(\omega)B(\lambda)=\tfrac{1}{2}\operatorname{Re}(C(\omega)C(-\lambda)-C(\omega)C(\lambda))$$

and

$$A(\omega)B(\lambda)=\tfrac{1}{2}\operatorname{Im}(C(\omega)C(-\lambda)-C(\omega)C(\lambda)).$$

We now have

$$\begin{aligned}E[A(\omega)A(\lambda)]&=\frac{1}{4\pi}\cos((\omega+\lambda)N/2)\int_{-\infty}^{\infty}\psi((\omega+x)N)\psi((\lambda-x)N)\phi(x)\,\mathrm{d}(Nx)\\&\quad+\frac{1}{4\pi}\cos((\omega-\lambda)N/2)\int_{-\infty}^{\infty}\psi((\omega+x)N)\psi((\lambda+x)N)\phi(x)\,\mathrm{d}(Nx)\\E[B(\omega)B(\lambda)]&=\frac{1}{4\pi}\cos((\omega-\lambda)N/2)\int_{-\infty}^{\infty}\psi((\omega+x)N)\psi((\lambda+x)N)\phi(x)\,\mathrm{d}(Nx)\\&\quad-\frac{1}{4\pi}\cos((\omega+\lambda)N/2)\int_{-\infty}^{\infty}\psi((\omega+x)N)\psi((\lambda-x)N)\phi(x)\,\mathrm{d}(Nx)\end{aligned}$$

and

$$\begin{aligned}E[A(\omega)B(\lambda)]&=\frac{1}{4\pi}\sin((\omega-\lambda)N/2)\int_{-\infty}^{\infty}\psi((\omega+x)N)\psi((\lambda+x)N)\phi(x)\,\mathrm{d}(Nx)\\&\quad-\frac{1}{4\pi}\sin((\omega+\lambda)N/2)\int_{-\infty}^{\infty}\psi((\omega+x)N)\psi((\lambda-x)N)\phi(x)\,\mathrm{d}(Nx).\end{aligned}$$

Thus, by the similar reason as in (A.19) and (A.20) we obtain for any fixed ω and λ

$$E[A(\omega)B(\lambda)] \to 0$$

and

$$\begin{aligned}\mathrm{Var}(A(\omega)) &= \frac{1}{4\pi}\int_{-\infty}^{\infty} \psi^2((\omega+x)N)\phi(x)\,\mathrm{d}(Nx)\\ &\quad + \frac{1}{4\pi}\cos(\omega N)\int_{-\infty}^{\infty} \psi((\omega+x)N)\psi((\omega-x)N)\phi(x)\,\mathrm{d}(Nx)\\ &\to \frac{1}{2}\phi(\omega,\theta_0)\end{aligned}$$

as $N \to \infty$.

Analogously,

$$\begin{aligned}\mathrm{Var}(B(\omega)) &= \frac{1}{4\pi}\int_{-\infty}^{\infty} \psi^2((\omega+x)N)\phi(x)\,\mathrm{d}(Nx)\\ &\quad - \frac{1}{4\pi}\cos(\omega N)\int_{-\infty}^{\infty} \psi((\omega+x)N)\psi((\omega-x)N)\phi(x)\,\mathrm{d}(Nx)\\ &\to \frac{1}{2}\phi(\omega,\theta_0)\end{aligned}$$

as $N \to \infty$.

Therefore, the proof of Theorem 2.5 follows from the fact that the normalized coefficients $A(\omega)/\sqrt{\phi(\omega,\theta_0)}$ and $B(\omega)/\sqrt{\phi(\omega,\theta_0)}$ are Gaussian, and asymptotically independent.

References

Anh, V., Heyde, C. (Eds.), 1999. Special issue on long-range dependence. J. Statist. Plann. Inference 80, 1.

Anh, V., Angulo, J., Ruiz-Medina, M., 1999a. Possible long-range dependence in fractional random fields. J. Statist. Plann. Inference 80, 95–110.

Anh, V., Heyde, C., Tieng, Q., 1999b. Stochastic models for fractal processes. J. Statist. Plann. Inference 80, 123–135.

Baillie, R., King, M. (Eds.), 1996. Special issue of journal of econometrics. Annals of Econometrics 73, 1.

Beran, J., 1994. Statistics for Long Memory Processes. Chapman & Hall, London.

Brockwell, P., Davis, R., 1990. Time Series Theory and Methods. Springer, New York.

Dahlhaus, R., 1989. Efficient parameter estimation for self-similar processes. Ann. Statist. 17, 1749–1766.

Dym, H., McKean, H.P., 1976. Gaussian Process, Function Theory and the Inverse Spectral Problem. Academic Press, New York.

Fan, J., Gijbels, I., 1996. Local Polynomial Modelling and Its Applications. Chapman & Hall, London.

Flandrin, P., 1989. On the spectrum of fractional Brownian motions. IEEE Trans. Inform. Theory 35, 197–199.

Frisch, U., 1995. Turbulence. Cambridge University Press, New York.

Gao, J., Anh, V., Heyde, C., Tieng, Q., 2001. Parameter estimation of stochastic processes with long-range dependence and intermittency. J. Time Ser. Anal. 22, 517–535.

Giraitis, L., Leipus, R., 1995. A generalized fractionally differencing approach in long-memory modelling. Liet. Mat. Rink. 35, 65–81.

FSC
www.fsc.org
MIX
Papier | Fördert
gute Waldnutzung
FSC® C083411

Kolibri 360 GmbH
Ferdinand-Jühlke-Straße 7
99095 Erfurt, Deutschland
produktsicherheit@kolibri360.de